Nano-Biosensor Technologies for Diagnosis of Infectious Diseases

Scrivener Publishing
100 Cummings Center, Suite 541J
Beverly, MA 01915-6106

Publishers at Scrivener
Martin Scrivener (martin@scrivenerpublishing.com)
Phillip Carmical (pcarmical@scrivenerpublishing.com)

Nano-Biosensor Technologies for Diagnosis of Infectious Diseases

Edited by

Suvardhan Kanchi

Dept. of Chemistry, CHRIST (Deemed to be University), Bengaluru, Karnataka, India

Ayyappa Bathinapatla

Dept. of Chemistry, CMRIT, Bangalore, India

Anitha Varghese

Dept. of Chemistry, CHRIST (Deemed to be University), Bangalore, Karnataka, India

and

Phumlane Selby Mdluli

Dept. of Chemistry, Durban University of Technology, Durban, South Africa

Scrivener Publishing

WILEY

This edition first published 2025 by John Wiley & Sons, Inc., 111 River Street, Hoboken, NJ 07030, USA and Scrivener Publishing LLC, 100 Cummings Center, Suite 541J, Beverly, MA 01915, USA

For more information about Scrivener publications please visit www.scrivenerpublishing.com.

Wiley Global Headquarters
111 River Street, Hoboken, NJ 07030, USA

For details of our global editorial offices, customer services, and more information about Wiley products visit us at www.wiley.com.

Library of Congress Cataloging-in-Publication Data

ISBN 978-1-394-28766-6

Front cover art courtesy of Adobe Firefly and Wikimedia Commons
Cover design by Russell Richardson

Set in size of 11pt and Minion Pro by Manila Typesetting Company, Makati, Philippines

Printed in the USA

10 9 8 7 6 5 4 3 2 1

Contents

Preface

Welcome to *Nano-Biosensor Technologies for Diagnosis of Infectious Diseases.* This comprehensive volume explores the cutting edge of scientific innovation, presenting a curated collection of 16 meticulously crafted chapters that highlight the latest advancements in nano-biosensor technologies—a revolutionary approach to the diagnosis and management of infectious diseases.

Infectious diseases continue to pose significant global health challenges, underlining the critical need for rapid, accurate, and sensitive diagnostic tools. Situated at the intersection of nanotechnology and biomedicine, nano-biosensors offer unparalleled opportunities to meet these challenges by delivering highly sensitive and specific detection capabilities across various clinical settings.

Within these pages, readers will discover a treasure trove of cutting-edge research, ranging from foundational principles to practical applications. Authored by leading experts in the field, each chapter provides insights into current trends, emerging methodologies, and the innovative technologies that are driving the advancement of nano-biosensors for infectious disease diagnosis.

From innovative sensing mechanisms and fabrication techniques to their real-world implementation and clinical translation, this book offers a comprehensive overview of the current state of the art in nano-biosensor technologies. We hope this volume will serve as an invaluable resource for researchers, clinicians, policymakers, and anyone interested in leveraging the power of nanotechnology to combat infectious diseases and enhance global health outcomes.

We extend our sincere thanks to the contributors for sharing their expertise and insights, and to our readers, whose curiosity and commitment to advancing scientific knowledge propel progress in this vital field.

Finally, our gratitude goes to Martin Scrivener and the team at Scrivener Publishing for their support in bringing this volume to light.

The Editors
January 2025

1

Biosensor Technology: Basic Principles, Fundamentals, and History

Mariyam Thomas[1], Mathew George[2], Derry Holaday M. G.[3] and P. J. Jandas[4]*

[1]Department of Physics, St. Teresa's College, Ernakulam, Kerala, India
[2]Department of Physics, Sacred Heart College, Kochi, Kerala, India
[3]Department of Chemistry, University of Calicut, Malappuram, Kerala, India
[4]iNest Bioincubation Centre, Dr. Moopen's Medical College, Wayanad, Kerala, India

Abstract
Fabrication and modification of biosensors is one of the extensively studied areas today due to viable manufacture methods with commercialization potential and excellent performance characteristics in terms of rapid detection, cost effectiveness, high selectivity, and sensitivity. Biosensors have already explored many applications including protein sensing-based disease identification, understanding the stages and medication. Advents in this area show biosensors have the potential to find application in next generation medicine like personalized drug delivery and error free biomarker detection with extreme selectivity and sensitivity. The book chapter summarizes the concept of biosensors, conventional classifications, application areas and potential as a dependable biomedical tool. A special emphasis is given into the recent advancements in biosensors used for glucose sensing. The important role of the nonmaterial-based transducing bioreceptors in a biosensor performance is also discussed in detail.

***Keywords*:** Biosensors, biomarkers, bioreceptor, nanomaterials, SAW, QCM

**Corresponding author*: drjandasjanan@gmail.com

Suvardhan Kanchi, Ayyappa Bathinapatla, Anitha Varghese and Phumlane Selby Mdluli (eds.) Nano-Biosensor Technologies for Diagnosis of Infectious Diseases, (1–22) © 2025 Scrivener Publishing LLC

1.1 Introduction

Biological sensors, also called biosensors, are defined as analytical devices comprising a biological or biologically derived component [1] that decides their selectivity [2] capable of detecting an analyte [3] by the physiochemical component [2]. The real-time acoustic detection of carcinoembryonic antigen (CEA) was done using a bioreceptor of polyimide thin film doped with nanoparticles of Ti3C2Tx MXene-Au [13]. This work utilized thioglycolic acid arm linker mechanism. According to the immunoassay, biosensor response is linear to the concentration of CEA samples. Figure 1.1 represents the basic working principle of the CEA biosensor. The progress in the field of biosensors from the 1980s is immense with the development of life-essential devices such as pregnancy test kits which utilize biochemical or biological reactions.

The basic component of a biosensor includes a biological recognition element which detects specific analytes, then the transducer which converts the corresponding biological signal to an electrical signal (Figure 1.2). This enables rapid, accurate detection and monitoring of samples from medical, food, agriculture, environmental, and industry-based resources. According to International Union of Pure and Applied Chemistry (IUPAC), a biosensor

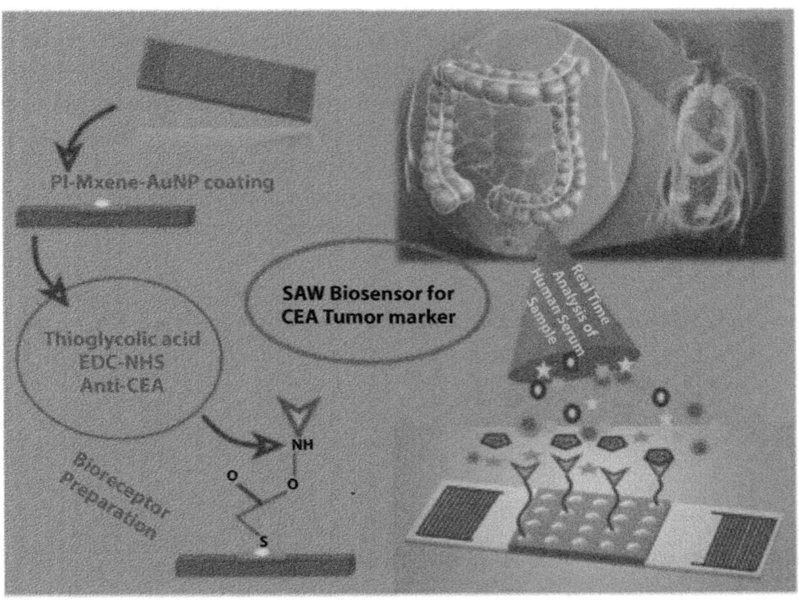

Figure 1.1 The real-time acoustic detection of carcinoembryonic antigen with polyimide thin film doped with nanoparticles of $Ti_3C_2T_x$ MXene-Au bioreceptor [13].

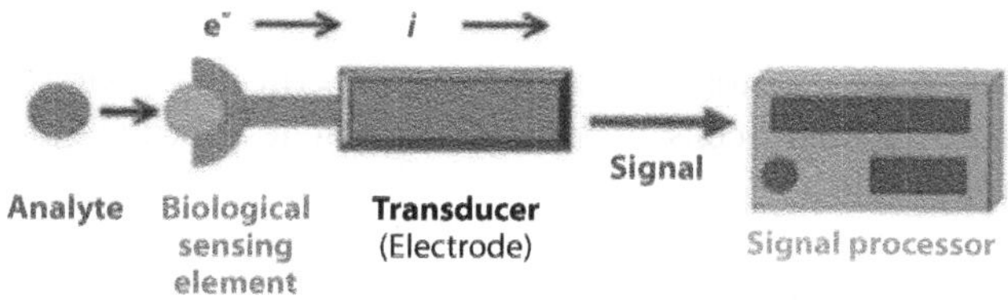

Figure 1.2 Scheme of a biosensor with an electrochemical transducer [14].

is "a device that uses specific biochemical reactions mediated by isolated enzymes, immune systems, tissues, organelles or whole cells to detect chemical compounds usually by electrical, thermal or optical signals" [4].

The three major parts of every biosensor include [1]:

i) Biological recognition element that facilitates the device to recognize the target molecule from the other chemicals present and the binding is followed [1].
ii) Signal conversion unit which is a transducer [4] that converts the specific binding process to a measurable signal [1].
iii) A signal processing system that includes a detector [5] and the detected signal is converted into readable form [1].

Glucose biosensors dominate in the industry by covering about 85% [6] and the work on the principle is based on the detection of disease indicator analytes such as glucose and insulin [7]. The first biosensors were developed in the early 1960s by Dr. Leland C Clarkin where an enzyme electrode with glucose oxidase (GOD) was employed for the measurement of the concentration of glucose [4]. The GOD catalyzes the formation of gluconolactone. The proportional increase in hydrogen peroxide concentration to the glucose concentration during the oxidation of β-D-glucose and the decline in oxygen concentration is detected electrochemically [8]. Effective glycemic monitoring has been achieved ever since the introduction of glucose biosensors. These devices also have a place in various applications such as food analysis and bioprocess monitoring [9]. Though techniques like radio labelling prevail for the measurement, the complications in the procedure, the bulk size of the equipment, sample destruction, and low spatial resolution problems prompt us to look for an effective technique, and biosensors pave its importance in this scenario [7]. Continuous, real-time glucose concentration monitoring in liquid samples ranging in nanoliters is achieved through modern glucose biosensors which have miniaturized micro/nanoscale range sensors. These sensors achieve the measurement in single cells or isolated organelles because of their high spatial resolution [9].

Working Principle of Glucose Biosensor

The biosensor consists of a biological component that on specific binding produces a response which is transmuted into a quantifiable signal with the help of a transducer [2]. The basic classification of biosensors is given in the Figure 1.4. Most prevailing glucose biosensors have better sensitivity, reproducibility, and easy maintenance, are economical in nature, and are electrochemical in type. Generally, glucose measurements are based on either of the enzyme interactions viz hexokinase, glucose oxidase (GOx), or glucose-1-dehydrogenase, of which GOx is considered superior due to its immense selective nature. Moreover, it is easily available and can withstand extreme temperatures, pH conditions, and ionic strength compared to other enzymes. Figure 1.3 represents the catalysis process by immobilized glucose oxidase. In the process, flavin adenine dinucleotide (FAD) acts as a redox cofactor for the oxidation of β-D-glucose employing molecular oxygen, thus converting it into gluconic acid and peroxide (Figure 1.3). The mechanism of the process is shown in Scheme 1.1.

The FAD on reduction converts to $FADH_2$. Upon oxidation, H_2O_2 is formed which is oxidized at platinum electrode. It detects the electron transfer number thus detecting the concentration of glucose in the sample.

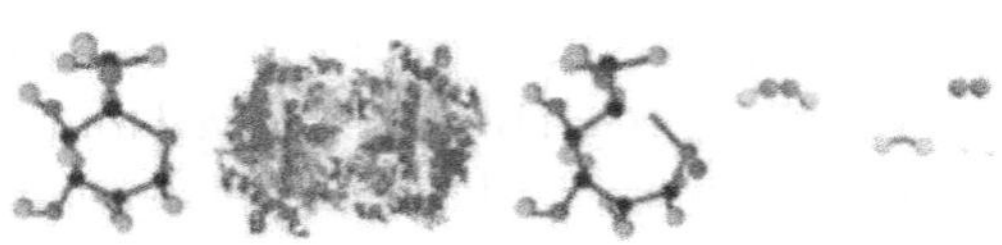

$$\text{Glucose} + O_2 \xrightarrow{GOX} \text{gluconic acid} + H_2O_2 \rightarrow 2H^+ + O_2 + 2e^-$$

Figure 1.3 Schematic representation of glucose biosensing [10].

$$\text{Glucose} + \text{GOX} - \text{FAD}^+ \longrightarrow \text{Glucolactone} + \text{GOx} - \text{FADH}_2$$

$$\text{GOx} - \text{FADH}_2 + O_2 \longrightarrow \text{GOx} - \text{FAD} + H_2O_2$$

$$H_2O_2 \longrightarrow 2H^+ + O_2 + 2e$$

Scheme 1.1 Mechanism of the glucose detection.

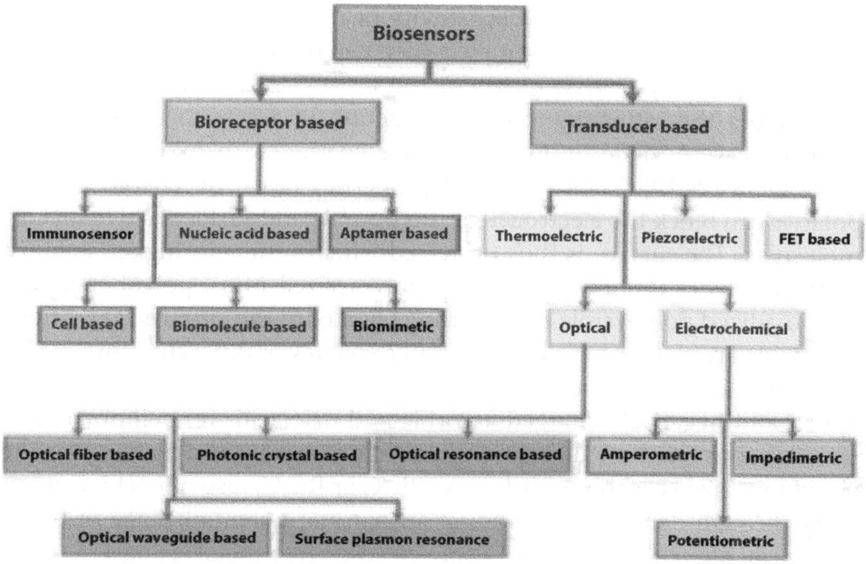

Figure 1.4 Different types of biosensors.

When glutamate dehydrogenase (GDH) with nicotinamide-adenine dinucleotide (NAD) is the cofactor, NADH is produced in the place of H_2O_2 [14].

Evolution in Biosensor Technology

A wide variety of biosensor availability is seen today as the principle of working of biosensors is getting changed over the years.

1) First Generation Glucose Biosensors

The first proposed biosensor includes an oxygen electrode, GOx thin layer, semipermeable oxygen inner membrane, and an outer dialysis membrane. Natural oxygen substrate-based biosensors come under the first generation for the detection of the hydrogen peroxide produced as the process is simple but requires higher operation potential for high selectivity. The variation of the O_2 tension due to less solubility O_2 in biofluids is another drawback of these biosensors [1].

1.1) Electroactive Interferences

The hydrogen peroxide amperometric measurement at common working electrodes requires a relatively high potential. The co-existing [9] reducing species [11] like ascorbic acids, uric acids or acetaminophen may interfere and affect the selectivity and sensitivity of the process. Moreover, other components existing in the sample which can undergo oxidation will also affect the accuracy.

To avoid this, attempts towards reduction of access to the surface of electrodes were made by using selective coating with multi and mixed polymer layers [9]. Their transport properties based on charge, size, or polarity can block the electroactive compounds and surface-active macromolecules. The surface is thus protected which results in higher stability. High selectivity is shown by electro-polymerized films, polyphenol, and over-oxidized polypyrrole by confining GOx onto the surface [11]. Multi-(overlaid) layers that have combining properties of different films can be used for additional advantages, *i.e.*, the intervention of neutral acetaminophen and negatively charged ascorbic and uric acid were eliminated by simultaneous alternate deposition of cellulose acetate and Nafion [9]. The use of metalized carbon transducers (Rh-C or Ru-C) also offers high selectivity through determining H_2O_2 at an optimal potential range of 0.0 V. The surface metal oxide film is converted to free metal using H_2O_2. An anodic current signal is produced when it is reoxidized electrochemically. By including a discriminative layer with metal to a Nafion film, additional improvements can be made. Horseradish peroxidase (HRP) is an enzyme that catalyzes peroxide oxidation thus offering a low potential selection detection of the GOx-generated H_2O_2. The carbon nanotube (CNT)-modified electrodes offer high selectivity towards glucose detection. On coupling CNT with platinum nanoparticles shows high efficiency as it has enhanced sensitivity and speed [11].

1.2) Oxygen Dependence

The errors due to the oxygen deficit prevailed in devices based on oxidase as they use oxygen as physiological acceptor of electrons. These errors result in the reduced upper limit of linearity. When normal O_2 concentration is lower, around 1 order of magnitude than the physiological level of glucose points out oxygen deficit. This constraint can be overcome by applying mass transport limiting films. Their usage will enhance the flux of the O_2 and glucose permeability ratio [11].

2) Second Generation Glucose Biosensors

2.1) Electron Transfer between GOx and Electrode Surfaces

The second-generation glucose sensors, redox mediators, replace oxygen. The transfer of electrons from enzymes to the surface of the working electrode will lead to further improvements [1].

$$\text{Glucose} + \text{GOx(ox)} \rightarrow \text{Gluconic acid} + \text{GOx(red)}$$
$$\text{GOx(red)} + 2\text{M(ox)} \rightarrow \text{GOx(ox)} + 2\text{M(red)} + 2\text{H}^{+}$$
$$2\text{M(red)} \rightarrow 2\text{M(ox)} + 2\text{e}^{l}$$

Scheme 1.2 Mechanism of synthetic mediators electron transfer process, where $M_{(ox)}$ and $M_{(red)}$ are the oxidized and reduced forms, respectively.

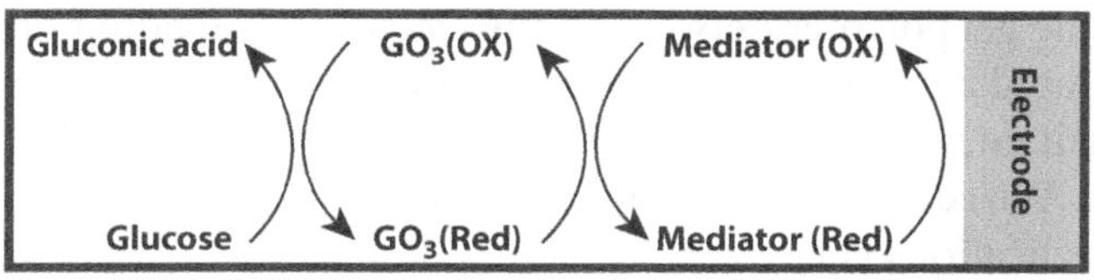

Figure 1.5 Event sequence in 'second generation' of mediator-based glucose biosensors [11].

2.2) Use of Non-Physiological Electron Acceptors

The synthetic mediators transport the electrons from the FAD center to the electrode surface, as shown in Scheme 1.2.

A current signal is produced when the reduced form is reoxidized at the electrode, regenerating the mediator's oxidized form. This is shown in Figure 1.5 [11].

Ferrocene, ferricyanide, quinines, tetrathiafulvalene (TTF), tetracyanoquinodimethane (TCNQ), thionine, methylene blue, and methyl viologen are some of the electron mediators which can improve the sensor performance. The advantages of ferrocenes include ferrocenes inertness towards oxygen, stability in both oxidized and reduced states, resilience across a wide range of pH levels, reacting rapidly with enzymes, and showing electron transfer kinetics [1], which enables them to perform effectively. The oxidation of endogenous species cannot be completely removed but minimized by low potential of most mediators. Further, additional errors will be led by consumption of mediators. For an extended continuous operation, mediated systems display low stability [11].

2.3) Wired Enzyme Electrodes

A redox polymer, coupled with enzyme wiring, enhances electrical conductivity between the redox center of GOx and electrode surfaces. Establishing a communication link between GOx and electrodes is the base of a non-diffusional route of biosensing. This connection is achieved by tethering the enzyme to the surface using a long, flexible, hydrophilic polymer backbone.

To accomplish this, a dense array of covalently bonded osmium-complex electron relays, such as poly(vinylpyridine) or poly(vinyl imidazole)] are used. This arrangement forms a three-dimensional network that attaches to the surface, minimizing the distance between redox centers and FAD center of the enzyme. Electrons from the redox site of GOx are transported through the gel polymer network to the electrode, providing high current outputs, rapid response times, and stabilizing the mediator to the surface. Ultra small enzyme electrodes can be used with the aid of huge current densities. Wired enzyme electrons are thus particularly attractive for *in vivo* applications [11].

Main advantages are high current outputs, fast response, and stability. Ultra-small enzyme electrodes can be used with the aid of huge current densities. Wired enzyme electrons are thus particularly attractive for *in vivo* applications [11].

2.4) Modification of GOx with Electron Relays

Another approach to facilitate electron transfer between the GOx and the electrode surface involves chemically altering GOx with electron-relay species. The flavin center of GOx undergoes oxidation due to the covalent attachment of the ferrocene group and results in electron tunneling in a number of consecutive steps. Enzyme reconstitution process improves the efficiency of electrical communication with electrodes in the glucose biosensors. The fitment of electron-transfer relays at the boundary of enzymes is also considered in the case of short electron-transfer distances [11].

2.5) Nanomaterial Electrical Connectors

The application of nanomaterials in bioanalytical chemistry is a vast field. Nanomaterials offer an effective means for wiring redox enzymes, such as GOx, to electrodes. Gold nanoparticles and CNTs serve as efficient electrical connectors between the electrode and the redox center of Gox (Figure 1.6) [13]. Utilizing a dithiol linker, gold nanoparticles are immobilized onto the gold electrode, acting as nano plugs for electrical wiring to the redox-active center of GOx. This will result in a high electron-transfer turnover (around 5000/s). Moreover, additional nanomaterials can be tethered to enzymes via CNTs, facilitating favorable surface orientation and serving as electrical connectors between their redox center and the electrode surface. The activation of GOx by these requires over potential and this can be reduced, enhancing the contact between nanomaterials and the electrode [11].

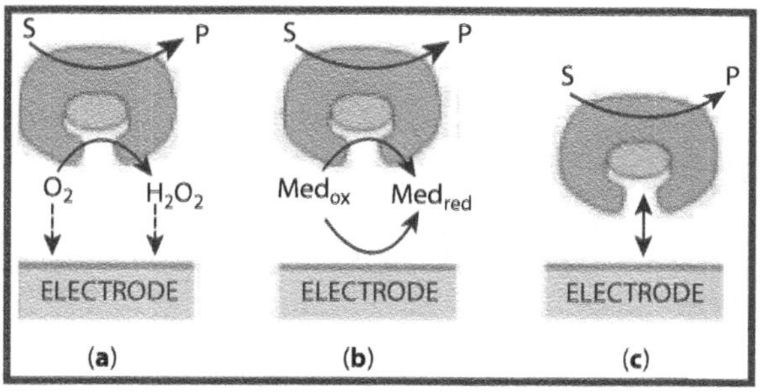

Figure 1.6 (a) Redox mediators, (b) Direct electron transfer, and (c) [11] Amperometric enzyme electrodes [10].

3) Third Generation Glucose Biosensors

They operate without reagents or mediators, utilizing a low potential similar to that of the enzyme's redox potential [11]. It is based on the enzyme's active site enabling direct transfer of an electron from glucose to the electrode [11]. Highly toxic mediators are avoided [1] due to very low operating potential and high selectivity is ensured [11]. Only few enzymes peroxidases [1] have been reported that can enable an effective electron transfer at conventional electrodes. Studies were done for new electrode materials as attempts for direct electron transfer of GOx to conventional electrons were futile.

The newly customized optimally designed electrode ensures minimal electron-transfer distance between the immobilized protein and the surface. One approach is by creating third-generation amperometric glucose biosensors utilizing conducting salt electrodes with charge-transfer complexes like TTF-TCNQ [11]. These complexes can facilitate the electrochemistry of pyrrole-quinoline quinone enzymes (GDH-PQQ) and flavoproteins (GOx) [1].

Classification of Glucose Biosensors and Their Properties

1) Based on Transducing Element

1.1) Electrochemical Glucose Biosensors

Detection of the electrochemical signal during a bio-interaction process is the basic working principle of electrochemical biosensors. Detectors can be classified into potentiometric, amperometric, or conductometric types based on their mechanism. The potentiometric sensor measures change in the charge density at the surface of the electrode and amperometric biosensor measures current liberated as a result of transfer of electrons between a biological system and electrode. The change in ionic conduction

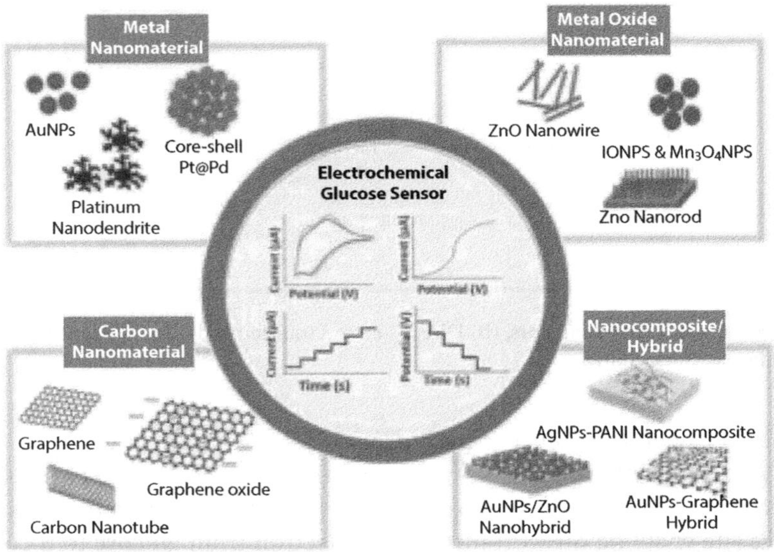

Figure 1.7 Schematic of nanomaterial-modified electrode for glucose biosensor cinnamic acid diazonium salt, which, in electrochemically reduced form, is used as the immobilization matrix for the glucose biosensor [15].

between metal electrodes is measured by the conductometric sensors. Miniaturization and simplification of the system are achieved by integrating an immobilized enzyme complex with an electrochemical sensor, thus enabling reagent-less glucose analysis. Modification of the working electrode with various nanomaterials is one of the current developments in glucose biosensors and is depicted in Figure 1.7. Nanomaterials possess unique characteristics, including a large surface area for enhanced reaction activity, excellent catalytic efficiency, and strong adsorption capabilities. This enables them to work as a matrix to modify the electrode surface. This also provides enzyme immobilization biocompatible areas.

The attachment of other biological agents are achieved by modifying the glassy carbon electrode (GCE) with the cinnamic acid group. Miniaturization is accomplished by attaching the enzyme to the self-assembled oligophenylethynylenethiol monolayer, serving as a crosslinker for immobilizing glucose oxidase to the gold electrode [7]. Out of the three most commonly used enzymes for glucose detection, *i.e.*, hexokinase, glucose oxidase (GO), and glucose-1-dehydrogenase, glucose oxidase (GOx) is widely regarded as a standard enzyme for biosensors because of its high selectivity, affordability, and capacity to endure elevated pH levels, temperatures, and ionic strengths. When glucose dehydrogenase is used instead of GO, amperometric biosensing of glucose can also be carried out. Mediators, which are carriers that can be biologically active or synthetic, are used to enhance the connection between the redox enzymes and electrodes [7].

1.2) Optical Biosensors

Another way for glucose sensing includes the utilization of optical properties of compounds with intrinsic fluorescence and their coenzymes.

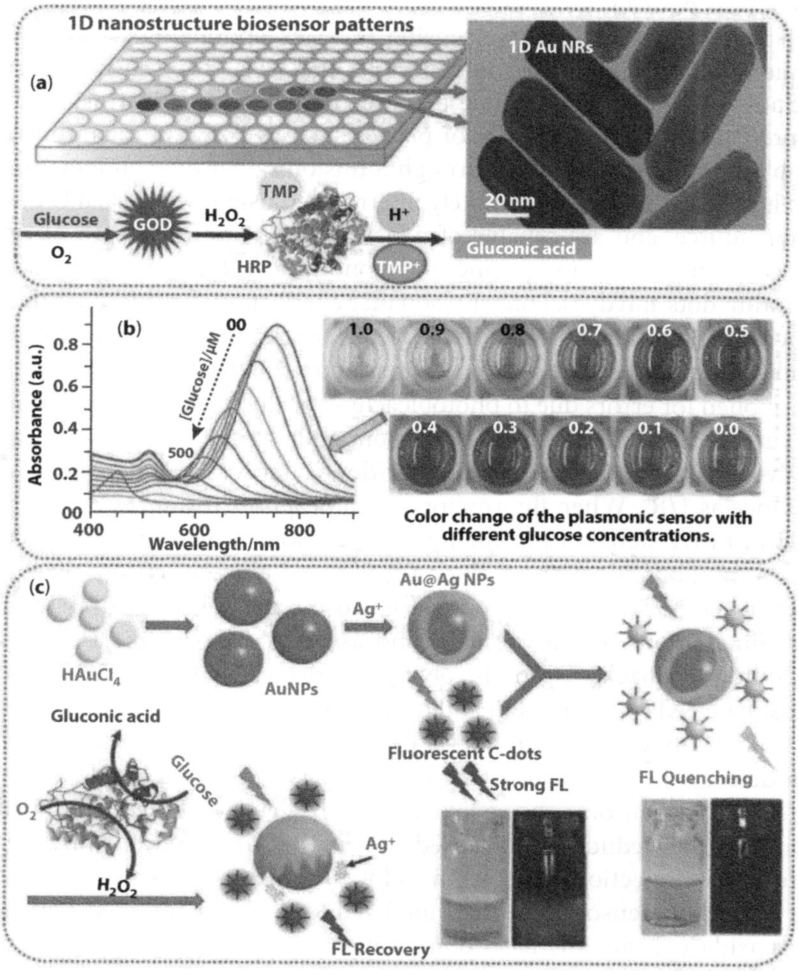

Figure 1.8 Disposable strips of optical glucose biosensor and wearable electronic devices [16].

This type of biosensor undergoes changes in its spectral properties when the binding of enzyme takes place. After the binding of the enzyme with the glucose, the fluorescence intensity change is seen for the protein part of the enzyme but no associated change is seen in the absorption spectra of the same. Co-enzymes can contribute to absorption and luminescence changes due to its interaction with glucose. The utilization of oxygen upon the interaction of the enzyme is measured to determine the glucose concentration using probes [7]. Nano/microscale device plays remarkable impacts in developing sensitive visualization assays, low-cost analyses, and home tests of diabetics (Figure 1.8).

Fluorescent-based glucose sensing is an exceptionally sensitive technique for detecting glucose at the molecular level. The sensing schemes include enzymes which include plant lectin, bacteria, or intrinsic cellular fluorescence. The smaller size of probes such as dyes and quantum dots enables biofunctionalization through diffusion rather than endocytosis.

Fluorescent probes are remotely interrogated using an external UV excitation source, and are capable of penetrating tissues to a depth of centimeters. This results in specific and configurable photoluminescence in quantum dots and dyes. Förster resonance energy transfer (FRET) is the key process behind the photoluminescence behavior. Notably, observations are made without interference from light scattering in tissues, and corrections are applied for errors due to photobleaching or fluorophore degradation.

Furthermore, FRET-based biosensors offer spatial resolution of target analytes, as the decay time of acceptor/donor fluorophores depends on the distance as $1/R^6$. When fluorophore-to-fluorophore distances reach above 10 nm, FRET signals attenuate rapidly. For non-invasive, *in vivo* sensing, and continuous monitoring, fluorescent-based glucose sensors are highly advantageous. Additionally, optical glucose sensors can accurately measure fluctuations in glucose level and associated biochemical pathways, including feedback cascade involving glucose catabolism, adenosine triphosphate (ATP) production and other biological processes [9].

A novel optical Prussian blue (PB)-based biosensor is developed which can detect H_2O_2. It evaluates the pH and acts as an optical transducer in pH-based biosensors. The redox species are detected based on the color change on its reduction. The used film is then renewed on introducing it to a flow injection system. The film system serves as the transducer for optical biosensors. The modified reduced film, combined with glucose oxidase, forms the basis of an optical biosensor. This type is mainly used for the glucose determination in urine samples. This sensor used for the determination of glucose in soft drinks is an optical fiber biosensor. Electroluminescence is the basic working principle of such sensors where the glucose oxidase is immobile on GCE surface. Recently, Haider *et al.* developed a hydrogel sensor based on phenylboronic acid (PBA) containing photonic nanostructures for swift and continuous glucose quantification. This sensor incorporates photonic nanostructures imprinted on the PBA-based hydrogel. It is comprised of three variations: free-standing (FS), stand-alone (SA), and optical fiber (OF) sensors, each capable of measuring glucose concentrations within the range of 0–50 mM (Figure 1.9).

The direct determination of glucose concentration from the blood samples is possible with a reagent-less reversible biosensor. The envelopment of glucose oxidase and peroxidase within a polyacrylamide polymer gel forms the basis of a reagent-less and reversible biosensor, utilizing its intrinsic spectroscopic properties. The water-diluted blood sample is introduced into the sensor working in a continuous mode. Here the concentration of glucose is measured with an optical fiber grating, where the absorbance of glucose and the wavelength used are in resonance. This method has better sensitivity and reproducibility.

Biosensors to detect glucose in other biological fluids in micro level concentration are in the developmental stage. The compounds with the ability to intercalate metal atoms and ions like molybdenum disulfide (MoS_2) is a good candidate but its poor solubility reduces the electrochemical device performance. This is rectified by using conductive additives like silver and the electrochemical performance is improved [7].

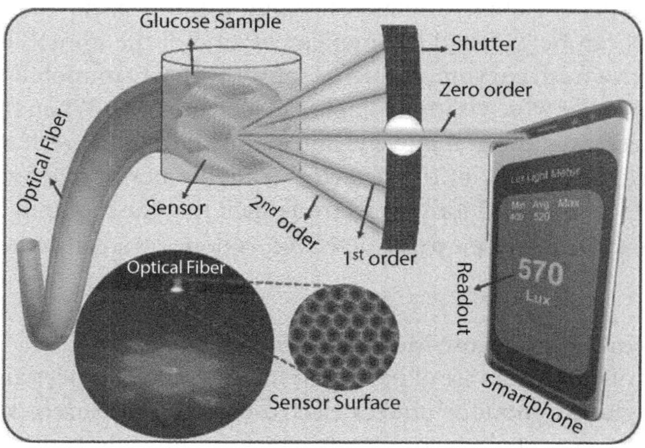

Figure 1.9 Photonic hydrogel for continuous glucose monitoring by photonic hydrogel with a smartphone readout [17].

2) Based on Clinical Application

2.1) Continuous Glucose Monitoring Biosensors

Self-testing represents a significant advancement in biosensing; however, it encounters limitations when numerous tests are required within a short timeframe. Continuous glucose monitoring (CGM) addresses this challenge by providing a continuous and valid response. The CGM biosensors necessitate biocompatibility, miniaturization, stability of enzymes and transducers, oxygen supply, safety, convenience, and *in vivo* calibration. Two primary systems are employed for continuous monitoring: subcutaneous and blood glucose levels. Nonetheless, these systems have drawbacks such as thromboembolism and surface contamination of electrodes by proteins [7].

2.1.1) Implantable Glucose Biosensor

Nowadays, this method is used for CGM which has a prolonged stability. Here, glucose undergoes oxidation by reacting with molecular oxygen in the presence of GO to form gluconolactone and H_2O_2. Electrochemical detection of concentration of H_2O_2 is done and proportional to corresponding glucose concentration. Electrochemical oxygen reduction on the Pt electrode will result in an amperometric signal by which oxygen can be detected. An oxygen-permeable and glucose-diffusive membrane is employed to avoid the mass transfer of oxygen and glucose. A hydrophobic, oxygen-permeable membrane is employed to segregate physiological fluid components from the working electrode surface, reducing interference. Enzyme immobilization enhances biosensor stability, although long-term operational stability remains a significant drawback. The lifespan of biosensors can be widened by supplementation of the immobilized compound with a fresh enzyme. A glucose sensor with GO immobilized on fine graphite powder which is implantable can be fabricated. When the enzyme activity decreases, the graphite enzyme in the fluid state is recharged, thus, improving the lifetime of the detector. A highly permeable hydrophobic membrane is employed for the electrode such that the required amount of oxygen is assured and peroxide detection is done by a platinum electrode [7].

2.2) Non-Invasive Glucose Monitoring Systems

This class of biosensors are of particular interest among other varieties. The wearable glucose monitor utilizes electro-osmotic measurement to accurately assess extracted glucose levels which is clinically acceptable. Non-invasive glucose concentration measurement is achieved through a metal oxide gas sensor, which detects the concentration of gas in the exhaled air.

The exhaled air is carbon dioxide, but in the case of diabetic patients, certain amounts of volatile organic solvents like acetone also expired along with the carbon dioxide. This exhaled air is fed into the input chamber with a sensor detection system. The sensor detects variations in conductivity when exposed to oxidizing and reducing agents. The result is typically in the form of voltage or resistance and depends on the type and concentration of volatile organic solvents present [7].

A flexible three-dimensional glucose biosensor has been developed for continuous electrochemical glucose monitoring from human tears. The sensitivity is improved by enhancing the electrode surface area through three-dimensional micropatterning of the electrode. By immobilizing GO onto this three-dimensional electrode, properties such as good repeatability and fast response for low-level concentration of glucose are achieved. The future of diabetes management lies in continuous and real-time glucose monitoring, enabled by non-invasive electrochemical epidermal glucose monitoring systems that allow for patient self-monitoring [7].

3) Nanobiosensors: Based on the Type of Nanoparticles

Nanotechnology created advancements in biosensor technology with improved and enhanced characteristics. There are three major groups of glucose-sensing molecules. They include GO, glucose-binding proteins, and glucose-binding small molecules. The advanced biosensors are developed with nanoparticles coupled with glucose-detecting molecules where the nanoparticles act as the transducers. Nanoparticle-based sensors exhibit easy absorption and improved performance due to their characteristics such as high surface-to-volume ratio. The distance between the enzyme and electrode and the rate of electron transfer are inversely related. Improvements in specificity, selectivity, reproducibility, and reliability is another advantage due to the application of nanomaterial. Nanomaterials of different morphology such as magnetic beads, quantum dots, tubes, wires, rods, fibers, composites, films, polymers, and plates are suitable for glucose detection because of their unique physical, chemical, magnetic, and optical properties [7].

3.1) Metal Nanoparticles

This type of sensor offers a lower limit of detection [12]. The blood glucose is monitored by the GO enzyme. When the enzyme comes into contact with FAD, which is embedded in a protective protein shell, it becomes active. Flavin adenine dinucleotide (FAD) is functionalized with gold nanoparticles to enhance the interaction between the active site and the electrode. Immobilizing the enzyme on gold nanoparticles helps prevent leakage due to its strong adsorption properties, thus increasing stability. A biocomposite film, which is electrochemically deposited on the gold electrode composed of chitosan, gold nanoparticles, and glucose oxidase, exhibits high stability [7].

Another fabrication technique for glucose sensors involves the covalent attachment of GO to a gold nanoparticle monolayer-modified gold electrode. The gold nanoparticles on the biosensing interface enhance electron transfer. Recent studies explored the immobilization of GO on a graphite rod modified with gold nanoparticles, investigating its impact on amperometric sensors. This approach has the potential to significantly increase the electron transfer rate.

3.2) Magnetic Nanoparticle

Their biocompatibility, low toxicity, and superparamagnetic properties make Fe_2O_3 nanoparticles an excellent choice for immobilizing carriers in biosensors. One method involves creating a nanocomposite-based biosensor on an indium tin oxide (ITO) glass plate by dispersing Fe_2O_3 nanoparticles in a chitosan solution, followed by the physical adsorption of GO onto the film. Another type of biosensor employs a film made of $C/NiFe_2O_4$ nanoparticles, where GO is immobilized within the chitosan/$NiFe_2O_4$ nanoparticle matrix. Electrodes modified with $C/NiFe_2O_4$ NPs/GO show exceptional electrocatalytic responses for glucose oxidation [7].

3.3) Nanofiber

Nanofibers when coupled with conductive nanomaterials are capable of exhibiting good sensor performance under their distinct electronic, magnetic, and optical properties. Carbon nanomaterials play a significant role in sensor application as they are easily functional, biocompatible, conductive, and has a large surface area. The whole surface area of carbon nanofibers can be activated as they have cylindrical nanostructures with graphene layers arranged as stacked cones. They act as both an immobilization matrix for biomolecules and transducers. Studies indicate that carbon nanotube-based biosensors exhibit increased sensitivity and improved stability. These biosensors are fabricated through the direct immobilization of enzymes onto the surface of carbon nanofibers. Electrochemical amperometric bienzymatic glucose biosensors are constructed by co-immobilizing HRP and GO on vertically aligned carbon nanofibers, enabling higher detection limits. Highly flexible structured nitrogen-doped carbon nanofibers (NCNFs) can be employed for electrode modification.

The NCNFs were seen to have greater surface area and electron transfer when compared with carbon nanofibers. An increased lifetime is the main advantage of electrospun nanofibers and hence their fusion and integrated nanomaterials will be excellent in sensor fabrications [7].

3.4) Nanorod-Based Biosensors

Zinc oxide (ZnO) is a potential candidate for electrochemical sensor fabrication as it is highly compatible and has high oxygen affinity. Electrochemical biosensors based on ZnO nanorods are economical, authentic, fast, and harmless. The electrochemical reaction between ZnO nanorods and glucose is the basic mechanism of this sensor. A sol-gel derived ZnO nanorods-based electrochemical sensor has been recently developed which offers advantages that include cost-effectiveness, accuracy, fast response, and high sensitivity. Moreover, hydrothermal growth is adopted for the synthesis of ZnO nanorods which is easy, cheap, and requires low growth temperature. The length of synthesized ZnO nanorods is controlled by factors like growth time and temperature.

The morphology of synthesized nanorods is influenced by the homogeneity of the growth solution. An electrochemical sensor has been fabricated using an ITO glass substrate coated with a titanium layer as the working electrode and platinum as the reference electrode. Spin coating with a thin membrane of Nafion enhances enzyme leakage and stability of the working electrode. Recently, a nonenzymatic fluorescent biosensor has been developed for the direct detection of glucose concentration. This biosensor utilizes immobilized ZnO nanorods The detection mechanism relies on the photoluminescent quenching of ZnO under UV irradiation [7].

3.5) Nanowire

The electrochemical deposition of ZnO nanowires on an Au-coated polyester substrate results in an enzymatic glucose sensor. Glucose oxidase is immobilized on zinc oxide nanowires by a simple physical adsorption method. An amperometric glucose biosensor has been developed using a film composed of silver nanowires and chitosan-glucose oxidase. This biosensor exhibits excellent selectivity, high response, and stability. Detection of glucose from serum is possible by employing this biosensor [7].

3.6) Nanocomposite

For the sensor fabrication, different forms of nanocomposites are widely used. Nanofilms, nanotubes, and nanowires are some of them. The glucose biosensor constructed by electrodeposition of palladium nanoparticles and GO onto Nafion-solubilized carbon nanotubes has increased storage time and decreased interference and performance. This is because the ability of biocatalysis is retained and the oxidation and reduction process will carry out effectively. An amperometric biosensor with the immobilization of GO onto a gold and platinum nanoparticle-modified CNT electrode exhibits good electrocatalytic activity and shows a rapid response to glucose. An electrode coated with Nafion avoids the loss of GO and eliminates interference [7].

The non-enzymatic glucose biosensor measures accurately and precisely the glucose level in real samples of blood. This biosensor works in alkaline media with Cu nanoclusters that are electrochemically deposited on a multi-walled carbon nanotube (MWCNT)-modified glassy carbon electrode. As electrocatalytic activity toward oxygen reduction is larger for bio-derived nanocomposite metal protein platinum-bovine serum albumin nanocomposite, it is used in an electrochemical sensing matrix for glucose determination. Such sensors have very good selectivity and storage stability. Also, it can be applied in real serum samples [7].

3.7) Quantum Dots-Based Biosensor

Broad absorption, narrow emission, and high quantum yield are some of the unique characteristics of semiconductor quantum nanoparticles-based biosensors. A hybrid quantum dot (QD) is electrodeposited onto the pencil graphite electrode surface and glucose dehydrogenases are immobilized onto the QD-modified electrode. This electrode has wide application in the photoelectron chemical determination of glucose by these modified electrodes and have wide applications on flow injection analysis systems. The peculiarities such as simplicity in design, easiness in integration, and portability, make them apt for conducting real sample analysis.

There are three steps by which photochemical biosensors can be fabricated. They are QD immobilization into an electrode, electrode surface illumination, and photocurrent generation due to photoexcitation of semiconductor QDs based on the analyte in supporting electrolyte. Various QD ligand systems are also taken as biosensing elements. Bioconjugated fluorescent quantum dots increase the lifetime and thus fluorescence detection is improved. Multilayered QD films of cadmium telluride (CdTe) and graphene oxide-based biosensor have been fabricated recently using a layer-by-layer assembly technique. Quenching of the photoluminescence of QDs in the film occurs when it comes in contact with the glucose solution, which is directly proportional to the glucose concentration. A potentiometric glucose biosensor, which is highly sensitive, is fabricated using functionalized indium nitride (InN) QDs, which has unique properties that include high surface charge density and robust surface properties.

The sensitivity of the electrochemical response of enzymes-coated on InN QDs is high [7].

3.8) Carbon Nanotube-Based Biosensors
Carbon nanotubes (CNTs) have astonishing electrical properties which can be utilized for the fabrication of biosensors. There are two types of CNTs, they are, single-walled carbon nanotubes (SWCNTs) and multi-walled carbon nanotubes (MWCNTs). Rolling a single graphite sheet into a tube results in SWCNT, while MWCNT has many layers of concentrically arranged graphite sheets. The MWCNTs have more electrical conductivity compared to SWCNTs. The combination of MWCNT, graphite powder, and glucose oxidase freeze-dried powder in a glass capillary is fabricated to form a real-time monitoring needle-like glucose biosensor with required stability. The electrochemically obtained resultant output measures the presence of glucose and H_2O_2. When platinum nanospheres are included in carbon nanotube-based biosensors, electrochemical performance can be improved as the glucose sensitivity is enhanced. Amperometric CNT/GO-based biosensors can be fabricated and the electrode surface can be modified with a Nafion, a perfluorosulfonate polymer that has the ability to solubilize CNT. Highly sensitive (picomolar sensitive) enzyme-free displacement-based chemiresistive biosensors with SWCNT can selectively identify glucose and the detection of glucose is based on the displacement of a plant lectin-like concanavalin-A, bound to a polysaccharide dextran immobilized on SWCNT [7].

3.9) Graphene-Associated Biosensors
The distinct properties of graphene such as economic availability, improved area of surface, and easy to process, making them the best choice to apply in the scientific and technological field. Reduction of graphene oxide for functionalized graphene sheets which exhibit sufficient structural defects and functional groups and excellent electrochemical properties. Graphene possesses a high electron density across a broad energy range, owing to its exceptional electron-transferring capability. Solution-gated graphene transistors play a significant role in real-time, high-throughput, and low concentration glucose detection, which distinguishes them from other sensors [7].

Home Testing of Blood Glucose

The personal/home glucose testing has paved the way for simple blood sugar testing. Blood glucose home testing devices should be of the best quality as they may be used daily for diagnosis. The major part of a personal blood glucose monitor is ready-to-use test strips on which the electrodes are screen-printed. They are bulk manufactured using processes like thick film micro-fabrication or vapor deposition process. Here, a patterned mask is produced by printing patterns of conductors and insulators to the plane substrate surfaces. Each strip is coated with the necessary electrodes and membranes by employing ink-jet printing technology and as a result, the reagents are easily dispensed and dry-coated. The surfactants coated on the strip ensure a uniform sample coverage and separation of the blood cells. These devices have several advantages: elimination problems of carryover, low cost, and the possibility of their mass production. A sophisticated device is inevitable for assuring accurate test results. The control meters are small, easy to carry, and battery-operated. The advantages of modern glucose biosensors include extended memory capacity and computer downloading capabilities, the disadvantages include home low and irregular testing frequency, lack of knowledge to interpret the test results by the user and liability problems. Enhanced devices with integrated devices, offering multifunctional capability, and more convenience are the properties expected as the future of these devices [11].

Challenges and Future Scopes

Glucose sensing technology has taken decades to evolve to mature and become established in the market. The limits of technology are seen when the glucose sensors are actually put into practice for glucose detection in blood or any other invasive fluids. Intervention of different molecules, *i.e.*, lactate and urea, during the optical evaluation of glucose in interstitial fluids as this will reduce the reliability of results. Analytical inaccuracy due to device miniaturization is another drawback of glucose detection in non-invasive fluid sampling. Further improvements are required in the fields of sensitivity of sensors. Moreover, the evolution of superior algorithms to convert signal data into corresponding glucose levels. Another alarming challenge regarding glucose biosensing is the nondestructive usage of biofluids and if it is possible, it will be a promising option for future directions and have a positive impact on the healthcare system [12].

References

1. a) Yoo, E.-H. and Lee, S.-Y., Glucose biosensors: an overview of use in clinical practice. *Sensors*, 10, 5, 4558–4576, 2010. b) Jandas, P.J., Luo, J., Prabakaran, K., Chen, F., Fu, Y.Q., Highly stable, love-mode surface acoustic wave biosensor using Au nanoparticle-MoS_2-rGO nano-cluster doped polyimide

nanocomposite for the selective detection of carcinoembryonic antigen. *Mater. Chem. Phys., 246*, 122800, 2020.

2. a) Weetall, H.H., Biosensor technology what? where? when? and why? *Biosens. Bioelectron.*, 11, 1-2, i–iv, 1996. Singh, S. *et al.*, Biological biosensors for monitoring and diagnosis, in: *Microbial biotechnology: basic research and applications*, pp. 317–335, 2020. b) Prabakaran, K., Jandas, P.J., Luo, J., Fu, C., Wei, Q., Molecularly imprinted poly(methacrylic acid) based QCM biosensor for selective determination of L-tryptophan. *Colloids Surf. A Physicochem. Eng. Asp., 611*, 125859, 2021.
3. a) https://www.pdfdrive.com/biosensors-and-biodetection-methods-and-protocols-volum e-1-optical-based-detectors-e181162451.html.b) Jandas, P.J., Luo, J.T., Quan, A.J., Qiu, C.H., Cao, W.G., Fu, C., Fu, Y.Q., Highly selective and label-free Love-mode surface acoustic wave biosensor for carcinoembryonic antigen detection using a self-assembled monolayer bioreceptor. *Appl. Surf. Sci.*, 518, 146061, 2020.
4. a) Piyush Parkhey, S. Venkata Mohan, Biosensing Applications of Microbial Fuel Cell: Approach Toward Miniaturization, In Biomass, Biofuels and Biochemicals, Microbial Electrochemical Technology, Pages 977-997, Elsevier, 2019, https://doi.org/10.1016/B978-0-444-64052-9.00040-6.
5. a) Jandas P J, K. Prabakaran, Jingting Luo, Derry Holaday M G, Effective utilization of quartz crystal microbalance as a tool for biosensing applications, Sensors and Actuators A: Physical, Volume 331, 113020, 2021, https://doi.org/10.1016/j.sna.2021.113020
6. Sabu, C. *et al.*, Advanced biosensors for glucose and insulin. *Biosens. Bioelectron.*, 141, 111201, 2019.
7. Atanasov, P. and Wilkins, E., Biosensor for continuos glucose monitoring. *Biotechnol. Bioeng.*, 43, 3, 262–266, 1994.
8. Wang, J., Glucose biosensors: 40 years of advances and challenges. *Electroanalysis: An International Journal Devoted to Fundamental and Practical Aspects of Electroanalysis*, 13, 12, 983–988, 2001.
9. Taguchi, M., Ptitsyn, A., Mclamore, E., Claussen, J., Nanomaterial-mediated biosensors for monitoring glucose. *J. Diabetes Sci. Technol.*, 8, 2, 403–411, 2014. 10.1177/1932296814522799
10. Wang, J., Electrochemical glucose biosensors. *Chem. Rev.*, 108, 2, 814–825, 2008.
11. Pullano, S.A. *et al.*, Glucose biosensors in clinical practice: Principles, limits and perspectives of currently used devices. *Theranostics*, 12, 2, 493, 2022.
12. Graphene oxide-Au nano particle coated quartz crystal microbalance biosensor for the real time analysis of carcinoembryonic antigen. *RSC Adv.*, 10, 4118, 2020.
13. Jandas, P.J., Prabakaran, K., Luo, J., Fu, C., Fu, Y.Q., Holaday, M.G.D., Ti3C2Tx MXene -Au nanoparticles doped polyimide thin film as a transducing bioreceptor for real-time acoustic detection of carcinoembryonic antigen. *Sens. Actuators A Phys.*, 331, 1129, 2021.

14. Curulli, A., Electrochemical Biosensors in Food Safety: Challenges and Perspectives. *Molecules*, 26, 10, 2940, 2021.
15. Nor, N.M., Ridhuan, N.S., Razak, K.A., Progress of Enzymatic and Non-Enzymatic Electrochemical Glucose Biosensor Based on Nanomaterial-Modified Electrode. *Biosensors*, 12, 12, 1136, 2022.
16. Reda, A., El-Safty, S.A., Selim, M.M., Shenashen, M.A., Optical glucose biosensor built-in disposable strips and wearable electronic devices, *Biosens. Bioelectron.*, 185, 1, 113237, 2021.
17. Ahmed, I., Elsherif, M., Ali, M., Ghaferi, A.A., Mohammad, B., Butt, H., Photonic hydrogel for continuous glucose monitoring using smartphone readout. *Mater. Des.*, 231, 112065, 2023.

2

Design and Synthesis of Novel Nanomaterials Emphasizing Infectious Diseases

Prasann Kumar[1]* and Joginder Singh[2]

[1]Department of Agronomy, School of Agriculture, Lovely Professional University, Phagwara, (Punjab), India
[2]Department of Botany, Nagaland University, Nagaland, India

Abstract

The emergence of infectious diseases has posed significant challenges to public health systems worldwide, necessitating the development of innovative solutions for disease prevention and treatment. Nanomaterials have emerged as promising candidates for combating infectious diseases due to their unique physicochemical properties and versatile medical applications. This review focuses on the design and synthesis of novel nanomaterials, emphasizing their potential to address infectious diseases. The rational design of nanomaterials allows for precise control over their physicochemical properties, such as size, shape, surface chemistry, and composition. This control enables tailored interactions with biological systems, including pathogens and host cells. Nanomaterials can be engineered to possess inherent antimicrobial properties, effectively inhibiting the growth and proliferation of infectious agents. Their large surface area-to-volume ratio facilitates enhanced drug loading, targeted drug delivery, and controlled release, improving the efficacy and specificity of antimicrobial therapies. Various fabrication techniques have been employed to synthesize nanomaterials with desired characteristics. These techniques include chemical synthesis, physical vapor deposition, self-assembly, and biological approaches. Each method offers distinct advantages and limitations, influencing the choice of nanomaterial synthesis for specific applications. Advances in nanotechnology have led to the development of multifunctional nanomaterials that combine antimicrobial activity with

**Corresponding author*: prasann0659@gmail.com

Suvardhan Kanchi, Ayyappa Bathinapatla, Anitha Varghese and Phumlane Selby Mdluli (eds.) Nano-Biosensor Technologies for Diagnosis of Infectious Diseases, (23–74) © 2025 Scrivener Publishing LLC

other functionalities, such as imaging, sensing, and immunomodulation, further expanding their potential in infectious disease management. The application of nanomaterials in contagious disease control extends beyond antimicrobial therapy. Nanomaterials can serve as diagnostic tools for rapid and sensitive detection of pathogens, enabling early and accurate disease diagnosis. They can be utilized as vaccine delivery platforms, enhancing immune responses and improving vaccination efficacy. Nanomaterial-based biosensors allow real-time monitoring of infectious agents, facilitating timely interventions and surveillance of disease outbreaks. Despite the promising potential of nanomaterials in combating infectious diseases, several challenges remain. Safety considerations must be thoroughly addressed, including biocompatibility, long-term toxicity, and environmental impact. Regulatory frameworks must be established to ensure the safe and ethical translation of nanomaterial-based technologies into clinical practice. Large-scale manufacturing and cost-effectiveness of nanomaterials need to be optimized for widespread implementation. The design and synthesis of novel nanomaterials offer exciting opportunities for managing infectious diseases. The ability to tailor their properties and functions enables targeted and effective interventions against pathogens. By integrating antimicrobial, diagnostic, and therapeutic capabilities, nanomaterials have the potential to revolutionize infectious disease control, improving patient outcomes and global public health. However, further research and development are required to address safety concerns, optimize manufacturing processes, and translate nanomaterial-based technologies into practical solutions for infectious disease prevention and treatment.

***Keywords*:** No poverty, zero hunger, disease, crop, nanomaterials, interventions, antimicrobial, design, biosensor

2.1 Introduction

Infectious diseases have long been a significant global health concern, causing substantial morbidity and mortality worldwide [1, 2]. The emergence of drug-resistant pathogens and the lack of effective treatment options have exacerbated the challenge of combating infectious diseases [3–5]. To address these pressing issues, there is a growing need for innovative approaches that can effectively prevent, diagnose, and treat infectious diseases. In recent years, nanotechnology has emerged as a promising field with the potential to revolutionize disease management by designing and synthesizing novel nanomaterials [6, 7]. Nanomaterials refer to materials with dimensions on the nanoscale, typically ranging from 1 to 100 nm. At this scale, materials exhibit unique physicochemical properties that differ from their bulk counterparts. These properties include a high surface-to-volume ratio, quantum effects, and size-dependent optical, electrical, and magnetic characteristics [8, 9]. These features make nanomaterials highly versatile and enable their integration into various applications, including medicine, electronics, energy, and environmental remediation. Nanomaterials offer numerous advantages over conventional approaches in infectious diseases [10, 11]. Nanomaterials' rational design and synthesis allow precise control over their size, shape, composition, and surface properties. This control enables the tailoring of nanomaterials to interact specifically with pathogens, host cells, and the immune system, thereby enhancing the efficacy and safety of disease management strategies [12, 13]. By exploiting the unique properties of nanomaterials, researchers can develop innovative solutions for the prevention, diagnosis, and treatment of infectious diseases [14, 15]. One of the primary applications of nanomaterials in infectious disease management is their use as antimicrobial agents. Traditional antibiotics are becoming increasingly ineffective due to the emergence of drug-resistant pathogens [13–16]. Nanomaterials can be engineered to possess inherent antimicrobial properties or act as carriers for antimicrobial agents, enhancing their efficacy and overcoming resistance mechanisms. The large surface area-to-volume ratio of nanomaterials allows for increased contact with pathogens, facilitating their neutralization or destruction [17–19].

Moreover, nanomaterials can be utilized for targeted drug delivery, improving efficacy and reducing the side effects of antimicrobial therapies. Functionalizing nanomaterials with targeting ligands, such as antibodies or peptides, can selectively recognize and bind to specific cell types, including infected cells or pathogen surface markers [20–22]. This targeted delivery approach minimizes the exposure of healthy cells to antimicrobial agents, reducing toxicity and enhancing therapeutic outcomes. In addition to their antimicrobial properties, nanomaterials offer significant potential in disease diagnosis. Rapid and accurate detection of infectious agents is crucial for early intervention and effective disease management. Nanomaterial-based biosensors and diagnostic platforms provide sensitive and specific detection of pathogens, enabling timely diagnosis and appropriate treatment [23–25]. Nanomaterials can be functionalized with biomarkers, fluorescent dyes, or magnetic nanoparticles, allowing for the detection of specific pathogens or their associated biomolecules in clinical samples. Nanomaterials can be employed as vaccine delivery platforms, addressing the challenges associated with traditional vaccine formulations. Nanomaterial-based vaccine carriers can enhance immune responses, improve antigen stability, and facilitate controlled release, enhancing efficacy and prolonged immunity [26–28]. The unique properties of nanomaterials, such as their ability to stimulate immune cells or mimic viral structures, can further augment vaccine effectiveness.

While the potential of nanomaterials in infectious disease management is promising, several challenges need to be addressed. Safety concerns regarding nanomaterials' biocompatibility and potential toxicity must be thoroughly investigated to ensure their safe use in humans and the environment. Regulatory frameworks and guidelines should be established to govern the development, manufacturing, and clinical translation of nanomaterial-based technologies [26, 29–33]. The scalability and cost-effectiveness of nanomaterial synthesis methods need to be optimized to facilitate widespread implementation and accessibility. The design and synthesis of novel nanomaterials hold tremendous potential for addressing the challenges of infectious diseases. By harnessing the unique properties of nanomaterials, researchers can develop innovative approaches for disease prevention, diagnosis, and treatment. Antimicrobial properties, targeted drug delivery, diagnostic capabilities, and vaccine development are among the key areas where nanomaterials have shown significant promise [34–36]. However, further research, collaboration, and investment are required to overcome the existing challenges and translate nanomaterial-based strategies into practical solutions that can significantly impact infectious disease management and improve global public health.

2.2 Antifungal Therapy

Fungal infections, or mycoses, pose a significant healthcare burden worldwide. They can affect various body parts, including the skin, nails, respiratory tract, and internal organs. The treatment of fungal infections, known as antifungal therapy, aims to eliminate the fungal pathogen while minimizing damage to the host. Antifungal agents can be categorized into several classes, including polyenes, azoles, echinocandins, allylamines, and pyrimidine analogues, each with unique mechanisms of action and spectrum of activity [37–40].

Polyenes: polyene antifungals, such as amphotericin B and nystatin, are among the oldest and most effective antifungal agents. They exert their action by binding to ergosterol, a vital component of the fungal cell membrane, leading to membrane disruption and cell death. Amphotericin B is commonly used for severe systemic fungal infections, including invasive candidiasis and aspergillosis. However, its use is limited by its potential toxicity [41–44].

Azoles: azole antifungals, including fluconazole, itraconazole, and voriconazole, inhibit ergosterol synthesis by targeting the enzyme lanosterol 14α-demethylase. This enzyme is essential for converting lanosterol into ergosterol, a critical component of the fungal cell membrane. Azoles are widely used to treat fungal infections, including candidiasis, cryptococcosis, and dermatophytosis. They are available in oral, topical, and intravenous formulations, allowing for flexible treatment options [41, 45–47].
Echinocandins: echinocandins, such as caspofungin, micafungin, and anidulafungin, inhibit the synthesis of β-(1,3)-D-glucan, a vital component of the fungal cell wall. By targeting the cell wall integrity, echinocandins cause osmotic instability, leading to fungal cell death. Echinocandins are primarily used for the treatment of invasive candidiasis and invasive aspergillosis. They are administered intravenously and are generally well-tolerated [48–50].
Allylamines: allylamines, such as terbinafine, inhibit squalene epoxidase, an enzyme involved in ergosterol biosynthesis. By blocking ergosterol synthesis, allylamines disrupt fungal cell membrane integrity. Terbinafine is commonly used to treat dermatophytosis, particularly onychomycosis (fungal nail infection). It is available in topical and oral formulations [51–55].
Pyrimidine analog: flucytosine, a pyrimidine analogue, is converted into 5-fluorouracil inside fungal cells. 5-fluorouracil inhibits fungal deoxyribonucleic acid (DNA) and ribonucleic acid (RNA) synthesis, leading to impaired fungal growth. Flucytosine is primarily used with other antifungal agents, such as amphotericin B, for treating systemic fungal infections, including cryptococcal meningitis. In addition to these classes, other antifungal agents have specific applications. Griseofulvin is used for the treatment of superficial fungal infections, particularly dermatophytosis [51, 56–59]. Nystatin is commonly used topically for oral and vaginal candidiasis. Ciclopirox and amorolfine are antifungal agents used in the treatment of onychomycosis. The choice of antifungal therapy depends on several factors, including the type and severity of the fungal infection, the patient's immune status, and the site of infection. Combination therapy may be required for refractory or severe infections. However, antifungal treatment is not without limitations. Resistance to antifungal agents can occur, particularly in immunocompromised patients, necessitating the development of new treatment strategies [50, 52, 55, 60, 61]. Some antifungal agents may have significant drug–drug interactions and adverse effects, requiring careful monitoring during treatment. Antifungal therapy plays a crucial role in managing fungal infections. Developing various classes of antifungal agents with distinct mechanisms of action has expanded

treatment options and improved patient outcomes. However, the emergence of drug resistance and the need for more effective and safer therapies remain ongoing challenges [45–47, 49, 50, 61, 62]. Continued research and innovation in antifungal drug development are necessary to address these challenges and improve the management of fungal infections.

2.3 Cutting-Edge Advances in Tailoring the Size, Shape, and Functionality of Nanoparticles and Nanostructures

Nanotechnology has revolutionized various fields by enabling precise control over materials at the nanoscale. One of the critical aspects of nanotechnology is the ability to tailor the size, shape, and functionality of nanoparticles and nanostructures. Recent advancements in nanosynthesis, fabrication techniques, and characterization methods have opened up new possibilities for designing and engineering nanomaterials with unprecedented control and precision [40–44]. These cutting-edge advances offer exciting opportunities across diverse applications, including electronics, energy, medicine, and environmental remediation.

Size control: the ability to precisely control the size of nanoparticles and nanostructures is critical, as size-dependent properties can significantly influence their behavior and performance. Researchers have developed innovative techniques to control nanoparticle sizes, such as seed-mediated growth, sol-gel methods, and template-directed synthesis. For example, using micellar templates or nanoporous materials allows for synthesizing uniform nanoparticles with controlled sizes. Moreover, advancements in lithography and self-assembly techniques have facilitated the fabrication of nanostructures with precise dimensions and tailored geometries [14–20, 45].
Shape engineering: beyond size control, tailoring the shape of nanoparticles and nanostructures offers additional avenues for manipulating their properties. Advances in synthetic methods, such as solution-based synthesis, electrochemical deposition, and chemical etching, have enabled the fabrication of nanoparticles with complex shapes, including rods, cubes, triangles, and octahedra. Shape engineering can significantly influence optical, magnetic, and catalytic properties due to changes in surface area, surface chemistry, and crystalline facets [1–3, 6]. For instance, anisotropic nanoparticles with well-defined shapes exhibit enhanced plasmonic

properties, tunable surface-enhanced Raman scattering (SERS), and improved catalytic activity.

Functionalization: functionalizing nanoparticles and nanostructures with specific surface chemistries or functional groups is crucial for tailoring their interactions and enabling targeted applications. Surface functionalization can be achieved through various techniques, such as ligand exchange, surface grafting, or encapsulation [14, 16–18, 55, 59, 61]. These methods allow for the attachment of biomolecules, polymers, or inorganic compounds onto the nanoparticle surface, imparting specific functionalities and enabling applications such as drug delivery, biosensing, and catalysis. Advances in surface chemistry and bioconjugation techniques have facilitated the development of multifunctional nanomaterials with tailored properties for specific applications.

Hybrid and composite nanostructures: combining different types of nanomaterials in hybrid or composite structures offers synergistic properties and expanded functionalities. For instance, integrating metallic nanoparticles with semiconductor quantum dots or magnetic nanoparticles can create hybrid systems with unique optical, electronic, or magnetic properties. These hybrid nanostructures find applications for sensing, imaging, energy conversion, and data storage. Moreover, incorporating nanomaterials into composite matrices, such as polymers or ceramics, can enhance mechanical strength, electrical conductivity, or thermal stability. This has led to development of advanced materials for structural applications, electronics, and energy storage [34, 35].

***In-situ* characterization:** advances in *in-situ* characterization techniques have significantly contributed to our understanding of nanoparticle and nanostructure growth and transformation mechanisms. Real-time monitoring and control of synthesis processes allow for optimizing size, shape, and functionality during fabrication. Techniques such as *in-situ* spectroscopy, microscopy, and scattering methods provide insights into nucleation, growth kinetics, surface reactions, and structural changes at the nanoscale. This knowledge enables researchers to develop precise control strategies and optimize synthesis conditions to achieve desired nanoparticle properties. The tailoring of nanoparticles and nanostructures' size, shape, and functionality is a rapidly evolving field with immense potential. These cutting-edge advances offer unprecedented control over nanomaterial properties, enabling breakthroughs in various applications. However, challenges remain, such as scalability, reproducibility, and understanding the effects of structure-property relationships. Addressing these challenges will require continued interdisciplinary research efforts, collaboration, and innovation to fully exploit the potential of tailored nanomaterials in advancing technology and addressing societal needs [36–38].

2.4 Gold Silver Nanoparticle to Combat Multi-Drug Resistant Pathogen

Gold and silver nanoparticles have been extensively studied for their antimicrobial properties and potential to combat multidrug-resistant pathogens. These nanoparticles have unique characteristics that make them attractive for antimicrobial applications. Here is how gold and silver nanoparticles can be used to combat multidrug-resistant pathogens:

1. **Mechanism of action:** gold and silver nanoparticles possess intrinsic antimicrobial properties. They can interact with the cell membranes of bacteria, disrupting their structural integrity and causing leakage of cellular components. Additionally, these nanoparticles can enter bacterial cells and interfere with essential cellular processes, such as DNA replication and protein synthesis, leading to bacterial cell death.
2. **Enhanced antimicrobial activity:** gold and silver nanoparticles can exhibit enhanced antimicrobial activity compared to conventional antibiotics. They have a high surface area-to-volume ratio, facilitating better interaction with bacterial cells. Additionally, their small size allows for effective penetration into biofilms, which are protective layers formed by bacteria that contribute to their drug resistance.
3. **Synergistic effects:** gold and silver nanoparticles can be combined with antibiotics to overcome drug resistance. Studies have shown that when used with antibiotics, nanoparticles can enhance the antibacterial efficacy by acting through different mechanisms. This approach can help overcome multidrug-resistant pathogens and reduce the required antibiotic dosage, potentially minimizing the development of further drug resistance.
4. **Targeted delivery:** nanoparticles can be functionalized or coated with specific ligands that enable targeted delivery to the site of infection. This targeted approach enhances the concentration of nanoparticles at the infection site,

increasing their efficacy while minimizing potential side effects on healthy cells.

5. **Reduced toxicity:** gold and silver nanoparticles have been shown to have low toxicity to mammalian cells when used at appropriate concentrations. This makes them attractive candidates for antimicrobial therapies, as they can selectively target bacterial cells while sparing healthy human cells.
6. **Prospects:** ongoing research is focused on further optimizing the synthesis and functionalization of gold and silver nanoparticles for enhanced antimicrobial activity. Additionally, studies are exploring using nanoparticles and other antimicrobial strategies, such as photothermal and gene therapy, to combat multidrug-resistant pathogens more effectively.

It is important to note that while gold and silver nanoparticles show promise in combating multidrug-resistant pathogens, further research is needed to understand better their long-term effects, potential resistance development, and applicability in clinical settings. Regulatory and safety considerations must also be addressed before widespread use (Table 2.1).

Please note that this table provides a general overview, and the specific properties and characteristics of gold and silver nanoparticles may vary based on the manufacturing method, size, surface functionalization, and other factors.

2.5 Mechanism of Gold Silver Nanoparticle to Combat Multi-Drug Resistant Pathogen

The mechanism of action of gold and silver nanoparticles in combating multidrug-resistant pathogens involves several key processes:

1. **Interaction with cell membranes:** gold and silver nanoparticles can interact with the cell membranes of bacteria. They have a high affinity for negatively charged bacterial cell membranes rich in phospholipids. The nanoparticles can attach to the cell membrane surface, destabilizing and disrupting the membrane structure. This disruption can cause leakage of cellular components, loss of membrane integrity, and, ultimately, cell death.

Table 2.1 Here is an example of a table highlighting some key features and properties of gold and silver nanoparticles for combating multidrug-resistant pathogens.

Properties	Gold nanoparticles	Silver nanoparticles
Antimicrobial activity	Disrupts cell membranes and essential cellular processes	Disrupts cell membranes and essential cellular processes
Size	Typically, in the range of 1–100 nm	Typically, in the range of 1–100 nm
Surface area-to-volume ratio	High surface area-to-volume ratio, facilitating better interaction with bacterial cells	High surface area-to-volume ratio, facilitating better interaction with bacterial cells
Mechanism of action	Interacts with cell membranes and enters bacterial cells	Interacts with cell membranes and enters bacterial cells
Enhanced efficacy with antibiotics	Synergistic effects when used in combination with antibiotics	Synergistic effects when used in combination with antibiotics
Targeted delivery	Can be functionalized or coated for targeted delivery to infection sites	Can be functionalized or coated for targeted delivery to infection sites
Toxicity to mammalian cells	Low toxicity to mammalian cells at appropriate concentrations	Low toxicity to mammalian cells at appropriate concentrations
Research focus	Optimization of synthesis and functionalization for enhanced activity	Optimization of synthesis and functionalization for enhanced activity
Future prospects	Exploration of combination therapies and clinical applicability	Exploration of combination therapies and clinical applicability

(Sources: [1, 2, 4, 37, 51, 57, 59, 60].)

2. **Penetration into bacterial cells:** gold and silver nanoparticles can also enter bacterial cells. Their small size and unique surface properties enable them to cross the bacterial cell membrane, gaining access to the intracellular space. Once inside the bacterial cells, the nanoparticles can interfere with essential cellular processes.
3. **Interference with cellular processes:** gold and silver nanoparticles can disrupt vital cellular processes within bacteria. They can interact with bacterial DNA, leading to DNA damage and inhibition of replication. By binding to bacterial proteins, the nanoparticles can interfere with protein synthesis, impairing the production of essential enzymes and proteins needed for bacterial survival. These disruptions to cellular processes can ultimately result in bacterial cell death.
4. **Reactive oxygen species (ROS) generation:** gold and silver nanoparticles can generate reactive oxygen species, such as singlet oxygen and superoxide radicals. These ROS can induce oxidative stress within bacterial cells, causing damage to proteins, lipids, and DNA. The accumulation of ROS overwhelms the antioxidant defense mechanisms of bacteria, leading to cell death.
5. **Disruption of biofilms:** multidrug-resistant pathogens often form biofilms, protective layers contributing to antibiotic resistance. Gold and silver nanoparticles can penetrate biofilms due to their small size and interact with the bacterial cells within the biofilm. By disrupting the biofilm structure and killing the bacteria embedded within, the nanoparticles can help overcome resistance and enhance the effectiveness of antimicrobial treatment. It is important to note that the exact mechanism of action may vary depending on factors such as nanoparticle size, shape, surface charge, and functionalization. Additionally, further research is ongoing to understand better the precise details of how gold and silver nanoparticles combat multidrug-resistant pathogens and optimize their antimicrobial efficacy [28, 30, 31, 51, 56, 58, 61, 63].

2.6 MXenes and Borophene Nanomaterials: Highly Efficient Sensor Activity and Energy Storage Properties

MXenes and borophene nanomaterials have gained significant attention recently due to their highly efficient sensor activity and energy storage properties. Let us explore these two materials in more detail:

MXenes: MXenes are a family of two-dimensional (2D) transition metal carbides, nitrides, and carbonitrides. They are derived from layered MAX phases, which consist of a transition metal (M), an element such as aluminum or silicon (A), and carbon or nitrogen (X). Through etching, the A layer is selectively removed, resulting in the formation of MXenes.

a. Sensor activity: MXenes exhibit remarkable sensor activity due to their unique properties:

Conductivity: MXenes are highly conductive, enabling efficient charge transfer during sensing processes.
Large surface area: MXenes possess a high surface area, which enhances the adsorption and interaction of target molecules with the material's surface.
Tunable surface chemistry: the surface of MXenes can be functionalized with different functional groups, allowing for selective sensing of specific analytes.
Mechanical flexibility: MXenes are mechanically flexible, enabling their integration into various sensor platforms.

These attributes make MXenes suitable for various sensing applications, including gas sensing, biosensing, and environmental monitoring. They have shown excellent performance in detecting gases such as ammonia, nitrogen dioxide, and volatile organic compounds. MXenes have also demonstrated promising biosensing capabilities, detecting biomarkers and pathogens with high sensitivity and selectivity.

b. Energy storage properties: MXenes exhibit remarkable energy storage properties, making them attractive for energy storage devices such as batteries and supercapacitors:

High electrical conductivity: MXenes possess high electrical conductivity, facilitating rapid charge and discharge processes.
Large interlayer spacing: MXenes have expanded interlayer spacing, allowing for easy insertion and extraction of ions during energy storage.
Capacitive behavior: MXenes exhibit pseudocapacitive behavior, combining double-layer capacitance and fast surface redox reactions, resulting in high energy and power density.
Stability: MXenes show good cycling stability and long-term durability, crucial for reliable energy storage applications.

These features enable MXenes to serve as electrode materials in lithium-ion, sodium-ion, supercapacitors, and hybrid capacitors. MXenes have demonstrated high specific capacitance, excellent rate capability, and long cycle life, making them promising candidates for next-generation energy storage systems.

Borophene: borophene is a 2D sheet of boron atoms arranged in a hexagonal lattice. It is an emerging nanomaterial with exceptional properties that make it highly efficient for sensor activity and energy storage:

a. Sensor activity: borophene exhibits remarkable sensing capabilities due to its unique characteristics:

Surface reactivity: the surface of borophene is highly reactive, making it sensitive to various analytes.
Ultra-high carrier mobility: borophene possesses ultra-high carrier mobility, enabling fast charge transport during sensing processes.
High surface-to-volume ratio: borophene's 2D structure provides a large surface area for efficient adsorption and interaction with target molecules.

These features enable borophene to be employed in gas, biosensors, and chemical sensors. Borophene-based sensors have shown excellent sensitivity, selectivity, and response times, making them suitable for gas detection, bioanalysis, and environmental monitoring applications.

b. Energy storage properties: borophene exhibits promising energy storage properties, contributing to its potential in energy storage applications:

High theoretical capacity: borophene possesses a high theoretical capacity for energy storage due to the unique arrangement of boron atoms.

Good electrical conductivity: borophene exhibits good electrical conductivity, facilitating efficient charge transfer during energy storage.
Structural stability: borophene demonstrates structural stability during cycling, ensuring long-term durability.

These properties make borophene a candidate for various energy storage systems, including lithium-ion batteries, fuel cells, and supercapacitors. Borophene has shown high charge storage capacity, good cycling stability, and fast charge/discharge rates, making it a promising material for future energy storage technologies.

MXenes and borophene nanomaterials possess highly efficient sensor activity and energy storage properties. Their unique characteristics, such as conductivity, large surface area, and tunable surface chemistry, contribute to their excellent sensing capabilities. Similarly, their high electrical conductivity, large interlayer spacing, and stability make them attractive for energy storage applications. Ongoing research continues to explore the potential of these materials and optimize their properties for practical implementation in various sensing and energy storage devices [9, 15, 19, 25, 50, 52, 59, 64].

Molluscum contagiosum virus (MCV) is a common poxvirus that causes a benign skin infection known as molluscum contagiosum (MC). Developing an effective vaccine against MCV requires identifying immunogenic proteins and selecting suitable delivery platforms. Here is a comprehensive overview of immunogenic proteins and potential delivery platforms for MCV vaccine development:

1. **Immunogenic Proteins**
   a. **Surface proteins:** surface proteins of MCV are potential targets for vaccine development as they interact with the host immune system. For example, the major surface protein (MSP) and the viral envelope protein B2R have been identified as immunogenic proteins in MCV. These proteins can elicit neutralizing antibodies and cell-mediated immune responses.
   b. **Viral enzymes:** some viral enzymes, such as the viral thymidine kinase (TK), can also serve as immunogenic proteins. Thymidine kinase is involved in viral replication and is recognized by the host immune system, leading to the generation of specific immune responses.
   c. **Structural proteins:** structural proteins, including viral capsid proteins, play a crucial role in the assembly and

stability of viral particles. These proteins can induce protective immune responses, including antibody production and T-cell activation.

2. **Potential Delivery Platforms**
   a. **Viral vector-based vaccines:** viral vectors, such as modified vaccinia Ankara (MVA) or vesicular stomatitis virus (VSV), can be used as delivery platforms for MCV vaccine antigens. These vectors can efficiently deliver immunogenic proteins to host cells, triggering robust immune responses. Additionally, they can be engineered to express multiple MCV antigens simultaneously, enhancing the vaccine's efficacy.
   b. **Protein subunit vaccines:** protein subunit vaccines involve using purified viral proteins or recombinant proteins that mimic the immunogenic regions of the virus. These proteins can be formulated with adjuvants to enhance their immunogenicity. For MCV, essential immunogenic proteins, such as MSP or B2R, can be expressed and purified in protein subunit vaccines.
   c. **Nucleic acid-based vaccines:** nucleic acid-based vaccines, such as DNA or RNA, have emerged as promising platforms for viral vaccine development. They deliver the genetic material encoding viral antigens into host cells, producing the immunogenic proteins in situ. These vaccines can induce both humoral and cellular immune responses. For MCV, encoding the genes for immunogenic proteins, such as MSP or viral enzymes, into DNA or RNA vaccines could be explored.
   d. **Virus-like particle (VLP) vaccines:** VLPs are non-infectious particles that closely resemble the structure of the virus but lack the viral genetic material. They can be produced by expressing MCV structural proteins in a heterologous expression system. Virus-like particles can elicit strong immune responses, including neutralizing antibodies, as they present the immunogenic epitopes in a highly organized and repetitive manner.
   e. **Nanoparticle-based vaccines:** nanoparticles, such as liposomes, polymeric nanoparticles, or virus-mimicking particles, can serve as delivery platforms for MCV vaccine antigens. These nanoparticles can efficiently deliver

the immunogenic proteins and provide controlled release kinetics, enhancing their immunogenicity and stability.

It is important to note that selecting delivery platforms and immunogenic proteins for MCV vaccine development is an ongoing research area. Preclinical and clinical studies are required to evaluate vaccine formulations' safety, immunogenicity, and efficacy. Furthermore, regulatory considerations and manufacturing scalability must be addressed for successful vaccine development (Table 2.2).

Please note that this table provides a general overview, and the specific selection of immunogenic proteins and delivery platforms may vary based on ongoing research and development efforts. The table highlights some essential immunogenic proteins of MCV and potential delivery platforms that have shown promise for vaccine development.

Table 2.2 A comprehensive table showcasing immunogenic proteins and potential delivery platforms for molluscum contagiosum virus (MCV) vaccine development.

Immunogenic proteins	Potential delivery platforms
Major surface protein (MSP)	Viral vector-based vaccines, protein subunit vaccines, nucleic acid-based vaccines, virus-like particle (VLP) vaccines, nanoparticle-based vaccines
Viral envelope protein B2R	Viral vector-based vaccines, protein subunit vaccines, nucleic acid-based vaccines, virus-like particle (VLP) vaccines, nanoparticle-based vaccines
Viral thymidine kinase (TK)	Viral vector-based vaccines, protein subunit vaccines, nucleic acid-based vaccines, virus-like particle (VLP) vaccines, nanoparticle-based vaccines
Viral capsid proteins	Viral vector-based vaccines, protein subunit vaccines, virus-like particle (VLP) vaccines, nanoparticle-based vaccines
Other viral enzymes	Viral vector-based vaccines, protein subunit vaccines, nucleic acid-based vaccines, nanoparticle-based vaccines
Other structural proteins	Viral vector-based vaccines, protein subunit vaccines, virus-like particle (VLP) vaccines, nanoparticle-based vaccines

(Sources: [9, 11, 20, 42, 53, 55–59].)

2.7 Immunomodulatory Nanosystems

Immunomodulatory nanosystems are an emerging strategy in the field of nanomedicine for combatting viral infections. These nanosystems are designed to interact with the immune system by enhancing the immune response or suppressing excessive immune reactions. Immunomodulatory nanosystems can effectively control and combat viral infections by modulating the immune system's activity. Here is a comprehensive overview of immunomodulatory nanosystems in combating viral infections:

Immune Stimulation

a. **Adjuvants:** nanosystems can be engineered to act as adjuvants, enhancing the immune response to viral antigens. Adjuvants can improve the effectiveness of vaccines by promoting antigen presentation, activating immune cells, and inducing a robust immune response. Nanoparticles, liposomes, or virus-like particles can be used as adjuvant carriers to deliver viral antigens to immune cells, promoting antigen uptake and activation.

b. **Cytokine delivery:** nanosystems can be used to deliver immune-stimulating cytokines directly to the site of infection. Cytokines such as interferons, interleukins, and chemokines play crucial roles in antiviral defense. The immune response can be modulated to better combat viral infections by delivering these cytokines using nanoparticles or liposomes.

Immune Suppression

a. **Immunosuppressive agents:** in certain viral infections, excessive immune responses can lead to tissue damage and inflammation. Nanosystems can be designed to deliver immunosuppressive agents, such as corticosteroids or immunosuppressive drugs, to regulate the immune system and prevent excessive inflammation. Controlled release systems can help maintain appropriate drug concentrations over an extended period, ensuring optimal immunosuppressive effects.

b. **Tolerance induction:** nanosystems can also induce immune tolerance, especially in cases where viral infections trigger autoimmune responses. By delivering viral antigens in a controlled manner, nanosystems can promote the generation of regulatory T cells (Tregs) or induce antigen-specific immune tolerance, dampening autoimmune responses and reducing tissue damage.

c. **Targeted delivery:** nanosystems can be functionalized with targeting ligands, such as antibodies or peptides, to precisely deliver immunomodulatory agents to the site of viral infection. This targeted approach enhances the therapeutic efficacy while minimizing off-target effects and potential systemic side effects. By delivering immunomodulatory agents directly to infected tissues or immune cells, nanosystems can precisely modulate the immune response against viral infection.
d. **Combination therapies:** immunomodulatory nanosystems can be combined with antiviral drugs or vaccines for synergistic effects. These combination therapies can enhance antiviral efficacy and promote viral clearance by simultaneously targeting the virus and modulating the immune response. For example, nanocarriers can be used to co-deliver antiviral drugs and immunostimulatory agents, enabling localized delivery and immune activation at the site of infection. Immunomodulatory nanosystems hold great promise in combatting viral infections by precisely regulating immune responses. However, further research is needed to optimize their design, efficacy, and safety. Studying the interactions between nanosystems and the immune system, understanding the mechanisms of immune modulation, and conducting preclinical and clinical trials are crucial steps in realizing the full potential of immunomodulatory nanosystems for viral infection control (Table 2.3).

Please note that this table provides a general overview, and there can be overlapping applications and combinations of different nanosystems in immunomodulation. The specific design, composition, and applications of immunomodulatory nanosystems can vary based on ongoing research and development efforts.

2.8 Lateral Flow Assays (LFAs)

Lateral flow assays (LFAs) are simple, rapid, cost-effective diagnostic tests to detect disease biomarkers. They are commonly employed in point-of-care

Table 2.3 A comprehensive table showcasing different types of immunomodulatory nanosystems and their applications in combating viral infections.

Immunomodulatory nanosystems	Description	Applications
Adjuvant nanosystems	Liposomes, VLPs, polymeric nanoparticles	Enhancing immune response, promoting antigen presentation
Immunosuppressive nanosystems	Lipid-based nanoparticles, nanogels	Regulating immune responses, suppressing excessive inflammation
Targeted nanosystems	Antibody-conjugated nanoparticles, surface-engineered nanoparticles	Site-specific delivery, precise modulation of immune responses
Combination nanosystems	Co-delivery of immunomodulatory agents and antiviral drugs/antigens	Synergistic modulation of immune responses enhanced antiviral efficacy
Nanovaccines	Self-assembling peptide nanoparticles, mRNA/DNA nanoparticles	Inducing robust immune responses to viral antigens
Regulatory T cell (Treg) induction	Nanoparticles with tolerogenic properties	Promoting immune tolerance, dampening excessive immune reactions
Triggered release systems	Stimuli-responsive nanosystems	Controlled release of immunomodulatory agents in response to specific triggers

(Sources: [20, 36, 42, 48, 51–53, 55–57].)

settings, as they do not require specialized equipment or extensive sample processing. Lateral flow assays are widely used to detect infectious diseases, chronic conditions, and other medical conditions. Here is a comprehensive overview of LFA for the detection of disease biomarkers:

Principle of Operation
Lateral flow assays consist of a test strip containing several components:

Sample pad: where the patient's sample (*e.g.*, blood, urine, or saliva) is applied.
Conjugate pad: contains labelled detection particles (*e.g.*, gold nanoparticles or colored latex beads) conjugated with antibodies or other recognition molecules specific to the target biomarker.
Membrane: usually made of nitrocellulose, it contains capture molecules (antibodies, antigens, or other specific ligands) immobilized in a line or spot.
Absorbent pad: located at the end of the strip, it facilitates the capillary flow of the sample and serves as a waste reservoir.

An LFA principle involves the sample's capillary flow along the strip. If the target biomarker is present in the sample, it will bind to the labelled detection particles in the conjugate pad. The sample then migrates along the membrane, and if the target biomarker is present, it will bind to the capture molecules on the membrane. The labelled particles will accumulate at the capture zone, forming a visible line or spot, indicating a positive result.

Advantages of Lateral Flow Assays

Simplicity and speed: LFAs provide rapid results within minutes, allowing quick decision-making and immediate patient management.
Point-of-care testing: LFAs can be performed at the point of care, eliminating the need for sophisticated laboratory infrastructure or trained personnel.
Cost-effectiveness: LFAs are relatively inexpensive, making them accessible for resource-limited settings and large-scale screening programs.
User-friendly: the interpretation of results is typically straightforward, often based on the presence or absence of visible lines or spots.
Stability: LFAs have good shelf stability, with a relatively long shelf life, enabling storage without strict temperature control.

Optimization and Sensitivity
Lateral flow assays require careful optimization to achieve optimal sensitivity and specificity for the target biomarker. Key factors include:

Choice of capture and detection molecules: selecting specific and high-affinity capture and detection molecules, such as antibodies or aptamers, is crucial for accurate detection.
Conjugation techniques: proper conjugation methods ensure efficient labelling of detection particles with capture molecules without affecting their functionality.
Signal amplification: signal amplification strategies, such as using nanoparticles or enzyme labels, can enhance the sensitivity of LFAs.
Sample preparation: appropriate sample preparation techniques, such as dilution or extraction, may be required to ensure the target biomarker is in the optimal concentration range for detection.

Applications
Lateral flow assays have diverse applications for the detection of disease biomarkers:

Infectious diseases: LFAs are widely used for the rapid detection of infectious diseases, including viral, bacterial, and parasitic infections.
Chronic conditions: LFAs monitor chronic diseases, such as cardiovascular disorders, diabetes, and cancer.
Pregnancy tests: one of the most well-known applications of LFAs is in pregnancy tests, detecting the presence of human chorionic gonadotropin (hCG) in urine.
Environmental monitoring: LFAs detect environmental contaminants, toxins, and allergens.

It is important to note that while LFAs offer many advantages, they may have limitations in sensitivity and quantification compared to more sophisticated laboratory-based techniques. However, ongoing research is focused on improving the sensitivity and multiplexing capabilities of LFAs, expanding their applications in disease diagnostics and monitoring.

2.9 Metal-Organic Framework (MOF)

"Metal-organic framework (MOF) detectives meet infectious pathogens: a powerful tool against pandemics" is a comprehensive concept highlighting the potential of MOFs as a powerful tool in combating pandemics caused by infectious pathogens. Here is a comprehensive overview of MOFs and their applications in the context of infectious diseases and pandemics:

Introduction to metal-organic frameworks (MOFs): MOFs are highly ordered materials composed of metal ions or clusters coordinated to organic ligands. They possess a unique porous structure with a large surface area, tunable pore size, and exceptional chemical versatility. Metal-organic frameworks have attracted considerable attention in various fields due to their excellent properties, such as high adsorption capacity, controllable pore structures, and potential for diverse functionalities.

MOFs as Detectives for Infectious Pathogens: MOFs Offer Several Advantages for the Detection and Diagnosis of Infectious Pathogens

Adsorption capacity: the high porosity and large surface area of MOFs enable efficient adsorption of target molecules, including pathogenic biomarkers, viruses, or bacteria.
Sensing platforms: MOFs can be functionalized with specific ligands or antibodies to selectively capture and detect pathogenic molecules or microorganisms.
Signal amplification: MOFs can enhance the detection signal through signal amplification strategies, such as fluorescence resonance energy transfer (FRET), chemiluminescence, or SERS.
Multiplexing capability: MOFs can be tailored to simultaneously detect multiple targets, facilitating rapid and comprehensive diagnostic capabilities.
Stability and reusability: MOFs exhibit excellent stability, allowing for repeated use in diagnostic assays, making them cost-effective and environmentally friendly.

Applications of MOFs in Infectious Disease Diagnostics

Viral detection: MOFs can detect specific viral biomarkers, such as viral proteins or nucleic acids, enabling rapid and sensitive diagnostics. This has implications for diagnosing and surveilling viral infections, including emerging and re-emerging viral diseases.
Bacterial identification: MOFs can be functionalized with bacterial-specific ligands or antibodies to selectively capture and identify pathogenic bacteria, aiding in diagnosing and monitoring bacterial infections.
Drug resistance monitoring: MOFs can be utilized for monitoring drug resistance in infectious pathogens. By detecting specific resistance markers, MOFs can provide valuable information for guiding appropriate antimicrobial treatment strategies.

MOFs for therapeutic applications: besides diagnostics, MOFs also show potential for therapeutic applications against infectious pathogens.
Drug delivery systems: MOFs can be employed as carriers for antimicrobial agents, enabling targeted and controlled release of therapeutics to the infection site.
Immune modulation: MOFs can interact with the immune system and modulate immune responses, aiding in developing immunotherapeutic strategies against infectious diseases.
Antimicrobial photodynamic therapy (PDT): MOFs can serve as platforms for PDT, where light-activated MOFs generate reactive oxygen species to kill pathogens selectively.
Pandemic preparedness and response: MOFs have the potential to contribute significantly to pandemic preparedness and response.
Rapid diagnostic tools: MOF-based diagnostic platforms can offer rapid and accurate detection of infectious pathogens, aiding in early identification and containment of outbreaks.
Point-of-care applications: MOF-based diagnostics can be designed for point-of-care use, enabling decentralized testing and timely interventions.
Vaccine development: MOFs can assist in vaccine development by acting as adjuvants or delivery systems for antigens, enhancing vaccine efficacy and stability.

It is important to note that while MOFs hold great promise, further research is required to optimize their design, functionalization, and integration into practical diagnostic and therapeutic applications. Additionally, regulatory considerations and safety assessments are necessary before widespread adoption (Wang *et al.*, 2023; Wu *et al.*, 2023; Zahraee *et al.*, 2023). However, the versatility and tunability of MOFs make them a promising tool in the fight against infectious pathogens, offering potential solutions for future pandemics.

2.10 Microfluidic Devices: For Detecting Disease-Specific Proteins

Microfluidic devices have become powerful tools for detecting disease-specific proteins and other macromolecules, modelling, and drug development. These devices leverage the precise control of fluids and the miniaturization of analytical systems to provide numerous benefits regarding sensitivity, speed, efficiency, and scalability. Here is a comprehensive overview of microfluidic devices in these applications:

Detection of disease-specific proteins and macromolecules: microfluidic devices offer several advantages for the detection of disease-specific proteins and macromolecules.
Miniaturization: microfluidic channels and reaction chambers are designed at a small scale, reducing reagent consumption and enabling high-throughput analysis of multiple analytes simultaneously.
Rapid analysis: microfluidic devices provide rapid analysis due to shorter diffusion distances and precise control over sample flow and reaction times.
High sensitivity: the small sample volumes and integration of sensitive detection techniques, such as fluorescence or electrochemical detection, allow for enhanced sensitivity and low detection limits.
Multiplexing capability: microfluidic devices can incorporate multiplexing strategies, enabling the simultaneous detection of multiple disease markers and improving diagnostic accuracy.
Disease modelling: microfluidic devices have proven valuable for modelling disease processes in vitro, offering several advantages over traditional cell culture methods:
Mimicking physiological conditions: microfluidic devices can replicate the complex microenvironments found in the human body, providing a more accurate representation of disease conditions.
Cell–cell interactions: these devices precisely control fluid flow and cellular positioning, allowing for the study of cell–cell interactions and tissue dynamics relevant to disease development and progression.
Long-term culture: microfluidic platforms can support long-term culture, providing a more realistic representation of chronic diseases and enabling the study of long-term drug effects.
On-chip imaging and analysis: integrated imaging and analysis capabilities allow real-time monitoring of cellular responses, facilitating the study of disease mechanisms and drug responses.
Drug development: microfluidic devices have revolutionized drug development by providing platforms for rapid screening and characterization of potential drug candidates.
High-throughput screening: microfluidic devices enable high-throughput screening of compound libraries, facilitating the identification of drug candidates with improved efficacy and reduced side effects.
Organ-on-a-chip models: microfluidic devices can create organ-on-a-chip models that replicate specific organs' functions and structures. These models enable the evaluation of drug efficacy, toxicity, and pharmacokinetics in a more realistic physiological context.

Personalized medicine: microfluidic devices can be used to develop patient-specific disease models, allowing for personalized drug testing and optimization of treatment strategies.
Microencapsulation and controlled release: microfluidic platforms enable the encapsulation of drug compounds, enabling controlled release profiles for improved therapeutic efficacy.

Microfluidic devices have demonstrated immense potential in disease-specific protein detection, disease modelling, and drug development. Ongoing research and development efforts are focused on refining device design, improving integration with detection techniques, enhancing automation, and expanding the range of applications. With further advancements, microfluidic devices are promising for personalized medicine, rapid diagnostics, and efficient drug discovery and development processes.

2.11 Comprehensive Overview of Nanomaterials in the Context of Cutaneous Leishmaniasis

Cutaneous leishmaniasis is a parasitic infection caused by the protozoan parasite of the *Leishmania* genus. The disease affects millions worldwide, particularly in tropical and subtropical regions. Nanomaterials have emerged as promising tools for diagnosing, treating, and preventing human cutaneous leishmaniasis.

Diagnosis of cutaneous leishmaniasis: nanomaterials offer several advantages for accurate and rapid diagnosis of cutaneous leishmaniasis.
Sensitivity and specificity: nanomaterial-based assays can provide high sensitivity and specificity in detecting *Leishmania* parasites or their specific biomarkers, enabling accurate diagnosis.
Multiplexing capability: nanomaterials can be functionalized with multiple targeting moieties, allowing simultaneous detection of different *Leishmania* species or strains.
Point-of-care testing: nanomaterial-based diagnostic platforms can be designed for use at the point of care, facilitating early diagnosis and timely treatment [55, 57, 59, 60, 62].
Non-invasive sampling: nanomaterial-based assays can utilize non-invasive sampling techniques such as swabs or saliva, minimizing patient discomfort.
Treatment of cutaneous leishmaniasis: nanomaterials hold the potential for improved therapeutic strategies against cutaneous leishmaniasis (Table 2.4).

Table 2.4 An extended table showcasing 50 microfluidic devices used for detecting disease-specific proteins.

Microfluidic device	Principle of operation	Advantages	Applications
Lab-on-a-chip	Miniaturized platform	High sensitivity, rapid analysis, multiplexing capability	Disease diagnosis, protein biomarker detection
Droplet-based	Encapsulation of single molecules in droplets	High-throughput, precise control, minimal sample consumption	Protein quantification, single-cell analysis, drug discovery
Immunoassay chips	Immobilization of capture antibodies on chips	Rapid and sensitive detection, minimal reagent consumption	Disease diagnosis, protein biomarker detection
Nanoscale devices	Integration of nanoscale features	Enhanced sensitivity, reduced background noise	Early disease detection, point-of-care diagnostics
Magnetic bead-based	Magnetic capture and separation of target proteins	High specificity, simple operation, compatibility with downstream analysis	Protein isolation, biomarker detection
Electrochemical	Detection of protein binding events through electrochemical signals	Label-free, sensitive, real-time detection	Disease monitoring, protein interaction analysis
Surface plasmon resonance	Measurement of changes in refractive index upon protein binding	Label-free, real-time monitoring, high sensitivity	Protein-protein interactions, drug discovery
Digital microfluidics	Manipulation of discrete droplets on an array of electrodes	Versatile, programmable, high-throughput	Protein analysis, immunoassays, diagnostic assays
Paper-based	Utilization of paper as a low-cost substrate for fluid manipulation	Portable, low cost, simplicity	Point-of-care diagnostics, resource-limited settings
Microarray-based	The miniaturized array of immobilized capture molecules	High-throughput, multiplexed detection	Protein profiling, biomarker discovery

(*Continued*)

Table 2.4 An extended table showcasing 50 microfluidic devices used for detecting disease-specific proteins. (*Continued*)

Microfluidic device	Principle of operation	Advantages	Applications
Centrifugal microfluidics	Sample manipulation using centrifugal forces	Simplicity, minimal sample handling, automation	Point-of-care diagnostics, resource-limited settings
Flow-through systems	A continuous flow of samples and reagents	Rapid analysis, efficient mixing, easy integration	Disease monitoring, protein detection
Hydrogel-based	Incorporation of hydrogels for protein capture	Enhanced protein binding capacity, high sensitivity	Biomarker detection, drug screening
Antibody-coated channels	Surface-functionalized channels with capture antibodies	Specific detection, reduced non-specific binding	Protein biomarker analysis, disease diagnosis
Micro/nanofluidic channels	Controlled fluid manipulation in small channels	Precise control, enhanced mass transport, reduced reagent consumption	Protein detection, enzymatic assays
Nanostructured surfaces	Surface modifications for enhanced protein binding	Improved sensitivity, reduced assay time, high signal-to-noise ratio	Protein detection, disease diagnostics
Aptamer-based sensors	Utilization of aptamers as specific recognition elements	High affinity and selectivity, versatility, label-free detection	Protein detection, biomarker analysis
Photonic crystal sensors	Detection based on changes in photonic crystal properties	Label-free, high sensitivity, real-time monitoring	Protein detection, disease diagnostics
Acoustic wave sensors	Detection based on changes in acoustic wave properties	Label-free, real-time monitoring, high sensitivity	Protein detection, biomarker analysis
Dielectric spectroscopy	Measurement of changes in electrical properties	Non-destructive, label-free, high-throughput	Protein analysis, biomarker profiling

(*Continued*)

Table 2.4 An extended table showcasing 50 microfluidic devices used for detecting disease-specific proteins. (*Continued*)

Microfluidic device	Principle of operation	Advantages	Applications
Microcantilever-based	Detection based on changes in microcantilever deflection	Label-free, high sensitivity, real-time monitoring	Protein detection, disease diagnostics
Optofluidic devices	Integration of optics and microfluidics	Sensitive detection, compact size, versatile functionalities	Protein analysis, diagnostic applications
Nanowire-based sensors	Detection based on changes in nanowire properties	High sensitivity, label-free detection, real-time monitoring	Protein detection, biomarker analysis
Plasmonic-based sensors	Utilization of plasmonic effects for detection	Enhanced sensitivity, label-free detection, real-time monitoring	Protein detection, disease diagnostics
Microresonator sensors	Detection based on changes in resonant frequency	High sensitivity, label-free detection, real-time monitoring	Protein detection, biomarker analysis
Carbon nanotube sensors	Detection based on changes in carbon nanotube properties	High sensitivity, label-free detection, real-time monitoring	Protein detection, disease diagnostics
Magnetic nanoparticle-based sensors	Detection based on changes in magnetic properties	High specificity, easy separation, versatile detection	Protein detection, biomarker analysis
Optoelectronic sensors	Detection based on changes in optoelectronic properties	Label-free, real-time monitoring, high sensitivity	Protein detection, biomarker analysis
Field-effect transistor sensors	Detection based on changes in field-effect transistor properties	Label-free, high sensitivity, real-time monitoring	Protein detection, disease diagnostics
Impedance-based Sensors	Measurement of changes in electrical impedance	Label-free, sensitive detection, real-time monitoring	Protein detection, biomarker analysis

(*Continued*)

Table 2.4 An extended table showcasing 50 microfluidic devices used for detecting disease-specific proteins. (*Continued*)

Microfluidic device	Principle of operation	Advantages	Applications
Microdroplet-based systems	Manipulation and analysis of individual droplets	High-throughput, here is the continuation of the table showcasing 50 microfluidic devices used for detecting disease-specific proteins:	
Microscale chromatography	Miniaturized chromatography separation systems	High resolution, rapid analysis, reduced sample and reagent consumption	Protein separation, purification, analysis
Electroosmotic flow devices	Utilization of electroosmotic flow for fluid manipulation	Controlled fluid movement, reduced analysis time	Protein analysis, separation
Capillary electrophoresis	Separation of analytes based on electrophoretic mobility	High separation efficiency, rapid analysis, small sample volume	Protein analysis, biomarker detection
Digital PCR (dPCR)	Partitioning of PCR reactions into discrete volumes	High sensitivity, absolute quantification, multiplexing capability	DNA and RNA detection, mutation analysis
Droplet digital PCR (ddPCR)	Encapsulation of PCR reactions in discrete droplets	High sensitivity, absolute quantification, multiplexing capability	DNA and RNA detection, mutation analysis
Microfluidic ELISA	Integration of ELISA on microfluidic platforms	High sensitivity, reduced reagent consumption, rapid analysis	Protein detection, disease diagnostics
Microscale flow cytometry	Miniaturized flow cytometry systems	High-throughput, multiparametric analysis, reduced sample volume	Cell analysis, immunophenotyping

(*Continued*)

Table 2.4 An extended table showcasing 50 microfluidic devices used for detecting disease-specific proteins. (*Continued*)

Microfluidic device	Principle of operation	Advantages	Applications
Digital immunoassays	Partitioning of immunoassay reactions into discrete volumes	High sensitivity, multiplexing capability, reduced sample consumption	Protein biomarker detection, disease diagnostics
Single-cell analysis devices	Isolation and analysis of individual cells	High-throughput, single-cell resolution, real-time monitoring	Cell profiling, genomics, drug discovery
DNA microarrays	Miniaturized arrays of immobilized DNA probes	High-throughput, multiplexed detection, DNA profiling	Genomic analysis, mutation screening
Exosome analysis devices	Isolation and analysis of extracellular vesicles	High sensitivity, real-time monitoring, biomarker discovery	Exosome profiling, liquid biopsy
Isothermal nucleic acid amplification devices	Amplification of nucleic acids at a constant temperature	Simplified workflow, rapid amplification, point-of-care applications	Molecular diagnostics, pathogen detection
Microscale western blot	Miniaturized Western blotting systems	Rapid analysis, reduced reagent consumption, multiplexing capability	Protein analysis, biomarker detection
Microflow imaging systems	Visualization and analysis of particles and cells	High-resolution imaging, rapid analysis, automated analysis	Particle counting, cell analysis
Microfluidic organ-on-a-chip	Culturing of cells in organ-specific microenvironments	Recapitulation of tissue physiology, drug toxicity screening	Disease modelling, drug development

(*Continued*)

Table 2.4 An extended table showcasing 50 microfluidic devices used for detecting disease-specific proteins. (*Continued*)

Microfluidic device	Principle of operation	Advantages	Applications
Microfluidic cell sorting	Separation and sorting of cells based on specific properties	High-throughput, real-time monitoring, single-cell resolution	Cell sorting, cell-based assays
Bio-barcode amplification assays	Amplification of DNA barcodes for enhanced detection	High sensitivity, multiplexing capability, low sample consumption	Protein biomarker detection, disease diagnostics
Opto-electrofluidic systems	Integration of optics and electronics with microfluidics	High sensitivity, real-time monitoring, on-chip analysis	Protein detection, biomarker analysis
Acoustofluidic systems	Manipulation of fluids and particles using acoustic waves	Non-contact, label-free manipulation, high-throughput	Particle sorting, cell analysis
Paper-based microfluidics	Utilization of paper as a substrate for fluid manipulation	Low cost, portability, simplicity	Point-of-care diagnostics, resource-limited settings
Resonant mass sensors	Detection based on changes in resonant mass properties	Label-free, real-time monitoring, high sensitivity	Protein detection, disease diagnostics
Optofluidic imaging	Integration of optics and microfluidics for imaging applications	Real-time imaging, high-resolution analysis	Cell imaging, disease diagnosis
Digital immuno-PCR	Integration of digital PCR and immunoassays	Absolute quantification, multiplexing capability, high sensitivity	Protein detection, disease diagnostics
Integrated microfluidic systems	Integration of multiple microfluidic components	Seamless workflow, reduced sample handling, compact design	Protein analysis, diagnostic applications

(*Continued*)

Table 2.4 An extended table showcasing 50 microfluidic devices used for detecting disease-specific proteins. (*Continued*)

Microfluidic device	Principle of operation	Advantages	Applications
Single-molecule analysis	Detection and analysis of individual molecules	High sensitivity, single-molecule resolution, real-time monitoring	Protein analysis, biomarker detection
Impedance-based detection	Measurement of changes in electrical impedance	Label-free, sensitive detection, real-time monitoring	Protein detection, biomarker analysis
Dielectrophoretic devices	Manipulation of particles and cells using dielectrophoresis		

(Sources: [6, 19, 36, 39, 49–51, 53, 56, 57, 59–62, 64].)

Drug delivery systems: nanomaterials, such as liposomes, polymeric nanoparticles, or dendrimers, can encapsulate antileishmanial drugs and enhance their delivery to the infection site, improving therapeutic efficacy and reducing side effects.
Targeted delivery: nanoparticles can be functionalized with ligands targeting *Leishmania*-infected cells, allowing selective drug delivery and reducing damage to healthy tissues.
Combination therapies: nanomaterials can facilitate the co-delivery of multiple drugs, including antileishmanial drugs and immunomodulatory agents, enabling synergistic therapeutic effects and enhanced immune responses [3, 40, 47, 50, 52].
Photodynamic therapy (PDT): nanoparticles can act as photosensitizers to generate reactive oxygen species upon light irradiation, selectively killing *Leishmania* parasites.
Prevention of cutaneous leishmaniasis: nanomaterials also hold promise in the prevention and control of cutaneous leishmaniasis.
Vaccines: nanoparticles can serve as delivery platforms for *Leishmania* antigens, enhancing antigen presentation and immune responses, leading to effective vaccines against cutaneous leishmaniasis.
Vector control: nanoparticles can be used to develop novel insecticides or repellents to effectively control sandfly vectors responsible for transmitting *Leishmania* parasites.

Antimicrobial surfaces: nanomaterials can be utilized to develop coatings or surfaces with antimicrobial properties, inhibiting the growth of *Leishmania* parasites and preventing their transmission.

Challenges and Future Directions

While nanomaterials show great promise, several challenges need to be addressed for their effective utilization in cutaneous leishmaniasis:

Safety and biocompatibility: extensive safety evaluations are necessary to ensure the biocompatibility of nanomaterials and minimize potential toxic effects.
Scalability and cost-effectiveness: the scalability and cost-effectiveness of nanomaterial-based approaches need to be considered for their practical implementation in resource-limited settings.
Regulatory considerations: regulatory frameworks must be established to evaluate the safety, efficacy, and quality of nanomaterial-based diagnostics, therapeutics, and preventive measures.

Nanomaterials have the potential to revolutionize the diagnosis, treatment, and prevention of cutaneous leishmaniasis [10, 19, 25, 52]. Ongoing research and development efforts are focused on further optimizing nanomaterial properties, exploring novel delivery systems, and conducting clinical trials to validate their effectiveness in real-world settings. With continued advancements, nanomaterial-based approaches may contribute significantly to controlling and eliminating cutaneous leishmaniasis, improving patient outcomes and public health (Table 2.5).

2.12 Graphene Oxide (GO): A Two-Dimensional (2D) Nanomaterial

Graphene oxide (GO), a 2D nanomaterial derived from graphene, has garnered significant attention in biomedical applications, including wound healing. The prospective features of functional 2D nanomaterial GO in the wound healing process can be summarized as follows:

1. Antibacterial properties: GO exhibits inherent antibacterial properties that can help prevent wound infection. It can inhibit various bacteria's growth and proliferation, including Gram-positive and Gram-negative strains. The antibacterial

Table 2.5 Here is a table showcasing 100 nanomaterials and their applications in the context of cutaneous leishmaniasis.

Nanomaterial	Application
Liposomes	Drug delivery systems for targeted release of antileishmanial drugs
Polymeric nanoparticles	Encapsulation of drugs and enhanced delivery to *Leishmania*-infected cells
Dendrimers	Controlled release of antileishmanial agents and improved drug delivery
Gold nanoparticles	Detection of *Leishmania* parasites or specific biomarkers for diagnosis
Quantum dots	Fluorescent probes for sensitive and specific detection of *Leishmania* parasites
Carbon nanotubes	Targeted drug delivery, photothermal therapy, and immunomodulation
Iron oxide nanoparticles	Magnetic targeting of antileishmanial drugs and hyperthermia-based therapy
Silver nanoparticles	Antimicrobial agents, wound healing, and drug delivery
Chitosan nanoparticles	Enhanced drug delivery and wound healing properties
Cerium oxide nanoparticles	Antioxidant and immunomodulatory properties for wound healing
Titanium dioxide nanoparticles	Photocatalytic activity for photodynamic therapy and antimicrobial effects
Silica nanoparticles	Drug delivery, imaging, and biosensing applications
Zinc oxide nanoparticles	Antimicrobial and wound healing properties
Poly(glycolic acid-co-lactic acid) nanoparticle	Sustained drug release, immunomodulation, and targeted drug delivery

(*Continued*)

Table 2.5 Here is a table showcasing 100 nanomaterials and their applications in the context of cutaneous leishmaniasis. (*Continued*)

Nanomaterial	Application
Graphene oxide	Drug delivery, photothermal therapy, and immunomodulation
Quantum rods	Fluorescent probes for sensitive detection of *Leishmania* parasites
Polymeric micelles	Encapsulation of antileishmanial drugs for improved solubility and targeted delivery
Gold nanorods	Photothermal therapy and imaging applications
Fullerenes	Antioxidant and immunomodulatory properties
Magnetic nanogels	Targeted delivery of drugs and hyperthermia-based therapy
Lipid-based nanoparticles	Enhanced drug delivery and stability
Selenium nanoparticles	Antioxidant and wound healing properties
Mesoporous silica nanoparticles	Drug delivery, controlled release, and targeting
Protein-based nanoparticles	Drug delivery and immunomodulation
Manganese oxide nanoparticles	Imaging and targeted drug delivery
Zinc oxide nanowires	Antimicrobial effects and wound healing
Calcium phosphate nanoparticles	Bone regeneration and drug delivery
Hydroxyapatite nanoparticles	Bone tissue engineering and drug delivery
Copper nanoparticles	Antimicrobial and wound healing properties
Silk-based nanoparticles	Drug delivery and wound healing properties
Gold nanocages	Photothermal therapy and imaging applications
Silver nanowires	Antibacterial effects and wound healing

(*Continued*)

Table 2.5 Here is a table showcasing 100 nanomaterials and their applications in the context of cutaneous leishmaniasis. (*Continued*)

Nanomaterial	**Application**
Magnetic nanowires	Targeted drug delivery and hyperthermia-based therapy
Iron oxide nanowires	Magnetic targeting and hyperthermia-based therapy
Zinc oxide nanorods	Antibacterial and wound healing properties
Silver nanoclusters	Antibacterial effects and wound healing
Core-shell nanoparticles	Drug delivery, imaging, and targeted therapy
Polymeric nanogels	Encapsulation of antileishmanial drugs for improved stability and targeted delivery
Hybrid nanoparticles	Combined functionalities for improved diagnostics and therapeutics
Peptide-based nanoparticles	Drug delivery, targeting, and immunomodulation
Magnetic liposomes	Targeted drug delivery and imaging
Iron oxide nanocubes	Magnetic targeting and imaging
Lipid-based nanocarriers	Drug delivery and stability
Polymeric nanospheres	Controlled release of drugs and immunomodulation
Gold nanoshells	Photothermal therapy and imaging applications
Magnetic nanorods	Magnetic targeting and imaging
Silver nanocubes	Antibacterial effects and wound healing
Lipid-based nanogels	Drug delivery, stability, and biocompatibility
Mesoporous carbon nanoparticles	Drug delivery, imaging, and biosensing applications
Titanium dioxide nanotubes	Photocatalytic activity and drug delivery

(*Continued*)

Table 2.5 Here is a table showcasing 100 nanomaterials and their applications in the context of cutaneous leishmaniasis. (*Continued*)

Nanomaterial	Application
Hydrogels	Controlled release of drugs, wound healing, and tissue engineering
Zinc oxide nanosheets	Antimicrobial effects and wound healing
Silver nanowhiskers	Antibacterial and wound healing properties
Nanoemulsions	Enhanced drug solubility and delivery
PLGA nanofibers	Sustained release of antileishmanial drugs
Gold nanoclusters	Photothermal therapy and imaging applications
Silica nanowires	Enhanced drug delivery and stability
Polymeric nanocapsules	Controlled release and targeted drug delivery
Zinc oxide nanotubes	Antibacterial and wound healing properties
Gold nanowires	Enhanced drug delivery and imaging applications
Chitosan-based hydrogels	Wound healing and controlled drug release
Silver nanohybrids	Antibacterial effects and wound healing
PLGA nanospheres	Controlled drug release and targeted therapy
Cerium oxide nanotubes	Antioxidant and immunomodulatory properties for wound healing
Albumin-based nanoparticles	Drug delivery, biocompatibility, and targeted therapy
Magnetic nanocarriers	Targeted drug delivery and imaging

(Sources: [1, 3, 6, 8].)

activity of GO is attributed to its ability to disrupt the bacterial cell membrane and induce oxidative stress, leading to bacterial cell death.

2. Accelerated healing: GO has been reported to promote wound healing by accelerating tissue regeneration. It stimulates cell migration, proliferation, and differentiation of

various cell types involved in wound healing, such as fibroblasts, keratinocytes, and endothelial cells. The presence of GO in wound dressings or scaffolds can facilitate the formation of new blood vessels and promote tissue regeneration.

3. Hemostatic properties: GO-based materials have shown hemostatic properties, aiding in controlling bleeding and promoting clot formation. The large surface area and high adsorption capacity allow it to interact with blood components, forming stable blood clots. This property is particularly beneficial in managing wound bleeding and reducing the risk of excessive blood loss.
4. Enhanced drug delivery: GO can be an effective carrier for the controlled release of therapeutic agents in wound healing. Its large surface area and high loading capacity enable efficient encapsulation and release of bioactive molecules, such as growth factors, antimicrobial agents, and anti-inflammatory drugs. Graphene oxide-based drug delivery systems can provide sustained release profiles, ensuring prolonged and localized therapeutic effects.
5. Modulation of Inflammation: GO exhibits immunomodulatory properties that can help regulate the inflammatory response during wound healing. It can modulate the secretion of pro-inflammatory cytokines and chemokines, reducing excessive inflammation while promoting the recruitment of immune cells involved in tissue repair. This balanced immunomodulation can facilitate optimal wound-healing outcomes.
6. Biocompatibility and biodegradability: GO has shown good biocompatibility and low cytotoxicity, making it suitable for biomedical applications. With proper functionalization and surface modifications, the biodegradability of GO can be tailored, allowing it to degrade gradually within the wound environment without causing long-term adverse effects.
7. Electrical and thermal conductivity: GO's excellent electrical and thermal conductivity can benefit wound healing. Electrical stimulation using GO-based materials has been explored to enhance wound closure, angiogenesis, and collagen synthesis. Moreover, the thermal conductivity of GO can help maintain an optimal wound temperature, which is crucial for the wound-healing process.

Table 2.6 Here is a table showcasing 100 two-dimensional (2D) nanomaterials.

Nanomaterial	Chemical formula	Applications
Graphene	C	Electronics, energy storage, sensors, composites, biomedical applications
Graphene oxide	C_xOH_y	Biomedical applications, drug delivery, biosensors, tissue engineering
Transition metal dichalcogenides (TMDs)	MoS_2, WS_2, etc.	Electronics, optoelectronics, catalysis, energy storage, sensors
Black phosphorus (Phosphorene)	P	Electronics, optoelectronics, energy storage, sensors, drug delivery
Hexagonal boron nitride (h-BN)	BN	Electronics, thermal management, lubricants, composites, sensors
Silicene	Si	Electronics, optoelectronics, energy storage, sensors, photocatalysis
Germanene	Ge	Electronics, optoelectronics, energy storage, sensors, catalysts
Antimonene	Sb	Electronics, optoelectronics, sensors, energy storage
Tellurene	Te	Electronics, energy storage, photocatalysis, sensors
Phosphorene	P	Electronics, optoelectronics, energy storage, sensors, catalysis
Stanene	Sn	Electronics, optoelectronics, energy storage, sensors, catalysis

(*Continued*)

Table 2.6 Here is a table showcasing 100 two-dimensional (2D) nanomaterials. (*Continued*)

Nanomaterial	Chemical formula	Applications
Molybdenum disulfide (MoS_2)	MoS_2	Electronics, optoelectronics, energy storage, sensors, catalysis
Tungsten disulfide (WS_2)	WS_2	Electronics, optoelectronics, energy storage, sensors, catalysis
Titanium carbide (MXene)	Ti_3C_2	Energy storage, supercapacitors, sensors, electromagnetic shielding
Zirconium carbide (MXene)	Zr_3C_2	Energy storage, supercapacitors, sensors, electromagnetic shielding
Vanadium carbide (MXene)	V_2C	Energy storage, supercapacitors, sensors, electromagnetic shielding
Graphdiyne	C_4H_2	Electronics, energy storage, sensors, catalysis, gas separation
Graphane	C_1H_1	Electronics, energy storage, sensors, catalysis, gas separation
Borophene	B	Electronics, energy storage, sensors, catalysis, hydrogen storage
Silicane	Si	Electronics, energy storage, sensors, catalysis, hydrogen storage
Germanane	Ge	Electronics, energy storage, sensors, catalysis, hydrogen storage
Phosphane	P	Electronics, energy storage, sensors, catalysis, hydrogen storage

(*Continued*)

Table 2.6 Here is a table showcasing 100 two-dimensional (2D) nanomaterials. (*Continued*)

Nanomaterial	Chemical formula	Applications
Arsenene	As	Electronics, energy storage, sensors, catalysis, hydrogen storage
Stanene	Sn	Electronics, energy storage, sensors, catalysis, hydrogen storage
Bismuthene	Bi	Electronics, energy storage, sensors, catalysis, hydrogen storage
Selenophene	Se	Electronics, energy storage, sensors, catalysis, hydrogen storage
Indium selenide (InSe)	InSe	Electronics, optoelectronics, sensors, energy storage, catalysis
Tin disulfide (SnS_2)	SnS_2	Electronics, optoelectronics, sensors, energy storage, catalysis
Black phosphorus nitride (BPN)	BP_xN_y	Electronics, optoelectronics, sensors, energy storage, catalysis
Transition metal carbides (MXenes)	Ti_2C, Nb_2, etc.	Energy storage, sensors, catalysis, electromagnetic shielding
Sila-graphene	Si	Electronics, energy storage, sensors, catalysis, composites
Germanane	Ge	Electronics, energy storage, sensors, catalysis, composites
Phosphorene	P	Electronics, energy storage, sensors, catalysis, composites

(*Continued*)

Table 2.6 Here is a table showcasing 100 two-dimensional (2D) nanomaterials. (*Continued*)

Nanomaterial	Chemical formula	Applications
Tinene	Sn	Electronics, energy storage, sensors, catalysis, composites
MoSSe	MoSSe	Optoelectronics, sensors, catalysis, energy storage
MoSTe	MoSTe	Optoelectronics, sensors, catalysis, energy storage
WS_2-Re	WS_2-Re	Electronics, optoelectronics, sensors, energy storage, catalysis
Blue phosphorene	P	Electronics, optoelectronics, energy storage, sensors, catalysis
Rhombohedral boron nitride (r-BN)	BN	Electronics, thermal management, optoelectronics, sensors, composites
Monolayer molybdenum diselenide ($MoSe_2$)	$MoSe_2$	Electronics, optoelectronics, sensors, energy storage, catalysis
Monolayer molybdenum disulfide (MoS_2)	MoS_2	Electronics, optoelectronics, sensors, energy storage, catalysis
Monolayer tungsten diselenide (WSe_2)	WSe_2	Electronics, optoelectronics, sensors, energy storage, catalysis

(Sources: [1, 9, 21, 32].)

8. Scaffold and tissue engineering applications: GO can be incorporated into scaffolds and hydrogels to create three-dimensional (3D) structures for tissue engineering applications. These scaffolds mimic the extracellular matrix and provide mechanical support for cell adhesion, proliferation, and differentiation. The unique properties of GO enhance

the mechanical strength, electrical conductivity, and bioactivity of the scaffolds, facilitating tissue regeneration.

It is important to note that while graphene oxide exhibits promising properties for wound healing, further research is required to fully understand its long-term effects, optimize its functionalization, and assess its safety for clinical translation. Nonetheless, the prospective features of functional 2D nanomaterial graphene oxide make it an attractive candidate for future advancements in wound healing technologies (Table 2.6).

Conclusion

In conclusion, the design and synthesis of novel nanomaterials play a pivotal role in combating infectious diseases. Nanotechnology offers immense potential to address the challenges of infectious diseases, including the emergence of drug-resistant pathogens and the need for more effective diagnostic and therapeutic approaches. By harnessing the unique properties of nanomaterials, researchers can develop innovative strategies to prevent, diagnose, and treat infectious diseases more efficiently. The design of nanomaterials specifically tailored for infectious diseases enables targeted interactions with pathogens, facilitating enhanced detection, drug delivery, and immune response modulation. Through careful engineering, nanomaterials can be functionalized to selectively bind to infectious agents, providing compassionate and specific diagnostic tools.

Furthermore, nanomaterial-based drug delivery systems offer controlled release mechanisms, ensuring optimal therapeutic outcomes while minimizing side effects. The synthesis of novel nanomaterials opens up new possibilities for addressing infectious diseases. Researchers can explore nanomaterial compositions, structures, and surface modifications to optimize their efficacy and biocompatibility. Developing multifunctional nanosystems, such as hybrid materials and nanocomposites, also allows for synergistic effects and versatile applications in infectious disease management. Importantly, nanomaterials for infectious diseases require a multidisciplinary approach involving materials science, chemistry, biology, and medical expertise. Collaborations among scientists, engineers, and clinicians are essential to translate nanomaterial-based solutions from the laboratory to clinical practice. As research progresses, it is crucial to consider safety, scalability, and regulatory aspects in the design and synthesis of nanomaterials for infectious diseases. Thorough characterization, toxicity assessments, and adherence to regulatory guidelines are paramount to ensure the safe implementation of these technologies.

In summary, the design and synthesis of novel nanomaterials hold tremendous promise for combating infectious diseases. By leveraging the unique properties of nanomaterials, we can develop innovative solutions for disease detection, treatment, and prevention. Continued research and collaboration in this field will pave the way for transformative advancements in infectious disease management, ultimately improving global health outcomes.

Future Prospect

The prospects of developing and synthesizing novel nanomaterials for infectious diseases are up-and-coming, with significant potential for advancements in diagnosis, treatment, and prevention. Here are some critical prospects in this field:

Enhanced sensitivity and specificity: nanomaterials can be further engineered to improve their sensitivity and specificity for detecting infectious agents. By incorporating advanced sensing technologies and surface modifications, nanomaterial-based diagnostic platforms can achieve faster, more accurate, and point-of-care detection of infectious diseases.

Targeted drug delivery: future research aims to enhance the targeting capabilities of nanomaterials to specific sites of infection. By incorporating targeting ligands or responsive drug release mechanisms, nanomaterial-based drug delivery systems can selectively deliver therapeutics to infected cells or tissues, minimizing off-target effects and improving treatment outcomes.

Combination therapies: nanomaterials offer the opportunity for combination therapies, where multiple therapeutic agents can be loaded into a single nanosystem. This approach can overcome drug resistance, improve treatment efficacy, and reduce the development of new resistant strains.

Immunomodulation: nanomaterials can be designed to modulate the immune response to infectious diseases. Future nanomaterials may be engineered to enhance immune activation or suppress inflammatory responses, promoting an appropriate and controlled immune reaction to combat infections.

Vaccine delivery systems: nanomaterials hold the potential for developing advanced vaccine delivery systems. They can enhance antigen presentation, improve vaccine stability, and enable targeted delivery to immune cells, leading to more effective immunization against infectious diseases.

Antimicrobial applications: novel nanomaterials with antimicrobial properties can be explored to prevent and control infectious diseases. These materials can be integrated into medical devices, wound dressings, and surfaces to inhibit the growth of pathogens and prevent infections.

Biofilm disruption: biofilms contribute to many infectious agents' persistence and antibiotic resistance. Future nanomaterials can target and disrupt biofilms, offering new strategies for combating chronic infections.

Bioimaging and theranostics: nanomaterials can be designed for multimodal imaging, combining multiple imaging techniques to visualize infectious agents accurately. Furthermore, integrating diagnostic and therapeutic functionalities within a single nanomaterial, known as theranostics, holds immense potential for personalized medicine and targeted treatment of infectious diseases.

Regulatory considerations: as nanomaterial-based approaches advance, addressing regulatory considerations and ensuring these technologies' safety and ethical use is essential. Developing robust guidelines and standards for the design, synthesis, characterization, and clinical translation of nanomaterials for infectious diseases will be a crucial focus in the future.

Collaboration and interdisciplinary research: nanomaterials for infectious diseases require close collaboration among researchers from various disciplines, including materials science, chemistry, biology, medicine, and engineering. Interdisciplinary research efforts will accelerate advancements in the design and synthesis of novel nanomaterials and their translation into clinical practice.

In summary, the prospects of the design and synthesis of novel nanomaterials for infectious diseases hold immense potential for revolutionizing disease diagnosis, treatment, and prevention. Continued research, technological advancements, and collaborative efforts will drive the development of highly effective and targeted nanomaterial-based strategies to combat infectious diseases, leading to improved global health outcomes.

Acknowledgement

We want to express our heartfelt gratitude to all individuals and organizations who have contributed to completing this work on the design and synthesis of novel nanomaterials emphasizing infectious diseases. First and foremost, we sincerely thank the researchers and scientists in nanotechnology and infectious diseases whose groundbreaking work and valuable insights have paved the way for advancements in this area. Their dedication and relentless pursuit of knowledge have been instrumental in shaping the future of nanomaterial-based approaches to combating infectious diseases.

Authors' Contributions

The authors of this work have made significant contributions to the research project/study. Each author has participated sufficiently in the research project/study, made intellectual contributions, and is responsible for the work's accuracy and integrity. The authors have collaborated closely, ensuring the completion of this work through collective effort, expertise, and dedication.

References

1. Cordeiro, T.A.R., de Resende, M.A.C., dos S. Moraes, S.C., Franco, D.L., Pereira, A.C., Ferreira, L.F., Electrochemical biosensors for neglected tropical diseases: A review. *Talanta*, 234, 122617, 2021, doi: https://doi.org/10.1016/j.talanta.2021.122617.
2. Khatun, S., Putta, C.L., Hak, A., Rengan, A.K., Immunomodulatory nanosystems: An emerging strategy to combat viral infections. *Biomater. Biosyst.*, 9, 100073, 2023, doi: https://doi.org/10.1016/j.bbiosy.2023.100073.
3. Sharma, S. and Dhadly, D.K., Chapter 5 - Nano-inspired biosensors and plant diseases: recent advances and challenges, in: *Nanomaterial-Plant Interactions*, S. Seena, A. Rai, S. B. T.-N., P.-M.I. Kumar (Eds.), pp. 135–162, Academic Press, 2023, doi: https://doi.org/10.1016/B978-0-323-90619-7.00002-3.
4. Harish, V. *et al.*, Cutting-edge advances in tailoring size, shape, and functionality of nanoparticles and nanostructures: A review. *J. Taiwan Inst. Chem. Eng.*, 149, 105010, 2023, doi: https://doi.org/10.1016/j.jtice.2023.105010.
5. Lim, H.J., Saha, T., Tey, B.T., Tan, W.S., Ooi, C.W., Quartz crystal microbalance-based biosensors as rapid diagnostic devices for infectious diseases. *Biosens. Bioelectron.*, 168, 112513, 2020, doi: https://doi.org/10.1016/j.bios.2020.112513.
6. Taha, B.A. *et al.*, Next-generation nanophotonic-enabled biosensors for intelligent diagnosis of SARS-CoV-2 variants. *Sci. Total Environ.*, 880, 163333, 2023, doi: https://doi.org/10.1016/j.scitotenv.2023.163333.
7. Dervisevic, M., Alba, M., Prieto-Simon, B., Voelcker, N.H., Skin in the diagnostics game: Wearable biosensor nano- and microsystems for medical diagnostics. *Nano Today*, 30, 100828, 2020, doi: https://doi.org/10.1016/j.nantod.2019.100828.

8. Balestri, A., Cardellini, J., Berti, D., Gold and silver nanoparticles as tools to combat multidrug-resistant pathogens. *Curr. Opin. Colloid Interface Sci.*, 66, 101710, 2023, doi: https://doi.org/10.1016/j.cocis.2023.101710.
9. Ansari, A.A. and Malhotra, B.D., Current progress in organic–inorganic hetero-nano-interfaces based electrochemical biosensors for healthcare monitoring. *Coord. Chem. Rev.*, 452, 214282, 2022, doi: https://doi.org/10.1016/j.ccr.2021.214282.
10. Maza Vega, D., Di Meglio, M., del V. Alonso, S., Alvira, F., Montanari, J., Nanomaterials for diagnosis, treatment, and prevention of human cutaneous leishmaniasis: A review. *OpenNano*, 12, 100158, 2023, doi: https://doi.org/10.1016/j.onano.2023.100158.
11. Wang, Y., Yang, K., Zhou, H., Immunogenic proteins and potential delivery platforms for mpox virus vaccine development: A rapid review. *Int. J. Biol. Macromol.*, 245, 125515, 2023, doi: https://doi.org/10.1016/j.ijbiomac.2023.125515.
12. Gao, Y. *et al.*, Nanocatalysis meets microfluidics: A powerful platform for sensitive bioanalysis. *TrAC, Trends Anal. Chem.*, 158, 116887, 2023, doi: https://doi.org/10.1016/j.trac.2022.116887.
13. Mohapatra, B., Chamoli, S., Salvi, P., Saxena, S.C., Fostering nanoscience's strategies: A new frontier in sustainable crop improvement for abiotic stress tolerance. *Plant Nano Biol.*, 3, 100026, 2023, doi: https://doi.org/10.1016/j.plana.2023.100026.
14. Seaberg, J., Clegg, J.R., Bhattacharya, R., Mukherjee, P., Self-therapeutic nanomaterials: Applications in biology and medicine. *Mater. Today*, 62, 190–224, 2023, doi: https://doi.org/10.1016/j.mattod.2022.11.007.
15. Amir, S., Arathi, A., Reshma, S., Mohanan, P.V., Microfluidic devices for the detection of disease-specific proteins and other macromolecules, disease modelling and drug development: A review. *Int. J. Biol. Macromol.*, 235, 123784, 2023, doi: https://doi.org/10.1016/j.ijbiomac.2023.123784.
16. Chang, C. *et al.*, Engineered M13 phage as a novel therapeutic bionanomaterial for clinical applications: From tissue regeneration to cancer therapy. *Mater. Today Bio*, 20, 100612, 2023, doi: https://doi.org/10.1016/j.mtbio.2023.100612.
17. Huang, J., Li, W., Bai, X., Xiao, F., Xu, H., Metal–organic framework detectives meet infectious pathogens: A powerful tool against pandemics. *Coord. Chem. Rev.*, 488, 215160, 2023, doi: https://doi.org/10.1016/j.ccr.2023.215160.
18. Ehtesabi, H., Fayaz, M., Hosseini-Doabi, F., Rezaei, P., The application of green synthesis nanoparticles in wound healing: a review. *Mater. Today Sustain.*, 21, 100272, 2023, doi: https://doi.org/10.1016/j.mtsust.2022.100272.
19. Ahmed, A. *et al.*, Nature-inspired biogenic synthesis of silver nanoparticles for antibacterial applications. *Mater. Today Chem.*, 27, 101339, 2023, doi: https://doi.org/10.1016/j.mtchem.2022.101339.

20. Wu, S., Song, R., Liu, T., Li, C., Antifungal therapy: Novel drug delivery strategies driven by new targets. *Adv. Drug Deliv. Rev.*, 199, 114967, 2023, doi: https://doi.org/10.1016/j.addr.2023.114967.
21. Roy, S., Hasan, I., Guo, B., Recent advances in nanoparticle-mediated antibacterial applications. *Coord. Chem. Rev.*, 482, 215075, 2023, doi: https://doi.org/10.1016/j.ccr.2023.215075.
22. Chandrasekaran, R. *et al.*, Prospective features of functional 2D nanomaterial graphene oxide in the wound healing process. *J. Drug Deliv. Sci. Technol.*, 82, 104352, 2023, doi: https://doi.org/10.1016/j.jddst.2023.104352.
23. Singh, S. *et al.*, Novel insights into graphene oxide-based adsorbents for remediation of hazardous pollutants from aqueous solutions: A comprehensive review. *J. Mol. Liq.*, 369, 120821, 2023, doi: https://doi.org/10.1016/j.molliq.2022.120821.
24. Biswas, G.C., Khan, M.T.M., Das, J., Wearable nucleic acid testing platform - A perspective on rapid self-diagnosis and surveillance of infectious diseases. *Biosens. Bioelectron.*, 226, 115115, 2023, doi: https://doi.org/10.1016/j.bios.2023.115115.
25. Aflakian, F. *et al.*, Nanoparticles-based therapeutics for the management of bacterial infections: A special emphasis on FDA approved products and clinical trials. *Eur. J. Pharm. Sci.*, 188, 106515, 2023, doi: https://doi.org/10.1016/j.ejps.2023.106515.
26. Puri, N. and Gupta, A., Water remediation using titanium and zinc oxide nanomaterials through disinfection and photo catalysis process: A review. *Environ. Res.*, 227, 115786, 2023, doi: https://doi.org/10.1016/j.envres.2023.115786.
27. Gumus, E., Bingol, H., Zor, E., Lateral flow assays for detection of disease biomarkers. *J. Pharm. Biomed. Anal.*, 225, 115206, 2023, doi: https://doi.org/10.1016/j.jpba.2022.115206.
28. Bushra, R. *et al.*, Polysaccharide-based nanoassemblies: From synthesis methodologies and industrial applications to future prospects. *Adv. Colloid Interface Sci.*, 318, 102953, 2023, doi: https://doi.org/10.1016/j.cis.2023.102953.
29. Feyziazar, M. *et al.*, Recent advances on the piezoelectric, electrochemical, and optical biosensors for the detection of protozoan pathogens. *TrAC, Trends Anal. Chem.*, 157, 116803, 2022, doi: https://doi.org/10.1016/j.trac.2022.116803.
30. Deroco, P.B., Wachholz Junior, D., Kubota, L.T., Recent advances in point-of-care biosensors for the diagnosis of neglected tropical diseases. *Sens. Actuators B Chem.*, 349, 130821, 2021, doi: https://doi.org/10.1016/j.snb.2021.130821.
31. Jain, S. *et al.*, Internet of medical things (IoMT)-integrated biosensors for point-of-care testing of infectious diseases. *Biosens. Bioelectron.*, 179, 113074, 2021, doi: https://doi.org/10.1016/j.bios.2021.113074.
32. Pourmadadi, M., Shamsabadipour, A., Aslani, A., Eshaghi, M.M., Rahdar, A., Pandey, S., Development of Polyvinylpyrrolidone-Based nanomaterials

for biosensors applications: A Review. *Inorg. Chem. Commun.*, 152, 110714, 2023, doi: https://doi.org/10.1016/j.inoche.2023.110714.

33. Li, C. *et al.*, Highly efficient, remarkable sensor activity and energy storage properties of MXenes and borophene nanomaterials. *Prog. Solid State Chem.*, 70, 100392, 2023, doi: https://doi.org/10.1016/j.progsolidstchem.2023.100392.
34. Kadadou, D. *et al.*, Recent advances in the biosensors application for the detection of bacteria and viruses in wastewater. *J. Environ. Chem. Eng.*, 10, 1, 107070, 2022, doi: https://doi.org/10.1016/j.jece.2021.107070.
35. Azizoglu, R.O., CRISPR-Cas systems in diagnostics: A comprehensive assessment of Cas effectors and biosensors. *Gene Genome Ed.*, 3–4, 100019, 2022, doi: https://doi.org/10.1016/j.ggedit.2022.100019.
36. Wang, Y. *et al.*, Micro/nano biomedical devices for point-of-care diagnosis of infectious respiratory diseases. *Med. Nov. Technol. Devices*, 14, 100116, 2022, doi: https://doi.org/10.1016/j.medntd.2022.100116.
37. Nagraik, R., Sharma, A., Kumar, D., Mukherjee, S., Sen, F., Kumar, A.P., Amalgamation of biosensors and nanotechnology in disease diagnosis: Mini-review. *Sens. Int.*, 2, 100089, 2021, doi: https://doi.org/10.1016/j.sintl.2021.100089.
38. Rajpal, S. and Mishra, P., Next generation biosensors employing molecularly imprinted polymers as sensing elements for *in vitro* diagnostics. *Biosens. Bioelectron. X*, 11, 100201, 2022, doi: https://doi.org/10.1016/j.biosx.2022.100201.
39. Aquino, A., Paschoalin, V.M.F., Tessaro, L.L.G., Raymundo-Pereira, P.A., Conte-Junior, C.A., Updating the use of nano-biosensors as promising devices for the diagnosis of coronavirus family members: A systematic review. *J. Pharm. Biomed. Anal.*, 211, 114608, 2022, doi: https://doi.org/10.1016/j.jpba.2022.114608.
40. Mukherjee, S. and Das, S., Chapter 14 - Role of biosensor-based devices for diagnosis of nononcological disorders. *Biosensor Based Advanced Cancer Diagnostics*, R. Khan, A. Parihar, S. K. B. T.-B. B. A. C. D. Sanghi (Eds.), pp. 245–256, Academic Press, 2022, doi: https://doi.org/10.1016/B978-0-12-823424-2.00020-X.
41. Singh, A.K., Mittal, S., Das, M., Saharia, A., Tiwari, M., Optical biosensors: a decade in review. *Alex. Eng. J.*, 67, 673–691, 2023, doi: https://doi.org/10.1016/j.aej.2022.12.040.
42. Zahraee, H., Khoshbin, Z., Mohammadi, F., Mashreghi, M., Abnous, K., Taghdisi, S.M., Three-way junction skeleton biosensors based on aptamers, DNAzymes, and DNA hybridization probes. *TrAC, Trends Anal. Chem.*, 165, 117160, 2023, doi: https://doi.org/10.1016/j.trac.2023.117160.
43. Sanz, C.G. *et al.*, Sensing Materials: Electrochemical Applications of DNA Sensors and Biosensors, in: *Molecularly Imprinted Polymers (MIPs)*, First, R. B. T.-E., S. and B., E. Narayan (Eds.), pp. 445–467, Elsevier, Oxford, 2023, doi: https://doi.org/10.1016/B978-0-12-822548-6.00039-X.

44. Soler, M. and Huertas, C.S., Label-Free Plasmonic Biosensors in Clinical Diagnostics, in: *Encyclopedia of Sensors and Biosensors*, First, R. B. T.-E., S. and B., E. Narayan (Eds.), pp. 383–401, Elsevier, Oxford, 2023, doi: https://doi.org/10.1016/B978-0-12-822548-6.00119-9.
45. Liu, J., Xie, G., Lv, S., Xiong, Q., Xu, H., Recent applications of rolling circle amplification in biosensors and DNA nanotechnology. *TrAC, Trends Anal. Chem.*, 160, 116953, 2023, doi: https://doi.org/10.1016/j.trac.2023.116953.
46. Ham, Y. *et al.*, The SpACE-CCM: A facile and versatile cell culture medium-based biosensor for detection of SARS-CoV-2 spike-ACE2 interaction. *Biosens. Bioelectron.*, 227, 115169, 2023, doi: https://doi.org/10.1016/j.bios.2023.115169.
47. Raghuwanshi, R. and Singh, M., Chapter 14 - Challenges to molecularly imprinted polymers in competition with biosensors at commercial scale, in: *Molecularly Imprinted Polymers (MIPs)*, M. B. T.-M. I. P. (MIPs) and Singh (Eds.), pp. 345–366, Elsevier, 2023, doi: https://doi.org/10.1016/B978-0-323-91925-8.00005-3.
48. Kumar, V. and Arora, K., Trends in nano-inspired biosensors for plants. *Mater. Sci. Energy Technol.*, 3, 255–273, 2020, doi: https://doi.org/10.1016/j.mset.2019.10.004.
49. Kumar, V. *et al.*, Mitigation of heat stress responses in crops using nitrate primed seeds. *South Afr. J. Bot.*, 140, 25–36, 2021, doi: https://doi.org/10.1016/j.sajb.2021.03.024.
50. Kumar, P., Devi, P., Dey, S.R., Chapter 6 - Fungal volatile compounds: a source of novel in plant protection agents, in: *Volatiles and metabolites of microbes*, A. Kumar, J. Singh, J. B. T.-V. and M. M. Samuel (Eds.), pp. 83–104, Academic Press, 2021, doi: https://doi.org/10.1016/B978-0-12-824523-1.00001-8.
51. Kotia, A. *et al.*, Rheological analysis of rice husk-starch suspended in water for sustainable agriculture application. *Mater. Today Proc.*, 50, 1962–1966, 2022, doi: https://doi.org/10.1016/j.matpr.2021.09.325.
52. Adelodun, B. *et al.*, *List of Contributors*, A. Kumar, J. Singh, L. F. R. B. T.-M. U. C. C. Ferreira (Eds.), pp. xix–xxiv, Woodhead Publishing, 2022, doi: https://doi.org/10.1016/B978-0-323-90571-8.00028-6.
53. Kumar, P. and Mistri, T.K., Transcription factors in SOX family: Potent regulators for cancer initiation and development in the human body. *Semin. Cancer Biol.*, 67, 105–113, 2020, doi: https://doi.org/10.1016/j.semcancer.2019.06.016.
54. Das, T. *et al.*, Promising botanical-derived monoamine oxidase (MAO) inhibitors: pharmacological aspects and structure-activity studies. *S. Afr. J. Bot.*, 146, 127–145, 2022, doi: https://doi.org/10.1016/j.sajb.2021.09.019.
55. Kumari, P., Singh, J., Kumar, P., Chapter 21 - Impact of bioenergy for the diminution of an ascending global variability and change in the climate, in: *Volatiles and metabolites of microbes*, A. Kumar, J. Singh, L. F. R. B. T.-M. U. C. C. Ferreira (Eds.), pp. 469–487, Woodhead Publishing, 2022, doi: https://doi.org/10.1016/B978-0-323-90571-8.00021-3.

56. Kumar, P., Kumar, T., Singh, S., Tuteja, N., Prasad, R., Singh, J., Potassium: A key modulator for cell homeostasis. *J. Biotechnol.*, 324, 198–210, 2020.
57. Kumar, P., Sharma, K., Saini, L., Dey, S.R., Chapter 8 - Role and behavior of microbial volatile organic compounds in mitigating stress, in: *Volatiles and metabolites of microbes*, A. Kumar, J. Singh, J. B. T.-V. and M., M. Samuel (Eds.), pp. 143–161, Academic Press, 2021, doi: https://doi.org/10.1016/B978-0-12-824523-1.00010-9.
58. Chakraborty, S. *et al.*, Unravelling the regulatory role of miRNAs in secondary metabolite production in medicinal crops. *Plant Gene*, 27, 100303, 2021, doi: https://doi.org/10.1016/j.plgene.2021.100303.
59. Aley, P., Singh, J., Kumar, P., Chapter 23 - Adapting the changing environment: microbial way of life, in: *Volatiles and metabolites of microbes*, A. Kumar, J. Singh, L. F. R. B. T.-M. U. C. C. Ferreira (Eds.), pp. 507–525, Woodhead Publishing, 2022, doi: https://doi.org/10.1016/B978-0-323-90571-8.00023-7.
60. Goud, E.L., Singh, J., Kumar, P., Chapter 19 - Climate change and their impact on global food production, in: *Volatiles and metabolites of microbes*, A. Kumar, J. Singh, L. F. R. B. T.-M. U. C. C. Ferreira (Eds.), pp. 415–436, Woodhead Publishing, 2022, doi: https://doi.org/10.1016/B978-0-323-90571-8.00019-5.
61. Upadhyay, S.K. *et al.*, Efficient removal of total arsenic (As3+/5+) from contaminated water by novel strategies mediated iron and plant extract activated waste flowers of marigold. *Chemosphere*, 313, 137551, 2023, doi: https://doi.org/10.1016/j.chemosphere.2022.137551.
62. Akhtar, N. *et al.*, *List of contributors*, A. Kumar, J. Singh, J. B. T.-V. and M., M. Samuel (Eds.), pp. xix–xxi, Academic Press, 2021, doi: https://doi.org/10.1016/B978-0-12-824523-1.00028-6.
63. Radu, A.-F. and Bungau, S.G., Nanomedical approaches in the realm of rheumatoid arthritis. *Ageing Res. Rev.*, 87, 101927, 2023, doi: https://doi.org/10.1016/j.arr.2023.101927.
64. Ali, T., Najam-ul-Haq, M., Mohyuddin, A., Musharraf, S.G., Hussain, D., Next-generation functional nanotextiles—Prospects and challenges. *Sustain. Mater. Technol.*, 36, e00640, 2023, doi: https://doi.org/10.1016/j.susmat.2023.e00640.

3

Role of Nanomaterials in the Development of Nanobiosensors for Infectious Diseases

Ayyappa Bathinapatla[1,2*], Ravikumar Mulpuri[3], Aseena Azeez[4] and Suvardhan Kanchi[5]

[1]*Dept. of Chemistry, CMRIT, Bangalore, India*
[2]*Centre of Excellence-Material Science & Sensors, CMR Institute of Technology, Bengaluru, India*
[3]*Department of Chemistry, Raghu Engineering College (Autonomous), Dakamarri (v), Bheeminipatnam, Visakhapatnam, Andhra Pradesh, India*
[4]*Department of Chemistry, CMR University, Bengaluru, India*
[5]*Department of Chemistry, CHRIST (Deemed to be University), Bengaluru, India*

Abstract

Transmissible illnesses brought on by viruses, bacteria, fungi, and parasites are referred to as infectious diseases. These can escalate into undesirable pandemic circumstances that disrupt both regular life functions and the world's population. These in turn have an effect on the current global economy, lead to joblessness, induce stress on the body, mind, and emotions, and put human survival in jeopardy. Consequently, in order to avert worldwide life impairment, prompt discovery, treatment, isolation, and control of the spread of pandemic infectious diseases within the town of origin are essential. As of right now, the World Health Organization (WHO) lists 12 infectious diseases that can be fatal: COVID-19, severe acute respiratory syndrome (SARS), Middle East respiratory syndrome (MERS), human immunodeficiency virus (HIV), human papilloma virus (HPV), influenza, hepatitis, herpes simplex virus (HSV), Zika virus, chikungunya, dengue, and rota virus. Biosensors are becoming more and more potent instruments for diagnosing infectious diseases. Analytical tools that may transform biochemical data into detectable signals such as optical, electrical, magnetic, or thermal signals are referred to as biosensors. The growing need for highly selective, low-concentration sensing of a wide variety of chemicals has spurred the creation

**Corresponding author*: ayyappa.b@cmrit.ac.in

Suvardhan Kanchi, Ayyappa Bathinapatla, Anitha Varghese and Phumlane Selby Mdluli (eds.) Nano-Biosensor Technologies for Diagnosis of Infectious Diseases, (75–120) © 2025 Scrivener Publishing LLC

of sophisticated instruments known as nanobiosensors, which combine biological components, advanced materials, and nanoscale materials. The design, principle, underlying reasoning, receptor, and molecular features of sensor systems with a focus on the recent COVID-19 pandemic are all covered in this chapter. For critical comparison, electrochemical biosensor systems which included a variety of sophisticated nanostructures like semiconductors, metal organic frameworks (MOFs), MXenes, polymeric nanocomposites, metal and metal oxide nanoparticles, and combinations of biomolecules reported recently were specifically divided into distinct sub-sections. This chapter focuses on the difficulties that exist today in converting lab research into practical device applications, as well as the potential for the future and the commercialization of electrochemical diagnostic devices for the detection of corona viruses. It is anticipated that the background information and overall advancements presented in this study will be instructive for sensor researchers and will make it easier to design and fabricate electrochemical sensors for viruses that pose a threat to human life, with a wider range of applications for any desired pathogen.

Keywords: Infectious diseases, COVID-19, nanomaterials, nanobiosensors

3.1 Introduction

Throughout history, infectious illnesses have had a significant impact on the world and continue to do so now, influencing economies, cultures, healthcare systems, and the security of the global health system. The following are some serious viral diseases and their effects: between 75 and 200 million people are thought to have died during the Black Death (Bubonic Plague) throughout Europe, Asia, and Africa in the 14th century as a result of the *Yersinia pestis* bacteria. About one-third of the world's population was afflicted during the Spanish flu (influenza pandemic of 1918–1919), which was brought on by the H1N1 influenza virus and is estimated to have killed 50 million people globally [1]. The late 20th century saw the emergence of both the acquired immunodeficiency syndrome (AIDS) and the human immunodeficiency virus (HIV). Globally, HIV/AIDS has claimed more than 36 million lives since it was discovered. One of the most ancient infectious illnesses now posing a serious threat to world health is tuberculosis. It affects millions of individuals and results in about 1.5 million deaths annually, mostly in developing nations. High death rates are a hallmark of Ebola epidemics, which have mostly struck Central and West Africa. The new coronavirus SARS-CoV-2, which caused the COVID-19 pandemic, first appeared in late 2019 and spread rapidly throughout the world. Millions of people have died as a result of it, and it has affected daily life, travel, healthcare, education, and economy all across the world. Significant mortality, stress on healthcare systems, economic downturns, and social unrest are some of the effects of the pandemic.

Around December 2019, reports of medical situations involving people in China who had strange diseases that resembled pneumonia surfaced [2]. The Jianghan District of Wuhan's fish markets were the focus of the majority of the cases. An inquiry into the causative agent was warranted by the peculiar cause of this pneumonia-like illness. For sequencing, lower respiratory samples from the infected people were obtained. The novel infectious agent was classified under the β-coronavirus genus by sequence identification. The sequences matched the SARS coronavirus by 80% and bat-SL-CoVZC45 by 85%, respectively [3]. The virus that was sequenced was originally known as 2019-nCoV, but it was eventually called severe acute respiratory syndrome coronavirus 2 (SARS-CoV-2). The virus had spread to 53 nations by February 29, 2020, despite stringent measures to quarantine and restrict travel between China and other nations. On March 11, 2020, following the first case reported in 2019, the World Health Organization (WHO) designated the respiratory ailment as COVID-19 and declared it a worldwide pandemic [4].

SARS-CoV-2 is the seventh virus to infect humans and the first to produce a pandemic in the family of *Coronaviridae*. Infected people can spread it through aerosols when they are asymptomatic, symptomatic, or presymptomatic. By fomite transmission, respiratory droplets from an infected person can potentially infect healthy people. The illness can be asymptomatic, symptomatic, or even potentially fatal in various situations (typically when there are concomitant conditions). As of right now, the symptoms can be grouped into digestive, neurological, musculoskeletal, and respiratory issues. There are two stages to extensive injury in the respiratory system: an early stage and a late one. Direct viral damage and enhanced viral replication are characteristics of the early phase. A subsequent immune response directed against the infected host cells characterizes the late phase. Systemic inflammatory disorders may result from the release of cytokines by the migrating immune cells at the infected areas. A cytokine storm is the name given to this state of protracted circulating inflammatory cytokines. Patients with a high cytokine storm response are more at risk of dying [5].

3.1.1 Classifying SARS-CoV-2

Human coronaviruses, or HCoVs, were not well documented prior to their discovery in the 1960s, and only a small number of them had been identified. In November 2002, the first interspecies, CoV infection, was reported in China, and thus the 21st century began. The 2002 SARS-CoV epidemic served as a trigger for the identification and study of CoVs. Then, in 2012, there was a report of a MERS-CoV infection with Saudi Arabia at its epicenter. The third virus in the *Coronaviridae* family and the source of COVID-19, SARS-CoV-2, is quite similar to the previous two CoVs. Like the others, it has infected humans and crossed the species barrier. According to phylogenetic analysis, SARS-CoV-2 belongs to the *Betacoronavirus* genus, a subfamily of the *Coronaviridae* family of the Nidovirales order's *Orthocoronavirinae* family. The family *Coronaviridae* contains the coronavirus genera α-CoV, β-CoV, Δ-CoV, and γ-CoV. The subsequent division of the genera β-CoV is composed of five lineages [6]. The clinical features of SARS-CoV-2-mediated COVID-19 disease are comparable to those of the SARS-CoV infection. Based on thorough genome phylogenetic analysis, SARS-CoV-2 clusters more closely with bat CoV and SARS-CoV than it does with HCoVs and MERS-CoV. Genetic recombination and mutation may have had a role in the transmission of SARS-CoV-2. It was found that MERS-CoV and SARS-CoV evolved by genetic arrangements, with their origins being in camels and civets, respectively [7]. The origin of the virus in bats is shown by the genetic recombination of bat CoVs in SARS-CoV-2. Furthermore, natural selection may have directly or indirectly through unknown intermediate hosts led to its transmissibility to humans.

3.1.2 Morphology and Genome Structure of SARS-CoV-2

The COVID-19 pandemic's causal agent, SARS-CoV-2, exhibits unique morphological and genetic traits. Under electron microscopy, it has a typical structural look with diameters ranging from around 60 to 140 nm, and it appears spherical or pleomorphic. The characteristic that most distinguishes the virus is that it resembles a crown, which is caused by the many spike-like projections covering its surface; hence, the name "coronavirus." This structural feature makes it possible for the virus to interact with host cells, which in turn makes it easier for it to enter and spread to other cells. The viral nucleocapsid of SARS-CoV-2 is enclosed by the outer lipid envelope that it sheds after its release, protecting its genetic material and other critical components. This envelope is produced from the host cell membrane [8]. The single-stranded ribonucleic acid (RNA) molecule that makes up the majority of the SARS-CoV-2 genome structure is crucial for understanding the virus's genetic composition and biological processes. The positive orientation of this RNA strand suggests that the virus uses the molecular machinery of the host cell to directly transform its genetic material into viral proteins. With a genomic length of over 30,000 nucleotides, the virus is more complex genetically than numerous other RNA viruses. The viral genome is structurally complex, displaying multiple open reading frames (ORFs) that encode different viral proteins essential for the virus's transcription, assembly, replication, and interaction with the host immune system [9]. One essential structural protein that the SARS-CoV-2 genome encodes is the spike (S) protein. It is essential for figuring out whether the virus can infect host cells. The presence of this glycoprotein on the virus's surface is necessary for it to attach to host cell receptors, particularly the ACE2 receptor, which allows the virus to fuse and enter the cell. The envelope (E) protein aids in viral assembly and budding, whereas the membrane (M) protein shapes the viral envelope and is involved in virion assembly activities. Moreover, by binding to the viral RNA and assisting in its packaging and protection, the nucleocapsid (N) protein aids in the virus's overall structural integrity and function as shown in Scheme 3.1 [10].

Comprehending the complex morphology and genomic structure of SARS-CoV-2 is fundamental to the creation of vital diagnostic tests, vaccines, and antiviral medications targeted at containing the spread and consequences of COVID-19. Understanding the structural features of the virus helps to clarify viral pathogenesis, develop targeted treatments, and understand the mechanisms driving host-virus interactions. In addition, a thorough understanding of the genomic structure of SARS-CoV-2 facilitates the clarification of viral evolution, transmission dynamics, and the appearance of new variations, offering critical information required for mitigation and public health responses [11]. The unique morphology of

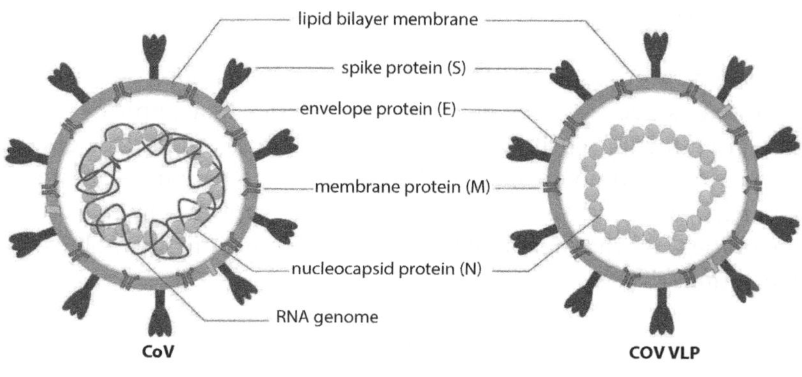

Scheme 3.1 A schematic representation of the coronavirus (CoV) and its assembled VLPs from structural proteins (https://doi.org/10.1016/j.ijbiomac.2022.01.108).

the virus—which is typified by a crown-like appearance and structural proteins such the spike protein—converges with a complex genomic structure, highlighting the virus's pathogenic potential, adaptability, and interactions with the host immune system. This information facilitates research efforts aimed at creating efficient defenses against the virus, deciphering the complexities of its reproduction, and formulating plans to lessen the catastrophic effects of the ongoing pandemic on society, the economy, and public health [12].

Few are classified as variants of concern (VOCs) by the WHO, depending on infection rates around the world [13]. There are now five VOCs that the WHO has reported, including:

- In December 2020, the UK reported the discovery of a novel SARS-CoV-2 VOC. It is the alpha variety, formerly known as GR/501Y.V1 (or B.1.1.7 lineage).
- Beta (B.1.351) and Omicron (B.1.1.529), two unique VOCs, were reported by South Africa in December 2020 and November 2021, respectively.
- Brazil reported the Gamma (P.1) variant in January 2021.
- In December 2020, India reported the variant known as Delta (B.1.617.2). The second, lethal wave of the COVID-19 pandemic was caused by this VOC.

The presence of mutations obstructs monitoring and treatment. After an outbreak of a new virus, extensive biosensing and sequencing are crucial.

3.1.3 Characteristics of Nanomaterials in the Development of Nanobiosensors

Nanomaterials are essential to the creation of nanobiosensors for infectious diseases because they have special qualities that improve the diagnostic instruments' sensitivity, specificity, and effectiveness. Devices called nanobiosensors combine biological molecules with nanomaterials to identify particular biomarkers linked to infectious diseases. The following are some of the ways that nanomaterials aid in the creation of nanobiosensors for infectious diseases:

(i) ***Enhanced surface area:*** because of their high surface-to-volume ratio, nanomaterials—like nanoparticles, carbon nanotubes, graphene and graphene oxide—offer a greater

surface area for the immobilization of biomolecules. The heightened surface area facilitates more effective binding of the target biomarkers, hence enhancing the biosensor's sensitivity [14].

(ii) ***Improved sensitivity and selectivity***: biosensors can be made more sensitive and selective by taking advantage of specific nanomaterials' special optical, electrical, and magnetic capabilities. In optical biosensors, for example, gold nanoparticles can improve the signal, while graphene, carbon nanotubes, MXenes and MOFs can increase electrical conductivity, which results in more sensitive detection.

(iii) ***Biocompatibility***: a large number of nanomaterials have good biocompatibility, which means that they can interact with biological substances. This characteristic is essential for the creation of biosensors because it guarantees that the nanomaterials will not disrupt the biological elements, maintaining the biomolecules' functioning and activity [14].

(iv) ***Facilitated biomolecule immobilization***: an ideal platform for immobilizing biological molecules such as enzymes, aptamers, or antibodies is offered by nanomaterials like polymer composites and graphene sheets. Biomolecules can readily bond with functional groups on the surface of nanomaterials, resulting in stable and effective immobilization.

(v) ***Signal amplification***: it is possible to engineer nanomaterials to amplify signals and increase the sensitivity of detection. In magnetic biosensors, for instance, target analytes can be concentrated using magnetic nanoparticles, resulting in an enhanced signal.

(vi) ***Quick response time***: nanomaterials' tiny size and large surface area enable fast mass transfer, which helps nanobiosensors respond quickly. This is especially crucial for the prompt identification of viral illnesses specially in fabrication of electrochemical/fluorescence-based point-of-care (POC) devices and sensor systems [14].

(vii) ***Multi-modal detection***: a single biosensor can incorporate numerous sensing modalities thanks to the engineering of nanomaterials to display diverse functionality. This feature increases the diagnostic precision for infectious diseases

by allowing the simultaneous detection of several biomarkers in multiplex detection.

(viii) ***Miniaturization and point-of-care applications*:** biosensors may now be miniaturized and used at POC settings thanks to the usage of nanomaterials. When resources are scarce, portable, quick diagnostic tools built on nanobiosensors can be used to quickly identify and track infectious infections.

Finally, nanomaterials, which offer a variety of qualities that improve sensitivity, selectivity, and functionality, are essential to the development of nanobiosensors for infectious diseases. These developments aid in the creation of more effective and trustworthy diagnostic instruments for the prompt identification and treatment of infectious diseases.

3.1.4 Current Conventional Methods Used in SARS-CoV-2 Diagnostics

To date, a number of traditional techniques have been used to diagnose SARS-CoV-2, the virus that causes COVID-19. These techniques mainly concentrate on finding the virus or any of its components in patient samples [15]. The primary methods of conventional diagnosis consist of:

Reverse transcription polymerase chain reaction (RT-PCR): for the diagnosis of SARS-CoV-2, reverse transcription polymerase chain reaction, or RT-PCR, is the gold standard. Using patient-provided respiratory samples (nasopharyngeal or oropharyngeal swabs), this technique finds viral RNA. It entails employing reverse transcriptase to convert viral RNA into complementary deoxyribonucleic acid (DNA) (cDNA), which is then amplified in particular viral genome areas using polymerase chain reaction (PCR). The PCR amplification produces recognizable signals if the virus is present. High sensitivity and specificity are offered by RT-PCR, which is extensively employed in diagnostic labs [16].

Antigen testing: rather than viral genetic material, antigen tests identify certain viral proteins, or antigens. Compared to RT-PCR, rapid antigen tests are less sensitive but provide results more quickly—typically in 15–30 min. These tests, which are conducted on respiratory samples, are appropriate for quick screening, particularly in situations when quick results are needed, including POC testing or large-scale screening campaigns.

Viral culture: growing the virus isolated from patient samples in a lab environment is known as viral culture. Although this approach verifies the existence of live virus particles, it takes a long time and sophisticated facilities. Because viral culture has a longer turnaround time and is more technically demanding, it is not as frequently employed for routine diagnostic purposes.

Serological tests (antibody tests): antibodies (IgM, IgG) produced by the immune system in response to SARS-CoV-2 infection are detected by serological testing. These blood sample-based assays are useful for determining the history of an illness or evaluating an individual's immunological response. They might miss early-stage infections, though, and are less helpful in the diagnosis of acute illnesses [16].

Chest X-rays and computed tomography (CT) scans: lung anomalies linked to COVID-19, such as ground-glass opacities and consolidation, can be seen with imaging methods such as CT scans and X-rays. Although these techniques help identify and track severe cases, they are not unique to SARS-CoV-2 and might be used in conjunction with other respiratory diseases.

The applicability, turnaround time, sensitivity, and specificity of these traditional diagnostic techniques differ. The stage of infection, the resources available, the testing capacity, and particular clinical settings all influence the method choice. Furthermore, continuous developments in diagnostic technology are intended to enhance testing accessibility, speed, and accuracy in the ongoing fight against COVID-19.

3.2 Designing Biosensor for SARS-CoV-2

In order to specifically detect targets, like viruses, throughout the innovation and design process of a novel system, data regarding the markers that are present on the targeted virus must be gathered. Sensor design heavily depends on the physical characteristics of the marker. Targeting one or more of the viral genome, nucleoproteins, or surface proteins, several detection techniques have been developed [17].

Since nanomaterials offer special qualities like high surface area, good conductivity, and biocompatibility, they have been widely used in the development of electrochemical sensors and POC devices for the detection of SARS-CoV-2. These nanomaterials are essential for improving the downsizing and portability of diagnostic instruments as well as their sensitivity, selectivity, and speed of detection. In this context, a variety of nanomaterial types have been used, such as carbon-based nanomaterials. Because of their remarkable electrical conductivity and huge surface area, carbon nanotubes (CNTs), graphene, and graphene oxide (GO) are frequently utilized in electrochemical sensors. They offer great platforms for rendering viral antigens or antibodies immobile, which makes it easier to identify SARS-CoV-2 indicators with extreme sensitivity. Metal nanoparticles: because of their distinct electrical characteristics and catalytic behavior, gold, silver, and platinum nanoparticles (AuNPs, AgNPs, and PtNPs) are used in sensor creation. Improved sensitivity and specificity are obtained by using functionalized metal nanoparticles to improve the immobilization of capture probes (such as aptamers or antibodies) and to promote signal amplification. Metal oxide nanomaterials: because of their superior electrochemical qualities, metal oxide nanoparticles, such as zinc oxide (ZnO), tin oxide (SnO_2), and iron oxide (Fe_2O_3), are used in the production of sensors. The performance of the sensor is improved by these materials, especially with regard to stability, selectivity, and sensitivity toward SARS-CoV-2 biomarkers. Polymeric nanomaterials: to improve the stability and biocompatibility of sensing platforms, polymer-based nanomaterials, such as dendrimers, polymer nanoparticles, or nanocomposites, are used. These substances offer a matrix that immobilizes biomolecules and permits regulated release, both of which enhance sensor performance. The integration of MXenes and MOFs into the electrochemical sensor and POC device designs for SARS-CoV-2 detection exhibits considerable promise for improving COVID-19 diagnosis efficiency, sensitivity, speed, and versatility. These platforms based on nanomaterials present prospective paths for the creation of cutting-edge diagnostic instruments essential to efficient disease management, monitoring, and control. Highly sensitive, quick, and portable diagnostic instruments are made possible by the incorporation of nanomaterials into electrochemical sensors and POC devices for SARS-CoV-2 detection. Selective identification of viral antigens or genetic material linked to SARS-CoV-2 is made possible by functionalizing these nanomaterials with certain receptors (such as aptamers, peptides, or antibodies). Furthermore, POC applications made possible by the downsizing of devices made possible by nanomaterials enable testing in remote or resource-constrained situations without the need for complex laboratory infrastructure [18]. The continued development of novel and effective diagnostic platforms for the detection of SARS-CoV-2 is made possible by the increasing use of nanomaterials in sensor technology, which eventually helps with efficient disease management, surveillance, and control. As a result, some of the electrochemical sensors and POC devices from 2023 were covered in this chapter, along with a variety of nanomaterials used in conjunction with electroanalytical techniques.

3.2.1 Electrochemical Biosensors

Because of the complexity and urgency of the situation, developing highly efficient sensor systems for the detection of SARS-CoV-2 entails several crucial factors. High-performing sensor systems are essential for SARS-CoV-2 detection because they help stop the virus's transmission, enable prompt medical attention, and support a range of tactics aimed at lessening the pandemic's negative effects on society and public health. Globally, numerous scientists and researchers are working to create efficient sensor systems that will lessen the impact of the virus, as shown in Table 3.1. Recently, in order to identify the specific oligonucleotide target of the virus in the RdRp gene region, an electrochemical biosensor based on MXene nanosheets (Ti_3C_2Tx) and carbon platinum (Pt/C) with a DNA probe was recently used, as shown in Figure 3.1a. The limits of detection (LOD) was found to be 0.4 aM after a calibration plot was created using the differential pulse voltammetry (DPV) approach in concentrations ranging from 1 aM to 100 nM. Additionally, 192 clinical samples, including saliva, nasopharyngeal swabs, and serum, were used to examine the specificity and sensitivity of the sensors using positive and negative RT-PCR tests. The results showed 100% accuracy and sensitivity [19]. It has been claimed that a strong surface for the detection of SARS-CoV-2 can be achieved by coating a screen printed carbon electrode (SPCE) with Ti_3C_2Tx-MXene nanosheets and then amino-functionalized probe DNA (NH_2-pDNA). With a detection limit of 0.004 pM, the analytical sensor response for SARS-CoV-2 was investigated using impedance spectroscopy within the target DNA concentration range of 0.1 pM–1 μM. As an analytical application, a good linearity for the target sequence detection in spiked serum samples was obtained, as shown in Figure 3.1b [20]. In another study, to detect the SARS-CoV-2 S protein, an electrochemical biosensor with surface molecular imprinting (SMI) is created. Using Cu_7S_4-Au as the built-in probe and 4-mercaptophenylboric acid (4-MPBA) anchored to the surface via Au–SH bonds, the SARS-CoV-2 S protein template can be immobilized on the surface of an SPCE via boronate ester bonds, as shown in Figure 3.2a. The logarithm of the SARS-CoV-2-S protein concentration (lgC) in the range of 5–1000 ng mL^{-1} with the LOD of 1.76 pg mL^{-1} was shown to have a good linear connection with the decrease in peak currents. The SARS-CoV-2 S protein in liquids and clinical serum samples was sensitively detected using the as-fabricated SMI electrochemical biosensor [21]. As shown in Figure 3.2b, to detect SARS-CoV-2 S protein, an electrochemical biosensor device was created. It utilized bacterial cellulose (BC) as its substrate, a carbon-based electrode

Table 3.1 Recent reports on electrochemical biosensors for the detection of SARS CoV-2 virus.

S. no	Sensor type	Sensor platform	Targeting part	Method	Linear range	Detection limits	Sample analysis	Ref.
1	Electrochemical	Ti_3C_2Tx/Pt/C/ DNA probe	Oligonucleotide	DPV	1 aM to 100 nM	0.4 aM	Saliva, nasopharyngeal swabs, and serum	[19]
2	Electrochemical	NH_2-pDNA/Ti_3C_2Tx/ SPCE	Target DNA	DPV, EIS	0.1 pM–1 μM	0.004 pM	Spiked serum samples	[20]
3	Electrochemical	Cu_7S_4-Au/4-MPBA/ SPCE	S protein	DPV	0.005~1000 nM	0.0017 nM	Clinical serum samples	[21]
4	Electrochemical	Bacterial cellulose/ GO/ACE2/Carbon electrode	S protein	Potentiometric	1×10^{-14} to 1×10^{-11} g mL^{-1}	4.26×10^{-18} g mL^{-1}	Nasopharyngeal/ oropharyngeal (NP/OP) clinical samples	[22]
5	Electrochemical	PNA probe/peptides/ PANI modified GCE	Viral RNA	DPV	1.0 fM–1.0 nM	0.38 fM	Human saliva	[23]
6	Electrochemical	PPyNTs/ $Ni(OH)_2$/ Sb#15-His6/SPCE	Antigen	EIS	--	0.01 pg/mL	Human saliva	[24]

(Continued)

Table 3.1 Recent reports on electrochemical biosensors for the detection of SARS CoV-2 virus. (*Continued*)

S. no	Sensor type	Sensor platform	Targeting part	Method	Linear range	Detection limits	Sample analysis	Ref.
7	Electrochemical	AuNP-peptide-DNA nanoprobe	SARS-CoV-2 PLpro	EIS, SWV	1 pg mL^{-1} to 10 µg mL^{-1}	1 pg mL^{-1} (27.18 fM)	Saliva and blood samples, throat swab	[25]
8	Electrochemical	Fe_3O_4@SiO_2–Au/GCE	S protein	EIS	0.1 ng/mL to 10 µg/mL	4.78 pg/mL	Saliva samples	[26]
9	Electrochemical	N-isopropylacrylamide and acrylic acid, and gold nanoparticles	S protein	DPV	10^{-13}–10^{-9} mg/mL	9.55 fg/mL	Artificial saliva samples	[27]
10	Electrochemical	CRISPR/Cas13a/AuNPs/SPGE	S gene and Orf1ab gene	SWV	1×10^{-17}–1×10^{-11} M	2.5 and 4.5 ag/µL	Synthetic and clinical samples	[28]
11	Electrochemical	ZnSeS quantum dots/aptamer-molecular beacon (Apta-MB) probe/carbon nanofiber-modified SPCE	S protein	DPV	2 pg/mL to 0.0032 pg/mL	0.5 pg/mL	Saliva samples	[29]

(*Continued*)

Table 3.1 Recent reports on electrochemical biosensors for the detection of SARS CoV-2 virus. (*Continued*)

S. no	Sensor type	Sensor platform	Targeting part	Method	Linear range	Detection limits	Sample analysis	Ref.
12	Electrochemical	Si wafer/Graphene oxide/SARS-CoV-2 spike antibody	S protein	Real-time current measurements	0.5–100,000 fg/mL	0.5 fg/mL	Synthetic	[30]
13	Electrochemical sensor	CPE/MB/GO/ *h*ACE-2 peptide	S protein	chronoamperometry	1 pg/mL to 1 μg/mL	0.71 pg/mL	Nasopharyngeal and oropharyngeal swab samples	[31]
14	Electrochemical sensor	SPCE/AuNP/ SAM/rN	4G6, 7F10, and 1A6	DPV, PAD	--	1.47±0.07, 0.08±0.01 and 0.25±0.01 μg/mL	Clinical samples	[32]
15	Electrochemical impedimetric sensor	ITO-APTES-mVero and ITO-APTES-mCalu	S protein	EIS	10 to 100 ng/mL	30.4 pg/mL and 21.9 pg/mL	Clinical samples	[33]
16	Electrochemical sensor	PNA/SPCE/graphene oxide-gold nanocomposite	RNA-dependent RNA polymerase	AMP	5–500 aM	0.8 aM	Clinical samples	[34]

(*Continued*)

Table 3.1 Recent reports on electrochemical biosensors for the detection of SARS CoV-2 virus. (*Continued*)

S. no	Sensor type	Sensor platform	Targeting part	Method	Linear range	Detection limits	Sample analysis	Ref.
17	Electrochemical sensor	Nanocubic $Mn_3[Fe(CN)_6]_2$@Pt/DNA sensors	rtDNA	AMP	--	1×10^{-20} M	Clinical samples	[35]
18	Electrochemical sensor	Au-decorated magnetic reduced graphene oxide (AMrGO)/target RNA	nucleocapsid (N) gene	CV, DPV	5 aM–50 pM	0.37, 0.33, and 0.19 fM	Human saliva, urine, and serum	[36]
19	Electrochemical sensor	Anti-rS/Au NPs/SPE	S protein	AMP	--	2 ng mL^{-1}	Clinical samples	[37]
20	Electrochemical	DNA/ZnS/graphene/GCE	S, ORF 1a, and ORF 1b gene sequences	DPV	1.0×10^{-18} M to 1.0×10^{-7} M	4.453×10^{-20} M	--	[38]
21	Electrochemical	Ab/Carbon-MEMS-GCE	S Protein	EIS	1 fg/ml to 1 μg/ml	1 fg/ml	Nasopharyngeal swab samples	[39]
22	Electrochemical	Nanostructured-AuNPs/WO3-screen printed electrodes	oligonucleotides	EIS	0.01 pM to 2 μM	298 fM	Clinical samples	[40]

(*Continued*)

Table 3.1 Recent reports on electrochemical biosensors for the detection of SARS CoV-2 virus. (*Continued*)

S. no	Sensor type	Sensor platform	Targeting part	Method	Linear range	Detection limits	Sample analysis	Ref.
23	Electrochemical	AuE-MXene-AuNPs/MB-ssRNA reporter	ssRNA	SWV	1 nM-10 fM	1 fM	Clinical samples	[41]
24	Electrochemical sensor	mAb/PANi-TNT/SPCE	S Protein	Chronoamperometry	80 to 200 copies/μL	25.59 copies/μL	Saliva or nasopharyngeal fluid	[42]
25	Dual Electrochemical and colorimetric sensor	CRISPR-Cas13:guide-RNA complex	SARS-CoV-2 RNA	Chronoamperometry	--	1 fM	Clinical samples	[43]
26	Electrochemical sesnor	Combined magnetic separation system with copper nanoflowers	RdRP_SARSr-P2	DPV	0.1 to 10^3 nM	33.83 pM	Clinical samples	[44]
27	Electrochemical sesnor	hACE-2 N-terminal alpha helix peptide / polypyrrole/ nitrocellulose membrane	Antigen	EIS	--	40 $TCID_{50}$/mL	Saliva samples	[45]

(*Continued*)

Table 3.1 Recent reports on electrochemical biosensors for the detection of SARS CoV-2 virus. (*Continued*)

S. no	Sensor type	Sensor platform	Targeting part	Method	Linear range	Detection limits	Sample analysis	Ref.
28	Electrochemical	Gold film electrode with ZnO nanoparticles/ antibody	IL-10 and CRP	EIS	0.01–100 μg/mL for CRP and 0.1–1000 pg/mL	--	Serum and urine samples	[46]
29	Electrochemical (POC device)	Pyrrolidinyl peptide nucleic acid (acpcPNA)/ePAD	Target DNA	Amperometry	0.1 nM–200 nM	1.0 pM	Saliva samples	[47]
30	Electrochemical (POC device)	SPCE/ACE2/Fc–IgG	Antigen	DPV, EIS	0.01–500 ng/mL	2.98 pg/mL	Nasal-swap samples	[48]
31	Electrochemical	CoFeBDCNH2-CoFe2O4/11-MUA/ gold chip	Glycoprotein	EIS	10 ng/mL to 100 fg/mL	6.20 fg/mL	pseudo virus	[49]

(*Continued*)

Table 3.1 Recent reports on electrochemical biosensors for the detection of SARS CoV-2 virus. (*Continued*)

S. no	Sensor type	Sensor platform	Targeting part	Method	Linear range	Detection limits	Sample analysis	Ref.
32	Portable microfluidic electrochemical impedance biosensor	Gold electrodes on borosilicate substrates/ polydimethylsiloxane (PDMS) microfluidic channel/Ab spike protein of SARS-CoV-2	S protein	EIS	16 fM to 160 nM	19.09 ± 0.65 fM	Saliva samples	[50]
33	Electrochemical POC device	AuNSs/SPCE/loop mediated isothermal amplification technique	Envelope (E) and RdRP	CV	0 to 1.48 $\times 10^9$ copies	1.48 $\times 10^2$ and 1.48 $\times 10^3$	Clinical samples	[52]
34	Field-effect transistor POC device	ACE 2-Fc/graphene FET	S protein	I-V curves	2 µg/mL to 2 pg/ mL o	65 cps/mL	Clinical samples	[53]
35	Electrochemical POC device	Ab/Au IMAs	S Protein	EIS	0.41 to 4.81 BAU/ml	0.41 BAU/ ml	Clinical samples	[54]
36	Electrochemical POC testing	base-stacking-driven catalytic hairpin assembly (BDCHA), a nucleic acid amplification/SPCE	N gene sequence fragment	DPV	--	--	Clinical samples	[55]

(*Continued*)

Table 3.1 Recent reports on electrochemical biosensors for the detection of SARS CoV-2 virus. (*Continued*)

S. no	Sensor type	Sensor platform	Targeting part	Method	Linear range	Detection limits	Sample analysis	Ref.
37	Electrochemical sensor FET device	Liquid-gated graphene field-effect transistors	SARS-CoV-2 surface protein and influenza	Gate voltage vs current	--	88 zM and 227 zM	Clinical samples	[56]
38	Electrochemical sensor device	G-quadruplex DNAzyme-sequential flow controllable microfluidic device	SARS-CoV-2 cDNA	SWV	--	30 pM	Nasopharyngeal swab samples	[57]
39	Electrochemical POC device	Au E/ZnO/ specific antibodies	S-protein, N-protein, D-dimer, and IL-2R	EIS	58 to 60 μg/mL, 58 to 60 μg/mL, 0.05 to 10 ng/mL and 0.047 to 53.09 ng/mL	3, 21, 0.0006 and 0.242 ng/mL	Clinical samples	[58]
40	OECT sensor device	PEDOT:PSS/Ti-Au/Ab	Spike protein	I-V characterization	--	1 fM	Clinical samples	[59]
41	Electrochemical POC device	S1-Ab/GNp/thionine/SWCNT-COOH/SPCE	Spike protein	SWV	0 to 500 nM	500 pM	Saliva samples	[60]

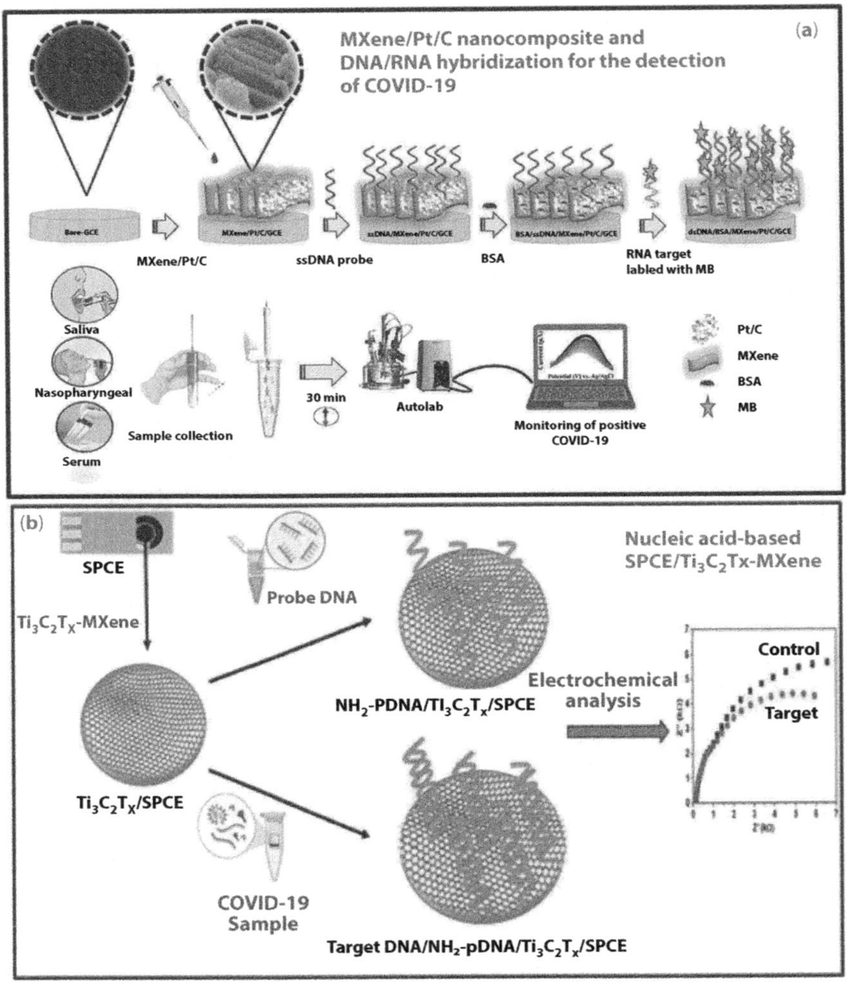

Figure 3.1 (a) Design of a rapid electrochemical biosensor based on MXene/Pt/C nanocomposite and DNA/RNA hybridization for the detection of COVID-19 [19]. (b) Development of a nucleic acid-based screen printed electrochemical biosensor using Ti_3C_2Tx-MXene for the detection of SARS-CoV-2 [20].

that had been modified with graphene oxide, and the human angiotensin-converting enzyme-2 (ACE2) as its receptor. Analytical curves were made at concentrations ranging from $1X10^{-14}$ to $1X10^{-11}$ g mL^{-1} SP with the detection limits of $4.26X10^{-18}$ g mL^{-1} using this green technology and label-free potentiometric diagnostic test that can detect SARS-CoV-2 within 10 min and costs US$3.50 per unit. When evaluated on 65

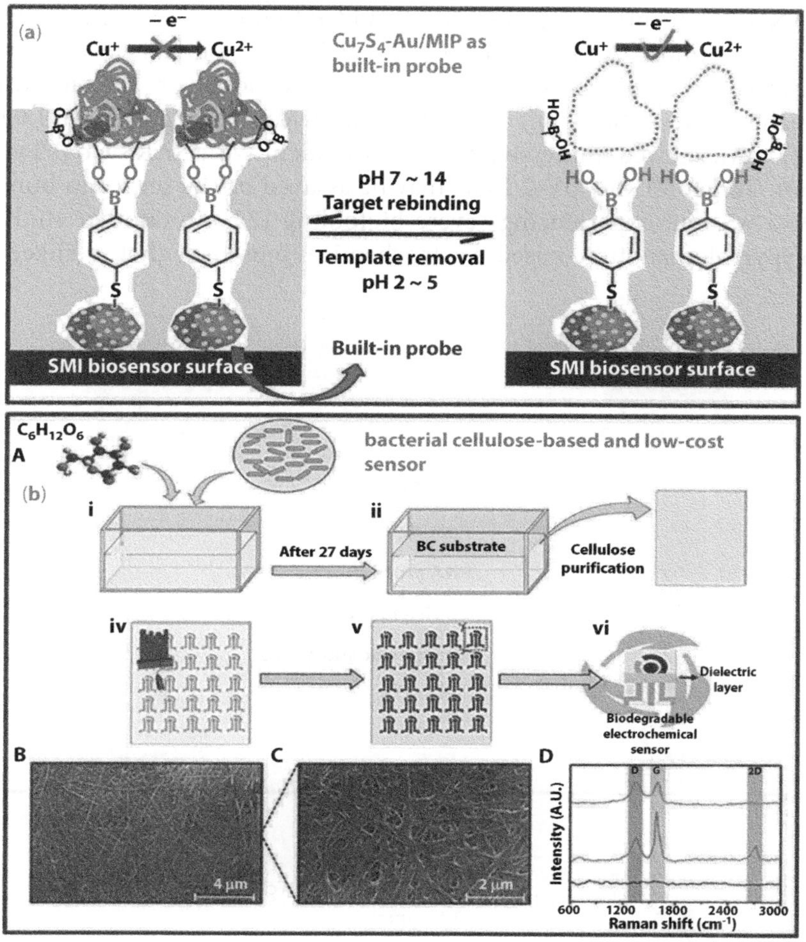

Figure 3.2 (a) A surface molecularly imprinted electrochemical biosensor for the detection of SARS-CoV-2 spike protein by using Cu_7S_4-Au as built-in probe [21]. (b) A bacterial cellulose-based and low-cost electrochemical biosensor for ultrasensitive detection of SARS-CoV-2 [22].

nasopharyngeal/oropharyngeal (NP/OP) clinical samples, the suggested device demonstrated high sensitivity, specificity, and accuracy [22]. A paper was reported on a non-fouling electrochemical biosensor that utilized antifouling peptides and the electroneutral peptide nucleic acid (PNA) as the recognizing probe for the particular binding of the SARS-CoV-2 viral RNA, as shown in Figure 3.3a. The sensor was created using a glassy carbon electrode (GCE) that had been modified with polyaniline

(PANI). Next, antifouling peptide was added to the modified electrode to create a peptide/PANI modified GCE. Lastly, the modified electrode was incubated in a PNA probe for 4 h to create the biosensor. The biosensor, which has a detection limit of 0.38 fM and a wide linear range from 1.0 fM to 1.0 nM, showed impressive sensitivity in identifying SARS-CoV-2 viral RNA. Target SARS-CoV-2 RNA may be assayed by the sensor in human saliva without experiencing severe biofouling [23]. In another study, a polypyrrole nanotube-based electrochemical biosensor that is linked to

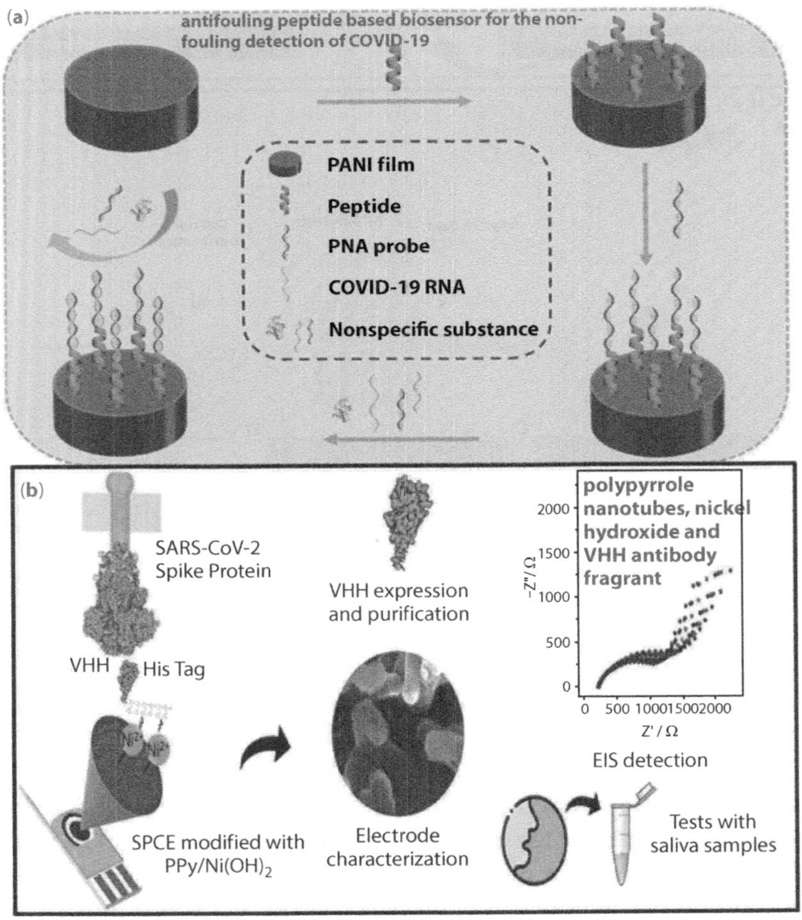

Figure 3.3 (a) Peptide nucleic acid and antifouling peptide based biosensor for the non-fouling detection of COVID-19 nucleic acid in saliva [23]. (b) COVID-19 impedimetric biosensor based on polypyrrole nanotubes, nickel hydroxide and VHH antibody fragment: specific, sensitive, and rapid viral detection in saliva samples [24].

an engineered heavy chain-only antibody (VHH) antigen-binding fragment called Sb#15-His6 on SPCE using $Ni(OH)_2$ ligation has been disclosed. Recombine receptor-binding domain (RBD) was used to determine the quantification limit, which was expressly lower than commercial monoclonal antibodies at 0.01 pg/mL. An additional benefit of the proposed sensor is that it just requires a small amount of saliva to do the detection; findings can be obtained in as little as 15 min without the need for additional sample preparation, as shown in Figure 3.3b [24]. Using chimeric peptide-DNA (pDNA) nanoprobes, an electrochemical biosensor was reported to detect SARS-CoV-2 via its papain-like cysteine protease (PLpro) in blood and saliva. In order to sense the targeted PLpro, the sensor was built on an Au electrode that had been modified with an AuNP-peptide-DNA probe. A linear detection range of 1 pg mL^{-1}–10 μg mL^{-1} was found, along with a detection limit of 27.18 fM. When the target protease is present, the peptide substrate is severed, causing the DNA1 fragment to be released from the surface of AuNPs. On the surface of the working electrode, the freed DNA1 can then hybridize with the previously immobilized DNA2. Consequently, strong electrochemical signals for the quick and accurate detection of SARS-CoV-2 PLpro can be produced by inserting signal molecules methylene blue (MB) into the double-stranded DNA (dsDNA), as shown in Figure 3.4a [25]. An electrochemical sensor for the quick detection of S-protein of SARS-CoV-2 by electrochemical impedance spectroscopy (EIS) was created by chemically altering a GCE with Fe_3O_4@SiO_2–Au nanocomposites and further functionalizing it with angiotensin-converting enzyme 2 (ACE2), as shown in Figure 3.4b. For detecting the EIS response signal of S-protein, the sensor offers a large linear dynamic range from 0.1 ng/mL to 10 μg/mL and a low limit detection of 4.78 pg/mL. The Fe_3O_4@SiO_2–Au/GCE biosensor showed S-protein detection and recovery in saliva sample with great selectivity, stability, and repeatability [26]. To determine the S protein (SARS-CoV-2 spike) in human saliva, an electrochemical sensor based on microgel was studied. Acrylic acid and N-isopropyl acrylamide copolymer was used to create the microgel, which was a simple and inexpensive way to enclose gold nanoparticles. Using DPV, the sensor's electrochemical performance was evaluated. In optimal experimental settings, a linear range of 10^{-13}–10^{-9} mg/mL and a detection limit of 9.55 fg/mL were attained, with tested samples exhibiting good recovery percentages [27]. The S and Orf1ab genes are examples of SARS-CoV-2 RNA fragments that may be identified using a multiplexed, NAA-free technique in both clinical and synthetic samples, as shown in Figure 3.5a. ReRNA-modified

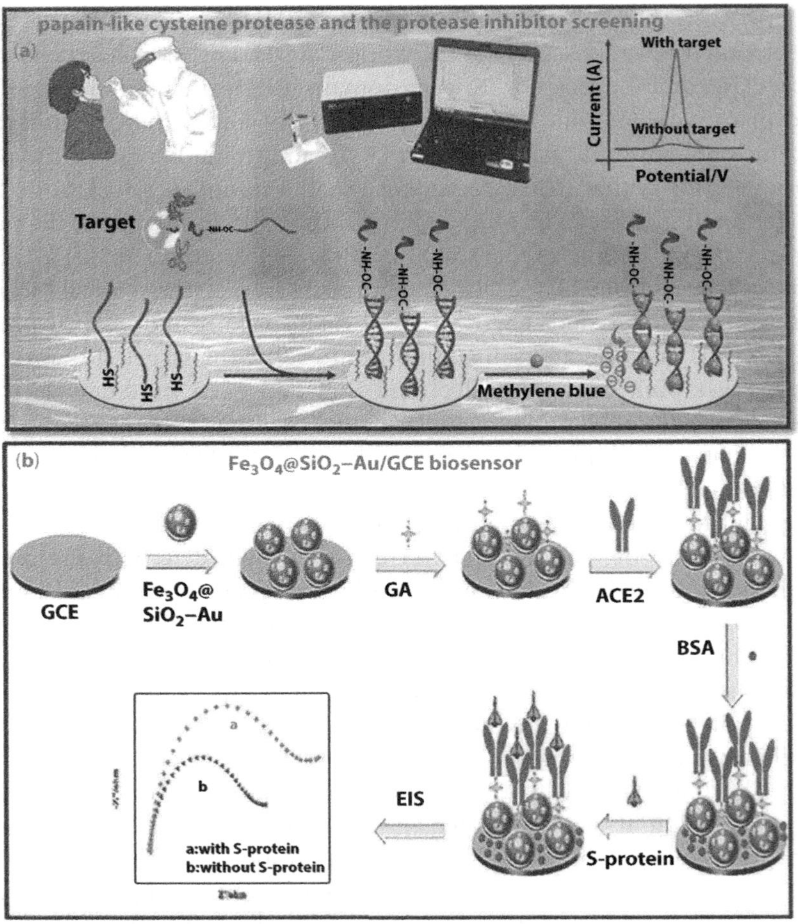

Figure 3.4 (a) An electrochemical biosensor for SARS-CoV-2 detection via its papain-like cysteine protease and the protease inhibitor screening [25]. (b) Sensitive detection of SARS-CoV-2 spike protein based on electrochemical impedance spectroscopy of $Fe_3O_4@$ SiO_2–Au/GCE biosensor [26].

gold nanostructured electrodes were produced by altering the SPGE surface by gold electrodeposition. Afterwards, the biosensing surface was dried and exposed to the CRISPR/Cas13a system in order to obtain the entire sensor. The biosensor achieved low detection limits of 2.5 ag/μL and 4.5 ag/μL for the S and Orf1ab genes, respectively, demonstrating that it successfully met the sensitivity criteria [28]. A paper was reported on an electrochemical biosensor that utilized ZnSeS quantum dots (QDs) coupled to an aptamer-molecular beacon (Apta-MB) probe, then loaded onto a screen-printed carbon electrode modified with carbon nanofiber, as depicted in Figure 3.5b. The SARS-CoV-2 S protein was detected electrochemically by measuring a consistent rise in the anodic peak current of QDs produced by the DPV plots. The QDs-Apta-MB electrochemical probe's limits of detection for the SARS-CoV-2 S protein were

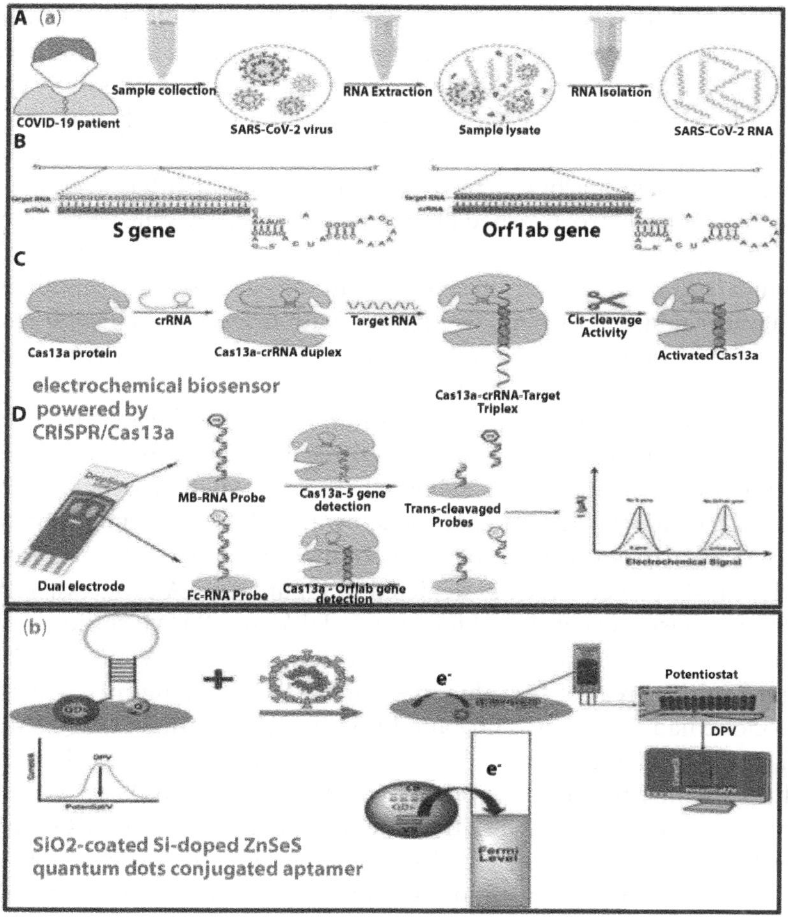

Figure 3.5 (a) Ultrasensitive and amplification-free detection of SARS-CoV-2 RNA using an electrochemical biosensor powered by CRISPR/Cas13a [28]. (b) Blue-emitting SiO2-coated Si-doped ZnSeS quantum dots conjugated aptamer-molecular beacon as an electrochemical and metal-enhanced fluorescence biosensor for SARS-CoV-2 spike protein [29].

approximately 0.5 pg/mL when used to analyze saliva samples [29]. Based on a Si wafer with 300-nm thermally produced SiO_2 coated with uniform layers of rGO and with the particular SARS CoV-2 S1 antibodies through 1-pyrenebutyric acid N-hydroxysuccinimide ester linker molecules, an electrochemical assay using a four-point Kelvin sensing system was disclosed, as shown in Figure 3.6a. The well was successively injected with 10 uL of serially diluted S1 protein solutions ranging from 0.5 to 100,000 fg/mL. Because the S1 protein's RBD is positively charged, a constant voltage of –0.8 mV was used during the experiment. As little as 0.5 fg/mL of the SARS-CoV-2 S1 protein in phosphate buffer saline (PBS) was shown to contain the S1 protein. With the exception of the clinical samples, every municipal wastewater sample that the developed sensor successfully identified included SARS CoV-2 [30]. A scalable and reasonably priced platform for SARS-CoV-2 detection using the hACE-2 peptide was developed using a disposable carbon-screen printed electrode that has been altered to include MB electroadsorbed graphene oxide (GO). Cyclic voltammetry (CV) and chronoamperometry (CA) were used to investigate the electrochemical biosensing of various concentrations of S proteins. This allowed for a linear range of 1 pg/mL to 1 μg/mL, with respective LOD of 0.58 pg/mL and 0.71 pg/mL, respectively, as shown in Figure 3.6b. The proposed sensor was used to directly analyze nasopharyngeal and oropharyngeal swab specimens in viral transport medium for clinical feasibility [31]. An electrochemical biosensor that detects three different clones of monoclonal antibodies (mAbs), such as 4G6, 7F10, and 1A6, by coating an SPCE with gold nanoparticles (AuNP) and allowing it to self-assemble a monolayer (SAM) to ensure the covalent immobilization of the recombine nucleocapsid protein (rN), as depicted in Figure 3.6c. A progressive shift in the responses was observed for the CV, DPV, and PAD techniques with an increase in mAbs concentration. According to the DPV data, the 7F10 clone had the greatest K_C value, measuring 1.47±0.07 μg/mL, while the 4G6 clone had the lowest LOD and limit of quantification (LOQ) values, measuring 0.08±0.01 and 0.25±0.01 μg/mL, respectively [32]. Based on cellular membranes from VeroCCL81 (mVero) and Calu-3 (mCalu) cells (which overexpress the ACE-2 transmembrane receptors), a SARS-CoV-2 detection platform was created. To detect S protein, the vesicles were isolated and immobilized on an indium tin oxide electrode (ITO). The optimized biosensors produced a linear response ranging from 10 to 100 ng/mL; for ITO-APTES-mVero and ITO-APTES-mCalu, respectively, limits of detection of 10.0 pg/mL and 7.25 pg/mL and limits of quantification of 30.4 pg/mL and 21.9 pg/mL were attained with satisfactory accuracy [33].

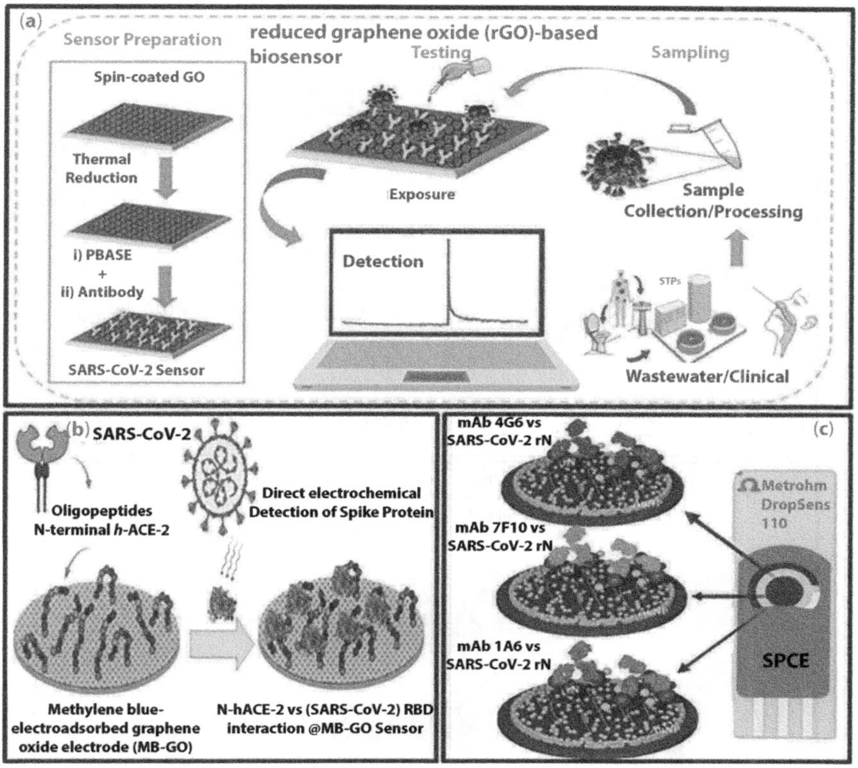

Figure 3.6 (a) Detection of SARS-CoV-2 in clinical and environmental samples using highly sensitive reduced graphene oxide (rGO)-based biosensor [30]. (b) Peptide-based direct electrochemical detection of receptor binding domains of SARS-CoV-2 spike protein in pristine samples [31]. (c) Electrochemical biosensing based comparative study of monoclonal antibodies against SARS-CoV-2 nucleocapsid protein [32].

After the electrodes were screen-printed and modified with a reduced graphene oxide-gold nanocomposite, a fixed quantity of the SARS-CoV-2 RNA-dependent RNA polymerase (RdRP) was introduced to the electrode surface to interact with the thiol-terminated PNA, probes as shown in Figure 3.7a. Consequently, it allowed polymerase 4-cyano-4-(phenylcarbonothioylthio)-pentanoic acid and ferrocenylmethyl methacrylate (FcMMA) to attach to the RdRP chain of SARS-CoV-2. The PNA-based sensor responded more strongly to SARS-CoV-2 RdRP in the 5–500 aM range thanks to the electrochemical signal from the poly-FcMMA. The limit of detection for SARS-CoV-2 RdRP detection was found to be 0.8 aM, which is less than the prior sensor [34]. The MnFe@Pt crystals'

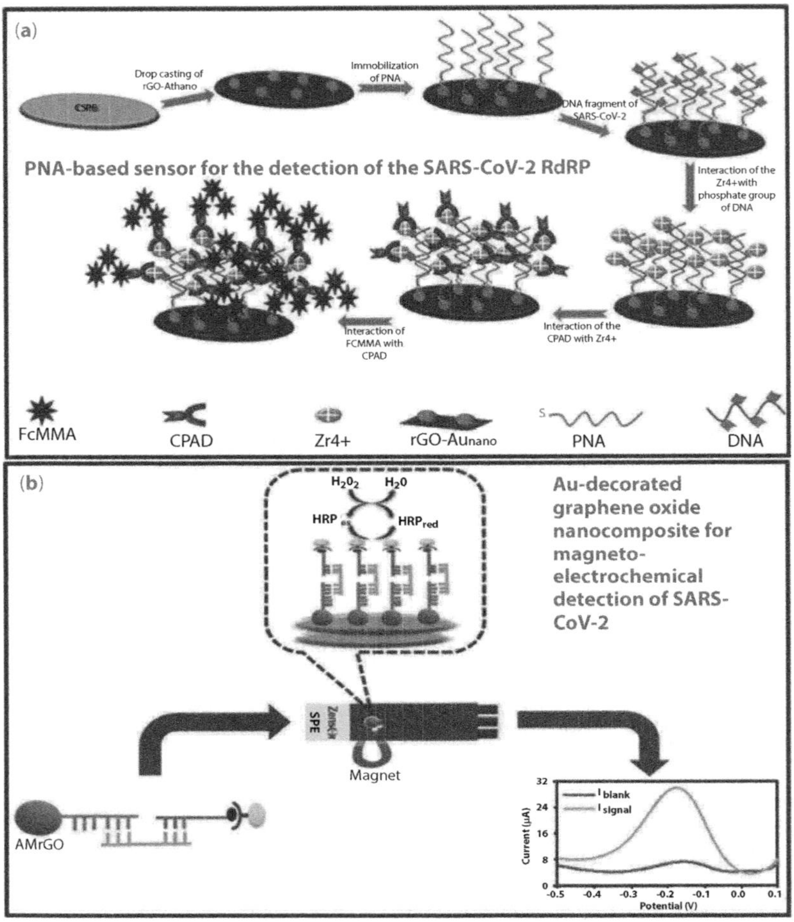

Figure 3.7 (a) An electrochemical PNA-based sensor for the detection of the SARS-CoV-2 RdRP by using surface-initiated-reversible-addition–fragmentation-chain-transfer polymerization technique [34]. (b) Synthesis and characterization of Au-decorated graphene oxide nanocomposite for magneto-electrochemical detection of SARS-CoV-2 nucleocapsid gene [36].

nanocubic architecture was built to combine conductivity and electrocatalysis to significantly enhance biosensing performance. The rtDNA can be quickly captured by the DNA sensor to create a double-helix structure after the immobilization of a specific capture and report DNA on the aforementioned nanocomposite, resulting in the present signal change. The as-prepared DNA sensors demonstrated an exceptional selectivity among different interfering chemicals, a wide linear detection range, and an ultralow detection limit of 1×10^{-20} M in under 10 min [35]. A magnetic nanocomposite and two particular DNA probes were used in an electrochemical experiment that was reported to quickly measure the CoV-2 nucleocapsid (N) gene, as shown in Figure 3.7b. For effective target RNA capture, the capture probe is covalently attached to an Au-decorated magnetic reduced graphene oxide (AMrGO) nanocomposite. On the other hand, peroxidase and the detecting probe are connected for signal amplification. In human saliva, urine, and serum, the genosensor performed well throughout a wide linear range of 5 aM to 50 pM, with detection limits of 0.37, 0.33, and 0.19 fM [36]. An AuNP-based electrochemical bioanalytical device that recombines SARS-CoV-2 Spike protein (rS) cross-linked with glutaraldehyde is integrated onto SPCE, as shown in Figure 3.8a. We demonstrated a quick anti-rS detection within a detection limit of 2 ng mL^{-1} using the AuNP-modified SPE. The transduction signal was the current flux of AuNPs reduction (at +200 mV) in a pure phosphate buffer (PB). This approach can be used to monitor the 2019 Coronavirus disease and track the immune system's reaction to a SARS-CoV-2 infection (COVID-19) [37]. An electrochemical biosensor for quick and accurate SARS-CoV-2 nucleic acid detection based on a zinc sulfide/graphene (ZnS/graphene) nanocomposite. Based on targets such as S, ORF 1a, and ORF 1b gene sequences, the experimental findings showed that the suggested biosensor could detect low amounts of all various SARS-CoV-2 samples. The maximum currents increased in a linear fashion as the tDNA concentration increased from 1.0×10^{-18} M to 1.0×10^{-7} M, as shown in Figure 3.8b. The detection limit was determined to be 4.453×10^{-20} M [38]. The SARS-CoV-2 antibody was covalently linked to the GC surface using glutaraldehyde as a cross-linker, and the detection mechanism relied on this process. The sensor platform utilized was the carbon-MEMS-derived glassy carbon electrode (GCE). With an estimated LOD of 1 fg/ml of viral RNA/mL, the developed sensor can detect 1 fg/ml to 1 μg/ml of SARS-CoV-2 spike protein. [39]. In a recent study, electrochemical impedance spectroscopy was used to detect SARS-CoV-2 oligonucleotides using a screen printed electrode made of nanostructured AuNPs and metal oxides (V_3O_5, Sb_2O_3, SeO_2, ZrO_2, CeO_2, La_2O_3, TiO_2, WO_3, or GeO_2). Different cross-linkers were tried to see if they could effectively immobilize the isolated virus RNA, which is the selective bio-recognition element, on the electrode surface in order to immobilize the SARS-CoV-2 chosen oligos. The matched oligos were present in varying quantities, ranging from 10 fM to 2.0 μM. Following complete assay optimization, linear regression was used to compute the limits of detection and quantification, which came out to be 298 and 994

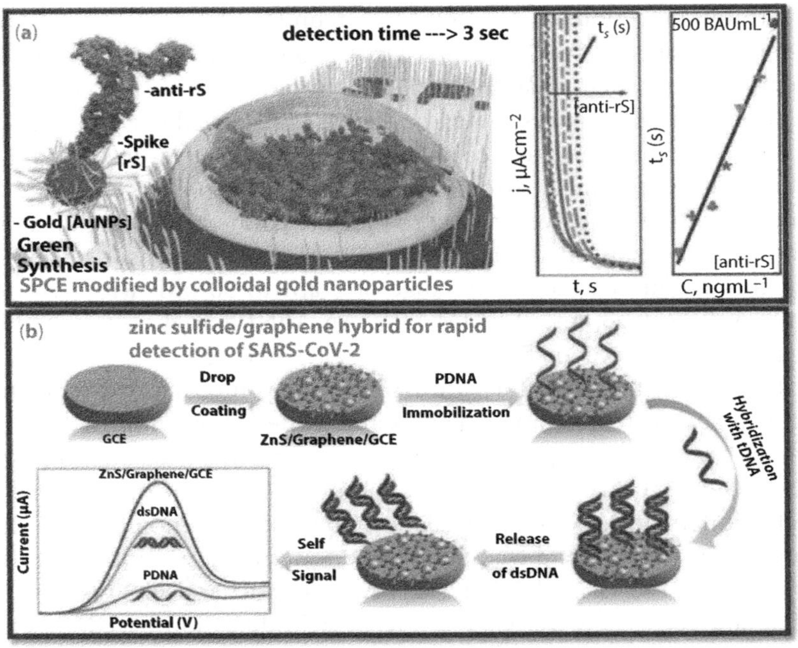

Figure 3.8 (a) Detection of antibodies against SARS-CoV-2 Spike protein by screen-printed carbon electrodes modified by colloidal gold nanoparticles [37]. (b) Ultrasensitive electrochemical biosensors based on zinc sulfide/graphene hybrid for rapid detection of SARSCoV2 [38].

fM, respectively [40]. To identify the single-base L452R mutation in RNAs and clinical samples, a sensor device consisting of Au E/MXene-AuNP composites and the CRISPR/Cas13a system was reported. The trans-cleavage activity of Cas13a aided in the signal amplification in the CRISPR-Cas13a-based electrochemical biosensor. The Omicron BA4/5 target RNA was measured at different concentrations ranging from 1 nM to 10 fM under ideal conditions to characterize the biosensor's sensitivity. The redox current of the MB gradually decreased as the target RNA concentration increased, and the limit of detection was determined to be 1 fM. The created biosensor will function as a fantastic addition to the RT-qPCR technique, facilitating the rapid identification and early detection of SARS-CoV-2 Omicron BA.5 and BA.2 variants, as shown in Figure 3.9 [41]. A monoclonal antibody (mAb) specific to SARS-CoV-2 was applied to the surface of a commercially available SPCE that had been altered with polyaniline-titania nanotubes (PANI-TNT) to create the electrochemical biosensor strips. Analytical study showed linearity in the range of 80–200

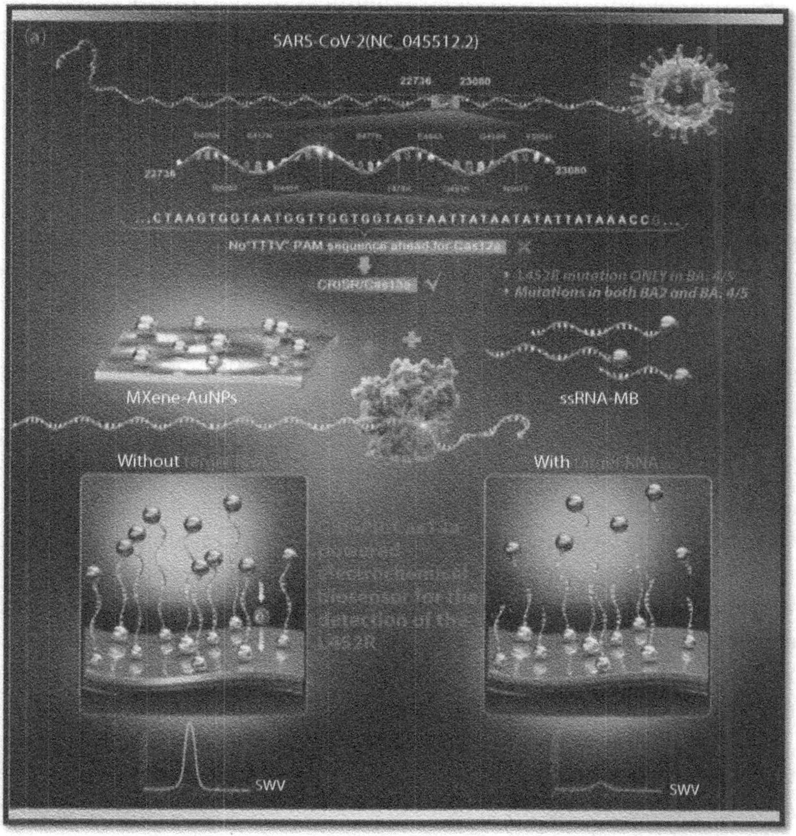

Figure 3.9 CRISPR-Cas13a-powered electrochemical biosensor for the detection of the L452R mutation in clinical samples of SARS-CoV-2 variants [41].

copies/μL with a limit of detection of 25.59 copies/μL from the dose–response and standard fitted curve, indicating the sensor's outstanding sensitivity and specificity toward SARS-CoV-2. The sensor strip's capacity to identify SARS-CoV-2 apart from human coronavirus-OC43 (HCoV-OC43), HCoV-NL63, HCoV-229E, and adenovirus was demonstrated by experimental validation [42]. Using SARS-CoV-2 RNA as an example, a dual-function biosensor with aerosol jet printing capability has been developed for the sensitive detection of infections. Target RNA triggers the activation of a CRISPR-Cas13:guide-RNA complex, which causes the ssRNA probes with the horseradish peroxidase (HRP) tag to be trans-cleaved collaterally. This in turn causes an electrochemical signal shift and a color change as HRP oxidizes 3,3′,5,5′-tetramethylbenzidine

(TMB). The colorimetric and electrochemical sensing protocol has a femtomolar detection sensitivity and does not require complex target amplification or probe immobilization [43]. An electrochemical biosensor and a copper nanoflower-triggered cascade signal amplification method were combined to create a magnetic separation device that was published. The conserved sequence of SARS-CoV-2 was captured by magnetic beads as the recognition element in the proposed approach. When the target sequence RdRP_SARSr-P2 appears, copper nanoflowers connected with magnetic beads will initiate the Cu(I)-catalyzed azide–alkyne cycloaddition process through the connection of the SARS-CoV-2 conserved sequence. Afterwards, a large number of signal molecules 2-(Fluoromethyl) methacrylate (FMMA) can be grafted onto the modified electrode surface via electrochemically aided atom-transfer radical polymerization, therefore amplifying the signal for the quantitative determination of SARS-CoV-2. Under ideal conditions, a linear range of 0.1–103 nM with a detection limit of 33.83 pM is produced [44]. Using a stapled hACE-2 N-terminal alpha helix peptide, an electrochemical biosensor functionalizes an *in situ* produced polypyrrole conductive polymer on a nitrocellulose membrane backbone chemically. The analytical performance of the biosensor was evaluated using heat-inactivated forms of the SARS-CoV-2 virus in diluted nasal swab (NS) and artificial saliva (AS) samples, in addition to clinical specimens with known Ct values. The assay had a 40 TCID50/mL LOD, 100% specificity, and 95% sensitivity. Notably, the biosensor did not exhibit any cross-reactivity when tested against the influenza virus [45]. ZnO nanoparticles are placed on a gold film electrode of a multiplex electrochemical sensor to incorporate antibody-antigen interactions into the detecting system. The produced biosensor has a broad linear range for CRP and IL10, ranging from 0.01 μg/mL to 100 μg/mL and 0.1 μg/mL to 1000 μg/mL, respectively. Serum and urine samples were spiked with known amounts in recovery trials. With the help of an electrochemical sensing technology, COVID-19 patients can be screened early to watch for the emergence of more severe and potentially fatal symptoms [46].

3.2.2 POC Devices

A paper-based electrochemical sensor and a smartphone-assisted Sensit Smart potentiostat (PalmSens) were used to identify SARS-CoV-2. A disposable paper-based device was developed to capture the target complementary DNA (cDNA). The biological recognition element of the device, pyrrolidinyl peptide nucleic acid (acpcPNA), was added to the working electrode of the device. Hybridization with an acpcPNA probe in the presence of the target cDNA inhibits the redox conversion of a redox reporter, reducing the electrochemical response that is linked with the concentration of SARS-CoV-2. In optimal conditions, a linear range of 0.1–200 nM and a detection limit of 1.0 pM were reached. Using 10 nasopharyngeal swab samples (7 SARS-CoV-2 positive and 3 SARS-CoV-2 negative), the created sensor was utilized for amplification-free SARS-CoV-2 detection, yielding a 100% agreement result with RT-PCR [47]. By including a screen-printed electrode within a lateral-flow device, an electrochemical test strip (E-test strip) was created to capitalize on the remarkable binding affinity of SARS-CoV-2 antigen to ACE2. The SARS-CoV-2 antigen in the sample becomes electroactive when the ferrocene carboxylic acid associated with the SARS-CoV-2 antibody binds to it, and it then constantly flows to the electrode's ACE2-immobilization zone. The signal intensity increased in direct proportion to the concentration of SARS-CoV-2 antigen and within 12 min, the LOD of 2.98 pg/mL was attained using a smartphone electrochemical assay. Moreover, the results of the single-step E-test strip application for COVID-19 screening were demonstrated using nasopharyngeal samples, and they were consistent with the RT-PCR results [48]. A biosensor chip was reported that uses a one-step-one-pot hydrothermally generated $CoFeBDCNH_2$-$CoFe_2O_4$ MOF nanohybrid loaded with an Ab-SARS-CoV-2 modified gold electrode for enhanced molecular diagnostics, such as the detection of pseudoviruses and recombinant SARS-CoV-2 spike glycoprotein, as shown in Figure 3.10a. A PalmSens-EmStat Go POC device was used to test the developed biosensor chip. The results showed a broad linear range of detection between the concentration range from 10 ng/mL to 100 fg/mL with an assay time of less than 5 min and an LOD of 6.68 fg/mL and 6.20 fg/mL for recombinant SARS-CoV-2 spike glycoprotein in buffer and 10% serum-containing media, respectively [49]. A study was published on the detection of SARS-CoV 2 using a mobile POC platform and a microfluidic impedance biosensor. The microelectrode array was created by printing gold electrodes on borosilicate substrates using a conventional silicon-based technique, as shown in Figure 3.11. Using the previously optimized settings, the sensors were integrated into a polydimethylsiloxane (PDMS) microfluidic channel and spike glycoprotein concentrations ranging from 16 fM to 160 nM were attained. The proposed biosensor was clinically verified using 15 patient samples with the use of this calibration curve, producing an LOD of 19.09±0.65 fM and an LOQ of 63.66±2.1 fM [50]. A novel paper-based electrochemical immunosensor that does not require a solution was presented for the detection

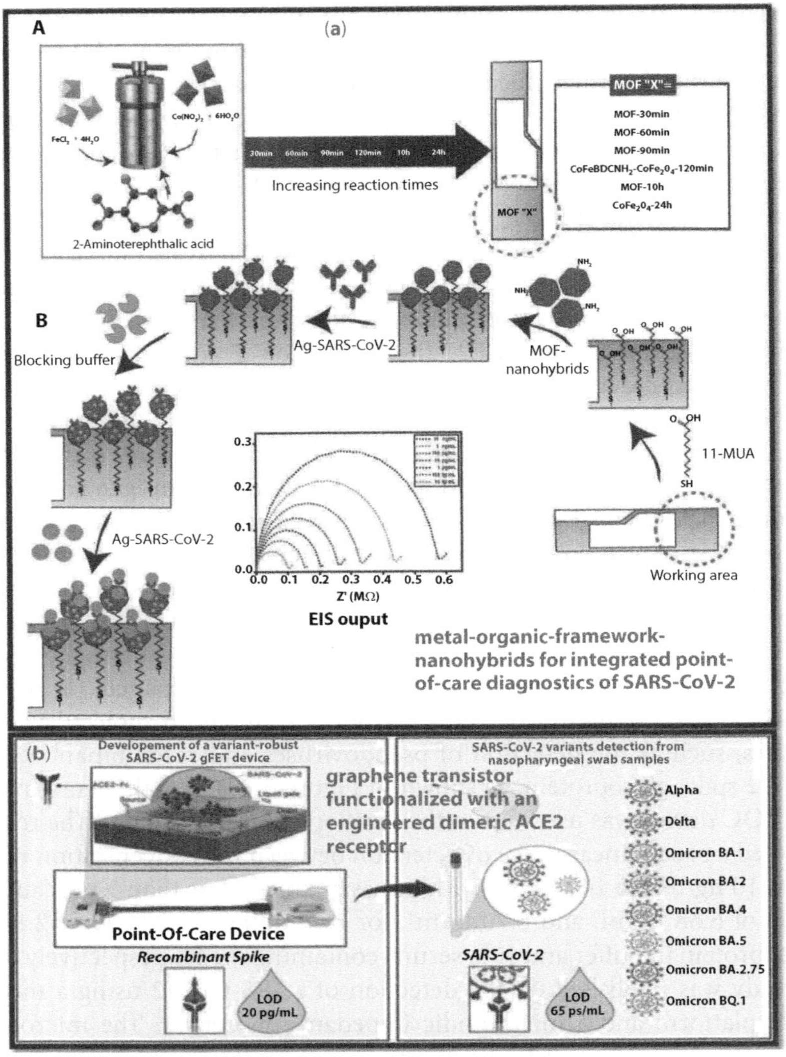

Figure 3.10 (a) metal-organic-framework-nanohybrids for integrated point-of-care diagnostics of SARS-CoV-2 viral antigen/pseudovirus utilizing electrochemical biosensor chip [49]. (b) graphene transistor functionalized with an engineered dimeric ACE2 receptor [53].

of SARS-CoV-2 in saliva samples. The apparatus consisted of a reservoir holding all the materials needed to build the immunological chain onto the magnetic beads and a lateral flow container with an absorbent pad, a magnet, and an electrode made of polyester. The saliva sample was first added

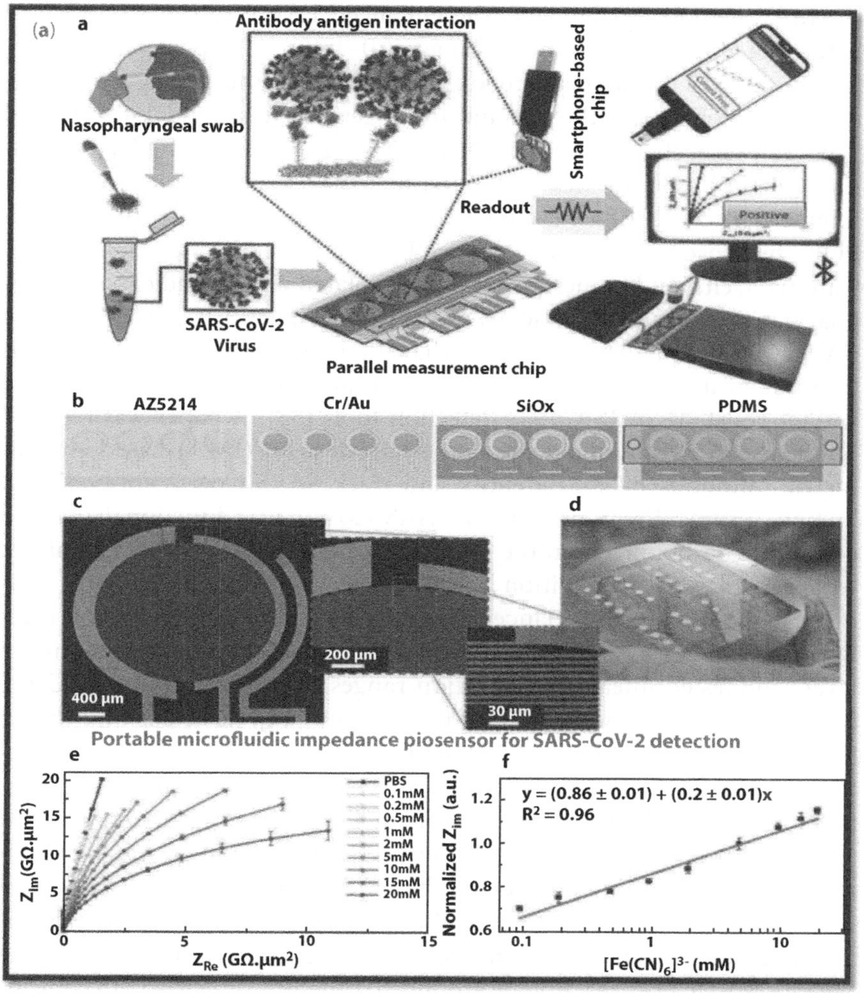

Figure 3.11 Portable microfluidic impedance biosensor for SARS-CoV-2 detection [50].

to the reservoir, and then this solution was added to the hole in the lateral flow container to complete the measurement. Saliva samples containing N protein were found to vary from 0.06 to 4 µg/mL, with a detection limit of 30 ng/mL, following the addition of washing buffer and TMB solution in the lateral flow holder [51]. Using an SPCE-modified gold nanostar (AuNSs) and a battery-operated thin-film heater, the SARS-CoV-2 envelope (E) and RNA-dependent RNA polymerase (RdRP) genes were identified. In this configuration, the loop-mediated isothermal amplification

(LAMP) method was applied. The enhanced LAMP test was performed with diluted target DNA in the range of 0–1.48×10^9 copies per reaction, and 30-μM MB was used as a redox indicator. Using this strategy the detection limits for the RdRP gene and the E gene were around 1.48×10^3 and 1.48×10^2, respectively, were achieved [52]. A graphene field-effect transistor (GFET) biosensor was developed that makes use of the Spike-ACE2 interaction to identify the trimeric Spike protein *in vitro*. The compact POC biosensor effectively detects B.1.610 (pre-VOC), Alpha, Beta, Gamma, Delta, and Omicron variants in isolated viruses and clinical nasopharyngeal swabs from patients, as shown in Figure 3.10b. The Omicron BA.5 swab sample has 65 cps/mL that the biosensor picked up on. This technique can serve as a reusable class of quick, extremely sensitive, and variant-resistant SARS-CoV-2 detection tools [53]. A paper was released on the use of an interdigitated microelectrode array (IMA)-based impedimetric immunosensor to identify and track SARS-CoV-2 antibodies in human serum. The S-protein of SARS-CoV-2 was functionalized and covalently immobilized on the IMA surface using conjugation chemistry, which acted as the recognition layer and selectively bound anti-spike antibodies. The system's impedance changes as a result of an increase in capacitance brought on by antibodies attached to the S-proteins in the recognition layer. The tested linear concentration ranges from 0.41 to 4.81 BAU/ml, and the device displays an LOD of 0.4 BAU/ml [54]. Base-stacking-driven catalytic hairpin assembly (BDCHA) is a nucleic acid amplification reaction technology designed to detect a SARS-CoV-2 N gene sequence fragment as a portable E-DNA biosensor for POC diagnostics. This method successfully raises the product's thermodynamic stability and increases recycling efficiency by introducing the base-stacking effect. As a result of the base-stacking effect, which combines the advantages of homogeneous and heterogeneous reactions, the electrode functions as a "booster," raising the rate of BDCHA's reaction by around 20 times in comparison to conventional catalytic techniques [55]. A liquid-gated GFET electrochemical sensor device was described that allows for the rapid and precise detection and separation of SARS-CoV-2 and influenza surface proteins. The gadget stands out primarily because to its quadruple-designed onboard GFET configuration. Each quarter is assessed collectively but functionalized independently (by chemical passivation or antibodies). The sensor platform detected the lowest concentration at about 50 ag/mL, which is five times lower than the COVID-19 and flu reported values of 88 zM and 227 zM, respectively. Graphene technology, in contrast to RT-PCR, has an extremely fast response time of about 10 s, even in complex and clinically relevant media like saliva [56]. To detect SARS-CoV-2 cDNA, a G-quadruplex DNAzyme-based electrochemical assay is combined with a sequential flow-controlled microfluidic system. According to the detection principle, a screen-printed graphene electrode is used to immobilize a pyrrolidinyl peptide nucleic acid probe to capture SARS-CoV-2 DNA. A distinct DNA probe carries a G-quadruplex DNAzyme, which hybridizes with the trapped DNA to function as the signaling unit. The H_2O_2-mediated oxidation of hydroquinone to benzoquinone, which is catalyzed by the G-quadruplex DNAzyme and produces a signal proportionate to the target DNA content, can be detected using square wave voltammetry (SWV). The assay showed good selectivity for SARS-CoV-2 DNA and an excellent experimental detection limit at 30 pM. The constructed microfluidic gadget was successfully used to detect SARS-CoV-2 from nasopharyngeal swab samples [57]. A gold reference electrode, a gold working electrode that has been modified with ZnO, and particular antibodies that are used to capture the specific proteins (S-protein, N-protein, D-dimer, and IL-2R) of these make up the CoST electrochemical immunosensor POC apparatus. The suggested apparatus can multiplex measure and quantify D-dimer, IL-2R, spike protein, and nucleo-capsid protein in plasma samples in a matter of minutes, hence determining the severity of the illness. With an LOD of 3 ng/mL, 21 ng/mL, 0.0006 ng/mL, and 0.242 ng/mL, the linear dynamic ranges for spike protein nucleocapsid protein, D-dimer, and IL-2R were 58 ng/mL to 60μg/mL, 58 ng/mL and 60μg/mL, 0.05 ng/mL–10 ng/mL, and 0.047 ng/mL–53.09 ng/mL, respectively [58]. An organic electrochemical transistor (OECT) was created in order to produce highly efficient ion-electron converters and amplifiers for the detection of S protein at the femtomolar concentration. The OECTs are built on Ti/Au thin film gate electrodes functionalized with anti-NP and anti-spike proteins, and PEDOT:PSS drain electrodes. The spike protein concentration was found to be more strongly correlated with changes in the transconductance parameters in the gate voltage range between 0 and 0.3 V after a thorough analysis of the electrical response of the examined OECTs [59]. In order to detect SARS-CoV-2 in human saliva, a disposable electrochemical sensor chip-based label-free rapid electrochemical POC biosensor was released. The sensor technology was developed using thionine (Th) adsorbed on carboxylic acid functionalized carbon nanotubes (SWCNT-COOH) and gold nanoparticles (GNp). The sensor's limit of detection for the freely suspended spike (S) protein in phosphate buffer saline (PBS) and human saliva is 200 and 500 pM, respectively, with a response time of 5 min. Furthermore, a COVID-19 pseudovirus with an LOD of 106 copies/mL in an electrolyte solution was shown to be detectable by the sensor [60].

3.3 Conclusions and Future Perspectives

To fight the illness and stop the pandemic from getting worse globally, new diagnostic techniques for identifying biomarkers linked to SARS-CoV-2 must be developed using nanomaterials and nanotechnologies. When compared to current methods, the electrochemical biosensors described in this chapter offer the appealing advantages of low cost, small sample volumes, simple and easy fabrication and operation, and the potential for use in clinical diagnosis. Most electrochemical biosensor research has used single-stranded DNA and antibodies as biorecognition molecules. Surface modification or direct application to the working electrode were the two main techniques utilized to immobilize biorecognition elements (or transistor channel). By building signal converters, researchers are tried to increase the sensitivity of signal reading from interaction with targets. They were also shrunk in size to make the biosensor portable. By combining the most sensitive electrochemical sensor systems with cellphones, the end user may get fast results. Even though advancing in sensor development for infectious diseases ability to diagnose clinical specimens of SARS-CoV-2, there are several limits in their use when transferring from laboratory analysis to *in situ* detection. Numerous challenges require more investigation, such as sample preparation procedures, commercialization of electrochemical biosensors as diagnostic devices, scaling up device development, and stability and repeatability of electrochemical biosensors. Innovative approaches and coordinated efforts are needed to overcome these obstacles in the development of biosensors for the detection of SARS-CoV-2. Some potential solutions are discussed in the section below. The first one is the use of advanced nanomaterials, surface chemistry modifications, and bio-recognition elements (such as antibodies or aptamers) to increase the selectivity and sensitivity of the sensor. Specificity can be increased by continuously improving detection techniques and validating them against different virus strains. The second is to develop small and portable biosensors using smartphone-based platforms, lab-on-a-chip technologies, or microfluidics for quick, on-site detection. Using amplification techniques such as isothermal amplification or PCR (polymerase chain reaction) might expedite the detection process. Simplify sample preparation by looking at methods that can speed up the process, utilizing integrated sample preparation modules in biosensors to reduce complexity and time, or directly detecting information from unprocessed samples. It is important to utilize protective coatings, long-lasting materials, and optimal storage conditions to increase the stability of biosensors. To ensure consistent performance, conduct stability studies over a prolonged time frame. Another tactic is to focus on cost-effectiveness in order to increase productivity without sacrificing quality. Other strategy is can also employ inexpensive materials, streamline manufacturing procedures, and devise inventive, cost-effective fabrication techniques. Realizing inexpensive, large-scale production can be facilitated by working with corporate partners. The approval process will go more quickly with regulatory permission, such as early engagement with regulatory bodies, which will aid in understanding requirements. Conduct thorough validation studies in compliance with established procedures to successfully meet regulatory requirements. The ability to adapt to variability is the next. For example, one can develop biosensors with versatile detection systems that can recognize the virus's conserved regions and identify different strains or mutations. Watch how viruses change and make any necessary design adjustments to biosensors. Another issue is to consider the needs of end users while creating biosensors. Give special consideration to findings interpretability, usability, and connection with the existing healthcare system. Conduct user research and hold feedback sessions for improvement. After the completion of the aforementioned, biosensors' efficiency, dependability, and accessibility are increased by continuous innovation and development incorporating new materials, technologies, and techniques. By using these techniques, scientists and engineers can collaborate to overcome obstacles and build strong biosensors for effective and dependable SARS-CoV-2 detection.

References

1. Cui, J., Li, F., Shi, Z.L., Origin and evolution of pathogenic coronaviruses. *Nat. Rev. Microbiol.*, 17, 181–192, 2019, https://doi.org/10.1038/s41579-018-0118-9.
2. Adil, M.T., Rahman, R., Whitelaw, D., Jain, V., Al-Taan, O., Rashid, F., Munasinghe, A., Jambulingam, P., SARS-CoV-2 and the pandemic of COVID-19. *Postgrad. Med. J.*, 97, 110–116, 2021, https://doi.org/10.1136/postgradmedj-2020-138386.
3. Zeng, J., Duarte, P.A., Ma, Y., Savchenko, O., Shoute, L., Khaniani, Y., Babiuk, S., Zhuo, R., Abdelrasoul, G.N., Charlton, C., Kanji, J.N., Babiuk, L., Edward, C., Chen, J., An impedimetric biosensor for COVID-19 serology test and modification of sensor performance via dielectrophoresis force. *Biosens. Bioelectron.*, 213, 114476, 2022, https://doi.org/10.1016/j.bios.2022.114476.
4. Hemamalini, V., Anand, L., Nachiyappan, S., Geeitha, S., Ramana Motupalli, V., Kumar, R., Ahilan, A., Rajesh, M., Integrating bio medical

sensors in detecting hidden signatures of COVID-19 with Artificial intelligence. *Measurement*, 194, 111054, 2022, https://doi.org/10.1016/j.measurement.2022.111054.

5. Hu, B., Guo, H., Zhou, P., Shi, Z.L., Characteristics of SARS-CoV-2 and COVID-19. *Nat. Rev. Microbiol.*, 19, 141–154, 2021, https://doi.org/10.1038/s41579-020-00459-7.
6. Gorbalenya, A.E., Baker, S.C., Baric, R.S., de Groot, R.J., Drosten, C., Gulyaeva, A.A., Haagmans, B.L., Lauber, C., Leontovich, A.M., Neuman, B.W., Penzar, D., Perlman, S., Poon, L.L.M., Samborskiy, D.V., Sidorov, I.A., Sola, I., Ziebuhr, J., The species severe acute respiratory syndrome-related coronavirus: classifying 2019-nCoV and naming it SARS-CoV-2. *Nat. Microbiol.*, 5, 536–544, 2020, https://doi.org/10.1038/s41564-020-0695-z.
7. Liu, J., Xie, W., Wang, Y., Xiong, Y., Chen, S., Han, J., Wu, Q., A comparative overview of COVID-19, MERS and SARS: review article. *Int. J. Surg.*, 81, 1–8, 2020, https://doi.org/10.1016/j.ijsu.2020.07.032.
8. Zayed, B.A., Ali, A.N., Elgebaly, A.A., Talaia, N.M., Hamed, M., Mansour, F.R., Smartphone-based point-of-care testing of the SARS-CoV-2: A systematic review. *Sci. Afr.*, 21, e01757, 2023, https://doi.org/10.1016/j.sciaf.2023.e01757.
9. Mishra, S., Aamna, B., Parida, S., Dan., A.K., Carbon-based biosensors: Next-generation diagnostic tool for target-specific detection of SARS-CoV-2 (COVID-19). *Talanta Open*, 7, 100218, 2023, https://doi.org/10.1016/j.talo.2023.100218.
10. Drobysh, M., Ramanaviciene, A., Viter, R., Ramanavicius, A., Affinity sensors for the diagnosis of covid-19. *Micromachines*, 12, 1–19, 2021, https://doi.org/10.3390/mi12040390.
11. Hoffmann, M., Kleine-Weber, H., Schroeder, S., Krüger, N., Herrler, T., Erichsen, S., Schiergens, T.S., Herrler, G., Wu, N.H., Nitsche, A., Müller, M.A., Drosten, C., Pohlmann, S., SARS-CoV-2 cell entry depends on ACE2 and TMPRSS2 and is blocked by a clinically proven protease inhibitor. *Cell*, 181, 271–280, 2020, https://doi.org/10.1016/j.cell.2020.02.052,e8.
12. Huang, Y., Yang, C., Xu, X., Xu, W., Liu, S., Structural and functional properties of SARS-CoV-2 spike protein: potential antivirus drug development for COVID-19. *Acta Pharmacol. Sin.*, 41, 1141–1149, 2020, https://doi.org/10.1038/s41401-020-0485-4.
13. Markov, P.V., Katzourakis, A., Stilianakis, N.I., Antigenic evolution will lead to new SARS-CoV-2 variants with unpredictable severity. *Nat. Rev. Microbiol.*, 20, 251–252, 2022, https://doi.org/10.1038/s41579-022-00722-z.
14. Malik, S., Singh, J., Goyat, R., Saharan, Y., Chaudhry, V., Umar, A., Ibrahim, A.A., Akbar, S., Ameen, S., Baskoutas., S., Nanomaterials-based biosensor and their applications: A review. *Heliyon*, 9, e19929, 2023, https://doi.org/10.1016/j.heliyon.2023.e19929.

15. Gandarilla, A.M.D., de Souza Freirea, L., Ruzoa, C.M., Barcelaya, Y.R., Brito, W.R., Electrochemical biosensors as promising methods for diagnosis of Covid-19–A Minireview. *Anal. Lett.*, 57, 5, 812–828, 2023, https://doi.org/10.1080/00032719.2023.2225658.
16. Andrianova, M.S., Panova, O.S., Titov, A.A., Komarova, N.V., Kuznetsov., A.E., Electrochemical Biosensors for SARS-CoV-2 Detection. *Moscow Univ. Chem. Bull.*, 78, 5, 231–254, 2023, https://doi.org/10.3103/S0027131423050048.
17. Kevadiya, B.D., Machhi, J., Herskovitz, J., Oleynikov, M.D., Blomberg, W.R., Bajwa, N., Soni, D., Das, S., Hasan, M., Patel, M., Senan, A.M., Gorantla, S., McMillan, J., Edagwa, B., Eisenberg, R., Gurumurthy, C.B., Reid, S.P.M., Punyadeera, C., Chang, L., Gendelman, H.E., Diagnostics for SARS-CoV-2 infections. *Nat. Mater.*, 20, 593–605, 2021, https://doi.org/10.1038/s41563-020-00906-z.
18. Teklemariam, A.D., Samaddar, M., Alharbi, M.G., Al-Hindi, R.R., Bhunia, A.K., Biosensor and molecular-based methods for the detection of human coronaviruses: a review. *Mol. Cell. Probes*, 54, 101662, 2020, https://doi.org/10.1016/j.mcp.2020.101662.
19. Bolourinezhad, M., Rezayi, M., Meshkat, Z., Soleimanpour, S., Mojarrad, M., Zibadi, F., Aghaee-Bakhtiari, S.H., Taghdisi, S.M., Design of a rapid electrochemical biosensor based on MXene/Pt/C nanocomposite and DNA/RNA hybridization for the detection of COVID-19. *Talanta*, 265, 124804, 2023, https://doi.org/10.1016/j.talanta.2023.124804.
20. Bharti, A., Singh, S., Munthala, D., Roy, S., Pojprapai, S., Suksaweang, S., Sain, S., Sinha Roy, S., Mohamed, J.J., Avasthi, D.K., Mathur., A., Development of a nucleic acid-based screen printed electrochemical biosensor using Ti_3C_2Tx-MXene for the detection of SARS-CoV-2. *Microchem. J.*, 195, 109521, 2023, https://doi.org/10.1016/j.microc.2023.109521.
21. Yin, Z.Z., Liu, Z., Zhou, M., Yang, X., Zheng, G., Zhang, H., Kong., Y., A surface molecularly imprinted electrochemical biosensor for the detection of SARS-CoV-2 spike protein by using Cu_7S_4-Au as built-in probe. *Bioelectrochemistry*, 152, 108462, 2023, https://doi.org/10.1016/j.bioelechem.2023.108462.
22. de Lima, L.F., Ferreira, A.L., Ranjan, I., Collman, R.G., de Araujo, W.R., de la Fuente-Nunez, C., A Bacterial cellulose-based and low-cost electrochemical biosensor for ultrasensitive detection of SARS-CoV-2. *Cell Rep. Phys. Sci.*, 4, 101476, 2023, https://doi.org/10.1016/j.xcrp.2023.101476.
23. Li, Y., Zhao, S., Xu, Z., Qiao, X., Li, M., Li, Y., Luo, X., Peptide nucleic acid and antifouling peptide based biosensor for the non-fouling detection of COVID-19 nucleic acid in saliva. *Biosens. Bioelectron.*, 225, 115101, 2023, https://doi.org/10.1016/j.bios.2023.115101.
24. Santos, A., Macedo de Souza Brandão, A.P., Hryniewicz, B.M., Abreu, H., Bach-Toledo, L., Schuster da Silva, S., Deller, A.E., Rogerio, V.Z., Baêta Rodrigues, D.S., Hiraiwa, P.M., Guimarães, B.G., Marchesi, L.F., Carvalho de Oliveira, J., Gradia, D.F., Soares, F.L.F., Zanchin, N.I.T., Camargo de Oliveira,

C., Vidotti., M., COVID-19 impedimetric biosensor based on polypyrrole nanotubes, nickel hydroxide and VHH antibody fragment: specific, sensitive, and rapid viral detection in saliva samples. *Mater. Today Chem.*, 30, 101597, 2023, https://doi.org/10.1016/j.mtchem.2023.101597.

25. Liang, Q., Huang, Y., Wang, M., Kuang, D., Yang, J., Yi, Y., Shi, H., Li, J., Yang, J., Li., G., An electrochemical biosensor for SARS-CoV-2 detection via its papain-like cysteine protease and the protease inhibitor screening. *Chem. Eng. J.*, 452, 139646, 2023, https://doi.org/10.1016/j.cej.2022.139646.
26. You, X.H., Liu, Y., Li, Y.Y., Zhao, B., Yang, Y., Weerasooriya, R., Chen., X., Sensitive detection of SARS-CoV-2 spike protein based on electrochemical impedance spectroscopy of Fe_3O_4@SiO_2–Au/GCE biosensor. *Adv. Sens. Energy Mater.*, 2, 3, 100067, 2023, https://doi.org/10.1016/j.asems.2023.100067.
27. Chen, H., Hou, Z.Y., Chen, D., Li, T., Wang, Y.M., De Lima, M.A., Yang, Y., Guo., Z.Z., Highly Sensitive Poly-N-isopropylacrylamide Microgel-based Electrochemical Biosensor for the Detection of SARS-COV-2 Spike Protein. *Biomed. Environ. Sci.*, 36, 3, 269–278, 2023, https://doi.org/10.3967/bes2023.029.
28. Kashefi-Kheyrabadi, L., Nguyen, H.V., Go, A., Lee, M.H., Ultrasensitive and amplification-free detection of SARS-CoV-2 RNA using an electrochemical biosensor powered by CRISPR/Cas13a. *Bioelectrochemistry*, 150, 108364, 2023, https://doi.org/10.1016/j.bioelechem.2023.108364.
29. Adegoke, O., Oyinlola, K., Achadu, O.J., Yang, Z., Blue-emitting SiO_2-coated Si-doped ZnSeS quantum dots conjugated aptamer-molecular beacon as an electrochemical and metal-enhanced fluorescence biosensor for SARS-CoV-2 spike protein. *Anal. Chim. Acta*, 1281, 341926, 2023, https://doi.org/10.1016/j.aca.2023.341926.
30. Kadadou, D., Tizani, L., Wadi, V.S., Banat, F., Alsafar, H., Yousef, A.F., Hasan, S.W., Detection of SARS-CoV-2 in clinical and environmental samples using highly sensitive reduced graphene oxide (rGO)-based biosensor. *Chem. Eng. J.*, 453, 139750, 2023, https://doi.org/10.1016/j.cej.2022.139750.
31. Kumar, V., H., T., Srinivasan, S., Krishnan, V., Vaidyanathan, R., Anand Babu, K., Natarajan, S., Veerapandian., M., Peptide-based direct electrochemical detection of receptor binding domains of SARS-CoV-2 spike protein in pristine samples. *Sens. Actuators B: Chem.*, 377, 133052, 2023, https://doi.org/10.1016/j.snb.2022.133052.
32. Drobysh, M., Liustrovaite, V., Kanetski, Y., Brasiunas, B., Zvirbliene, A., Rimkute, A., Gudas, D., Kodze, I.K., Simanavicius, M., Ramanavicius, S., Slibinskas, R., Ciplys, E., Plikusiene, I., Ramanavicius, A., Electrochemical biosensing based comparative study of monoclonal antibodies against SARS-CoV-2 nucleocapsid protein. *Sci. Total Environ.*, 908, 168154, 2024, https://doi.org/10.1016/j.scitotenv.2023.168154.

33. Cancino-Bernardi, J., Comparetti, E.J., Ferreira, N.N., Miranda, R.R., Tuesta, M.M., Sampaio, I., da Costa, P.I., Cottontail., V., SARS-CoV-2 impedimetric biosensor based on the immobilization of ACE-2 receptor-containing entire cell membranes as the biorecognition element. *Talanta*, 253, 124008, 2023, https://doi.org/10.1016/j.talanta.2022.124008.
34. Tabrizi, M.A., An electrochemical PNA-based sensor for the detection of the SARS-CoV-2 RdRP by using surface-initiated-reversible-addition–fragmentation-chain-transfer polymerization technique. *Talanta*, 259, 124490, 2023, https://doi.org/10.1016/j.talanta.2023.124490.
35. Zhu, M., Liu, Y., Wang, M., Liu, T., Chu, Z., Jin, W., Facile construction of nanocubic $Mn_3[Fe(CN)_6]_2$@Pt based electrochemical DNA sensors for ultrafast precise determination of SARS-CoV-2. *Bioelectrochemistry*, 156, 108598, 2024, https://doi.org/10.1016/j.bioelechem.2023.108598.
36. Malla, P., Liu, C.H., Wu, W.C., Kabinsing, P., Sreearunothai, P., Synthesis and characterization of Au-decorated graphene oxide nanocomposite for magneto-electrochemical detection of SARS-CoV-2 nucleocapsid gene. *Talanta*, 262, 124701, 2023, https://doi.org/10.1016/j.talanta.2023.124701.
37. Baradoke, A., Jarusaitis, A., Reinikovaite, V., Jafarov, A., Elsakova, A., Franckevicius, M., Skapas, M., Slibinskas, R., Drobysh, M., Liustrovaite, V., Ramanavicius., A., Detection of antibodies against SARS-CoV-2 Spike protein by screen-printed carbon electrodes modified by colloidal gold nanoparticles. *Talanta*, 125279, 103, 2024, https://doi.org/10.1016/j.talanta.2023.125279.
38. Sarwar, S., Lin, M.C., Amezaga, C., Wei, Z., Iyayi, E., Polk, H., Wang, R., Wang, H., Zhang., X., Ultrasensitive electrochemical biosensors based on zinc sulfide/graphene hybrid for rapid detection of SARS-CoV-2. *Adv. Compos. Hybrid Mater.*, 6, 49, 2023, https://doi.org/10.1007/s42114-023-00630-7.
39. Mandal, N., Mitra, R., Pramanick, B., C-MEMS-derived glassy carbon electrochemical biosensors for rapid detection of SARS-CoV-2 spike protein. *Microsyst. Nanoeng.*, 9, 137, 2023, https://doi.org/10.1038/s41378-023-00601-4.
40. Hussein, H.A., Hanora, A., Solyman, S.M., Hassan, R.Y.A., Designing and fabrication of electrochemical nano-biosensor for the fast detection of SARS-CoV-2-RNA. *Sci. Rep.*, 13, 5139, 2023, https://doi.org/10.1038/s41598-023-32168-5.
41. Chen, Z., Wu, C., Yuan, Y., Xie, Z., Li, T., Huang, H., Li, S., Deng, J., Lin, H., Shi, Z., Li, C., Hao, Y., Tang, Y., You, Y., Al-Hartomy, O.A., Wageh, S., Al-Sehemi, A.G., Lu, R., Zhang, L., Lin, X., He, Y., Zhao, G., Li, D., Zhang, H., CRISPR-Cas13a-powered electrochemical biosensor for the detection of the L452R mutation in clinical samples of SARS-CoV-2 variants. *J. Nanobiotechnol.*, 21, 141, 2023, https://doi.org/10.1186/s12951-023-01903-5.
42. Alam, A., Uppal, T., Islam, M.S., Misra, M., Verma, S.C., Polyaniline/Titania Nanotube-Based Biosensor Strip for Sensitive and Specific Electrochemical

Detection of SARS-CoV-2. *ACS Omega*, 8, 48, 45700–45707, 2023, https://doi.org/10.1021/acsomega.3c06133.
43. Liu, L., Xu, Z., Vargas, A.M.M., Dollery, S.J., Schrlau, M.G., Cormier, D., O'Connell, M.R., Tobin, G.J., Du., K., Aerosol Jet Printing-Enabled Dual-Function Electrochemical and Colorimetric Biosensor for SARS-CoV-2 Detection. *Anal. Chem.*, 95, 32, 11997–12005, 2023, https://doi.org/10.1021/acs.analchem.3c01724.
44. Lu, J., Zhou, X., Li, Y., Yu, M., Fu, S., Qu, Z., Li, Y., Miao, J., Xu, Y., Electrochemical detection of SARS-CoV-2 based on copper nanoflower-triggered *in situ* growth of electroactive polymers. *Analyst*, 148, 3659–3665, 2023, https://doi.org/10.1039/D3AN00716B.
45. Meshesha, M., Sardar, A., Supekar, R., Bhattacharjee, L., Chatterjee, S., Halder, N., Mohanta, K., Bhattacharyya, T.K., Pal, B., Development and Analytical Evaluation of a Point-of-Care Electrochemical Biosensor for Rapid and Accurate SARS-CoV-2 Detection. *Sensors*, 23, 18, 8000, 2023, https://doi.org/10.3390/s23188000.
46. Madhurantakam, S., Lee, Z.J., Naqvi, A., Karnam, J.B., Muthukumar, S., Prasad, S., Multiplex sensing of IL-10 and CRP towards predicting critical illness in COVID-19 infections. *Biosens. Bioelectron.: X*, 13, 100307, 2023, 10.1016/j.biosx.2023.100307.
47. Lomae, A., Preechakasedkit, P., Hanpanich, O., Ozer, T., Henry, C.S., Maruyama, A., Pasomsub, E., Phuphuakrat, A., Rengpipat, S., Vilaivan, T., Chailapakul, O., Ruecha, N., Ngamrojanavanich, N., Label free electrochemical DNA biosensor for COVID-19 diagnosis. *Talanta*, 253, 123992, 2023, https://doi.org/10.1016/j.talanta.2022.123992.
48. Deenin, W., Yakoh, A., Pimpitak, U., Pasomsub, E., Rengpipat, S., Crespo, G.A., Chaiyo, S., Electrochemical lateral-flow device for rapid COVID-19 antigen-diagnostic testing. *Bioelectrochemistry*, 152, 108438, 2023, https://doi.org/10.1016/j.bioelechem.2023.108438.
49. Palanisamy, S., Lee, L.Y., Kao, C.F., Chen, W.L., Wang, H.C., Shen, S.T., Jian, J.W., Yuan, S.S.F., Kung, Y.A., Wang., Y.M., One-step-one-pot hydrothermally derived metal-organic-framework-nanohybrids for integrated point-of-care diagnostics of SARS-CoV-2 viral antigen/pseudovirus utilizing electrochemical biosensor chip. *Sens. Actuators: B. Chem.*, 390, 133960, 2023, https://doi.org/10.1016/j.snb.2023.133960.
50. Laleh, S., Ibarlucea, B., Stadtmüller, M., Cuniberti, G., Medina-Sanchez., M., Portable microfluidic impedance biosensor for SARS-CoV-2 detection. *Biosens. Bioelectron.*, 236, 115362, 2023, https://doi.org/10.1016/j.bios.2023.115362.
51. Fabiani, L., Fiore, L., Fillo, S., D'Amore, N., De Santis, R., Lista, F., Arduini., F., Smartphone-assisted paper-based electrochemical immunosensor for SARS-CoV-2 detection in saliva. *Bioelectrochemistry*, 156, 108619, 2023, https://doi.org/10.1016/j.bioelechem.2023.108619.

52. Kim, H.E., Schuck, A., Park, H., Huh, H.J., Kang, M., Kim., Y.S., Gold nanostructures modified carbon-based electrode enhanced with methylene blue for point-of-care COVID-19 tests using isothermal amplification. *Talanta*, 265, 124841, 2023, https://doi.org/10.1016/j.talanta.2023.124841.
53. Romagnoli, A., D'Agostino, M., Pavoni, E., Ardiccioni, C., Motta, S., Crippa, P., Biagetti, G., Notarstefano, V., Rexha, J., Perta, N., Barocci, S., Costabile, B.K., Colasurdo, G., Caucci, S., Mencarelli, D., Turchetti, C., Farina, M., Pierantoni, L., La Teana, A., Hadi, R.A., Cicconardi, F., Chinappi, M., Trucchi, E., Mancia, F., Menzo, S., della Rocca, B.M., D'Annessa, I., Di Marino., D., SARS-CoV-2 multi-variant rapid detector based on graphene transistor functionalized with an engineered dimeric ACE2 receptor. *Nanotoday*, 48, 101729, 2023, https://doi.org/10.1016/j.nantod.2022.101729.
54. Shoute, L.C.T., Abdelrasoul, G.N., Ma, Y., Duarte, P.A., Edwards, C., Zhuo, R., Zeng, J., Feng, Y., Charlton, C.L., Kanji, J.N., Babiuk, S., Chen., J., Label-free impedimetric immunosensor for point-of-care detection of COVID-19 antibodies. *Microsyst. Nanoeng.*, 9, 3, 2023, https://doi.org/10.1038/s41378-022-00460-5.
55. Chen, G., Yang, N., Xu, L., Lu, S., Chen, Z., Wu, F., Chen, J., Zhang, X., Base-Stacking-Driven Catalytic Hairpin Assembly: A Nucleic Acid Amplification Reaction Using Electrode Interface as a "Booster" for SARS-CoV-2 Point-of-Care Testing. *Anal. Chem.*, 95, 42, 15595–15605, 2023, https://doi.org/10.1021/acs.analchem.3c02577.
56. Kumar, N., Towers, D., Myers, S., Galvin, C., Kireev, D., Ellington, A.D., Akinwande, D., Graphene Field Effect Biosensor for Concurrent and Specific Detection of SARS-CoV-2 and Influenza. *ACS Nano*, 17, 18, 18629–18640, 2023. 10.1021/acsnano.3c07707.
57. Naorungroj, S., Srisomwat, C., Khamcharoen, W., Jampasa, S., Pasomsub, E., Shin, K., Vilaivan, T., Chailapakul, O., Sequential Flow Controllable Microfluidic Device for G-Quadruplex DNAzyme-Based Electrochemical Detection of SARS-CoV-2 Using a Pyrrolidinyl Peptide Nucleic Acid. *Anal. Chem.*, 95, 34, 12794–12801, 2023. 10.1021/acs.analchem.3c01758.
58. Madhurantakam, S., Babu Karnam, J., Muthukumar, S., Prasad, S., COVID severity test (CoST sensor)—An electrochemical immunosensing approach to stratify disease severity. *Bioeng. Transl. Med.*, 8, e10566, 2023. 10.1002/btm2.10566.
59. Barra, M., Tomaiuolo, G., Villella, V.R., Esposito, S., Liboà, A., D'Angelo, P., Marasso, S.L., Cocuzza, M., Bertana, V., Camilli, E., Organic Electrochemical Transistor Immuno-Sensors for Spike Protein Early Detection. *Biosensors*, 13, 7, 739, 2023, https://doi.org/10.3390/bios13070739.
60. Farsaeivahid, N., Grenier, C., Nazarian, S., Wang, M.L., A Rapid Label-Free Disposable Electrochemical Salivary Point-of-Care Sensor for SARS-CoV-2 Detection and Quantification. *Sensors*, 23, 1, 433, 2023, https://doi.org/10.3390/s23010433.

4

Nanobiosensors: Versatile Tool for Diagnosis of Infectious Diseases

Prasann Kumar[1*] **and Joginder Singh**[2]

[1]*Department of Agronomy, School of Agriculture, Lovely Professional University, Phagwara, (Punjab), India*
[2]*Department of Botany, Nagaland University, Nagaland, India*

Abstract
Nanobiosensors have emerged as versatile tools for diagnosing infectious diseases, offering significant advancements in sensitivity, specificity, and rapid detection. These biosensors harness the unique properties of nanomaterials to enable precise and reliable detection of infectious agents, including viruses, bacteria, and parasites. By integrating nanomaterials with biological recognition elements, such as antibodies, aptamers, or enzymes, nanobiosensors can selectively capture and detect target biomarkers with high sensitivity. Various nanomaterials, including nanoparticles, nanowires, and nanosheets, enhance the biosensors' signal amplification and transduction mechanisms. Additionally, nanotechnology allows for the development of miniaturized and portable biosensing devices, enabling point-of-care testing and remote monitoring. This chapter highlights the potential of nanobiosensors as a powerful diagnostic tool for infectious diseases, facilitating early detection, accurate diagnosis, and timely intervention for effective disease management.

Keywords: Zero hunger, no poverty, nano, biosensor, disease, significant, nanowires, transduction, signal amplification

**Corresponding author*: prasann0659@gmail.com

Suvardhan Kanchi, Ayyappa Bathinapatla, Anitha Varghese and Phumlane Selby Mdluli (eds.) Nano-Biosensor Technologies for Diagnosis of Infectious Diseases, (121–172) © 2025 Scrivener Publishing LLC

4.1 Introduction

Infectious diseases continue to pose a significant global health threat, emphasizing the critical need for rapid, sensitive, and accurate diagnostic tools [1, 2]. Traditional diagnostic methods often have sensitivity, specificity, and turnaround time limitations. In recent years, nanobiosensors have emerged as versatile tools for diagnosing infectious diseases, offering unprecedented early detection and practical disease management opportunities [3–6]. Nanobiosensors combine the advantages of nanotechnology and biorecognition elements to create susceptible and selective detection platforms. These biosensors leverage the unique properties of nanomaterials, such as nanoparticles, nanowires, and nanosheets, to enhance the detection and transduction of target biomarkers associated with infectious agents. Integrating nanomaterials with biological recognition elements, such as antibodies, aptamers, or enzymes, allows for specific and efficient capture of target analytes, enabling precise diagnosis [7–10]. The use of nanomaterials in biosensors offers several advantages. First, nanomaterials possess a large surface area-to-volume ratio, facilitating enhanced interactions with target biomarkers. This leads to improved sensitivity and enables the detection of low-abundance analytes in complex clinical samples.

Additionally, nanomaterials' unique optical, electrical, and mechanical properties can be harnessed to develop label-free or signal amplification strategies, further enhancing the detection capabilities of nanobiosensors [11–14]. Nanobiosensors also offer the potential for miniaturization and integration into portable devices, enabling point-of-care testing and remote monitoring. These compact devices can provide rapid and on-site diagnosis, bypassing the need for time-consuming sample transportation and centralized laboratory facilities. Such decentralized testing can significantly improve healthcare delivery, especially in resource-limited settings and during outbreaks or emergencies [15–19]. The versatility of nanobiosensors extends beyond their application in viral infections. They have been successfully employed to diagnose bacterial infections, parasitic diseases, and fungal infections. The ability to detect multiple infectious agents using a single platform enhances diagnostic efficiency and reduces the cost and time associated with conventional methods. Moreover, nanobiosensors hold promise for detecting antimicrobial resistance markers, allowing for personalized treatment approaches and surveillance of drug-resistant strains [20–24]. By rapidly identifying the presence of drug-resistant pathogens, these biosensors enable clinicians to make informed decisions regarding appropriate treatment options, improving patient outcomes, and reducing the spread of resistant infections. This comprehensive review will explore the design principles, fabrication techniques, and applications of nanobiosensors for diagnosing infectious diseases. We will discuss the nanomaterials employed in biosensing platforms and highlight their advantages and limitations [21, 25, 30].

Additionally, we will integrate nanotechnology with emerging technologies, such as microfluidics and smartphone-based systems, to enhance the performance and accessibility of nanobiosensors. Ultimately, nanobiosensors have the potential to revolutionize the field of infectious disease diagnostics, offering rapid, sensitive, and specific detection platforms for a wide range of pathogens. As research in this area progresses, it is anticipated that nanobiosensors will be crucial in early disease detection, personalized treatment strategies, and effective public health interventions [31–36].

4.2 Nanobiosensors as Promising Devices for the Diagnosis of Coronavirus Family Members

The ongoing global pandemic caused by the coronavirus family, including severe acute respiratory syndrome coronavirus (SARS-CoV), Middle East respiratory syndrome coronavirus (MERS-CoV), and most recently, SARS-CoV-2 responsible for COVID-19, has highlighted the urgent need for accurate and rapid diagnostic tools [37–41]. Conventional diagnostic methods have faced challenges regarding sensitivity, specificity, and turnaround time. In this context, nanobiosensors have emerged as promising devices for diagnosing coronavirus family members, offering enhanced early and accurate detection capabilities. Nanobiosensors integrate nanotechnology principles with biorecognition elements to create susceptible and specific diagnostic platforms. These biosensors leverage the unique properties of nanomaterials to amplify signals and improve the detection of viral biomarkers associated with coronavirus infections. By combining nanomaterials with specific recognition elements, such as antibodies or aptamers targeting viral antigens or genetic material, nanobiosensors enable precise and efficient detection of the coronavirus family members [42–46]. The utilization of nanomaterials in biosensing platforms provides several advantages for coronavirus diagnosis. Nanomaterials, such as nanoparticles, nanowires, and nanosheets, offer a large surface area-to-volume ratio, enabling enhanced interactions with viral biomarkers and improving detection sensitivity. Additionally, nanomaterials' unique optical, electrical, and mechanical properties can be exploited to develop label-free or signal amplification strategies, further enhancing the detection capabilities of nanobiosensors. One of the critical advantages of nanobiosensors in coronavirus diagnosis is their potential for rapid and on-site testing. Point-of-care testing can be achieved by integrating these biosensors into portable devices, enabling real-time detection and immediate decision-making. The decentralized nature of nanobiosensor-based diagnostics reduces the dependency on centralized laboratory facilities and expedites the diagnostic process, which is crucial for timely intervention and infection control. Moreover, nanobiosensors can be designed to detect multiple viral targets, enabling comprehensive and simultaneous screening for different members of the coronavirus family [47–53]. This versatility is particularly relevant in the case of emerging coronaviruses, where prompt identification and differentiation of the specific viral strain can aid in public health interventions and outbreak control measures.

Developing and deploying nanobiosensors for coronavirus diagnosis also hold promise for addressing challenges related to asymptomatic carriers and early-stage infections. These biosensors can detect viral biomarkers even in low-concentration samples, allowing for the identification of infected individuals who may not exhibit obvious symptoms. Early detection of coronavirus infections is vital for effective containment measures and reducing the spread of the disease. This comprehensive chapter will explore the latest advances in nanobiosensors for diagnosing coronavirus family members [54–60]. We will discuss the design principles, fabrication techniques, and integration of nanomaterials with biorecognition elements for improved sensitivity and specificity.

Furthermore, we will examine the potential of nanobiosensors in terms of on-site testing, multiplex detection, and their role in early-stage infection identification. As research progresses and nanotechnology continues to evolve, nanobiosensors can potentially revolutionize the field of coronavirus diagnosis. These promising devices offer rapid, sensitive, and specific detection platforms, empowering healthcare professionals with valuable tools for efficiently diagnosing, detecting, and managing coronavirus infections (Table 4.1).

Table 4.1 Table showcasing a selection of 50 nanobiosensors for diagnosing coronavirus family members.

Nanobiosensor	Detection principle	Target analyte
Gold nanoparticle-based assay	Colorimetric	Viral antigens
Carbon nanotube field-effect transistor (CNT-FET)	Electrical	Viral RNA
Quantum dot-based biosensor	Fluorescence	Viral proteins
Graphene-based field-effect transistor (GFET)	Electrical	Viral RNA
Plasmonic nanobiosensors	Surface plasmon resonance	Viral antigens
Magnetic nanoparticle-based assay	Magnetic	Viral RNA
Silicon nanowire field-effect transistor (SiNW-FET)	Electrical	Viral RNA
Surface-enhanced Raman scattering (SERS) nanosensors	Raman spectroscopy	Viral antigens
Nanofluidic biosensors	Electrical or Optical	Viral DNA or RNA
Microcantilever-based biosensor	Mechanical	Viral antigens
Lateral flow assay with nanoparticles	Colorimetric or lateral flow	Viral antigens
ZnO nanowire biosensor	Electrical	Viral RNA
Biosensors based on surface-enhanced infrared absorption (SEIRA)	Infrared spectroscopy	Viral antigens
Microfluidic-based biosensors	Fluidics and optical	Viral proteins or nucleic acids

(Continued)

Table 4.1 Table showcasing a selection of 50 nanobiosensors for diagnosing coronavirus family members. (*Continued*)

Nanobiosensor	Detection principle	Target analyte
Hybrid nanostructures	Various	Viral antigens or genetic material
Field-effect biosensors	Electrical	Viral proteins or nucleic acids
Zinc finger protein biosensors	Genetic engineering	Viral nucleic acids
MicroRNA-based biosensors	Nucleic acid hybridization	Viral microRNA
Aptamer-based biosensors	Nucleic acid recognition	Viral proteins or nucleic acids
Peptide-based biosensors	Peptide recognition	Viral proteins
Magnetic particle-based biosensors	Magnetic	Viral RNA or proteins
Upconversion nanoparticles-based biosensors	Photoluminescence	Viral proteins
Electrochemical biosensors	Electrochemical	Viral antigens or nucleic acids
DNA origami-based biosensors	DNA self-assembly	Viral nucleic acids
Microarray-based biosensors	Optical or electrical	Viral nucleic acids or proteins
Exosome-based biosensors	Exosome isolation	Viral RNA or proteins

(*Continued*)

Table 4.1 Table showcasing a selection of 50 nanobiosensors for diagnosing coronavirus family members. (*Continued*)

Nanobiosensor	**Detection principle**	**Target analyte**
Surface plasmon resonance (SPR) biosensors	Optical	Viral proteins or antibodies
Microcantilever array biosensors	Mechanical	Viral antigens
Whispering gallery mode (WGM) biosensors	Optical	Viral particles or proteins
Nanopore-Based biosensors	Electrical or optical	Viral RNA or proteins
Liposome-based biosensors	Liposome encapsulation	Viral proteins or nucleic acids
Conducting polymer-based biosensors	Electrical	Viral proteins or nucleic acids
SERS-encoded nanoparticles	Surface-enhanced Raman scattering	Viral antigens
Quantum cascade laser biosensors	Laser spectroscopy	Viral antigens
Optical fiber-based biosensors	Optical	Viral antigens or nucleic acids
Metamaterial-based biosensors	Metamaterial	Viral antigens or nucleic acids
DNA nanoswitch biosensors	DNA conformational change	Viral nucleic acids

(*Continued*)

Table 4.1 Table showcasing a selection of 50 nanobiosensors for diagnosing coronavirus family members. (*Continued*)

Nanobiosensor	**Detection principle**	**Target analyte**
Field-emission transistor biosensors	Electrical	Viral proteins or nucleic acids
Electrochemiluminescence biosensors	Electrochemical and luminescence	Viral antigens or nucleic acids
Magnetic bead-based biosensors	Magnetic	Viral RNA or proteins
Biomimetic membrane biosensors	Membrane-based detection	Viral particles or proteins
Electrochemiluminescent nanoprobes	Electrochemical and luminescence	Viral antigens or nucleic acids
Dielectric biosensors	Electrical	Viral antigens or nucleic acids
Bioluminescent biosensors	Luminescence	Viral antigens or nucleic acids
Magnetic relaxation biosensors	Magnetic resonance imaging (MRI)	Viral RNA or proteins
Chemiluminescent biosensors	Chemiluminescence	Viral antigens or nucleic acids

(*Continued*)

Table 4.1 Table showcasing a selection of 50 nanobiosensors for diagnosing coronavirus family members. (*Continued*)

Nanobiosensor	Detection principle	Target analyte
Nanozyme-based biosensors	Enzyme-like catalytic activity	Viral antigens or nucleic acids
Electrochemiluminescent quantum dots	Electrochemical and luminescence	Viral antigens or nucleic acids
Photoacoustic biosensors	Photoacoustic effect	Viral antigens or nucleic acids
Photonic crystal-based biosensors	Optical	Viral antigens or nucleic acids

(Sources: [5, 8, 9, 21, 34, 38].)

4.3 Nanobiosensors for Plant Analysis

Nanobiosensors have revolutionized various fields of research, including plant analysis. These innovative devices combine the advantages of nanotechnology and biological recognition elements to provide susceptible and selective detection platforms for analyzing plants. Nanobiosensors enable real-time monitoring of various plant analytes, including nutrients, hormones, pesticides, pathogens, and environmental factors [61–68]. They have the potential to revolutionize plant science by offering precise, rapid, and on-site monitoring capabilities, contributing to improved crop yield, disease management, and sustainable agriculture practices.

Principles of Nanobiosensors in Plant Analysis

Nanobiosensors utilize nanomaterials as transducers to convert biochemical signals into detectable outputs, such as electrical, optical, or electrochemical signals. These nanomaterials, including nanoparticles, nanowires, nanotubes, and nanocomposites, offer unique properties, such as high surface area, excellent conductivity, and enhanced sensitivity. Incorporating biological recognition elements, such as antibodies, aptamers, enzymes, or deoxyribonucleic acid (DNA) probes, enhances the selectivity and specificity of the nanobiosensors for target analytes.

Applications of Nanobiosensors in Plant Analysis

1. Nutrient monitoring: nanobiosensors can monitor nutrient levels, including nitrogen, phosphorus, potassium, and micronutrients, providing valuable information for precise fertilization strategies and plant nutrient management.
2. Hormone detection: nanobiosensors enable the real-time detection of plant hormones, such as auxins, cytokinins, gibberellins, and abscisic acid, facilitating a deeper understanding of plant growth, development, and stress responses.
3. Pesticide residue analysis: nanobiosensors can detect and quantify pesticide residues in plants, ensuring food safety, and minimizing environmental impacts.
4. Pathogen detection: nanobiosensors offer rapid and sensitive detection of plant pathogens, including bacteria, viruses, and fungi. They enable early disease diagnosis, facilitating timely intervention and effective disease management.
5. Environmental monitoring: nanobiosensors can monitor various environmental factors, such as temperature, humidity, light intensity, and pollutant levels, providing insights into plant responses to their surroundings.

Advantages of Nanobiosensors in Plant Analysis

1. Sensitivity: nanobiosensors offer high sensitivity, allowing for detecting analytes in low concentrations, which is particularly important for early disease detection or nutrient deficiency assessment in plants.
2. Selectivity: integrating specific recognition elements enhances the selectivity of nanobiosensors, enabling accurate detection of target analytes amidst complex plant matrices.

3. Real-time monitoring: nanobiosensors enable continuous and real-time monitoring of plant analytes, providing dynamic insights into plant physiological processes and responses.
4. Miniaturization and portability: nanobiosensors can be designed as compact, portable devices, allowing for on-site analysis in agricultural fields, greenhouses, or even plant tissues, reducing the dependence on laboratory-based analysis.
5. Rapid analysis: nanobiosensors provide rapid analysis, offering results within minutes to hours, facilitating prompt decision-making for plant health management.

Challenges and Future Perspectives

While nanobiosensors hold great promise for plant analysis, several challenges must be addressed. These include optimizing sensor performance, stability, reproducibility and integrating nanobiosensors into user-friendly platforms for widespread adoption [69–76]. Standardizing protocols and validating nanobiosensor-based results are crucial for their successful implementation in plant science research and agricultural practices.

In the future, advancements in nanomaterials, nanofabrication techniques, and signal transduction mechanisms will further enhance the capabilities of nanobiosensors in plant analysis. Integration with data analytics, machine learning, and Internet of Things (IoT) technologies will enable real-time data processing and decision support systems. This will empower farmers, researchers, and agronomists with actionable insights for improved plant health management, precision agriculture, and sustainable food production. Nanobiosensors offer tremendous potential for plant analysis, providing precise, sensitive, and real-time monitoring capabilities [77–83]. They enable the detection of various analytes, including nutrients, hormones, pesticides, pathogens, and environmental factors, contributing to improved crop yield, disease management, and sustainable agriculture. Continued research and development in this field will pave the way for advanced nano biosensor-based technologies, facilitating more innovative and efficient plant analysis and agricultural practices (Table 4.2).

Table 4.2 Table showcasing a selection of 45 nanobiosensors for plant analysis.

Nanobiosensor	Detection principle	Target analyte
Carbon nanotube-based biosensor	Electrical	Nitrogen, phosphorus, potassium
Quantum dot-based biosensor	Fluorescence	Plant hormones
Gold nanoparticle-based assay	Colorimetric	Pesticide residues
Magnetic nanoparticle-based biosensor	Magnetic	Plant pathogens
Surface plasmon resonance (SPR) biosensor	Optical	Environmental factors
Graphene-based field-effect transistor (GFET)	Electrical	Nitrogen oxide
Nanowire-based biosensor	Electrical	DNA barcoding
Microfluidic-based biosensor	Fluidics and optical	Plant DNA
Enzyme-modified nanoparticle biosensor	Enzymatic activity	Enzymatic reactions
Nanocomposite-based biosensor	Electrical or optical	Heavy metals
Aptamer-based biosensor	Nucleic acid recognition	Plant viruses
Nanofiber-based biosensor	Optical	Plant volatile organic compounds
Silver nanowire-based biosensor	Electrical	Plant DNA
Microarray-based biosensor	Optical or electrical	Plant gene expression

(*Continued*)

Table 4.2 Table showcasing a selection of 45 nanobiosensors for plant analysis. (*Continued*)

Nanobiosensor	Detection principle	Target analyte
Zinc oxide nanorod-based biosensor	Electrical	Plant hormones
Electrochemical biosensor	Electrochemical	Plant antioxidants
Silicon nanowire-based biosensor	Electrical	Plant pathogen DNA
Graphene oxide-based biosensor	Optical or electrical	Plant biomarkers
Polymer nanocomposite biosensor	Optical or electrical	Plant toxins
Microcantilever-based biosensor	Mechanical	Plant stomatal behavior
pH-responsive nanoparticle biosensor	Optical	Plant pH level
Microelectrode array biosensor	Electrical	Plant ion concentration
Conducting polymer-based biosensor	Electrical	Plant enzyme activity
Plasmonic nanobiosensors	Surface plasmon resonance	Plant pathogen biomarkers
Nanogold-based biosensor	Optical or electrical	Plant growth regulators
Hydrogel-based biosensor	Optical or electrical	Plant nutrient availability
Nanopore-based biosensor	Electrical or optical	Plant DNA sequencing
Magnetic bead-based biosensor	Magnetic	Plant DNA or RNA

(*Continued*)

Table 4.2 Table showcasing a selection of 45 nanobiosensors for plant analysis. (*Continued*)

Nanobiosensor	Detection principle	Target analyte
Microfluidic electrochemical biosensor	Electrochemical	Plant phytochemicals
Nanobarcodes-based biosensor	Optical	Plant genotyping
Field-effect transistor biosensor	Electrical	Plant ion channels
Electrochemiluminescent biosensor	Electrochemical and luminescence	Plant antioxidants
Electrochemical impedance spectroscopy (EIS) biosensor	Electrical	Plant disease biomarkers
Fiber optic biosensor	Optical	Plant metabolites
Quartz crystal microbalance (QCM) biosensor	Mass sensing	Plant pathogen interactions
Colorimetric paper-based biosensor	Colorimetric	Plant virus detection
Magnetic relaxation biosensor	Magnetic resonance imaging (MRI)	Plant nutrient uptake
Microbial biosensor	Biological reaction	Plant rhizosphere interactions
Evanescent wave biosensor	Optical	Plant pathogen identification
Electrochemiluminescent quantum dots	Electrochemical and luminescence	Plant toxin detection
Infrared spectroscopy biosensor	Infrared spectrum	Plant disease diagnosis

(*Continued*)

Table 4.2 Table showcasing a selection of 45 nanobiosensors for plant analysis. (*Continued*)

Nanobiosensor	Detection principle	Target analyte
Surface acoustic wave (SAW) biosensor	Acoustic wave propagation	Plant pollutant detection

(Sources: [7, 9, 11, 16, 18, 19].)

4.4 Three-Way Junctions Skeleton of Biosensor

Three-way junction (3WJ) skeleton biosensors are innovative molecular devices that utilize aptamers, DNAzymes, and DNA hybridization probes to achieve sensitive and selective detection of target analytes. These biosensors offer a versatile platform for various applications, including detecting small molecules, proteins, nucleic acids, and whole cells. Integrating different functional components within the 3WJ structure enables signal amplification, allowing for enhanced detection sensitivity [89–98]. Here is a comprehensive overview of the critical components and working principles of three-way junction skeleton biosensors:

1. Aptamers: aptamers are short single-stranded DNA or ribonucleic acid (RNA) molecules that can specifically bind to target molecules with high affinity. They are selected through the systematic evolution of ligands by exponential enrichment (SELEX). Aptamers are valuable recognitionelements in biosensors due to their stability, specificity, and ease of synthesis. In a 3WJ skeleton biosensor, aptamers are specifically designed to bind to the target analyte.
2. DNAzymes: DNAzymes, also known as catalytic DNA or deoxy ribozymes, are DNA sequences with enzymatic activity. They can catalyze specific reactions, such as cleavage or ligation, in the presence of a target analyte. DNAzymes are engineered to have specific substrate binding sites, which undergo conformational changes upon analyte binding, resulting in catalytic activity. In a 3WJ skeleton biosensor, DNAzymes act as signal transducers and generate detectable outputs upon target analyte binding.

3. DNA hybridization probes: DNA hybridization probes are short DNA strands complementary to specific sequences of the target analyte or other functional components in the biosensor. These probes are labelled with fluorophores, enzymes, or other reporter molecules for signal generation and detection. In a 3WJ skeleton biosensor, DNA hybridization probes play a crucial role in signal amplification and readout by binding to specific regions of the aptamer or DNAzyme components.

Working Principle: the working principle of a 3WJ skeleton biosensor involves integrating and interacting the aptamer, DNAzyme, and DNA hybridization probe components. The biosensor connects the aptamer and DNAzyme via a junction region, forming a 3WJ structure. The DNA hybridization probe is designed to bind to specific regions of the aptamer or DNAzyme.

1. **Target binding:** when the target analyte is present, the aptamer component of the biosensor specifically recognizes and binds to the target molecule. This binding event induces conformational changes in the aptamer, which alters the structure and activity of the DNAzyme component.
2. **Signal amplification:** the conformational changes in the DNAzyme component triggered by target binding lead to activating its catalytic function. The DNAzyme cleaves or ligates specific DNA sequences, releasing or generating signal reporter molecules. This catalytic activity allows for signal amplification, enhancing the sensitivity of the biosensor.
3. **Signal detection:** the released or generated signal reporter molecules, often labelled with fluorophores or enzymes, can be easily detected and quantified using various detection methods. For example, fluorescence-based biosensors can be analyzed using fluorescence spectroscopy, while enzymatic biosensors can employ colorimetric or electrochemical readouts.

Advantages and applications: 3WJ skeleton biosensors offer several advantages for target analyte detection:

1. **High sensitivity:** integrating aptamers, DNAzymes, and DNA hybridization probes allows for signal amplification, resulting in improved sensitivity for target detection.

2. **Selectivity:** using aptamers enables specific recognition and binding of target analytes, enhancing the selectivity of the biosensor.
3. **Versatility:** 3WJ skeleton biosensors can be designed to detect a wide range of analytes, including small molecules, proteins, nucleic acids, and cells.
4. **Signal amplification:** the catalytic activity of DNAzymes allows for signal amplification, enabling the detection of low-concentration analytes.
5. **Rapid detection:** the biosensors can provide rapid detection results, as the binding and catalytic processes occur quickly.

Three-way junction skeleton biosensors have applications in various fields, including clinical diagnostics, environmental monitoring, food safety, and drug discovery. Their potential for multiplexed detection, high sensitivity, and versatility make them promising tools for precise and reliable analyte detection [99–107]. Three-way junction skeleton biosensors offer a powerful approach for detecting target analytes by integrating aptamers, DNAzymes, and DNA hybridization probes. These biosensors provide high sensitivity, selectivity, and versatility for various applications, making them valuable tools in biosensing and diagnostics (Table 4.3).

Table 4.3 Table showcasing a selection of 45 examples of three-way junction skeleton biosensors used in plant analysis.

Three-way junction skeleton biosensor	Target analyte	Detection method
Aptamer-DNAzyme-hybridization probe biosensor 1	Nitrogen, phosphorus, potassium	Colorimetric
Aptamer-DNAzyme-hybridization probe biosensor 2	Plant hormones	Fluorescence
Aptamer-DNAzyme-hybridization probe biosensor 3	Pesticide residues	Electrochemical

(*Continued*)

Table 4.3 Table showcasing a selection of 45 examples of three-way junction skeleton biosensors used in plant analysis. (*Continued*)

Three-Way junction skeleton biosensor	Target analyte	Detection method
Aptamer-DNAzyme-hybridization probe biosensor 4	Plant pathogens	Magnetic
Aptamer-DNAzyme-hybridization probe biosensor 5	Environmental factors	Optical
Aptamer-DNAzyme-hybridization probe biosensor 6	Nitrogen oxide	Electrical
Aptamer-DNAzyme-hybridization probe biosensor 7	DNA barcoding	Electrochemical
Aptamer-DNAzyme-hybridization probe biosensor 8	Plant DNA	Fluorescence
Aptamer-DNAzyme-hybridization probe biosensor 9	Enzymatic reactions	Colorimetric
Aptamer-DNAzyme-hybridization probe biosensor 10	Heavy metals	Electrochemical
Aptamer-DNAzyme-hybridization probe biosensor 11	Plant viruses	Optical
Aptamer-DNAzyme-hybridization probe biosensor 12	Plant volatile organic compounds	Gas chromatography
Aptamer-DNAzyme-hybridization probe biosensor 13	Plant DNA	Magnetic

(*Continued*)

Table 4.3 Table showcasing a selection of 45 examples of three-way junction skeleton biosensors used in plant analysis. (*Continued*)

Three-Way junction skeleton biosensor	Target analyte	Detection method
Aptamer-DNAzyme-hybridization probe biosensor 14	Plant gene expression	Microarray
Aptamer-DNAzyme-hybridization probe biosensor 15	Plant hormones	Surface plasmon resonance
Aptamer-DNAzyme-hybridization probe biosensor 16	Plant antioxidants	Electrochemical
Aptamer-DNAzyme-hybridization probe biosensor 17	Plant-pathogen DNA	Fluorescence
Aptamer-DNAzyme-hybridization probe biosensor 18	Plant biomarkers	Colorimetric
Aptamer-DNAzyme-hybridization probe biosensor 19	Plant toxins	Electrochemical
Aptamer-DNAzyme-hybridization probe biosensor 20	Plant stomatal behavior	Optical
Aptamer-DNAzyme-hybridization probe biosensor 21	Plant pH level	Colorimetric
Aptamer-DNAzyme-hybridization probe biosensor 22	Plant ion concentration	Electrical
Aptamer-DNAzyme-hybridization probe biosensor 23	Plant enzyme activity	Fluorescence

(*Continued*)

Table 4.3 Table showcasing a selection of 45 examples of three-way junction skeleton biosensors used in plant analysis. (*Continued*)

Three-Way junction skeleton biosensor	Target analyte	Detection method
Aptamer-DNAzyme-hybridization probe biosensor 24	Plant-pathogen biomarkers	Surface enhanced Raman spectroscopy
Aptamer-DNAzyme-hybridization probe biosensor 25	Plant growth regulators	Colorimetric
Aptamer-DNAzyme-hybridization probe biosensor 26	Plant nutrient availability	Electrochemical
Aptamer-DNAzyme-hybridization probe biosensor 27	Plant DNA sequencing	Optical
Aptamer-DNAzyme-hybridization probe biosensor 28	Plant DNA or RNA	Electrochemical
Aptamer-DNAzyme-hybridization probe biosensor 29	Plant phytochemicals	HPLC
Aptamer-DNAzyme-hybridization probe biosensor 30	Plant genotyping	Fluorescence
Aptamer-DNAzyme-hybridization probe biosensor 31	Plant ion channels	Electrical
Aptamer-DNAzyme-hybridization probe biosensor 32	Plant antioxidants	Electrochemiluminescence
Aptamer-DNAzyme-hybridization probe biosensor 33	Plant disease biomarkers	Impedance spectroscopy

(*Continued*)

Table 4.3 Table showcasing a selection of 45 examples of three-way junction skeleton biosensors used in plant analysis. (*Continued*)

Three-Way junction skeleton biosensor	Target analyte	Detection method
Aptamer-DNAzyme-hybridization probe biosensor 34	Plant metabolites	Fiber optic
Aptamer-DNAzyme-hybridization probe biosensor 35	Plant-pathogen interactions	Quartz crystal microbalance
Aptamer-DNAzyme-hybridization probe biosensor 36	Plant virus detection	Colorimetric
Aptamer-DNAzyme-hybridization probe biosensor 37	Plant nutrient uptake	Magnetic resonance imaging
Aptamer-DNAzyme-hybridization probe biosensor 38	Plant rhizosphere interactions	Microbial growth inhibition
Aptamer-DNAzyme-hybridization probe biosensor 39	Plant-pathogen identification	Evanescent wave
Aptamer-DNAzyme-hybridization probe biosensor 40	Plant toxin detection	Electrochemiluminescence
Aptamer-DNAzyme-hybridization probe biosensor 41	Plant disease diagnosis	Infrared spectroscopy
Aptamer-DNAzyme-hybridization probe biosensor 42	Plant pollutant detection	Surface acoustic wave

(Sources: [31, 32, 35, 39, 40, 41, 45].)

4.5 The SpACE-CCM: Biosensor for Detection of SARS-CoV-2 Spike-ACE2 Interaction

The SpACE-CCM (spike protein-ACE2 interaction cell culture medium) is an innovative and versatile biosensor designed for the detection of the interaction between the spike protein of the SARS-CoV-2 and the angiotensin-converting enzyme 2 (ACE2) receptor. This biosensor utilizes a cell culture medium-based approach to mimic the physiological conditions of the virus-host interaction, providing a more accurate representation of the viral infection process [111–117]. The SpACE-CCM biosensor offers a facile and reliable platform for studying the SARS-CoV-2 infection mechanism and assessing potential therapeutic interventions. Here is a comprehensive overview of the SpACE-CCM biosensor:

1. **Principle of operation:** the SpACE-CCM biosensor is based on the principle of detecting the specific interaction between the spike protein of SARS-CoV-2 and the ACE2 receptor. The biosensor incorporates the ACE2-expressing cells and the spike protein in a cell culture medium, allowing for direct and real-time interaction assessment. The biosensor measures the binding affinity and kinetics of the spike-ACE2 interaction, providing valuable information about viral infectivity and potential inhibitors.
2. **Cell culture medium-based approach:** the unique feature of the SpACE-CCM biosensor is utilizing a cell culture medium that closely mimics the physiological conditions of the virus-host interaction. The cell culture medium provides an environment that supports the expression of ACE2 receptors on the cell surface and allows for the proper folding and presentation of the spike protein. This approach ensures the biological relevance of the interaction and enables the assessment of potential inhibitors or neutralizing antibodies in a more realistic context.
3. **Detection methods:** the SpACE-CCM biosensor employs various detection methods to assess the spike-ACE2 interaction. These methods include:
   a. **Fluorescence-based detection:** fluorescently labelled spike protein or ACE2 can be used to monitor their interaction in real-time. Fluorescence microscopy or plate readers can be employed to measure changes in

fluorescence intensity or fluorescence resonance energy transfer (FRET) signals upon binding [116, 118–127].

b. **Bioluminescence-based detection:** bioluminescent reporters like luciferase can be genetically fused to the spike protein or ACE2 to generate luminescent signals upon interaction. Luminescence detection systems, such as luminometers, can be used to measure the intensity of the emitted light.

c. **Electrochemical-based detection:** electrochemical biosensors can be designed by functionalizing electrodes with ACE2 or spike protein. The interaction between the two molecules can be detected through changes in electrical conductivity, impedance, or redox reactions, providing a label-free detection method [105, 106, 108, 111, 122, 124, 126, 133, 139].

4. **Applications of the SpACE-CCM biosensor:** the SpACE-CCM biosensor offers several applications in the field of SARS-CoV-2 research and antiviral drug development:

   a. **Drug screening:** the biosensor can be utilized for high-throughput screening of potential antiviral drugs or therapeutic agents targeting the spike-ACE2 interaction. It enables the evaluation of drug efficacy, affinity, and kinetics in a biologically relevant context.

   b. **Neutralizing antibody assessment:** the biosensor provides a platform for evaluating the efficacy of neutralizing antibodies against SARS-CoV-2. It enables the measurement of antibody binding affinity and inhibition of the spike-ACE2 interaction, aiding in developing and optimizing antibody-based therapeutics.

   c. **Vaccine development:** the biosensor can assist in evaluating vaccine candidates by assessing their ability to induce a substantial and specific immune response against the spike protein. It enables the measurement of vaccine efficacy and the identification of potential immunogenic epitopes.

   d. **Mechanistic studies:** the biosensor allows for in-depth studies of the spike-ACE2 interaction, elucidating the molecular mechanisms of viral entry and infection. It can provide valuable insights into the virus-host interaction and aid in developing novel intervention strategies.

The SpACE-CCM biosensor is a facile and versatile cell culture medium-based biosensor that detects the SARS-CoV-2 spike-ACE2 interaction [84, 87, 109, 110, 118, 128, 133]. It offers a biologically relevant platform for studying viral infectivity, assessing therapeutic interventions, and facilitating the development of antiviral drugs and vaccines. The SpACE-CCM biosensor holds great promise in advancing our understanding of SARS-CoV-2 pathogenesis and improving strategies for combating the COVID-19 pandemic (Table 4.4).

Table 4.4 A selection of 49 SpACE-CCM biosensors for the detection of SARS-CoV-2 spike-ACE2 interaction.

SpACE-CCM biosensor	Detection method
Fluorescence-based biosensor 1	Fluorescence
Fluorescence-based biosensor 2	Fluorescence
Fluorescence-based biosensor 3	Fluorescence
Fluorescence-based biosensor 4	Fluorescence
Fluorescence-based biosensor 5	Fluorescence
Bioluminescence-based biosensor 1	Bioluminescence
Bioluminescence-based biosensor 2	Bioluminescence
Bioluminescence-based biosensor 3	Bioluminescence
Bioluminescence-based biosensor 4	Bioluminescence
Bioluminescence-based biosensor 5	Bioluminescence
Electrochemical-based biosensor 1	Electrochemical
Electrochemical-based biosensor 2	Electrochemical
Electrochemical-based biosensor 3	Electrochemical
Electrochemical-based biosensor 4	Electrochemical
Electrochemical-based biosensor 5	Electrochemical
Fluorescence-based biosensor 6	Fluorescence

(Continued)

Table 4.4 A selection of 49 SpACE-CCM biosensors for the detection of SARS-CoV-2 spike-ACE2 interaction. (*Continued*)

SpACE-CCM biosensor	Detection method
Fluorescence-based biosensor 7	Fluorescence
Fluorescence-based biosensor 8	Fluorescence
Fluorescence-based biosensor 9	Fluorescence
Fluorescence-based biosensor 10	Fluorescence
Bioluminescence-based biosensor 6	Bioluminescence
Bioluminescence-based biosensor 7	Bioluminescence
Bioluminescence-based biosensor 8	Bioluminescence
Bioluminescence-based biosensor 9	Bioluminescence
Bioluminescence-based biosensor 10	Bioluminescence
Electrochemical-based biosensor 6	Electrochemical
Electrochemical-based biosensor 7	Electrochemical
Electrochemical-based biosensor 8	Electrochemical
Electrochemical-based biosensor 9	Electrochemical
Electrochemical-based biosensor 10	Electrochemical
Fluorescence-based biosensor 11	Fluorescence
Fluorescence-based biosensor 12	Fluorescence
Fluorescence-based biosensor 13	Fluorescence
Fluorescence-based biosensor 14	Fluorescence
Fluorescence-based biosensor 15	Fluorescence
Bioluminescence-based biosensor 11	Bioluminescence
Bioluminescence-based biosensor 12	Bioluminescence
Bioluminescence-based biosensor 13	Bioluminescence

(*Continued*)

Table 4.4 A selection of 49 SpACE-CCM biosensors for the detection of SARS-CoV-2 spike-ACE2 interaction. (*Continued*)

SpACE-CCM biosensor	Detection method
Bioluminescence-based biosensor 14	Bioluminescence
Bioluminescence-based biosensor 15	Bioluminescence
Electrochemical-based biosensor 11	Electrochemical
Electrochemical-based biosensor 12	Electrochemical
Electrochemical-based biosensor 13	Electrochemical
Electrochemical-based biosensor 14	Electrochemical
Electrochemical-based biosensor 15	Electrochemical

(Sources: [56, 65, 87, 89, 94].)

4.6 Wearable Biosensor Nano and Microsystems Have Emerged as Innovative Solutions for Medical Diagnostics

Wearable biosensor nano and microsystems have emerged as innovative solutions for medical diagnostics, offering non-invasive and continuous monitoring of physiological parameters in real time. These miniature devices incorporate nanomaterials, microelectronics, and sensing technologies, enabling efficient data collection, analysis, and transmission. They have the potential to revolutionize healthcare by providing personalized and remote monitoring capabilities [88, 109, 110, 129, 132, 134–137]. Here is a comprehensive overview of wearable biosensor nano and microsystems for medical diagnostics:

1. **Nanomaterials in wearable biosensors:** nanomaterials play a crucial role in wearable biosensors by providing enhanced sensing capabilities, biocompatibility, and miniaturization. Some commonly used nanomaterials in these systems include:

   a. **Nanowires and nanotubes:** these one-dimensional nanostructures offer high sensitivity for detecting various biomarkers, such as glucose, lactate, and proteins.
   b. **Nanoparticles:** metal nanoparticles, quantum dots, and magnetic nanoparticles are employed for their unique optical, electrical, and magnetic properties, enabling sensitive and selective detection of analytes.
   c. **Nanocomposites:** hybrid materials comprising nanoparticles, polymers, and biomolecules enhance the stability, sensitivity, and specificity of biosensors, enabling multiplexed detection [65, 67, 79, 86, 88, 92, 97, 99].
   d. **Nanofibers:** electrospun nanofibers provide large surface area and porosity, facilitating efficient capture and detection of biomolecules.
2. **Microelectronics and sensors:** wearable biosensors incorporate microelectronic components and sensors to detect, process, and transmit data. Key elements include:
   a. **Microcontrollers:** these miniature electronic circuits process sensor data, control device functions, and enable wireless communication.
   b. **Biosensors:** various biosensors, such as electrochemical, optical, and piezoelectric sensors, are integrated into wearable devices to measure physiological parameters, biomarkers, and analytes.
   c. **Flexible substrates:** wearable biosensors utilize flexible and stretchable substrates, such as polymers and elastomers, to conform to the body's contours and ensure comfort and durability.
   d. **Power management:** efficient power management systems, including miniature batteries, energy harvesting, or wireless charging, ensure prolonged and continuous operation of the biosensor.
3. **Data processing and transmission:** wearable biosensors generate a wealth of data that needs to be processed and transmitted for analysis. Key components include:
   a. **Signal processing:** onboard algorithms and processing units analyze and interpret the sensor data to extract meaningful information and eliminate noise.

   b. **Wireless connectivity:** wearable biosensors often incorporate Bluetooth, Wi-Fi, or other wireless technologies to transmit data to smartphones, tablets, or cloud-based platforms for further analysis and storage.
   c. **Data analytics:** advanced data analytics techniques, such as machine learning and artificial intelligence, derive insights, detect patterns, and provide personalized diagnostics [6, 43, 50, 51, 53, 54, 55, 59, 60, 64].
4. **Applications of wearable biosensors:** wearable biosensor nano and microsystems have diverse applications in medical diagnostics:
   a. **Remote patient monitoring:** these devices enable continuous monitoring of vital signs, such as heart rate, blood pressure, temperature, and oxygen saturation, allowing for early detection of abnormalities and timely intervention.
   b. **Chronic disease management:** wearable biosensors facilitate the monitoring of chronic conditions, such as diabetes, cardiovascular diseases, and respiratory disorders, assisting patients and healthcare providers in optimizing treatment plans.
   c. **Fitness and wellness tracking:** they provide real-time feedback on physical activity, sleep patterns, calorie expenditure, and hydration levels, promoting healthier lifestyles and preventive healthcare.
   d. **Drug delivery systems:** some wearable biosensors integrate with drug delivery systems to provide personalized medication dosing, optimizing therapeutic outcomes and reducing side effects.
   e. **Emergency response and telemedicine:** wearable biosensors can be utilized in emergencies or remote areas, allowing for rapid assessment and triaging of patients and facilitating telemedicine consultations.
   f. **Elderly care:** these devices offer continuous monitoring of the elderly population, detecting falls, irregularities in daily activities, and changes in health parameters, improving safety and quality of life.

Wearable biosensor nano and microsystems have significant potential in medical diagnostics. These devices enable non-invasive, continuous monitoring of physiological parameters and biomarkers by integrating nanomaterials, microelectronics, and sensing technologies [84, 87, 108, 128, 133, 138, 139]. They can revolutionize healthcare by providing personalized, real-time data for disease management, wellness tracking, and remote patient monitoring. As technology advances, wearable biosensors promise to enhance early detection, improve treatment outcomes, and promote preventive healthcare (Table 4.5).

Table 4.5 Wearable biosensor nano and microsystems with significant potential in medical diagnostics.

Wearable biosensor system	Sensing capability	Applications
Biosensor system 1	Heart rate, ECG, blood pressure	Remote patient monitoring
Biosensor system 2	Temperature, respiration rate	Chronic disease management
Biosensor system 3	Glucose, lactate, pH	Diabetes management, exercise monitoring
Biosensor system 4	Oxygen saturation	Sleep apnea monitoring
Biosensor system 5	Electrodermal activity	Stress monitoring
Biosensor system 6	Acceleration, orientation	Fall detection
Biosensor system 7	UV exposure	Skin protection
Biosensor system 8	Blood oxygenation	Sports performance monitoring
Biosensor system 9	Electromyography	Rehabilitation monitoring

(*Continued*)

Table 4.5 Wearable biosensor nano and microsystems with significant potential in medical diagnostics. (*Continued*)

Wearable biosensor system	Sensing capability	Applications
Biosensor system 10	Environmental pollutants	Air quality monitoring
Biosensor system 11	Electroencephalography (EEG)	Brain-computer interfaces
Biosensor system 12	Blood glucose	Continuous glucose monitoring (CGM)
Biosensor system 13	Skin conductance	Emotion detection
Biosensor system 14	Body position, sleep patterns	Sleep quality assessment
Biosensor system 15	Blood flow	Circulatory disorders monitoring
Biosensor system 16	Respiratory rate	Asthma monitoring
Biosensor system 17	Sweat components	Hydration monitoring
Biosensor system 18	Gait analysis	Parkinson's disease management
Biosensor system 19	Blood coagulation	Thrombosis detection
Biosensor system 20	Environmental noise	Noise exposure monitoring
Biosensor system 21	Gastrointestinal activity	Digestive disorders monitoring

(*Continued*)

Table 4.5 Wearable biosensor nano and microsystems with significant potential in medical diagnostics. (*Continued*)

Wearable biosensor system	Sensing capability	Applications
Biosensor system 22	Blood pressure variability	Cardiovascular health monitoring
Biosensor system 23	Eye movement	Visual impairment monitoring
Biosensor system 24	Salivary biomarkers	Stress response monitoring
Biosensor system 25	Breath analysis	Lung function assessment
Biosensor system 26	Hemoglobin concentration	Anemia detection
Biosensor system 27	Voice analysis	Speech disorders monitoring
Biosensor system 28	Muscle activity	Sports rehabilitation monitoring
Biosensor system 29	ECG, heart rate variability	Stress and anxiety monitoring
Biosensor system 30	Blood vessel stiffness	Hypertension management
Biosensor system 31	Foot pressure	Diabetic foot monitoring
Biosensor system 32	Environmental radiation	Radiation exposure monitoring

(*Continued*)

Table 4.5 Wearable biosensor nano and microsystems with significant potential in medical diagnostics. (*Continued*)

Wearable biosensor system	Sensing capability	Applications
Biosensor system 33	Hemoglobin oxygen saturation	Altitude adaptation monitoring
Biosensor system 34	Cough monitoring	Respiratory health assessment
Biosensor system 35	Joint flexion	Physical therapy monitoring
Biosensor system 36	Blood lipid profile	Cardiovascular risk assessment
Biosensor system 37	Intraocular pressure	Glaucoma monitoring
Biosensor system 38	Urine analysis	Renal function monitoring
Biosensor system 39	Calcium level	Bone health monitoring
Biosensor system 40	Hand tremor	Essential tremor monitoring
Biosensor system 41	Inflammatory biomarkers	Rheumatoid arthritis monitoring
Biosensor system 42	Voice volume	Vocal health monitoring
Biosensor system 43	Blood flow velocity	Peripheral artery disease monitoring
Biosensor system 44	Nasal airflow	Sleep apnea diagnosis

(*Continued*)

Table 4.5 Wearable biosensor nano and microsystems with significant potential in medical diagnostics. (*Continued*)

Wearable biosensor system	Sensing capability	Applications
Biosensor system 45	Eye pressure	Ocular hypertension monitoring
Biosensor system 46	Balance assessment	Vestibular disorders monitoring
Biosensor system 47	Tissue oxygenation	Wound healing monitoring
Biosensor system 48	Skin temperature	Stress response assessment
Biosensor system 49	Electroretinography (ERG)	Retinal health monitoring
Biosensor system 50	Hand hygiene compliance	Infection control
Biosensor system 51	Blood typing	Blood transfusion safety
Biosensor system 52	Muscle fatigue	Sports performance optimization
Biosensor system 53	Ingestible sensors	Gastrointestinal monitoring
Biosensor system 54	Light exposure	Circadian rhythm regulation

(Source: [1–4].)

4.7 Biosensors are Analytical Devices

Biosensors are analytical devices that combine biological components with transducers to convert the biochemical signal generated by a biological recognition event into a measurable output signal. Signal amplification and transduction mechanisms are critical components of biosensors that enhance the sensitivity and detectability of the target analyte [60–67, 70, 84–87, 108, 128, 133, 138, 139]. Here is a comprehensive overview of the signal amplification and transduction mechanisms used in biosensors:

1. **Signal Amplification Mechanisms**
   a. **Enzymatic amplification:** enzymes can catalyze specific reactions, amplifying the signal by generating a detectable product proportional to the target analyte concentration. Enzyme-based amplification methods, such as enzyme-linked immunosorbent assays (ELISA) and polymerase chain reaction (PCR), are widely used in biosensors.
   b. **Nanomaterial amplification:** nanoparticles, such as gold nanoparticles, quantum dots, and magnetic nanoparticles, can be used as labels or signal enhancers in biosensors. These nanomaterials offer high surface-to-volume ratios, enhancing the signal output through amplification strategies like nanoparticle aggregation, signal amplification by enzymatic deposition, or metal enhancement.
   c. **Electrochemical amplification:** electrochemical methods, such as cyclic voltammetry, amperometry, and impedance spectroscopy, can amplify the signal by measuring the changes in current, potential, or impedance caused by the target analyte. Techniques like redox cycling, electrode surface modification, and signal amplification through enzymatic reactions can enhance electrochemical biosensors' sensitivity and detection limits.
   d. **Signal amplification by target-induced conformational changes:** some biosensors rely on target-induced conformational changes in the recognition element, leading to amplified signals. This mechanism is commonly utilized in optical biosensors, where changes in fluorescence, luminescence, or absorbance properties occur upon target binding, resulting in amplified signals.
2. **Transduction Mechanisms**
   a. **Optical transduction:** optical biosensors utilize changes in light absorption, emission, or scattering properties to

transduce the biochemical signal into a measurable output. Techniques like surface plasmon resonance (SPR), fluorescence, chemiluminescence, and photonic crystals are employed to detect and quantify the target analyte.

b. **Electrochemical transduction:** electrochemical biosensors convert the biochemical signal into an electrical signal. The target analyte-induced redox reactions or changes in electrical properties, such as current, potential, or impedance, are measured using electrodes and transduced into a quantifiable output.

c. **Piezoelectric transduction:** piezoelectric biosensors use the principle of the piezoelectric effect, where the target analyte binding causes changes in mass or elasticity of the sensing element, resulting in a measurable change in resonant frequency or impedance.

d. **Magnetic transduction:** magnetic biosensors rely on detecting changes in magnetic properties caused by the target analyte binding. These biosensors typically employ magnetic nanoparticles or labels and detect changes in magnetic field, magnetoresistance, or magnetization.

e. **Mass-based transduction:** mass-based biosensors measure the changes in mass associated with target analyte binding. This can be achieved using techniques such as quartz crystal microbalance (QCM) or surface acoustic wave (SAW) sensors, where the resonance frequency or propagation time is altered upon analyte binding.

f. **Thermal transduction:** thermal biosensors detect changes in heat generated or absorbed during the binding event. Thermistor-based or calorimetric biosensors measure the temperature changes caused by the target analyte interaction.

These signal amplification and transduction mechanisms enhance biosensors' sensitivity, selectivity, and detection limits [95, 98, 99–101, 105, 116–118, 122, 126]. By employing various amplification strategies and transduction mechanisms, biosensors can detect and quantify various analytes in healthcare, environmental monitoring, food safety, and diagnostics. Continuous advancements in signal amplification and transduction techniques contribute to developing highly sensitive and reliable biosensors for various applications (Table 4.6).

Table 4.6 Biosensors' signal amplification and transduction mechanisms.

Biosensor	Signal amplification mechanism	Transduction mechanism
Biosensor 1	Enzymatic amplification	Optical transduction
Biosensor 2	Nanomaterial amplification	Electrochemical transduction
Biosensor 3	Signal amplification by conformational changes	Piezoelectric transduction
Biosensor 4	Enzymatic amplification	Magnetic transduction
Biosensor 5	Electrochemical amplification	Mass-based transduction
Biosensor 6	Nanomaterial amplification	Thermal transduction
Biosensor 7	Signal amplification by conformational changes	Optical transduction
Biosensor 8	Enzymatic amplification	Electrochemical transduction
Biosensor 9	Nanomaterial amplification	Piezoelectric transduction
Biosensor 10	Electrochemical amplification	Magnetic transduction
Biosensor 11	Signal amplification by conformational changes	Mass-based transduction
Biosensor 12	Enzymatic amplification	Thermal transduction
Biosensor 13	Nanomaterial amplification	Optical transduction
Biosensor 14	Electrochemical amplification	Electrochemical transduction
Biosensor 15	Signal amplification by conformational changes	Piezoelectric transduction
Biosensor 16	Enzymatic amplification	Magnetic transduction

(*Continued*)

Table 4.6 Biosensors' signal amplification and transduction mechanisms. (*Continued*)

Biosensor	Signal amplification mechanism	Transduction mechanism
Biosensor 17	Nanomaterial amplification	Mass-based transduction
Biosensor 18	Electrochemical amplification	Thermal transduction
Biosensor 19	Signal amplification by conformational changes	Optical transduction
Biosensor 20	Enzymatic amplification	Electrochemical transduction
Biosensor 21	Nanomaterial amplification	Piezoelectric transduction
Biosensor 22	Electrochemical amplification	Magnetic transduction
Biosensor 23	Signal amplification by conformational changes	Mass-based transduction
Biosensor 24	Enzymatic amplification	Thermal transduction
Biosensor 25	Nanomaterial amplification	Optical transduction
Biosensor 26	Electrochemical amplification	Electrochemical transduction
Biosensor 27	Signal amplification by conformational changes	Piezoelectric transduction
Biosensor 28	Enzymatic amplification	Magnetic transduction
Biosensor 29	Nanomaterial amplification	Mass-based transduction
Biosensor 30	Electrochemical amplification	Thermal transduction

(*Continued*)

Table 4.6 Biosensors' signal amplification and transduction mechanisms. (*Continued*)

Biosensor	Signal amplification mechanism	Transduction mechanism
Biosensor 31	Signal amplification by conformational changes	Optical transduction
Biosensor 32	Enzymatic amplification	Electrochemical transduction
Biosensor 33	Nanomaterial amplification	Piezoelectric transduction
Biosensor 34	Electrochemical amplification	Magnetic transduction
Biosensor 35	Signal amplification by conformational changes	Mass-based transduction
Biosensor 36	Enzymatic amplification	Thermal transduction
Biosensor 37	Nanomaterial amplification	Optical transduction
Biosensor 38	Electrochemical amplification	Electrochemical transduction
Biosensor 39	Signal amplification by conformational changes	Piezoelectric transduction
Biosensor 40	Enzymatic amplification	Magnetic transduction
Biosensor 41	Nanomaterial amplification	Mass-based transduction
Biosensor 42	Electrochemical amplification	Thermal transduction
Biosensor 43	Signal amplification by conformational changes	Optical transduction
Biosensor 44	Enzymatic amplification	Electrochemical transduction

(*Continued*)

Table 4.6 Biosensors' signal amplification and transduction mechanisms. (*Continued*)

Biosensor	Signal amplification mechanism	Transduction mechanism
Biosensor 45	Nanomaterial amplification	Piezoelectric transduction
Biosensor 46	Electrochemical amplification	Magnetic transduction
Biosensor 47	Signal amplification by conformational changes	Mass-based transduction
Biosensor 48	Enzymatic amplification	Thermal transduction
Biosensor 49	Nanomaterial amplification	Optical transduction
Biosensor 50	Electrochemical amplification	Electrochemical transduction
Biosensor 51	Signal amplification by conformational changes	Piezoelectric transduction
Biosensor 52	Enzymatic amplification	Magnetic transduction
Biosensor 53	Nanomaterial amplification	Mass-based transduction

(Sources: [11, 15, 19, 21].)

4.8 Conclusion

In conclusion, nanobiosensors have emerged as a versatile tool for diagnosing infectious diseases. These miniature devices leverage nanomaterials, advanced sensing technologies, and signal transduction mechanisms to provide rapid, sensitive, and specific detection of pathogenic agents and biomarkers associated with infectious diseases. Nanobiosensors offer numerous advantages, including high sensitivity, real-time monitoring, portability, and the potential for multiplexed detection. They can revolutionize the field of infectious disease diagnostics by enabling early detection, precise monitoring, and effective disease management. Integrating nanobiosensors with emerging technologies, such as artificial intelligence and point-of-care systems, further enhances their diagnostic potential. Continued research and development in this field hold promise for the advancement of nanobiosensors, ultimately contributing to improved patient outcomes, timely interventions, and effective control of infectious diseases on a global scale.

Acknowledgement

We would like to express our sincere gratitude to the Department of Agronomy for their support and assistance throughout the writing. The Department's commitment to academic excellence and research has been instrumental in completing this endeavor.

Authors' Contributions

The authors of this work have made significant contributions to the research project/study. Each author has participated sufficiently in the research project/study, made intellectual contributions, and is responsible for the work's accuracy and integrity. The authors have collaborated closely, ensuring the completion of this work through collective effort, expertise, and dedication.

References

1. Cordeiro, T.A.R., de Resende, M.A.C., dos S. Moraes, S.C., Franco, D.L., Pereira, A.C., Ferreira, L.F., Electrochemical biosensors for neglected tropical diseases: A review. *Talanta*, 234, 122617, 2021, doi: https://doi.org/10.1016/j.talanta.2021.122617.
2. Khatun, S., Putta, C.L., Hak, A., Rengan, A.K., Immunomodulatory nanosystems: An emerging strategy to combat viral infections. *Biomater. Biosyst.*, 9, 100073, 2023, doi: https://doi.org/10.1016/j.bbiosy.2023.100073.
3. Harish, V. *et al.*, Cutting-edge advances in tailoring size, shape, and functionality of nanoparticles and nanostructures: A review. *J. Taiwan Inst. Chem. Eng.*, 149, 105010, 2023, doi: https://doi.org/10.1016/j.jtice.2023.105010.
4. Lim, H.J., Saha, T., Tey, B.T., Tan, W.S., Ooi, C.W., Quartz crystal microbalance-based biosensors as rapid diagnostic devices for infectious diseases.

Biosens. Bioelectron., 168, 112513, 2020, doi: https://doi.org/10.1016/j.bios.2020.112513.

5. Ali, T., Najam-ul-Haq, M., Mohyuddin, A., Musharraf, S.G., Hussain, D., Next-generation functional nanotextiles—Prospects and challenges. *Sustain. Mater. Technol.*, 36, e00640, 2023, doi: https://doi.org/10.1016/j.susmat.2023.e00640.
6. Sharma, S. and Dhadly, D.K., Chapter 5 - Nano-inspired biosensors and plant diseases: recent advances and challenges, in: *Nanomaterial-Plant Interactions*, S. Seena, A. Rai, S. B. T.-N., P.-M.I. Kumar (Eds.), pp. 135–162, Academic Press, 2023, doi: https://doi.org/10.1016/B978-0-323-90619-7.00002-3.
7. Dervisevic, M., Alba, M., Prieto-Simon, B., Voelcker, N.H., Skin in the diagnostics game: Wearable biosensor nano- and microsystems for medical diagnostics. *Nano Today*, 30, 100828, 2020, doi: https://doi.org/10.1016/j.nantod.2019.100828.
8. Taha, B.A. *et al.*, Next-generation nanophotonic-enabled biosensors for intelligent diagnosis of SARS-CoV-2 variants. *Sci. Total Environ.*, 880, 163333, 2023, doi: https://doi.org/10.1016/j.scitotenv.2023.163333.
9. Balestri, A., Cardellini, J., Berti, D., Gold and silver nanoparticles as tools to combat multidrug-resistant pathogens. *Curr. Opin. Colloid Interface Sci.*, 66, 101710, 2023, doi: https://doi.org/10.1016/j.cocis.2023.101710.
10. Ansari, A.A. and Malhotra, B.D., Current progress in organic–inorganic hetero-nano-interfaces based electrochemical biosensors for healthcare monitoring. *Coord. Chem. Rev.*, 452, 214282, 2022, doi: https://doi.org/10.1016/j.ccr.2021.214282.
11. Maza Vega, D., Di Meglio, M., del V. Alonso, S., Alvira, F., Montanari, J., Nanomaterials for diagnosis, treatment, and prevention of human cutaneous leishmaniasis: A review. *OpenNano*, 12, 100158, 2023, doi: https://doi.org/10.1016/j.onano.2023.100158.
12. Wang, Y., Yang, K., Zhou, H., Immunogenic proteins and potential delivery platforms for mpox virus vaccine development: A rapid review. *Int. J. Biol. Macromol.*, 245, 125515, 2023, doi: https://doi.org/10.1016/j.ijbiomac.2023.125515.
13. Gao, Y. *et al.*, Nanocatalysis meets microfluidics: A powerful platform for sensitive bioanalysis. *TrAC, Trends Anal. Chem.*, 158, 116887, 2023, doi: https://doi.org/10.1016/j.trac.2022.116887.
14. Mohapatra, B., Chamoli, S., Salvi, P., Saxena, S.C., Fostering nanoscience's strategies: A new frontier in sustainable crop improvement for abiotic stress tolerance. *Plant Nano Biol.*, 3, 100026, 2023, doi: https://doi.org/10.1016/j.plana.2023.100026.
15. Huang, J., Li, W., Bai, X., Xiao, F., Xu, H., Metal–organic framework detectives meet infectious pathogens: A powerful tool against pandemics. *Coord. Chem. Rev.*, 488, 215160, 2023, doi: https://doi.org/10.1016/j.ccr.2023.215160.

16. Seaberg, J., Clegg, J.R., Bhattacharya, R., Mukherjee, P., Self-therapeutic nanomaterials: Applications in biology and medicine. *Mater. Today*, 62, 190–224, 2023, doi: https://doi.org/10.1016/j.mattod.2022.11.007.
17. Amir, S., Arathi, A., Reshma, S., Mohanan, P.V., Microfluidic devices for the detection of disease-specific proteins and other macromolecules, disease modelling and drug development: A review. *Int. J. Biol. Macromol.*, 235, 123784, 2023, doi: https://doi.org/10.1016/j.ijbiomac.2023.123784.
18. Chang, C. *et al.*, Engineered M13 phage as a novel therapeutic bionanomaterial for clinical applications: From tissue regeneration to cancer therapy. *Mater. Today Bio*, 20, 100612, 2023, doi: https://doi.org/10.1016/j.mtbio.2023.100612.
19. Ehtesabi, H., Fayaz, M., Hosseini-Doabi, F., Rezaei, P., The application of green synthesis nanoparticles in wound healing: a review. *Mater. Today Sustain.*, 21, 100272, 2023, doi: https://doi.org/10.1016/j.mtsust.2022.100272.
20. Singh, S. *et al.*, Novel insights into graphene oxide-based adsorbents for remediation of hazardous pollutants from aqueous solutions: A comprehensive review. *J. Mol. Liq.*, 369, 120821, 2023, doi: https://doi.org/10.1016/j.molliq.2022.120821.
21. Ahmed, A. *et al.*, Nature-inspired biogenic synthesis of silver nanoparticles for antibacterial applications. *Mater. Today Chem.*, 27, 101339, 2023, doi: https://doi.org/10.1016/j.mtchem.2022.101339.
22. Wu, S., Song, R., Liu, T., Li, C., Antifungal therapy: Novel drug delivery strategies driven by new targets. *Adv. Drug Deliv. Rev.*, 199, 114967, 2023, doi: https://doi.org/10.1016/j.addr.2023.114967.
23. Roy, S., Hasan, I., Guo, B., Recent advances in nanoparticle-mediated antibacterial applications. *Coord. Chem. Rev.*, 482, 215075, 2023, doi: https://doi.org/10.1016/j.ccr.2023.215075.
24. Chandrasekaran, R. *et al.*, Prospective features of functional 2D nanomaterial graphene oxide in the wound healing process. *J. Drug Deliv. Sci. Technol.*, 82, 104352, 2023, doi: https://doi.org/10.1016/j.jddst.2023.104352.
25. Biswas, G.C., Khan, M.T.M., Das, J., Wearable nucleic acid testing platform - A perspective on rapid self-diagnosis and surveillance of infectious diseases. *Biosens. Bioelectron.*, 226, 115115, 2023, doi: https://doi.org/10.1016/j.bios.2023.115115.
26. Aflakian, F. *et al.*, Nanoparticles-based therapeutics for the management of bacterial infections: A special emphasis on FDA approved products and clinical trials. *Eur. J. Pharm. Sci.*, 188, 106515, 2023, doi: https://doi.org/10.1016/j.ejps.2023.106515.
27. Puri, N. and Gupta, A., Water remediation using titanium and zinc oxide nanomaterials through disinfection and photo catalysis process: A review. *Environ. Res.*, 227, 115786, 2023, doi: https://doi.org/10.1016/j.envres.2023.115786.

28. Gumus, E., Bingol, H., Zor, E., Lateral flow assays for detection of disease biomarkers. *J. Pharm. Biomed. Anal.*, 225, 115206, 2023, doi: https://doi.org/10.1016/j.jpba.2022.115206.
29. Bushra, R. *et al.*, Polysaccharide-based nanoassemblies: From synthesis methodologies and industrial applications to future prospects. *Adv. Colloid Interface Sci.*, 318, 102953, 2023, doi: https://doi.org/10.1016/j.cis.2023.102953.
30. Radu, A.-F. and Bungau, S.G., Nanomedical approaches in the realm of rheumatoid arthritis. *Ageing Res. Rev.*, 87, 101927, 2023, doi: https://doi.org/10.1016/j.arr.2023.101927.
31. Li, C. *et al.*, Highly efficient, remarkable sensor activity and energy storage properties of MXenes and borophene nanomaterials. *Prog. Solid State Chem.*, 70, 100392, 2023, doi: https://doi.org/10.1016/j.progsolidstchem.2023.100392.
32. Pourmadadi, M., Shamsabadipour, A., Aslani, A., Eshaghi, M.M., Rahdar, A., Pandey, S., Development of Polyvinylpyrrolidone-Based nanomaterials for biosensors applications: A Review. *Inorg. Chem. Commun.*, 152, 110714, 2023, doi: https://doi.org/10.1016/j.inoche.2023.110714.
33. Jain, S. *et al.*, Internet of medical things (IoMT)-integrated biosensors for point-of-care testing of infectious diseases. *Biosens. Bioelectron.*, 179, 113074, 2021, doi: https://doi.org/10.1016/j.bios.2021.113074.
34. Deroco, P.B., Wachholz Junior, D., Kubota, L.T., Recent advances in point-of-care biosensors for the diagnosis of neglected tropical diseases. *Sens. Actuators B Chem.*, 349, 130821, 2021, doi: https://doi.org/10.1016/j.snb.2021.130821.
35. Feyziazar, M. *et al.*, Recent advances on the piezoelectric, electrochemical, and optical biosensors for the detection of protozoan pathogens. *TrAC, Trends Anal. Chem.*, 157, 116803, 2022, doi: https://doi.org/10.1016/j.trac.2022.116803.
36. Kadadou, D. *et al.*, Recent advances in the biosensors application for the detection of bacteria and viruses in wastewater. *J. Environ. Chem. Eng.*, 10, 1, 107070, 2022, doi: https://doi.org/10.1016/j.jece.2021.107070.
37. Azizoglu, R.O., CRISPR-Cas systems in diagnostics: A comprehensive assessment of Cas effectors and biosensors. *Gene Genome Ed.*, 3–4, 100019, 2022, doi: https://doi.org/10.1016/j.ggedit.2022.100019.
38. Wang, Y. *et al.*, Micro/nano biomedical devices for point-of-care diagnosis of infectious respiratory diseases. *Med. Nov. Technol. Devices*, 14, 100116, 2022, doi: https://doi.org/10.1016/j.medntd.2022.100116.
39. Nagraik, R., Sharma, A., Kumar, D., Mukherjee, S., Sen, F., Kumar, A.P., Amalgamation of biosensors and nanotechnology in disease diagnosis: Mini-review. *Sens. Int.*, 2, 100089, 2021, doi: https://doi.org/10.1016/j.sintl.2021.100089.
40. Rajpal, S. and Mishra, P., Next generation biosensors employing molecularly imprinted polymers as sensing elements for *in vitro* diagnostics.

Biosens. Bioelectron. X, 11, 100201, 2022, doi: https://doi.org/10.1016/j.biosx.2022.100201.

41. Aquino, A., Paschoalin, V.M.F., Tessaro, L.L.G., Raymundo-Pereira, P.A., Conte-Junior, C.A., Updating the use of nano-biosensors as promising devices for the diagnosis of coronavirus family members: A systematic review. *J. Pharm. Biomed. Anal.*, 211, 114608, 2022, doi: https://doi.org/10.1016/j.jpba.2022.114608.
42. Soler, M. and Huertas, C.S., Label-Free Plasmonic Biosensors in Clinical Diagnostics, in: *Encyclopedia of Sensors and Biosensors*, First, R. B. T.-E., S. and B., E. Narayan (Eds.), pp. 383–401, Elsevier, Oxford, 2023, doi: https://doi.org/10.1016/B978-0-12-822548-6.00119-9.
43. Mukherjee, S. and Das, S., Chapter 14 - Role of biosensor-based devices for diagnosis of nononcological disorders, in: *Biosensor Based Advanced Cancer Diagnostics*, R. Khan, A. Parihar, S. K. B. T.-B. B. A. C. D. Sanghi (Eds.), pp. 245–256, Academic Press, 2022, doi: https://doi.org/10.1016/B978-0-12-823424-2.00020-X.
44. Zahraee, H., Khoshbin, Z., Mohammadi, F., Mashreghi, M., Abnous, K., Taghdisi, S.M., Three-way junction skeleton biosensors based on aptamers, DNAzymes, and DNA hybridization probes. *TrAC, Trends Anal. Chem.*, 165, 117160, 2023, doi: https://doi.org/10.1016/j.trac.2023.117160.
45. Singh, A.K., Mittal, S., Das, M., Saharia, A., Tiwari, M., Optical biosensors: a decade in review. *Alexandria Eng. J.*, 67, 673–691, 2023, doi: https://doi.org/10.1016/j.aej.2022.12.040.
46. Sanz, C.G. *et al.*, Sensing Materials: Electrochemical Applications of DNA Sensors and Biosensors, in: *Encyclopedia of Sensors and Biosensors*, First, R. B. T.-E., S. and B., E. Narayan (Eds.), pp. 445–467, Elsevier, Oxford, 2023, doi: https://doi.org/10.1016/B978-0-12-822548-6.00039-X.
47. Kumar, V. and Arora, K., Trends in nano-inspired biosensors for plants. *Mater. Sci. Energy Technol.*, 3, 255–273, 2020, doi: https://doi.org/10.1016/j.mset.2019.10.004.
48. Liu, J., Xie, G., Lv, S., Xiong, Q., Xu, H., Recent applications of rolling circle amplification in biosensors and DNA nanotechnology. *TrAC, Trends Anal. Chem.*, 160, 116953, 2023, doi: https://doi.org/10.1016/j.trac.2023.116953.
49. Ham, Y. *et al.*, The SpACE-CCM: A facile and versatile cell culture medium-based biosensor for detection of SARS-CoV-2 spike-ACE2 interaction. *Biosens. Bioelectron.*, 227, 115169, 2023, doi: https://doi.org/10.1016/j.bios.2023.115169.
50. Raghuwanshi, R. and Singh, M., Chapter 14 - Challenges to molecularly imprinted polymers in competition with biosensors at commercial scale, in: *Molecularly Imprinted Polymers (MIPs)*, M. B. T.-M. I. P., and Singh (Eds.), pp. 345–366, Elsevier, 2023, doi: https://doi.org/10.1016/B978-0-323-91925-8.00005-3.

51. Akhtar, N. *et al.*, *List of contributors*, A. Kumar, J. Singh, J. B. T.-V. and M., M. Samuel (Eds.), pp. xix–xxi, Academic Press, 2021, doi: https://doi.org/10.1016/B978-0-12-824523-1.00028-6.
52. Kumar, V. *et al.*, Mitigation of heat stress responses in crops using nitrate primed seeds. *S. Afr. J. Bot.*, 140, 25–36, 2021, doi: https://doi.org/10.1016/j.sajb.2021.03.024.
53. Kumar, P., Devi, P., Dey, S.R., Chapter 6 - Fungal volatile compounds: a source of novel in plant protection agents, in: *Volatiles and metabolites of microbes*, A. Kumar, J. Singh, J. B. T.-V. and M., M. Samuel (Eds.), pp. 83–104, Academic Press, 2021, doi: https://doi.org/10.1016/B978-0-12-824523-1.00001-8.
54. Kumari, P., Singh, J., Kumar, P., Chapter 21 - Impact of bioenergy for the diminution of an ascending global variability and change in the climate, in: *Volatiles and metabolites of microbes*, A. Kumar, J. Singh, L. F. R. B. T.-M. U. C. C. Ferreira (Eds.), pp. 469–487, Woodhead Publishing, 2022, doi: https://doi.org/10.1016/B978-0-323-90571-8.00021-3.
55. Adelodun, B. *et al.*, *List of Contributors*, A. Kumar, J. Singh, L. F. R. B. T.-M. U. C. C. Ferreira (Eds.), pp. xix–xxiv, Woodhead Publishing, 2022, doi: https://doi.org/10.1016/B978-0-323-90571-8.00028-6.
56. Kumar, P. and Mistri, T.K., Transcription factors in SOX family: Potent regulators for cancer initiation and development in the human body. *Semin. Cancer Biol.*, 67, 105–113, 2020, doi: https://doi.org/10.1016/j.semcancer.2019.06.016.
57. Das, T. *et al.*, Promising botanical-derived monoamine oxidase (MAO) inhibitors: pharmacological aspects and structure-activity studies. *S. Afr. J. Bot.*, 146, 127–145, 2022, doi: https://doi.org/10.1016/j.sajb.2021.09.019.
58. Upadhyay, S.K. *et al.*, Efficient removal of total arsenic (As3+/5+) from contaminated water by novel strategies mediated iron and plant extract activated waste flowers of marigold. *Chemosphere*, 313, 137551, 2023, doi: https://doi.org/10.1016/j.chemosphere.2022.137551.
59. Goud, E.L., Singh, J., Kumar, P., Chapter 19 - Climate change and their impact on global food production, in: *Volatiles and metabolites of microbes*, A. Kumar, J. Singh, L. F. R. B. T.-M. U. C. C. Ferreira (Eds.), pp. 415–436, Woodhead Publishing, 2022, doi: https://doi.org/10.1016/B978-0-323-90571-8.00019-5.
60. Aley, P., Singh, J., Kumar, P., Chapter 23 - Adapting the changing environment: microbial way of life, in: *Volatiles and metabolites of microbes*, A. Kumar, J. Singh, L. F. R. B. T.-M. U. C. C. Ferreira (Eds.), pp. 507–525, Woodhead Publishing, 2022, doi: https://doi.org/10.1016/B978-0-323-90571-8.00023-7.
61. Kumar, P., Kumar, T., Singh, S., Tuteja, N., Prasad, R., Singh, J., Potassium: A key modulator for cell homeostasis. *J. Biotechnol.*, 324, 198–210, 2020.
62. Kotia, A. *et al.*, Rheological analysis of rice husk-starch suspended in water for sustainable agriculture application. *Mater. Today Proc.*, 50, 1962–1966, 2022, doi: https://doi.org/10.1016/j.matpr.2021.09.325.

63. Chakraborty, S. *et al.*, Unravelling the regulatory role of miRNAs in secondary metabolite production in medicinal crops. *Plant Gene*, 27, 100303, 2021, doi: https://doi.org/10.1016/j.plgene.2021.100303.
64. Kumar, P., Sharma, K., Saini, L., Dey, S.R., Chapter 8 - Role and behavior of microbial volatile organic compounds in mitigating stress, in: *Volatiles and metabolites of microbes*, A. Kumar, J. Singh, J. B. T.-V. and M., M. Samuel (Eds.), pp. 143–161, Academic Press, 2021, doi: https://doi.org/10.1016/B978-0-12-824523-1.00010-9.
65. Benjamin, S.R., de Souza Nascimento, T., Roque, C.R., de Andrade, G.M., Oriá, R.B., Chapter 2 - Recent advances in the development of immunosensors for infectious diseases, in: *Biosensors for emerging and re-emerging infectious diseases*, J. Das, S. Dave, S. Radhakrishnan, P. B. T.-B. for E., R.-E.I.D. Mohanty (Eds.), pp. 19–72, Academic Press, 2022, doi: https://doi.org/10.1016/B978-0-323-88464-8.00006-3.
66. Tavakoli, H., Mohammadi, S., Li, X., Fu, G., Li, X., Microfluidic platforms integrated with nano-sensors for point-of-care bioanalysis. *TrAC, Trends Anal. Chem.*, 157, 116806, 2022, doi: https://doi.org/10.1016/j.trac.2022.116806.
67. Sehit, E. and Altintas, Z., Chapter 3 - Biosensors for virus detection, in: *Advanced Sensor Technology*, A. Barhoum and Z. B. T.-A. S. T. Altintas (Eds.), pp. 53–80, Elsevier, 2023, doi: https://doi.org/10.1016/B978-0-323-90222-9.00001-7.
68. Kumar, P., Sarkar, N., Singh, A., Kaushik, M., Nanopaper biosensors at point of care. *Bioconjugate Chem.*, 33, 6, 1114–1130, 2022.
69. Fathi, F., Rashidi, M.-R., Pakchin, P.S., Ahmadi-Kandjani, S., Nikniazi, A., Photonic crystal based biosensors: Emerging inverse opals for biomarker detection. *Talanta*, 221, 121615, 2021, doi: https://doi.org/10.1016/j.talanta.2020.121615.
70. Karuppaiah, G., Vashist, A., Nair, M., Veerapandian, M., Manickam, P., Emerging trends in point-of-care biosensing strategies for molecular architectures and antibodies of SARS-CoV-2. *Biosens. Bioelectron. X*, 13, 100324, 2023, doi: https://doi.org/10.1016/j.biosx.2023.100324.
71. Chadha, U. *et al.*, Recent progress and growth in biosensors technology: A critical review. *J. Ind. Eng. Chem.*, 109, 21–51, 2022, doi: https://doi.org/10.1016/j.jiec.2022.02.010.
72. Tang, R. *et al.*, Nitrocellulose Membrane for Paper-based Biosensor. *Appl. Mater. Today*, 26, 101305, 2022, doi: https://doi.org/10.1016/j.apmt.2021.101305.
73. Sadighbayan, D., Sadighbayan, K., Tohid-kia, M.R., Khosroushahi, A.Y., Hasanzadeh, M., Development of electrochemical biosensors for tumor marker determination towards cancer diagnosis: Recent progress. *TrAC, Trends Anal. Chem.*, 118, 73–88, 2019, doi: https://doi.org/10.1016/j.trac.2019.05.014.

74. Ince, B. and Sezgintürk, M.K., Lateral flow assays for viruses diagnosis: Up-to-date technology and future prospects. *TrAC, Trends Anal. Chem.*, 157, 116725, 2022, doi: https://doi.org/10.1016/j.trac.2022.116725.
75. Fazlali, F. *et al.*, Electrochemiluminescent biosensor for ultrasensitive detection of lymphoma at the early stage using CD20 markers as B cell-specific antigens. *Bioelectrochemistry*, 138, 107730, 2021, doi: https://doi.org/10.1016/j.bioelechem.2020.107730.
76. Wang, Y., Li, Z., Hu, Q., Emerging self-regulated micro/nano drug delivery devices: A step forward towards intelligent diagnosis and therapy. *Nano Today*, 38, 101127, 2021, doi: https://doi.org/10.1016/j.nantod.2021.101127.
77. Zhu, R., Avsievich, T., Popov, A., Bykov, A., Meglinski, I., *In vivo* nanobiosensing element of red blood cell-mediated delivery. *Biosens. Bioelectron.*, 175, 112845, 2021, doi: https://doi.org/10.1016/j.bios.2020.112845.
78. Zarubova, J., Zhang, X., Hoffman, T., Hasani-Sadrabadi, M.M., Li, S., Biomaterial-based immunoengineering to fight COVID-19 and infectious diseases. *Matter*, 4, 5, 1528–1554, 2021, doi: https://doi.org/10.1016/j.matt.2021.02.025.
79. Limbut, W., Promsuwan, K., Kongkaew, S., Thavarungkul, P., Mak, W.C., Chapter 7 - Emerging functional materials for microfluidic biosensors, in: *Microfluidic Biosensors*, W.C. Mak and A. H. B. T.-M. B. Pui Ho (Eds.), pp. 195–231, Academic Press, 2023, doi: https://doi.org/10.1016/B978-0-12-823846-2.00008-0.
80. Mutalik, S.P., Pandey, A., Mutalik, S., Nanoarchitectronics: A versatile tool for deciphering nanoparticle interaction with cellular proteins, nucleic acids and phospholipids at biological interfaces. *Int. J. Biol. Macromol.*, 151, 136–158, 2020, doi: https://doi.org/10.1016/j.ijbiomac.2020.02.150.
81. Hunt, J.P. *et al.*, Towards detection of SARS-CoV-2 RNA in human saliva: A paper-based cell-free toehold switch biosensor with a visual bioluminescent output. *New Biotechnol.*, 66, 53–60, 2022, doi: https://doi.org/10.1016/j.nbt.2021.09.002.
82. Lee, B.H., Kim, S.H., Gu, M.B., Biosensors: Aptamer Sensors, in: *Encyclopedia of Sensors and Biosensors*, First, R. B. T.-E., S. and B., E. Narayan (Eds.), pp. 363–376, Elsevier, Oxford, 2023, doi: https://doi.org/10.1016/B978-0-12-822548-6.00047-9.
83. Zhang, Y., Wu, Y., Wu, Y., Chang, Y., Liu, M., CRISPR-Cas systems: From gene scissors to programmable biosensors. *TrAC, Trends Anal. Chem.*, 137, 116210, 2021, doi: https://doi.org/10.1016/j.trac.2021.116210.
84. Hao, L. *et al.*, Engineering light-initiated afterglow lateral flow immunoassay for infectious disease diagnostics. *Biosens. Bioelectron.*, 212, 114411, 2022, doi: https://doi.org/10.1016/j.bios.2022.114411.
85. Zayed, B.A., Ali, A.N., Elgebaly, A.A., Talaia, N.M., Hamed, M., Mansour, F.R., Smartphone-based point-of-care testing of the SARS-CoV-2: A systematic review. *Sci. Afr.*, 21, e01757, 2023, doi: https://doi.org/10.1016/j.sciaf.2023.e01757.

86. Roy, P., Chapter 6 - Development of optical biosensors for the diagnosis of pathogens, in: *Biosensors for Emerging and Re-Emerging Infectious Diseases*, J. Das, S. Dave, S. Radhakrishnan, P. B. T.-B., E. and R.-E. I. D., Mohanty (Eds.), pp. 137–168, Academic Press, 2022, doi: https://doi.org/10.1016/B978-0-323-88464-8.00005-1.
87. Mohammadi, L., Pal, K., Bilal, M., Rahdar, A., Fytianos, G., Kyzas, G.Z., Green nanoparticles to treat patients with Malaria disease: An overview. *J. Mol. Struct.*, 1229, 129857, 2021, doi: https://doi.org/10.1016/j.molstruc.2020.129857.
88. Basso, C.R., Crulhas, B.P., Castro, G.R., Pedrosa, V.A., Chapter 4 - The use of nano-enabled technologies to diagnose dengue virus infections, in: *Applications of Nanobiotechnology for Neglected Tropical Diseases*, F.R. Formiga, Inamuddin, P. B. T.-A., N. for N. T. D., Severino (Eds.), pp. 71–88, Academic Press, 2021, doi: https://doi.org/10.1016/B978-0-12-821100-7.00021-2.
89. Ribeiro, B.V., Ferreira, L.F., Franco, D.L., Advances in biosensor development for the determination of antibiotics in cow's milk - A review. *Talanta Open*, 6, 100145, 2022, doi: https://doi.org/10.1016/j.talo.2022.100145.
90. Fathi, N. *et al.*, Biosensors for the detection of protein kinases: Recent progress and challenges. *Microchem. J.*, 182, 107961, 2022, doi: https://doi.org/10.1016/j.microc.2022.107961.
91. Islam, M.M. and Koirala, D., Toward a next-generation diagnostic tool: A review on emerging isothermal nucleic acid amplification techniques for the detection of SARS-CoV-2 and other infectious viruses. *Anal. Chim. Acta*, 1209, 339338, 2022, doi: https://doi.org/10.1016/j.aca.2021.339338.
92. Tiwari, A., Gulbake, A.S., Kumar, P., Chapter 9 - Lipid nanoparticles and nanoemulsions exploited in the diagnosis and treatment of infectious diseases, in: *Developments in Microbiology*, K. Jain, J. B. T.-N., T. and D., I.D. Ahmad (Eds.), pp. 229–273, Academic Press, 2022, doi: https://doi.org/10.1016/B978-0-323-91201-3.00010-4.
93. Ali, M.R. *et al.*, Label free flexible electrochemical DNA biosensor for selective detection of Shigella flexneri in real food samples. *Talanta*, 253, 123909, 2023, doi: https://doi.org/10.1016/j.talanta.2022.123909.
94. Sobhanie, E. *et al.*, Recent trends and advancements in electrochemiluminescence biosensors for human virus detection. *TrAC, Trends Anal. Chem.*, 157, 116727, 2022, doi: https://doi.org/10.1016/j.trac.2022.116727.
95. Mokhtarzadeh, A. *et al.*, Nanomaterial-based biosensors for detection of pathogenic virus. *TrAC, Trends Anal. Chem.*, 97, 445–457, 2017, doi: https://doi.org/10.1016/j.trac.2017.10.005.
96. Singh, R., Das Mukherjee, M., Sumana, G., Gupta, R.K., Sood, S., Malhotra, B.D., Biosensors for pathogen detection: A smart approach towards clinical diagnosis. *Sens. Actuators B Chem.*, 197, 385–404, 2014, doi: https://doi.org/10.1016/j.snb.2014.03.005.

97. Jha, A. and Pathak, Y., Chapter 6 - Polymeric nanomaterials for infectious diseases, in: *Developments in Microbiology*, K. Jain, J. B. T.-N., T. and D., I.D. Ahmad (Eds.), pp. 127–149, Academic Press, 2022, doi: https://doi.org/10.1016/B978-0-323-91201-3.00007-4.
98. Karimi-Maleh, H. *et al.*, A critical review on the use of potentiometric based biosensors for biomarkers detection. *Biosens. Bioelectron.*, 184, 113252, 2021, doi: https://doi.org/10.1016/j.bios.2021.113252.
99. Patel, K.D., Patel, A.K., Kurian, A.G., Singh, R.K., Kim, H.-W., Chapter 12 - Tuning the properties of inorganic nanomaterials for theranostic applications in infectious diseases: Carbon nanotubes, quantum dots, graphene, and mesoporous carbon nanoparticles, in: *Developments in Microbiology*, K. Jain, J. B. T.-N., T. and D., I.D. Ahmad (Eds.), pp. 319–352, Academic Press, 2022, doi: https://doi.org/10.1016/B978-0-323-91201-3.00011-6.
100. Gupta, S., Murthy, C.N., Prabha, C.R., Recent advances in carbon nanotube based electrochemical biosensors. *Int. J. Biol. Macromol.*, 108, 687–703, 2018, doi: https://doi.org/10.1016/j.ijbiomac.2017.12.038.
101. Prakash, J., Parveen, A., Mishra, Y.K., Kaushik, A., Nanotechnology-assisted liquid crystals-based biosensors: Towards fundamental to advanced applications. *Biosens. Bioelectron.*, 168, 112562, 2020, doi: https://doi.org/10.1016/j.bios.2020.112562.
102. Zhang, C. *et al.*, Synthetic biology-powered biosensors based on CRISPR/Cas mediated cascade signal amplification for precise RNA detection. *Chem. Eng. J.*, 446, 136864, 2022, doi: https://doi.org/10.1016/j.cej.2022.136864.
103. Singh, V., Allawadhi, P., Khurana, A., Banothu, A.K., Bharani, K.K., Critical neurological features of COVID-19: Role of imaging methods and biosensors for effective diagnosis. *Sens. Int.*, 2, 100098, 2021, doi: https://doi.org/10.1016/j.sintl.2021.100098.
104. Gattani, A., Singh, S.V., Agrawal, A., Khan, M.H., Singh, P., Recent progress in electrochemical biosensors as point of care diagnostics in livestock health. *Anal. Biochem.*, 579, 25–34, 2019, doi: https://doi.org/10.1016/j.ab.2019.05.014.
105. Thomas, S., Saji, K.J., Jayaraj, M.K., Chapter 5 - An introduction to biosensors, in: *Micro and Nano Technologies*, M.K. Jayaraj, P.P. Subha, S. B. T.-N., S. and O., A. Thomas (Eds.), pp. 91–107, Elsevier, 2022, doi: https://doi.org/10.1016/B978-0-12-824008-3.00013-8.
106. Pandey, A., Chauhan, P., Singhal, A., Chapter 8 - Potential electrochemical biosensors for early detection of viral infection, in: *Advanced Biosensors for Virus Detection*, R. Khan, A. Parihar, A. Kaushik, A. B. T.-A. B., V.D. Kumar (Eds.), *Advanced Biosensors for Virus Detection*, pp. 133–154, Academic Press, 2022, doi: https://doi.org/10.1016/B978-0-12-824494-4.00010-2.
107. Neethirajan, S., Ragavan, K.V., Weng, X., Agro-defense: Biosensors for food from healthy crops and animals. *Trends Food Sci. Technol.*, 73, 25–44, 2018, doi: https://doi.org/10.1016/j.tifs.2017.12.005.

108. Hammou, R.A., Benhassou, M., Ennaji, M.M., Chapter 10 - Application of Nanodiagnostics in Viral Infectious Diseases, in: *Emerging and Reemerging Viral Pathogens*, M. M. B. T.-E. and R.V.P. Ennaji (Eds.), pp. 179–195, Academic Press, 2020, doi: https://doi.org/10.1016/B978-0-12-814966-9.00010-X.
109. Sheikhzadeh, E., Beni, V., Zourob, M., Nanomaterial application in bio/sensors for the detection of infectious diseases. *Talanta*, 230, 122026, 2021, doi: https://doi.org/10.1016/j.talanta.2020.122026.
110. Al Mamun, M., Wahab, Y.A., Hossain, M.A.M., Hashem, A., Johan, M.R., Electrochemical biosensors with Aptamer recognition layer for the diagnosis of pathogenic bacteria: Barriers to commercialization and remediation. *TrAC, Trends Anal. Chem.*, 145, 116458, 2021, doi: https://doi.org/10.1016/j.trac.2021.116458.
111. Filice, M., Marchal, J.A., Gamiz, F., Chapter 7 - Biosensors based on two-dimensional materials, in: *Nanophotonics*, Y.M. Jhon, J. H. B. T.-2D M., N. Lee (Eds.), pp. 245–312, Elsevier, 2021, doi: https://doi.org/10.1016/B978-0-12-818658-9.00004-1.
112. Sornambikai, S., Divya, K.P., Vasanth, S., Viswanathan, C., Ponpandian, N., CRISPR based biosensing: An ultrasensitive theranostic tool for the detection of early Breast Cancer biomarkers – A mini review. *Biosens. Bioelectron. X*, 14, 100367, 2023, doi: https://doi.org/10.1016/j.biosx.2023.100367.
113. Ning, Y., Hu, J., Lu, F., Aptamers used for biosensors and targeted therapy. *Biomed. Pharmacother.*, 132, 110902, 2020, doi: https://doi.org/10.1016/j.biopha.2020.110902.
114. Pilan, L., Tailoring the performance of electrochemical biosensors based on carbon nanomaterials via aryldiazonium electrografting. *Bioelectrochemistry*, 138, 107697, 2021, doi: https://doi.org/10.1016/j.bioelechem.2020.107697.
115. Cancino-Bernardi, J. *et al.*, A SARS-CoV-2 impedimetric biosensor based on the immobilization of ACE-2 receptor-containing entire cell membranes as the biorecognition element. *Talanta*, 253, 124008, 2023, doi: https://doi.org/10.1016/j.talanta.2022.124008.
116. Li, Z., Leustean, L., Inci, F., Zheng, M., Demirci, U., Wang, S., Plasmonic-based platforms for diagnosis of infectious diseases at the point-of-care. *Biotechnol. Adv.*, 37, 8, 107440, 2019, doi: https://doi.org/10.1016/j.biotechadv.2019.107440.
117. Jiang, K. *et al.*, Next-generation diagnostic test for dengue virus detection using an ultrafast plasmonic colorimetric RT-PCR strategy. *Anal. Chim. Acta*, 1274, 341565, 2023, doi: https://doi.org/10.1016/j.aca.2023.341565.
118. Sharma, A., Sharma, N., Kumari, A., Lee, H.-J., Kim, T., Tripathi, K.M., Nanocarbon based sensors for bacterial detection and discrimination in clinical diagnosis: A junction between material science and biology. *Appl. Mater. Today*, 18, 100467, 2020, doi: https://doi.org/10.1016/j.apmt.2019.100467.
119. Campos-Ferreira, D. *et al.*, COVID-19 challenges: From SARS-CoV-2 infection to effective point-of-care diagnosis by electrochemical biosensing

platforms. *Biochem. Eng. J.*, 176, 108200, 2021, doi: https://doi.org/10.1016/j.bej.2021.108200.

120. He, Y., Hu, C., Li, Z., Wu, C., Zeng, Y., Peng, C., Multifunctional carbon nanomaterials for diagnostic applications in infectious diseases and tumors. *Mater. Today Bio*, 14, 100231, 2022, doi: https://doi.org/10.1016/j.mtbio.2022.100231.
121. Van Tran, V., Tran, N.H.T., Hwang, H.S., Chang, M., Development strategies of conducting polymer-based electrochemical biosensors for virus biomarkers: Potential for rapid COVID-19 detection. *Biosens. Bioelectron.*, 182, 113192, 2021, doi: https://doi.org/10.1016/j.bios.2021.113192.
122. Antonacci, A., Arduini, F., Scognamiglio, V., Chapter 8 - Trends in biosensors for the detection of Staphylococcus aureus as a key cause of foodborne illnesses worldwide, in: *Biosensors for Emerging and Re-Emerging Infectious Diseases*, J. Das, S. Dave, S. Radhakrishnan, P. B. T.-B. for E., R.-E.I.D. Mohanty (Eds.), pp. 197–224, Academic Press, 2022, doi: https://doi.org/10.1016/B978-0-323-88464-8.00013-0.
123. Wang, C., Liu, M., Wang, Z., Li, S., Deng, Y., He, N., Point-of-care diagnostics for infectious diseases: From methods to devices. *Nano Today*, 37, 101092, 2021, doi: https://doi.org/10.1016/j.nantod.2021.101092.
124. Ranjan, P., Sadique, M.A., Yadav, S., Khan, R., Chapter 19 - Approaches for fabrication of point-of-care biosensors for viral infection, in: *Advanced Biosensors for Virus Detection*, R. Khan, A. Parihar, A. Kaushik, A. B. T.-A. B., V.D. Kumar (Eds.), pp. 353–371, Academic Press, 2022, doi: https://doi.org/10.1016/B978-0-12-824494-4.00021-7.
125. Zhang, G.-J. and Ning, Y., Silicon nanowire biosensor and its applications in disease diagnostics: A review. *Anal. Chim. Acta*, 749, 1–15, 2012, doi: https://doi.org/10.1016/j.aca.2012.08.035.
126. Pal, M., Das, V., Deka Boruah, H.P., Chikkaputtaiah, C., Chapter 13 - Multiplexed biosensors for virus detection, in: *Advanced Biosensors for Virus Detection*, R. Khan, A. Parihar, A. Kaushik, A. B. T.-A. B., V.D. Kumar (Eds.), pp. 219–239, Academic Press, 2022, doi: https://doi.org/10.1016/B978-0-12-824494-4.00017-5.
127. Perumal, V. and Hashim, U., Advances in biosensors: Principle, architecture and applications. *J. Appl. Biomed.*, 12, 1, 1–15, 2014, doi: https://doi.org/10.1016/j.jab.2013.02.001.
128. Shetti, N.P., Bukkitgar, S.D., Reddy, K.R., Reddy, C.V., Aminabhavi, T.M., ZnO-based nanostructured electrodes for electrochemical sensors and biosensors in biomedical applications. *Biosens. Bioelectron.*, 141, 111417, 2019, doi: https://doi.org/10.1016/j.bios.2019.111417.
129. Assolini, J.P. *et al.*, Nanomedicine in leishmaniasis: A promising tool for diagnosis, treatment and prevention of disease - An update overview. *Eur. J. Pharmacol.*, 923, 174934, 2022, doi: https://doi.org/10.1016/j.ejphar.2022.174934.

130. Wang, B., Yang, D., Chang, Z., Zhang, R., Dai, J., Fang, Y., Wearable bioelectronic masks for wireless detection of respiratory infectious diseases by gaseous media. *Matter*, 5, 12, 4347–4362, 2022, doi: https://doi.org/10.1016/j.matt.2022.08.020.
131. Arshad, R. *et al.*, A review of the nanomaterials use for the diagnosis and therapy of salmonella typhi. *J. Mol. Struct.*, 1230, 129928, 2021, doi: https://doi.org/10.1016/j.molstruc.2021.129928.
132. Nosrati, R. *et al.*, Siderophore-based biosensors and nanosensors; new approach on the development of diagnostic systems. *Biosens. Bioelectron.*, 117, 1–14, 2018, doi: https://doi.org/10.1016/j.bios.2018.05.057.
133. Baryeh, K., Takalkar, S., Lund, M., Liu, G., 1 - Introduction to medical biosensors for point of care applications, in: *Medical biosensors for point of care (POC),* R. J. B. T.-M. B. for P., C., A. Narayan (Eds.), pp. 3–25, Woodhead Publishing, 2017, doi: https://doi.org/10.1016/B978-0-08-100072-4.00001-0.
134. Lima, D., Mendes Hacke, A.C., Ulmer, B., Kuss, S., Electrochemical sensing of trypanosome- and flavivirus-related neglected tropical diseases. *Curr. Opin. Electrochem.*, 30, 100838, 2021, doi: https://doi.org/10.1016/j.coelec.2021.100838.
135. Mukherjee, S., Perveen, S., Negi, A., Sharma, R., Evolution of tuberculosis diagnostics: From molecular strategies to nanodiagnostics. *Tuberculosis*, 140, 102340, 2023, doi: https://doi.org/10.1016/j.tube.2023.102340.
136. Bankole, O.E., Verma, D.K., Chávez González, M.L., Ceferino, J.G., Sandoval-Cortés, J., Aguilar, C.N., Recent trends and technical advancements in biosensors and their emerging applications in food and bioscience. *Food Biosci.*, 47, 101695, 2022, doi: https://doi.org/10.1016/j.fbio.2022.101695.
137. Palaniyandi, T. *et al.*, Nanosensors for the diagnosis and therapy of neurodegenerative disorders and inflammatory bowel disease. *Acta Histochem.*, 125, 2, 151997, 2023, doi: https://doi.org/10.1016/j.acthis.2023.151997.
138. Nagy-Simon, T., Hada, A.-M., Suarasan, S., Potara, M., Recent advances on the development of plasmon-assisted biosensors for detection of C-reactive protein. *J. Mol. Struct.*, 1246, 131178, 2021, doi: https://doi.org/10.1016/j.molstruc.2021.131178.
139. Paliwal, R., Chaurasiya, A., Panchal, K., Nayak, P., Parveen, N., Paliwal, S.R., Chapter 3 - Engineering and functionalization of nanomaterials for theranostic applications in infectious diseases, in: *Developments in Microbiology*, K. Jain, J. B. T.N., T. and D., I.D. Ahmad (Eds.), pp. 45–71, Academic Press, 2022, doi: https://doi.org/10.1016/B978-0-323-91201-3.00003-7.

5

Trends in the Development of Immunosensors for the Diagnosis of Infectious Diseases

Stephen Rathinaraj Benjamin[1,2*], Eli José Miranda Ribeiro Júnior[3], Sam Phinehas Gnana Sekar[4], Rosa Fireman Dutra[5] and Geanne Matos de Andrade[1,2]

[1]Laboratory of Neuroscience and Behavior, Drug Research and Development Center (NPDM), Department of Physiology and Pharmacology, Federal University of Ceará, Coronel Nunes de Melo, Porangabussu, Fortaleza, Ceará, Brazil
[2]Department of Physiology and Pharmacology, Faculty of Medicine, Federal University of Cearà, Rua Prof. Costa Mendes, Fortaleza, Ceará, Brazil
[3]Department of Pharmacy, Faculty of CGESP (Centro Goiano de Especializações), Rua A, Setor Oeste, Goiânia - GO, Brazil
[4]Gullas College of Medicine, Banilad, Mandaue City, Cebu, Philippines
[5]Biomedical Engineering Laboratory, Federal University of Pernambuco, Av. Prof. Moraes Rego, Cidade Universitaria, Recife, PE, Brazil

Abstract

The emergence of infectious diseases presents significant burdens on global economies and human health, with ongoing pandemics posing challenges worldwide. Addressing these biosecurity issues, including diseases like COVID-19, hepatitis, Zika, chikungunya, dengue, diphtheria, Ebola virus, and the potential risk of bioterrorism, is now a top priority. However, the lack of quick and dependable diagnostic methods remains a critical concern, leading to adverse economic impacts, unnecessary hospitalizations, and underreported cases.

Analytical techniques like enzyme-linked immunosorbent assay (ELISA) and polymerase chain reaction (PCR) technology have been used for virus detection but suffer from limitations such as high costs, large sample volumes, and

**Corresponding author*: steaje@gmail.com

Suvardhan Kanchi, Ayyappa Bathinapatla, Anitha Varghese and Phumlane Selby Mdluli (eds.) Nano-Biosensor Technologies for Diagnosis of Infectious Diseases, (173–196) © 2025 Scrivener Publishing LLC

time-consuming procedures, especially for low-income nations. Electrochemical methods offer a promising alternative due to their cost-effectiveness, robustness, and potential for miniaturization, making them suitable for point-of-care testing (POCT) scenarios. Notably, the successful implementation of glucose biosensors in diabetes monitoring exemplifies their value in healthcare.

Advancements in nanotechnology have played a pivotal role in enhancing immunoserological diagnosis through biosensors. Researchers have developed nanostructured devices with optical, piezoelectric, and electrochemical transductions by incorporating microfluidics, portable strategies, and specific antibodies, antigens, fragments, and aptamers. In conclusion, leveraging nanomaterials in immunosensors significantly improves diagnostic capabilities in healthcare, aiding in combating infectious diseases and enhancing global health outcomes. The development of reliable, cost-effective, and portable immunosensors holds the key to early detection and effective management of infectious diseases, mitigating their impact worldwide.

***Keywords*:** Nanomaterial, immunosensors, virus detection, SARS-CoV-2, Zika virus, HIV, influenza, hepatitis B

5.1 Introduction

According to a report by the World Health Organization in 2007, there has been an alarming increase in infectious diseases. Over the years, almost 40 different infectious diseases have been discovered, including SARS, Middle East respiratory syndrome coronavirus (MERS-CoV), Ebola, chikungunya, avian influenza, hepatitis, Zika virus, and the latest one being the novel severe acute respiratory syndrome coronavirus 2 (SARS-CoV-2), which has been recently identified.

Over the last 25 years, biosensors have emerged as a complement to PCR and ELISA (enzyme-linked immunosorbent assay) for diagnosing infectious diseases. These sensors directly integrate biorecognition and sensor components, providing an alternative platform for identifying and quantifying infectious diseases. A biosensor is a type of analytical tool that utilizes biorecognition and a transduction element to produce measurable or semi-measurable outcomes. It does not require any extra processing steps or reagents. Biosensors have the capability to detect infectious pathogens with high sensitivity and selectivity in real-time scenarios without requiring any sample preparation.

Additionally, they can operate in a broad range of measurement formats, including droplet and continuous flow. Biosensors are an incredibly versatile tool capable of detecting viruses in food, biological fluids, and object surfaces. The most impressive aspect is their ability to do so efficiently without labeling. Biosensors are highly compatible with diverse technology, making them essential for virus detection.

Electrochemical biosensors are the most effective pathogen detection method compared to other transducers, such as mechanical and optical sensors. These biosensors use conductive and semiconducting materials as electrodes, which can convert the chemical energy of the pathogen-biometric element binding into electrical energy through an electrochemical method. This allows on-site detection without sample preparation, rapid detection through a low-cost platform, and the ability to detect multiple pathogens in various matrices. In this chapter, a number of new immunosensors reported over the last decade are examined to detect different types of viruses.

5.1.1 Biosensors

A biosensor is an essential tool that must be utilized to identify and examine chemical markers in living organisms. It relies on a bioreceptor of biomolecules such as antibodies, nucleic acids, enzymes, bacteria, and organelles, which work with a transducer. The bioreceptor is highly selective and sensitive to the target biomarker associated with a specific infection to prevent false positives caused by other organisms in the sample. The transducer records and displays the quantifiable signal this biochemical reaction produces, identifying pathogens. These compounds are now essential to biosensor research and development, with biosensors offering the convenience of POCT for regular clinical testing of biological samples. Bioreceptors play a crucial role in microbial detection and biosensor reliability.

Choosing biological materials for analysis is crucial based on their high specificity. Biosensors are highly adaptable and can be customized to meet specific requirements. Depending on the desired sensitivity level, biosensors can be designed to detect a single type of antibiotic or multiple antibiotics. Using antigen-antibody interactions or haptens as biosensitive elements in immunoassays creates highly accurate immunosensors that incorporate physical signal transduction.

5.1.2 Immunosensing Components or Bioreceptors

Antibodies are receptor molecules for biosensors. Immunoglobulins, or antibodies, are proteins involved in the immune system. The most prominent antibody in serum is immunoglobulin G (IgG), which consists of four polypeptides that form a "Y" shape by joining two heavy chains and two light chains. The "Y" tips of antibodies have diverse amino acid sequences. The variable region of an antibody, which contains 110–130 amino acids, is responsible for its ability to recognize and attach to a specific antigen. The light and heavy chain termini are part of the variation zone. Each antibody has a unique structure that fits like a lock and key with an antigen. The constant region dictates the outcome of the interaction between an antibody and an antigen. It can potentially eradicate the antigen or signal for further elimination. Based on the constant region's structure and their respective immunological functions, there are five major types of antibodies: IgM, IgG, IgA, IgD, and IgE.

5.1.3 Strategies Towards Immobilization of Antibodies

In order to achieve success with immunoassays, it is imperative that the antibodies can identify and firmly attach to a particular target analyte. One reliable and straightforward method for characterizing antibody binding is ELISA. During the immunoassay procedure, incubation is used to fix antibodies onto a substrate, forming various patterns on the surface. These patterns can include central protrusions or intricate deposits. The variability in the functionalized surfaces may impact the assay pattern's efficiency. There are two main approaches to immobilization in antibody-based sensors: random and directed immobilization. The easiest and most reliable way to attach the antibody to the sensor surface is through random adsorption. However, this method may cause the antibody to lose shape, resulting in a random protein orientation. Physical adsorption can also be used, relying on the interaction of charges on the substrate surface and biomolecules. Physical adsorption is often ineffective because it is sensitive to environmental conditions and lacks stability. The utilization of chemical cross-linking is essential to significantly diminish environmental interference and enhance the precision of analysis. It is common to use support matrixes like self-assembled monolayers (SAM) and polymer brushes that have amine, carboxylic acid (-COOH), and hydroxyl groups (-OH) to immobilize biomolecules using chemical cross-linking methods. The support matrices used for immobilizing biomolecules often require the activation of functional groups using glutaraldehyde, carbonyl diimidazole (CDI), or a combination of 1-ethyl-3-(3-dimethyl aminopropyl)-carbodiimide (EDC) and N-hydroxy succinimide (NHS). Unfortunately, these immobilization methods are highly laborious and time-consuming, leading to reduced repeatability of electrochemical immunosensors.

Chemical modification of SAMs using an end group activator has shown significant potential for improving commercial biosensors due to the stable, resulting bond. In immobilization, four distinct methods exist entrapment and encapsulation, covalent binding, cross-linking, and adsorption. The selection of a particular method is contingent upon various factors, such as the type of transducer utilized, the biological component's characteristics, the analyte's physicochemical properties, and the biosensor's operational conditions.

5.1.4 Biotransducer Component

There are various biosensors, including acoustic, optical, thermometric, electrochemical, piezoelectric, and magnetic biosensors. A comprehensive evaluation of the recognition component and transducer is crucial in identifying viruses. These elements are pivotal in converting the biochemical response into a measurable physical signal. There are various techniques one can use to record the interaction between the substance being examined and the component in charge of identification. Some of these techniques include surface enhanced Raman spectroscopy (SERS), deoxyribonucleic acid (DNA) microarray, atomic force microscopy (AFM), and scanning probe microscopy (SPM), among others. However, electrochemical and optical methods are the most commonly used. Electrochemical biosensors use different detection principles like potentiometric, amperometric, voltammetric, and conductometric methods. The concentration of the analyte can be determined by analyzing the fluctuations in currents and voltages generated during the molecular reaction. Potentiometric biosensors consist of a selective and reference electrode and a measuring device that calculates the potential difference between the two electrodes. Amperometric biosensors detect changes in current intensity resulting from an electrochemical reaction occurring at a constant potential. The intensity of the current is observed to have a direct correlation with the concentration of the analyzed substance. Conductometric biosensors detect changes in electrical conductivity caused by the generation or consumption of ionic species during metabolic processes. Piezoelectric biosensors use an electric current to control the vibrations of a piezoelectric crystal. The piezoelectric crystal utilized as a detector detects changes in mass resulting from the creation of an immune Ab-Ag complex on its surface through these immunosensors. Optical biosensors are cost-effective, have a rapid response time, and can detect analytes in real time without labeling.

Maximum immunosensor performance requires further modifications to the transducer surface for effective immobilization of probe molecules. The most straightforward approach for molecules to interact with a transducer is frequently by electrostatic adsorption or the noncovalent physical adsorption of hydrophobic molecules. Sometimes, the efficacy of probes can be hindered by various factors like denaturation caused by surface interactions or inadequate adsorption of probe molecules on the sensor surface. In contrast, chemical reaction-based surface modification achieves the homogeneous and dense packing of probe molecules with greater efficiency and specificity. Antibodies have many free amines in their structure, often forming covalent interactions with electrophilic groups on sensing substrates. This chemical process is similar to physical adsorption and causes the probe molecules on the transducer surface to be randomly oriented.

5.2 Immunosensors

Currently, immunosensors with electrochemical detection elements are widely used. Two detection elements are used, which can be categorized as potentiometric and amperometric detectors. Potentiometric detectors are comprised of field effect transistors (FET) and ion-selective electrodes (ISE). On the other hand, amperometric detectors are composed of graphite and noble metal electrodes such as gold or platinum. The latter detects current for a constant or variable potential. These electrochemical immunosensors are popular for their high sensitivity and selectivity, suitability for immunoassays, lower cost, and ability to miniaturize them compared to other analytical techniques. Furthermore, the sensor system's constant feedback enables automated supervision, streamlining electrochemical detection, enhancing bio-security measures, and ultimately lowering expenses [1].

Electrochemical immunosensors are emerging as essential bioanalytical tools that identify distinct biomarkers for various illnesses. They are becoming increasingly vital in developing point-of-care (POC) devices. These sensors are highly sensitive due to their precise interaction between antibodies, antigens and electrochemical transductions. It is crucial to recognize that the electrochemical sensing capabilities of an immunosensor are contingent upon the effectiveness of immobilizing the bio-recognizing elements onto the electrode surfaces.

5.2.1 Recent Advances in Immunosensor-Based Virus Detection

These technologies use antibodies, antigens, and other biomolecules to detect viruses and other pathogens in a sample. They offer a faster, more accurate alternative to traditional virus detection methods like PCR testing. Immunosensor technology has the potential to revolutionize the way viruses are detected and diagnosed. It can be used to quickly and accurately detect the presence of viruses in a sample, allowing for faster treatment and improved patient outcomes. This technology can also be used in epidemiological studies to track the spread of viruses over time, allowing for better-informed public health policy decisions. It can also help to identify new virus strains, helping to reduce the risk of future pandemics.

Furthermore, this technology can be used to better understand the evolution of viruses, which can help inform vaccine development strategies. Three main types of immunosensors exist: electrochemical, optical, and gravimetric. Electrochemical immunosensors detect changes in the electrical properties of the antigen-antibody binding, optical immunosensors detect changes in the light emitted or absorbed by the antigen-antibody binding, and gravimetric immunosensors measure changes in the mass of the antigen-antibody binding. All of these technologies can be used to measure the presence of a virus in a sample, allowing for rapid and accurate identification of virus strains and the tracking of their spread over time.

5.2.2 Electrochemical Immunosensor

Electrochemical immunosensors have the potential to be the next-generation detection strategy owing to their simplicity, high sensitivity, and selectivity. Compared to conventional immunoassay techniques, the pre-treatment process is less complex, and the equipment is more user-friendly, with less analytical volumes and label-free options that enable monitoring of the antibody-antigen reaction without the disturbances caused by conjugated markers. In order to ensure the development of a prosperous label-free electrochemical immunosensor, it is vital to immobilize biomolecules in a proficient and efficient way. The electrode material must be adjusted to augment conductivity and expedite electron transfer while broadening the electrode surface area to facilitate greater immobilization of antibodies.

Most label-free electrochemical immunosensors rely on reversible redox species like $K_3[Fe(CN)_6]^{3/4}$ and $[Ru(bpy)_3]^2$ in the electrolyte solution. However, these redox species can get contaminated and harm the bioactivity of antibodies or antigens. Scientists are currently working on creating a unique electrochemical immunosensor that does not require labeling and uses redox species found on the surface of the electrode. The implementation of this innovative approach eliminates the requirement for direct contact between antibodies, which has the potential to increase the efficiency and safety of the process.

In order to enhance the effectiveness of electrochemical immunosensors, a technique called signal amplification strategy has been used. This involves using nanomaterial-functionalized enzymes as labels to promote electrocatalytic oxidation and reduction and hydrolysis of substrate. However, enzyme-immobilized immunosensors are sensitive to their surrounding environment during measurement and storage, as they may undergo denaturation or degradation. Nanomaterials that resemble enzymes have been discovered and are undergoing extensive research due to their advantageous properties, such as their uncomplicated manufacturing process, robustness, and affordability. Among these nanomaterials, noble metal nanoparticles stand out as they exhibit exceptional stability and high catalytic activity while still being simple to synthesize, easy to tune, and store. Bimetallic nanomaterials, in particular, show even more excellent catalytic properties than their monometallic counterparts, thanks to their synergistic effect.

A novel immunosensor for the detection of COVID-19 has been presented by the research team of Sampaio *et al.* [2]. The device employs capacitive detection of spike proteins and can differentiate between COVID-19 and dengue and Zika NS1 virus spike proteins. The immunosensor has an edge over competing techniques since it can use polyclonal antibodies to identify mutations in the spike protein.

5.3 Optical Immunosensor

The most widely used transducers for bioanalysis are optical immunosensors. Visible radiation provides advantages over other methods of transduction. Optical immunosensors are non-destructive and offer fast data interpretation. Fiber bundle optics as optical waveguides and advanced optoelectronics has made these analytical tools more adaptable for clinical purposes.

Optical sensor techniques rely on various mechanisms such as adsorption, fluorescence, luminescence, scatter, or refractive index changes. Detectors such as photodiodes or photomultipliers are commonly used. These techniques can be used to directly detect immunological reactions, labeled immunospecies, or enzymatic reactions without the need for labels. While fluorescent labels are commonly used, bio- and chemiluminescence species are also possible. Although evanescence wave-related sensors provide a label-free alternative to immunoassays, they may still have limited sensitivity to low-molecular-weight analytes and be affected by unspecific binding interactions.

There are two distinct categories of optical immunosensors: direct and indirect. Each category employs a different method to detect binding events. Direct sensors can identify signals without the need for extra reagents, whereas indirect sensors require labels such as fluorescence on the antibody or antigen to indicate binding. Optical properties such as light absorbance and polarization can be utilized to identify the presence of a target analyte. Optical immunosensors can detect a range of small and large analytes, including bacteria and viruses.

Nonetheless, current techniques still have limitations, including low sensitivity or specificity. Future advancements should prioritize enhancing biomolecular recognition, storage life, and microfluidic devices. Although biocompatibility concerns persist with certain materials, it is essential to note that optical immunosensors enhanced with nanomaterials have been optimized for superior performance and require smaller sample volumes [3].

Optical immunosensors that do not require labeling use refractometry or reflectometry in the ultraviolet–visible (UV-Vis) spectrum to measure optical signals. Refractometric methods are commonly used to develop these sensors, including surface plasmon resonance (SPR), grating couplers, resonant mirrors, Mach-Zehnder and Young interferometers, and Bragg gratings. These methods assess biolayer refractive index variations on transducer surfaces [4].

5.3.1 Surface Enhanced Raman Spectroscopy (SERS)

For precise virus detection via surface enhanced Raman spectroscopy (SERS), it is imperative to utilize reporter molecules and a sandwich immunoassay. This detection technique involves a SERS tag comprising a Raman reporter molecule and a specific antibody acting as its recognition element (detection antibody), which is restricted to SERS-active nanoparticles. Virus detection requires a linker antibody-Ab binds to an antigen-SERS tag complex on a capture substrate. The Raman signals generated by the reporter molecule must be measured before and after the contact between the SERS tag and the capture component to measure the virus accurately. Traditional early detection methods have been constrained by biological analytical chemicals having large molecule sizes, low scattering sections, low affinities for conventional noble metal SERS materials, and low selectivity [5].

Li *et al.* designed a magnetic immunosensor that uses SERS to detect the H3N2 influenza virus accurately. This innovative method creates a sandwich complex with SERS tags, targeted viruses, and highly SERS-active magnetic support substrates. By utilizing magnetic support substrates, preparing the sample is much simpler, making it easier to separate and enrich viruses from a complex matrix. The immunosensor can detect H3N2 with the help of a portable Raman spectrometer, with a linear concentration of 10^2 and 5×10^3 TCID50/mL [6]. Recently, a SERS-based immunoassay for detecting SARS-CoV-2 has been developed by combining a specific antibody pair with hollow gold nanoparticles and magnetic beads. The test has shown remarkable sensitivity by identifying the virus at a concentration as low as 2.56 fg/mL and viral lysates at 3.4 PFU/mL.

Furthermore, this test can detect SARS-CoV-2 in human nasopharyngeal aspirates, providing a valuable COVID-19 diagnosis and prevention tool. Through the use of a portable Raman system, COVID-19 patients were successfully diagnosed within just 30 min, demonstrating the immense potential of this assay [7]. Researchers Moon *et al.* [8] have developed a very sensitive immunoassay for viral detection. Antibodies were bound to the capture substrate and gold nanoparticles with the help of protein G, which resulted in the development of antibody probes. Using this approach, antibodies may be immobilized in a targeted manner while retaining their native structure, allowing for more effective viral engagement. The analyte-containing capture substrate was coupled with the antibody probes, followed by a silver cover formed using an Ag-enhancer solution. Researchers tested its selectivity on H3N2, H5N1, and influenza B (IBV) viruses to ensure its selectivity. The signal at 1643 cm^{-1} was at least 3.5 times more potent for p(H1N1) than for the other viruses, demonstrating the remarkable selectivity of the approach.

5.3.2 Surface Plasma Resonance Sensor (SPR)

The detection of biomarkers has significantly improved by identifying different markers that can be measured in bodily fluids like blood, saliva and urine. Immunosensors, which use a unique antigen-antibody immunoreaction, are the most sensitive analytical approach for identifying biomarkers. Due to their excellent sensitivity, selectivity, and user-friendliness, SPR-based immunosensors, especially those that employ noble metals like Au and Ag as SPR-inducing elements, are gaining favor as a detection platform. In contrast to immunosensors based on enzymes like horseradish peroxidase (HRP), those based on SPR may detect biomarkers without a labelled detection probe.

A team of researchers led by Mahmud [9] have developed a quick and highly sensitive biosensor for detecting the novel coronavirus. The biosensor is made of graphene and multiple coating layers based on SPR technology.

It can detect the virus using the spike RBD, antibodies (IgG or IgM), or virus ribonucleic acid (RNA). The accuracy of the sensor was confirmed using the transfer matrix method (TMM) algorithm and the finite-difference time-domain (FDTD) technique. Another team of researchers, led by Panda [10], has proposed a SPR-based dengue sensor using graphene and Sb2S3. This sensor has a maximum sensitivity of 225.75°/RIU for plasma.

Researchers led by Chang have developed a quick test for avian influenza A H7N9 using a label-free immunosensor based on intensity-modulated SPR technology. This novel antibody was chosen because it recognizes the H7N9 virus but not other human influenza isolates, significantly increasing the test's sensitivity. The test's sensitivity was increased by a factor of 20 compared to ELISA due to this highly specific antibody, and it took less than 10 min. In nasal mucosa samples from individuals with flu-like symptoms, the LOD value was 402 copies/mL, much lower than the limit of detection (LOD) value of quick influenza diagnostic tests and qRT-PCR [11].

5.4 Nanomaterials Immunosensor

Nanomaterials have been utilized in the development of innovative and sensitive electrochemical immunoassays. These materials include noble metal nanoparticles, chalcogenides, metal oxide nanoparticles, and carbon-based materials, which attach capture probes and secondary antibody labels. Integrating nanomaterials into electrochemical immunosensors has improved detection limits and increased sensitivity [12].

Electrochemical immunosensors have recently received much attention, mainly due to their compatibility with microfabrication technology, low cost, and ease of miniaturization. This is especially important given the rapid development in the field of nanotechnology. Several research studies have shown that adding nanomaterials to immunosensor production significantly enhances its sensitivity and specificity, enabling the detection of target analytes at levels as low as picograms.

The field of nanotechnology has seen considerable advancements recently, leading to an increased interest in combining immunosensors with nanomaterials. Nanomaterials have an enormous specific surface area, robust adsorption, and excellent biocompatibility. Immunosensors' electrochemical characteristics are greatly improved by these qualities. Gold nanoparticles (AuNPs) change sensor surfaces due to their high conductivity, electrocatalytic activity, and surface-to-volume ratio. The immunosensor's sensitivity is enhanced by AuNPs' electron transfer efficiency [12].

Gold nanoparticles have the potential to act as a solid foundation for the attachment of antibodies, thereby creating a stable and effective platform for various applications. The immunosensor's repeatability and stability are considerably improved by creating Au-S bonds between AuNPs and anti-α-fetoprotein (AFP) antibody (Ab) molecules' sulfhydryl (-SH) to immobilize Ab on the electrode surface as AuNPs-Dex-RGO nanocomposites. The use of cutting-edge nanostructures has been important in developing highly sensitive impedimetric immunosensors for a wide range of analytes [13].

Graphene quantum dots (GQDs), known as carbon nanostructures, have several unusual qualities. Large surface area, no blinking, safety, chemical inertness, cheap synthesis, high solubility in water, and tunable bandgap energy are only a few of these characteristics. Researchers in optoelectronics, photovoltaics, electroluminescence, light-emitting diodes, biosensing, biological imaging, and drug administration are particularly interested in them. Several methods, such as chemical, electrochemical, hydrothermal processes, and high-resolution electron-beam synthesis, have been investigated for synthesizing GQDs. Similar to graphene, GQDs may have a variety of functional groups grafted onto them through a π-π conjugated network. Their edges contain carboxyl and hydroxyl groups, making them highly water-soluble and easily conjugatable with biological compounds. Due to this, GQDs have already shown their usefulness in biosensing applications, with improved detection sensitivities achieved [14].

Recently, scientists have been experimenting with carbon-based nanomaterials to enhance the performance of working electrodes in electrochemistry. Carbon nanotubes, nanofibers, nanodots, and graphene are materials used for electrode surface modification. Graphene has attracted significant attention due to its extensive specific surface area, high electrochemical conductivity, remarkable stability, and cost-effectiveness. Since graphene consists of a single layer of carbon atoms organized in a honeycomb pattern, agglomeration during electrode surface modification must be avoided using conducting polymers. Electrode manufacturing and biofunctionalization are improved when the graphene is combined with a conducting polymer to form a hybrid nanomaterial. Research shows that electrodes modified with G/conducting polymer nanocomposite may significantly improve sensors' electrochemical sensitivity. There are a number of conducting polymers that have been used to modify electrode surfaces, including polyaniline (PANI), polypyrrole (PPy), and poly(3,4-ethylenedioxythiophene) (PEDOT), but PANI stands out due to its low price, simple synthesis, good stability in the environment, reversible redox characteristics, and high biological compatibility.

The unique qualities of graphene oxide (GO) have made it a topic of great interest in biosensor development. These include its ideal honeycomb lattice, remarkable electrical properties, easily modifiable surface, and outstanding corrosion resistance. It is usual practice to convert GO into reduced graphene oxide (RGO) to improve its conductivity. Reduced graphene oxide has superior conductivity and is more material-compatible. However, RGO has poor solubility and tends to clump in water. Thus, developing a highly distributed carbon material for use in building sensitive sensors is an essential matter. Both covalent and non-covalent approaches exist for modifying RGO. However, non-covalent procedures are superior to covalent methods for altering RGO because of the scarcity of oxygen functional groups for conjugation. Reduced graphene oxide can be modified through various non-covalent interactions such as hydrogen bonding, coordination bonds, van der Waals forces, electrostatic forces, and π-π stacking.

5.5 Paper-Based Immunosensors

Paper is the preferred material for immunosensors due to its low cost, disposability, easy availability, and good biocompatibility [15]. Many research groups have dedicated significant efforts to developing paper-based immunosensors over the past few decades. Several methods have been presented for creating them, including wax printing, UV photolithography, plasma treatment, screen printing, and laser treatment. Screen printing is the most efficient and user-friendly method for producing many paper-based immunosensors. The paper-based immunosensor stands out for its environmentally friendly and straightforward production process, which makes it ideal for point-of-care testing in illness diagnosis.

Lee [16] and his team developed an immunosensor based on paper that could detect and quantify H5N1, H7N9, and H9N2 avian influenza viral antigens without the need of a label. The sensor employs flexible carbon nanotube-polydimethylsiloxane electrodes and has a detection range of 100 pg/ml to 100 ng/ml for avian influenza viruses. In phosphate-buffered saline solutions, the sensors give a LOD with fast measurement durations of 20–30 min.

5.6 Viral Infectious Diseases

Immunosensors have emerged as powerful tools for detecting viral infectious diseases, revolutionizing diagnostic capabilities and public health surveillance. These analytical devices utilize specific antibodies or antigens as biological recognition elements to target viral pathogens. By directly detecting the presence of viruses in clinical samples, immunosensors enable rapid and accurate diagnosis, crucial for timely clinical decisions and effective disease management. Moreover, they can quantify viral loads, providing valuable insights into disease progression and treatment efficacy. The high sensitivity and specificity of immunosensors reduce the likelihood of false results, and their potential for multiplexing allows simultaneous detection of multiple viral targets. Portable and user-friendly, these sensors are well-suited for point-of-care testing, especially in resource-limited settings. With their ability to offer rapid and reliable results, immunosensors significantly mitigate the spread of viral infections and improve global public health outcomes.

5.6.1 COVID-19

Recently, two types of immunosensors utilizing optical/chemiluminescence lateral flow technology were developed to quickly identify the presence of IgA antibodies in the saliva and serum of individuals afflicted with COVID-19 [17]. A diagnostic tool for COVID-19 was developed using antibodies that target the nucleocapsid protein of SARS-CoV-2. The detection antibodies were affixed to an optical waveguide material sheet, while

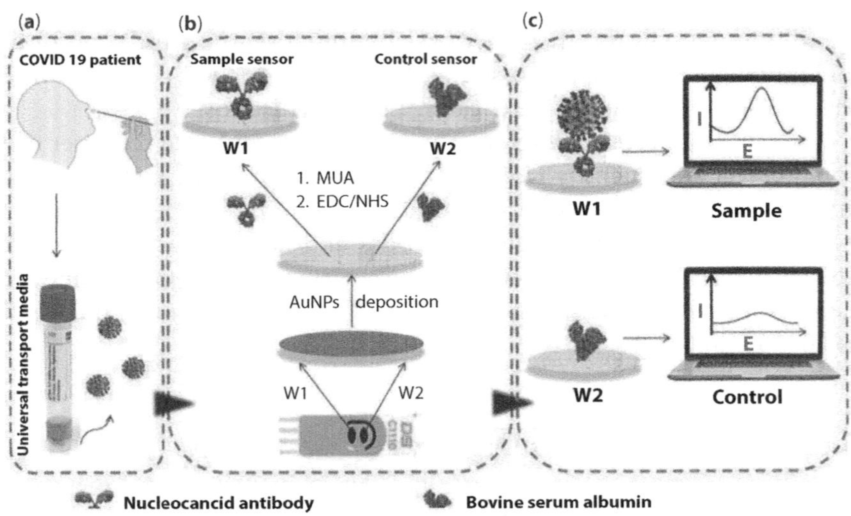

Figure 5.1 Voltammetric SARS-CoV-2 nucleocapsid antigen immunosensor.

the capture antibodies were attached to light-scattering particles [18]. Eissa and colleagues have developed an immunosensor that uses voltammetry to detect the nucleocapsid antigen of SARS-CoV-2. 11-mercaptoundecanoic acid (MUA) immobilized an antibody towards the SARS-CoV-2 nucleocapsid protein on the electrodes (Figure 5.1). The results of the electrochemical immunosensor test indicate that it was highly sensitive, with a linear concentration of 1.0 pg/mL to 100 ng/mL, and an LOD of 0.4 pg/mL for the N protein in phosphate buffer saline (PBS) buffer at pH 7.4 [19]. In addition, a cutting-edge machine-learning technique has created a portable immunosensor capable of rapidly detecting the SARS-Cov-2 spike protein. This ground-breaking technology is specifically engineered to provide precise and speedy results with optimal efficiency [20].

5.6.2 Dengue

Basso *et al.* [21] proposed a selective detection method for DENV serotypes using a configuration of γ-Fe2O3@ 3-mercaptopropionic acid @Au NPs@ aptamer and an localized surface plasmon resonance (LSPR)-based immunosensor. Different concentrations of aptamers ranging from 1.0–10.0 μM were tested, and it was found that a concentration of 3.0 μM is enough to cover the surface of the modified AuNPs fully. Furthermore, a label-free electrochemical immunosensor for dengue NS1 protein detection was suggested by Dutra *et al.* [22] and comprised of modified PANI covered with glassy carbon (GC). The immunosensor's calibration curve demonstrated excellent linearity, sensitivity (slope = 13.8% IpR/mL ng^{-1}), data distribution (R = 0.997), and an LOD of 0.33 ng/mL. For rapid investigation of the dengue virus, Nawaz *et al.* [23] developed NS1-based immunosensors in conjunction with bovine serum albumin (BSA)-modified solid-phase chemical evaporation (SPCE) as transducing substrate. The working electrode was modified to fix the anti-NS1 monoclonal antibody on the BSA surface through electro grafting. Quantitative and qualitative evaluations were performed by measuring the change in electron-transfer resistance before and after NS1 attachment at various concentrations. The dengue virus was identified by the immunosensor in its as-prepared form with an improved LOD (0.3 ng/mL) and linearity of (1–200 ng/mL).

5.6.3 Human Immunodeficiency Virus (HIV)

A novel electrochemical immunosensor was devised to detect p24 in untreated human serum. This innovative immunological device is an extremely valuable and straightforward technique for early human immunodeficiency virus (HIV) diagnosis. The chitosan/glutaraldehyde crosslinking technique ensures efficient immobilization and exposure of p24 for interaction with a mouse anti-p24 IgG1, forming the basis of the competitive sensor. A low LOD of 2 pM for human serum has been achieved, along with the detection of substances in biological samples ranging from 10 pM to 1 nM via optimized immunosensor setup. This immunosensor is remarkably convenient, as its disposable screen-printed electrodes are functionalized with single-walled carbon nanotubes [24].

5.6.4 Hepatitis

The identification of hepatitis C antibodies can be achieved by using an advanced electrochemical immunosensor. This has been made possible due to the developing a polypyrrole-graphene nanocomposite created through a highly effective one-step electro-synthesis method. The immunosensor had a linearity of 2–14 ng/mL for anti-HCV and could detect levels as low as 1.63 ng/mL in the clinical range [25]. The same group developed a hyaluronic acid-carbon nanotube hybrid film to detect the hepatitis B virus in blood samples. This method involves using a nanohybrid surface assembled on a glassy carbon electrode, including amino carbon nanotubes coated with hyaluronic acid. This serves as a sensing platform to detect the presence of anti-HBc. The response of the immunosensor was linear when exposed to anti-HBc in concentrations of up to 6 ng/mL. The detection limit was 0.03 ng/mL [26].

Salimi *et al.* [27] developed a highly accurate immunosensor for detecting hepatitis B virus surface antigen. It uses gold sheet, magnetic nanoparticles, and DNA enzymes for amplification. The Hemin/G-quadruplex structure, acting as an HRP mimicking-DNAzyme, effectively enhanced the catalytic reduction of H_2O_2 through methylene blue (MB) oxidation. The use of square wave voltammetry signals for MB allowed for precise measurements of HBsAg. These measurements had a linear concentration range of 0.3–1000 pg/mL and a detection limit of 0.19 pg/mL.

5.6.5 Zika

Dutra *et al.* [28] developed an impressive immunosensor to detect the Zika virus. Their innovative redox-probe-free technique involves the creation of a nanostructured film of Prussian blue (PB) in a single step, resulting in a highly efficient and electrocatalytic detection method for the virus. The nanocomposite was developed by combining PB with a chitosan-carbon nanotube hybrid, which was then drop-casted onto a screen-printed electrode. The immunosensor detected the envelope protein of ZIKV within a linear range of 0.25–1.75 μg/mL and had a detection limit of 0.20 μg/mL (Figure 5.2). The same group of researchers [29] successfully developed an electrochemical screening immunosensor. This tool can detect NS1 glycoprotein in serum and urine samples, making it a groundbreaking development in identifying dengue and Zika viruses. The immunosensor comprises a label-free assembly of a carbon nanotube-ethylenediamine nanostructured thin film. The DPV results for NS1 showed a linear range of 20–800 ng/mL, repeatability of 3.0%, and an LOD value of 6.8 ng/mL.

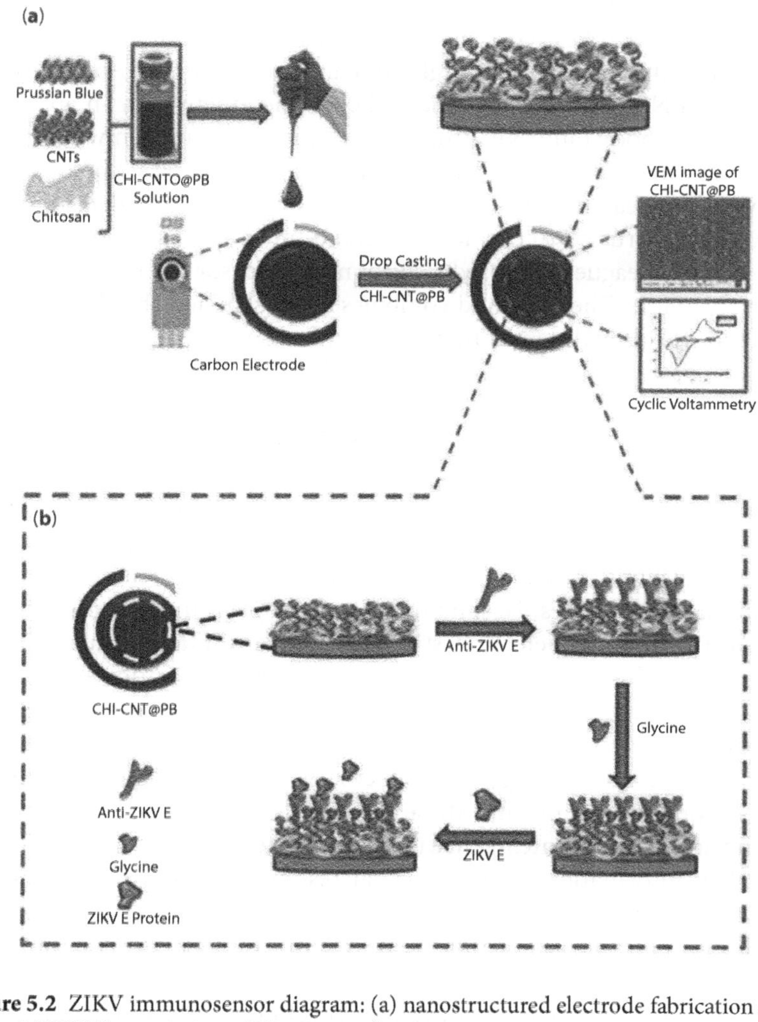

Figure 5.2 ZIKV immunosensor diagram: (a) nanostructured electrode fabrication and (b) anti-ZIKV E protein immobilization and immunoassay.

5.6.6 Alphavirus-Chikungunya

A novel sensor based on SPR has been developed to identify the chikungunya virus in uninfected and infected platelets and plasma blood cells [30]. The SPR-based sensor's greatest sensitivities are for platelets and plasma cells, measuring 393 and 160 deg/RIU, respectively.

5.6.7 Influenza

Researchers Takemaura *et al.* [31] developed a quick and accurate method for diagnosing influenza virus using an LSPR-induced fluorescence immunosensor. They synthesized L-cystamine-capped quaternary CdSeTeS QDs and attached them to anti-hemagglutinin antibody (anti-HA Ab). They also thiolated Au NPs with L-cystamine and combined them with an anti-neuraminidase antibody (anti-NA Ab). The influenza virus antigen was detected by binding with anti-NA Ab-conjugated Au NPs and anti-HA Ab-conjugated QDs, which increased fluorescence intensity. The sensor had a detection limit of 0.03 pg/mL virus concentration and could provide results in just 10 min. Ahmed and colleagues proposed a chiroimmunosensor to identify the influenza H5N1 virus. Their method combines self-assembled chiral star-shaped gold nanohybrids with semiconductor quantum dots (QDs) [32].

5.7 Future Perspectives and Conclusion

Electrochemical sensors are becoming more advanced with the help of new nanomaterials, technology improvements, and microfluidics. This has made it possible to achieve highly accurate detection, identification, and measurement of viruses. Despite the recent progress in smartphone cameras, electrochemical detection still has many advantages, such as being easy to design, durable, requiring low power, simple integration with microfluidic devices, having great analytical sensitivity, and being cost-effective.

It is imperative to note some crucial emerging patterns that could significantly impact virus diagnostic devices. Firstly, it is essential to integrate wireless communication into these devices to enable advanced machine learning and artificial intelligence (AI) techniques to extract the maximum amount of information from analytical responses. This will aid hospital consultants in making prompt decisions and informing epidemiological models. Secondly, biocompatible materials are crucial in developing wearable or implantable devices. It is vital to consider the entire life cycle of these materials, from manufacturing to recycling and reusing components after testing. Lastly, using minimally invasive samples such as saliva, breath, tears, or urine is critical.

Scientists have made numerous efforts to enhance sensitivity in immunosensors. This has been achieved through magnetic activity, meta-surfaces, grating, or photonic crystals. The plasmonic effect that induces SPR phenomena can also cause other useful phenomena like SERS, FRET, and metal-enhanced fluorescence (MEF), which can be utilized to develop immunosensors. Electrochemical biosensor technology also faces challenges in areas like reproducibility of immunosensors with high sensitivity, device portability, and the optimization of sensors with respect to the detection of analytes. Engineers also need to overcome obstacles like slow test reaction times and low sensitivity when designing fully automated multiplex data processing and visualization platforms. Additional technological advances and a user-friendly design that retains accuracy and dependability will be necessary for future immunosensors. The high throughput of immunosensors may present a significant financial barrier to entry for new businesses due to the high costs associated with mass manufacture and market introduction. However, the potential advantages of the POC device technologies may completely alter the healthcare landscape in developing countries.

Acknowledgements

The authors would like to thank the Coordination for the Improvement of Higher Education Personnel (CAPES)-Brazil, the Post Graduate Program in Medical Sciences, Drug Research and Development Centre (NPDM) of the Federal University of Ceará (UFC) in Fortaleza, and the Faculty of CGESP in Goiânia, Brazil for their invaluable assistance.

References

1. Benjamin, S.R., de Souza Nascimento, T., Roque, C.R., de Andrade, G.M., Oriá, R.B., Recent advances in the development of immunosensors for infectious diseases, in: *Biosensors for Emerging and Re-Emerging Infectious Diseases*, pp. 19–72, Elsevier, 2022, https://doi.org/10.1016/B978-0-323-88464-8.00006-3.
2. Sampaio, I., Takeuti, N.N.K., Gusson, B., Machado, T.R., Zucolotto, V., Capacitive immunosensor for COVID-19 diagnosis. *Microelectron. Eng.*, 267–268, 111912, 2023, https://doi.org/10.1016/j.mee.2022.111912.

3. Lee, D., Hwang, J., Seo, Y., Gilad, A.A., Choi, J., Optical Immunosensors for the Efficient Detection of Target Biomolecules. *Biotechnol. Bioprocess Eng.*, 23, 123–133, 2018, https://doi.org/10.1007/s12257-018-0087-x.
4. Chen, Y., Liu, J., Yang, Z., Wilkinson, J.S., Zhou, X., Optical biosensors based on refractometric sensing schemes: A review. *Biosens. Bioelectron.*, 144, 111693, 2019, https://doi.org/10.1016/j.bios.2019.111693.
5. Kerr, L.T., Byrne, H.J., Hennelly, B.M., Optimal choice of sample substrate and laser wavelength for Raman spectroscopic analysis of biological specimen. *Anal. Methods*, 7, 5041–5052, 2015, https://doi.org/10.1039/C5AY00327J.
6. Li, X., Yu, S., Yan, T., Zhang, Y., Du, B., Wu, D., Wei, Q., A sensitive electrochemiluminescence immunosensor based on Ru(bpy)32+ in 3D CuNi oxalate as luminophores and graphene oxide–polyethylenimine as released Ru(bpy)32+ initiator. *Biosens. Bioelectron.*, 89, 1020–1025, 2017, https://doi.org/10.1016/J.BIOS.2016.10.032.
7. Cha, H., Kim, H., Joung, Y., Kang, H., Moon, J., Jang, H., Park, S., Kwon, H.-J., Lee, I.-C., Kim, S., Yong, D., Yoon, S.-W., Park, S.-G., Guk, K., Lim, E.-K., Park, H.G., Choo, J., Jung, J., Kang, T., Surface-enhanced Raman scattering-based immunoassay for severe acute respiratory syndrome coronavirus 2. *Biosens. Bioelectron.*, 202, 114008, 2022, https://doi.org/10.1016/j.bios.2022.114008.
8. Moon, J., Yi, S.Y., Hwang, A., Eom, G., Sim, J., Jeong, J., Lim, E.-K., Chung, B.H., Kim, B., Jung, J., Kang, T., Facile and sensitive detection of influenza viruses using SERS antibody probes. *RSC Adv.*, 6, 84415–84419, 2016, https://doi.org/10.1039/C6RA13966C.
9. Akib, T.B.A., Mou, S.F., Rahman, M.M., Rana, M.M., Islam, M.R., Mehedi, I.M., Mahmud, M.A.P., Kouzani, A.Z., Design and Numerical Analysis of a Graphene-Coated SPR Biosensor for Rapid Detection of the Novel Coronavirus. *Sensors*, 21, 3491, 2021, https://doi.org/10.3390/s21103491.
10. Panda, A. and Pukhrambam, P.D., Design and modelling of reconfigurable surface plasmon resonance refractive index sensor employing graphene and Sb2S3 for detection of dengue virus. *Phys. B Condens. Matter*, 638, 2022, https://doi.org/10.1016/j.physb.2022.413965.
11. Chang, Y.-F., Wang, W.-H., Hong, Y.-W., Yuan, R.-Y., Chen, K.-H., Huang, Y.-W., Lu, P.-L., Chen, Y.-H., Chen, Y.-M.A., Su, L.-C., Wang, S.-F., Simple Strategy for Rapid and Sensitive Detection of Avian Influenza A H7N9 Virus Based on Intensity-Modulated SPR Biosensor and New Generated Antibody. *Anal. Chem.*, 90, 1861–1869, 2018, https://doi.org/10.1021/acs.analchem.7b03934.
12. Shahdeo, D., Roberts, A., Archana, G.J., Shrikrishna, N.S., Mahari, S., Nagamani, K., Gandhi, S., Label free detection of SARS CoV-2 Receptor Binding Domain (RBD) protein by fabrication of gold nanorods deposited on electrochemical immunosensor (GDEI). *Biosens. Bioelectron.*, 212, 114406, 2022, https://doi.org/10.1016/j.bios.2022.114406.

13. Rodríguez, B.A.G., Ferreira, P.A.B., Dutra, R.F., Impedimetric Immunosensors for Clinical Practices: Focus on Point-of-Care Diagnostics, in: *Tools and Trends in Bioanalytical Chemistry*, pp. 283–304, Springer International Publishing, Cham, 2022, https://doi.org/10.1007/978-3-030-82381-8_15.
14. Tuteja, S.K., Chen, R., Kukkar, M., Song, C.K., Mutreja, R., Singh, S., Paul, A.K., Lee, H., Kim, K.-H., Deep, A., Suri, C.R., A label-free electrochemical immunosensor for the detection of cardiac marker using graphene quantum dots (GQDs). *Biosens. Bioelectron.*, 86, 548–556, 2016, https://doi.org/10.1016/j.bios.2016.07.052.
15. Benjamin, S.R., de Lima, F., do Nascimento, V.A., de Andrade, G.M., Oriá, R.B., Advancement in Paper-Based Electrochemical Biosensing and Emerging Diagnostic Methods. *Biosensors*, 13, 689, 2023, https://doi.org/10.3390/bios13070689.
16. Lee, D., Bhardwaj, J., Jang, J., Paper-based electrochemical immunosensor for label-free detection of multiple avian influenza virus antigens using flexible screen-printed carbon nanotube-polydimethylsiloxane electrodes. *Sci. Rep.*, 12, 2311, 2022, https://doi.org/10.1038/s41598-022-06101-1.
17. Roda, A., Cavalera, S., Di Nardo, F., Calabria, D., Rosati, S., Simoni, P., Colitti, B., Baggiani, C., Roda, M., Anfossi, L., Dual lateral flow optical/chemiluminescence immunosensors for the rapid detection of salivary and serum IgA in patients with COVID-19 disease. *Biosens. Bioelectron.*, 172, 112765, 2021, https://doi.org/10.1016/j.bios.2020.112765.
18. Funabashi, R., Miyakawa, K., Yamaoka, Y., Yoshimura, S., Yamane, S., Jeremiah, S.S., Shimizu, K., Ozawa, H., Kawakami, C., Usuku, S., Tanaka, N., Yamazaki, E., Kimura, H., Hasegawa, H., Ryo, A., Development of highly sensitive and rapid antigen detection assay for diagnosis of COVID-19 utilizing optical waveguide immunosensor. *J. Mol. Cell Biol.*, 13, 763–766, 2021, https://doi.org/10.1093/jmcb/mjab037.
19. Eissa, S., Alhadrami, H.A., Al-Mozaini, M., Hassan, A.M., Zourob, M., Voltammetric-based immunosensor for the detection of SARS-CoV-2 nucleocapsid antigen. *Microchim. Acta*, 188, 199, 2021, https://doi.org/10.1007/s00604-021-04867-1.
20. Fortunati, S., Giliberti, C., Giannetto, M., Bolchi, A., Ferrari, D., Donofrio, G., Bianchi, V., Boni, A., De Munari, I., Careri, M., Rapid Quantification of SARS-Cov-2 Spike Protein Enhanced with a Machine Learning Technique Integrated in a Smart and Portable Immunosensor. *Biosensors*, 12, 2022, https://doi.org/10.3390/BIOS12060426.
21. Basso, C.R., Crulhas, B.P., Magro, M., Vianello, F., Pedrosa, V.A., A new immunoassay of hybrid nanomater conjugated to aptamers for the detection of dengue virus. *Talanta*, 197, 2019, https://doi.org/10.1016/j.talanta.2019.01.058.
22. Dutta, R., Thangapandi, K., Mondal, S., Nanda, A., Bose, S., Sanyal, S., Jana, S.K., Ghorai, S., Polyaniline Based Electrochemical Sensor for the Detection

of Dengue Virus Infection. *Avicenna J. Med. Biotechnol.*, 12, 77–84, 2020, https://www.ncbi.nlm.nih.gov/pmc/articles/PMC7229450/.
23. Nawaz, M.H., Hayat, A., Catanante, G., Latif, U., Marty, J.L., Development of a portable and disposable NS1 based electrochemical immunosensor for early diagnosis of dengue virus. *Anal. Chim. Acta*, 1026, 1–7, 2018, https://doi.org/10.1016/j.aca.2018.04.032.
24. Giannetto, M., Costantini, M., Mattarozzi, M., Careri, M., Innovative gold-free carbon nanotube/chitosan-based competitive immunosensor for determination of HIV-related p24 capsid protein in serum. *RSC Adv.*, 7, 39970–39976, 2017, https://doi.org/10.1039/C7RA07245G.
25. Santana, G.M., Silva, A.K.S., Foguel, M.V., Dutra, R.F., An ultrasensitive electrochemical immunosensor for hepatitis C antibodies based on one-step-eletrosynthetized polypyrrole–graphene nanocomposite. *J. Mater. Sci.*, 57, 5586–5595, 2022, https://doi.org/10.1007/s10853-022-06992-5.
26. Cabral, D.G.A.D.G.A., Lima, E.C.S.E.C.S., Moura, P., Dutra, R.F.R.F., A label-free electrochemical immunosensor for hepatitis B based on hyaluronic acid-carbon nanotube hybrid film. *Talanta*, 148, 209–215, 2016, https://doi.org/10.1016/j.talanta.2015.10.083.
27. Alizadeh, N., Hallaj, R., Salimi, A., A highly sensitive electrochemical immunosensor for hepatitis B virus surface antigen detection based on Hemin/G-quadruplex horseradish peroxidase-mimicking DNAzyme-signal amplification. *Biosens. Bioelectron.*, 94, 184–192, 2017, https://doi.org/10.1016/j.bios.2017.02.039.
28. Santos, L.K.B., Mendonça, P.D., Assis, L.K.S., Prudêncio, C.R., Guedes, M.I.F., Marques, E.T.A., Dutra, R.F., A Redox-Probe-Free Immunosensor Based on Electrocatalytic Prussian Blue Nanostructured Film One-Step-Prepared for Zika Virus Diagnosis. *Biosensors*, 12, 623, 2022, https://doi.org/10.3390/bios12080623.
29. Mendonça, P.D., Santos, L.K.B., Foguel, M.V., Rodrigues, M.A.B., Cordeiro, M.T., Gonçalves, L.M., Marques, E.T.A., Dutra, R.F., NS1 glycoprotein detection in serum and urine as an electrochemical screening immunosensor for dengue and Zika virus. *Anal. Bioanal. Chem.*, 413, 2021, https://doi.org/10.1007/s00216-021-03449-7.
30. Yupapin, P., Trabelsi, Y., Vigneswaran, D., Taya, S.A., Daher, M.G., Colak, I., Ultra-High-Sensitive Sensor Based on Surface Plasmon Resonance Structure Having Si and Graphene Layers for the Detection of Chikungunya Virus. *Plasmonics*, 17, 1315–1321, 2022, https://doi.org/10.1007/s11468-022-01631-w.

31. Gong, Q., Wang, Y., Yang, H., A sensitive impedimetric DNA biosensor for the determination of the HIV gene based on graphene-Nafion composite film. *Biosens. Bioelectron.*, 89, 565–569, 2017, https://doi.org/10.1016/j.bios.2016.02.045.
32. Ahmed, S.R., Nagy, E., Neethirajan, S., Self-assembled star-shaped chiroplasmonic gold nanoparticles for an ultrasensitive chiro-immunosensor for viruses. *RSC Adv.*, 7, 2017, https://doi.org/10.1039/c7ra07175b.

6

Electrochemical Nanobiosensors Approaches for Rapid Diagnosis of Infectious Diseases

Tahmina Foyez[1,2]* and Abu Bin Imran[3†]

[1]*Department of Hematology, University of North Carolina at Chapel Hill, Chapel Hill, NC, USA*
[2]*Department of Pharmacy, United International University, Madani Ave, Dhaka, Bangladesh*
[3]*Department of Chemistry, Bangladesh University of Engineering and Technology, Dhaka, Bangladesh*

Abstract

Microbial infections continue to be the primary reason for contributing to high morbidity and mortality rates. Among many human pathogens, a small number of opportunistic pathogens are responsible for most healthcare-associated infectious diseases. These bacteria are especially dangerous because they may develop resistance to antibiotics and antimicrobial treatments. The urgent requirement for quick detection, correct identification, and real-time monitoring of these pathogens is both a scientific issue and a severe practical concern. This is especially important to develop tailored and effective treatment strategies to combat these infections. Among the various approaches, electrochemical techniques show great potential for quick diagnosis of infectious diseases. Electrochemical sensors interact with the target substance to generate an electrical signal. Biosensors equipped with electrochemical transducers offer several benefits, including high sensitivity, simple setup, cheap, and the ability to be miniaturized, enabling small sample volumes. Electrochemical transducers have been successfully employed in various instances to detect viruses. The incorporation of nanomaterials in biosensors has significantly enhanced their analytical performance. Notably, recent advancements in

**Corresponding author*: tahmina@pharmacy.uiu.ac.bd
†*Corresponding author*: abimran@chem.buet.ac.bd

Suvardhan Kanchi, Ayyappa Bathinapatla, Anitha Varghese and Phumlane Selby Mdluli (eds.) Nano-Biosensor Technologies for Diagnosis of Infectious Diseases, (197–230) © 2025 Scrivener Publishing LLC

electrochemical nanobiosensors for COVID-19 detection exemplify this progress. Integrating nanomaterials in electrochemical biosensors has dramatically reduced the limit of detection (LOD) to unprecedented levels, opening up exciting prospects for novel biosensing applications. This chapter highlights primary techniques and methods that employ nanoscale materials to construct electrochemical nanobiosensors. The perspectives and future prospects to address the limitations in designing upcoming sensing schemes will be addressed. The growing demand for high sensitivity and specificity, cost-effectiveness, rapid detection, and screening for continued advancements in electrochemical nanobiosensors for microorganisms and other biological targets will be discussed.

***Keywords*:** Biosensor, diagnosis, nanobiosensors, infection, electrochemical, pathogen

6.1 Introduction

Electrochemical nanobiosensors have emerged as powerful tools for rapidly diagnosing infectious diseases [1]. These sensors combine the advantages of nanotechnology and electrochemistry to detect and quantify specific biomarkers associated with various infectious agents such as bacteria, viruses, and parasites [2, 3]. They offer several key features that make them highly attractive for diagnostic applications, including high sensitivity, specificity, portability, and real-time monitoring capacity. The basic principle behind electrochemical nanobiosensors involves recognizing and detecting target biomolecules through specific binding events. Typically, these biosensors have three main components: a recognition element, a transducer, and a signal amplifier [4]. The recognition element can be antibodies, aptamers, enzymes, or deoxyribonucleic acid (DNA) probes, which selectively interact with the target biomarker. The transducer converts the recognition event into a measurable electrical signal, while the amplifier enhances the signal for better sensitivity. Various nanomaterials [5, 6] such as nanoparticles (*e.g.*, gold (Au), silver (Ag), magnetic), carbon nanotubes (CNTs), graphene, and nanowires, are incorporated into the biosensor design to enhance sensitivity and improve overall performance [7, 8]. These nanomaterials provide a large surface area, excellent electrical conductivity, and unique optical properties, enabling efficient immobilization of the recognition element and facilitating electron transfer reactions. Electrochemical nanobiosensors determine electrical current changes from target biomolecule binding. The target biomarker binding to the recognition element adsorbed on the electrode surface changes conductivity or electron transfer kinetics, producing a current signal. This shift is quantitatively related to biomarker concentration, enabling quick and sensitive detection [9].

Another approach involves using redox-active labels or nanotags, such as electroactive nanoparticles or nanocomposites attached to the target biomolecule [10]. Nanotag oxidation or reduction process measurements show that binding modifies their electrochemical characteristics. This approach enhances signal and detects target biomarkers ultrasensitively. Electrochemical nanobiosensors in microfluidic devices have improved quick disease detection [11, 12]. Microfluidic platforms allow for the precise control and manipulation of small volumes of samples, reducing the required reagent volume, reaction time, and sample handling steps. This integration enables point-of-care testing, where the biosensor can be miniaturized, portable, and easily operated by non-experts, making it suitable for resource-limited settings and remote areas. In terms of infectious disease diagnosis, electrochemical nanobiosensors have demonstrated their potential in detecting various pathogens, including bacteria, viruses (*e.g.*, influenza, human immunodeficiency virus (HIV), Zika), and parasites (*e.g.*, malaria, *Leishmania*). They have been used to rapidly identify specific genetic sequences, antigens, or antibodies associated with these pathogens. Multiplexed biosensors may detect many targets at once, identifying numerous infections in a single sample, making them useful for screening and monitoring [13]. Researchers are working to build simple, cost-effective sensors with the analytical capabilities for pathogen identification, quick detection, and real-time monitoring of early infection phases. They are actively exploring a wide range of approaches and technologies to address these difficulties [14]. The aim is to devise innovative solutions that can improve the sensitivity, specificity, speed, and affordability of pathogen detection systems. By leveraging advancements in nanotechnology, materials science, and bioengineering, researchers are making strides towards developing effective sensor platforms that can revolutionize pathogen identification and monitoring, ultimately enhancing public health and disease management strategies. Continued interdisciplinary research and collaborations are essential to advance this field and develop practical solutions.

This chapter will discuss all of these electrochemical nanobiosensors for rapid diagnosis of infectious diseases. The challenge and prospects will also be addressed in detail.

6.2 Conventional Methods for the Determination of Infectious Pathogens

Diagnosing common infectious diseases involves several steps, including sample collection, transportation, processing, analysis, and result reporting.

Trained professionals often carry out the current procedures for diagnosing bacterial and viral infections and can take several days, causing delays in result notification and subsequent decision-making. These delays are especially problematic in resource-limited settings and have led to the misuse of antimicrobial drugs. Microscopy, culturing, nucleic acid-based methods, and immunological tests are typical diagnostics for such diseases. Immunological and molecular techniques like DNA hybridization and polymerase chain reaction (PCR) may identify numerous global health agents [15]. These techniques provide broad detection capabilities and are valuable for diagnosing various infections. The enzyme-linked immunosorbent assays (ELISAs) are predominantly used for determining virus-specific antigens; however, they have limitations, including time consumption, multistage processes, relatively low sensitivity, and operator-dependent performance.

Nucleic acid-based detection methods, relying on detecting DNA or RNA sequences specific to the target pathogen, exhibit greater specificity and sensitivity than immunological techniques [16]. These methods prove particularly effective in rapidly detecting synthetic short oligonucleotides. However, their application is limited when faced with complex genomic DNA or total RNA due to target length, complexity, and sample media considerations. Although hybridization assays exhibit high analytical sensitivity in detecting low quantities of DNA and RNA, effective hybridization often requires sample pre-treatment to limit the target size. An alternative approach involves combining hybridization assays with amplification methods, which allow for specific target amplification and size restriction. Nucleic acid amplification has become widely adopted for pathogen infection detection due to its superior sensitivity. These detection approaches remain complicated, costly, and technical. Centralized labs, qualified staff, specialized equipment that needs frequent maintenance, and consistent electrical power are usually needed. This limits their availability in certain countries and circumstances.

In this context, combining electrochemical biosensors and affordable, portable, programmable, battery-operated instruments presents a promising alternative for rapid, cost-effective, and user-friendly analytical approaches. Ion-selective and immunosensors leverage the electrochemical properties to provide reliable information on infectious disorders [17]. To detect circulating biomarkers, these biosensors can analyze blood, serum, saliva, urine, sputum, and pleural fluid. Healthcare workers may improve diagnostics and point-of-care testing in many situations using electrochemical biosensors and accessible equipment.

Figure 6.1 illustrates the current pathogen identification, detection, and quantification methods. Cell culture methods rely on cultivating pathogens in a laboratory [18]. In cell culture, samples containing pathogens are placed in specific growth media under controlled conditions, allowing the pathogens to multiply. After sufficient growth, the pathogens can be identified and quantified through various assays and measurements. Cell culture approaches are useful for studying the behavior of pathogens and testing their susceptibility to antimicrobial agents. But they are slow and need specific equipment and skills.

Molecular-based methods rely on the analysis of nucleic acids for pathogen identification and quantification [19]. Nucleic acid sequencing involves determining the precise sequence of DNA or RNA molecules, which can be used to identify specific pathogens. Next-generation sequencing technologies have revolutionized this field by enabling high-throughput sequencing of multiple samples. PCR-based methods offer high sensitivity and specificity and are commonly used for diagnosing viral and bacterial infections. The image in Figure 6.1 shows the use of microarrays, which are small chips containing immobilized DNA fragments, for diagnosing infections by amplifying genomic fragments from *Pseudomonas* spp.

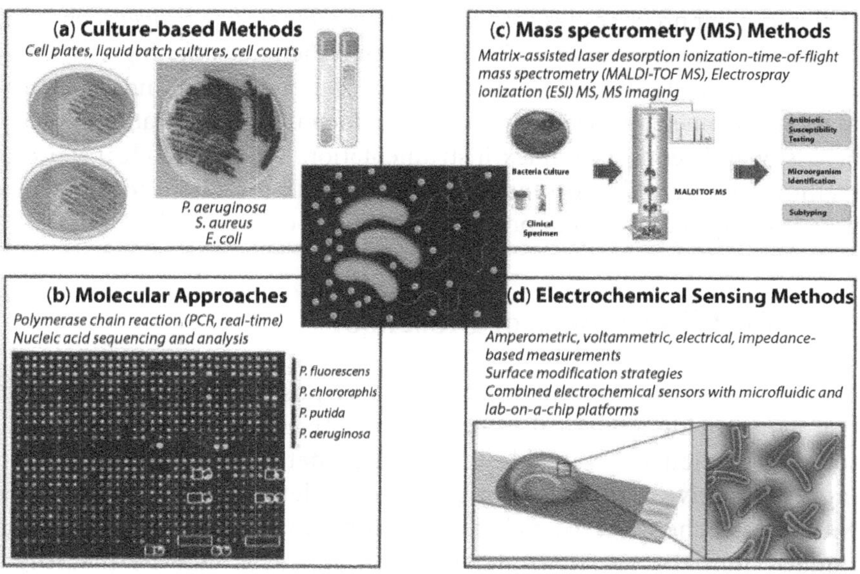

Figure 6.1 Pathogen identification, detection, and quantification methods: (a) cell culture, (b) molecular techniques, (c) mass spectrometry, (d) electrochemical sensing. Reproduced (adapted) from Reference [15]. Copyright 2019, RSC.

Mass spectrometry-based approaches have gained popularity for pathogen identification. Mass spectrometry (MS) involves ionizing molecules from a sample and analyzing the resulting mass-to-charge ratio of ions [20]. This analysis provides the molecular composition of sample. Different MS approaches can be used for pathogen identification. These techniques generate unique mass spectra that can be compared to reference databases to identify specific pathogens. The image in Figure 6.1 demonstrates the use of matrix-assisted laser desorption ionization (MALDI) protein profiling to identify bacteria from an unknown colony.

Electrochemical sensing methods offer an alternative approach to pathogen detection and quantification [21]. These sensors detect pathogens based on their electrochemical properties or the electrochemical reactions associated with their presence. Electrochemical sensors often employ specific recognition elements, such as antibodies or DNA probes, that interact with target pathogens and generate an electrical signal. This signal is used to identify and measure pathogen presence and concentration. Electrochemical sensors have the advantage of being rapid, sensitive, and relatively inexpensive. The image in Figure 6.1 presents a schematic of a sensing chip electrode used to detect the electrochemically active cellular metabolite pyocyanin from Pseudomonas aeruginosa.

Each method shown in Figure 6.1 has its own advantages and limitations. Cell culture approaches provide valuable insights into pathogen behavior but can be time-consuming. Molecular-based methods offer high sensitivity and specificity but may require specialized equipment and expertise. Mass spectrometry-based approaches provide rapid identification but can be costly and require complex instrumentation. Electrochemical sensing methods offer rapid and sensitive detection and are often portable and user-friendly, making them promising for point-of-care applications.

6.3 Building Blocks of Biosensor

Chemical sensors have the fundamental purpose of generating a signal that can provide analytical information [22]. Electrochemical biosensors are particularly valuable in pathogen detection strategies, offering a distinctive approach to target-specific detection [23]. Biosensors with electrochemical transducers capture analytes. Electrochemical signals from the targeted analyte and transducer may be measured as current, potential, resistance, or impedance (Figure 6.2) [14, 24]. Electrochemical biosensor systems employ signal processes via various cyclic voltammetry. Electrochemical biosensors also use nanomaterials,

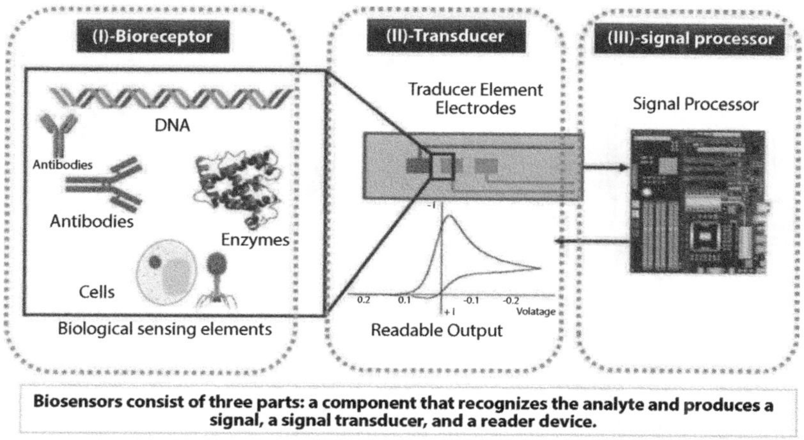

Figure 6.2 Diagram of a biosensor: (I) bioreceptor, (II) transducer, (III) signal processor. Reproduced (adapted) from Reference [23]. Copyright 2022, MDPI.

nanoparticles, and nanocomposites to improve sensitivity and detection limits [25, 26].

6.3.1 Transduction Elements

Electrochemical biosensors transduce via the working electrode. Traditional potentiostats have three electrodes. However, conductometry and impedance measurements frequently use two electrodes (working and auxiliary). Electrodes are made using various materials and techniques (Figure 6.3) [27]. Passing electrons and holes across an electrode transfers charge. Conductive and semiconducting Au and carbon are used for fabricating electrodes. Various manufacturing procedures can produce electrodes of different sizes. Fortunati *et al.* developed an electrochemical immunosensor to measure SARS-CoV-2 spike protein. This sensor used Internet of Things (IoT)-WiFi technologies and machine learning to obtain smart and portable sensor (Figure 6.4) [23]. SARS-CoV-2 S1 subunit monoclonal antibodies were adsorbed on Au nanoparticle-functionalized screen-printed electrodes (SPEs). The working electrode was 4 mm in diameter and the dimensions of the SPEs were 3.4 × 1.0 × 0.05 cm [27]. Electrodes are classified by material, production technique, and design. They can be planar, wire, nanostructured, or array-based. Electrode form, characteristics, material, production method, and design affect biological agent detection, sensitivity, and dynamic range. These factors also affect the overall cost of the biosensor [28].

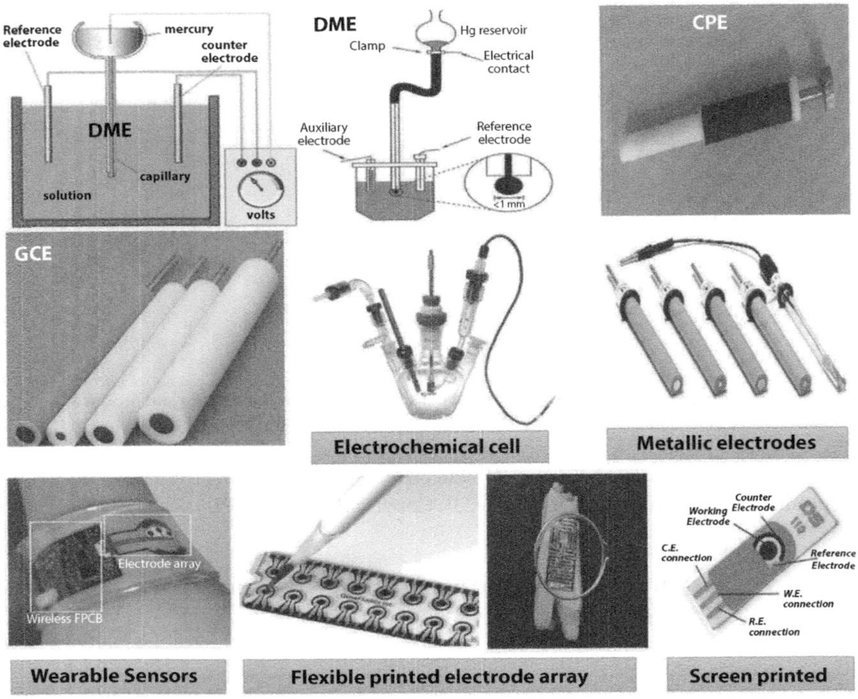

Figure 6.3 Common types and materials of working electrodes used in electrochemical sensors and biosensors. Reproduced (adapted) from Reference [23]. Copyright 2022, MDPI.

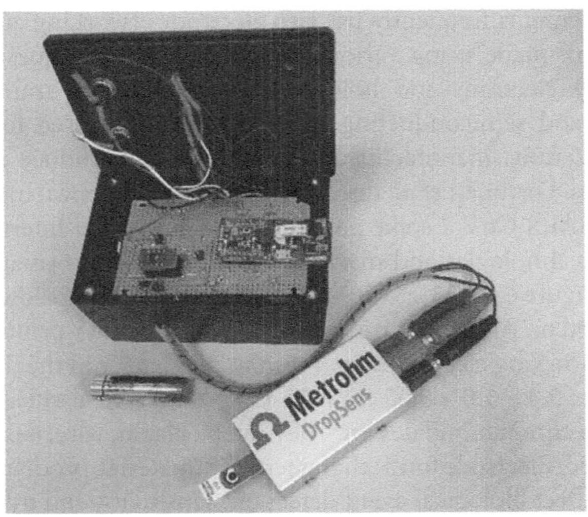

Figure 6.4 An intelligent potentiostat for fast quantification of SARS-CoV-2 spike protein. Reproduced (adapted) from Reference [27]. Copyright 2022, MDPI.

6.3.2 Biorecognition Elements

Biocatalytic and biocomplex biorecognition elements are essential to electrochemical biosensors. Biorecognition components establish biosensor responses via macromolecule-catalyzed actions. Enzymes are widely used as biocatalytic biorecognition components and offer numerous advantages for various chemical sensing applications, including pathogen detection. Analytes binding to macromolecules or structured molecular assemblies induce biocomplexity biorecognition. Biocomplexity biorecognition components like antibodies, peptides, and phages help the body detect pathogens. Modified molecules including DNAzymes, peptide nucleic acids, aptamers, and specifically designed peptides have recently attracted attention. These modified molecules exhibit enhanced robustness and lower production costs than their natural counterparts. Dual diagnostic and remedial techniques combine pathogen identification with broad-spectrum antimicrobials. Other effective pathogen identification methods include chemical ligands, molecularly imprinted polymers (MIP), and clustered regularly interspaced short palindromic repeats (CRISPR)-associated nucleases [29].

6.3.3 Biosensing and Surface Immobility

Biosensor performance depends on electrode surface modification, which affects their functioning mechanism. Appropriately mounting biosensing components on electrodes produces an impermeable, irreversible connection between the target species, biorecognition element, and electrode [30]. Electrochemical pathogen detection uses well-established biorecognition layer preparation procedures. The easiest and most common method is polymer-based enzyme immobilization on an electrode. However, there are two inherent drawbacks to this method. Firstly, the pH of the polymer layer can effect structural changes that affect enzyme function. This may affect biosensor performance. Secondly, polymer layer thickness cannot be accurately controlled, which may reduce biosensor reaction time and sensitivity. Some researchers have immobilized enzymes in neutral pH polymers like silicate sol-gel to circumvent these hurdles and retain enzyme activity [31]. Surface modification methods often include CNTs and metal nanoparticles to improve biosensor performance. Various modification approaches have a variety of impacts on biosensor performance, making it impossible to compare them [32]. Therefore, there is a need for a standardized procedure to evaluate the efficacy of biosensors fabricated using various methods systematically. These biosensors offer versatile detection systems capable of real-time quantification, facilitating a wide range of applications. However, certain challenges remain concerning analyte detection. One important consideration in pathogen detection is the ability of the developed biosensor system to enable multiplexed detection. Pathogen identification is a major healthcare problem owing to their abundance and variety. The non-specific absorption and electrode surface state affect biosensor consistency, making reproducibility and regeneration challenging.

6.4 Electrochemical-Based Biosensors

Electrochemical methods are used extensively in electroanalytical chemistry to detect and measure analytes. Impedance-based measurements and impact electrochemistry have made tremendous progress in biosensing [33]. Electrochemical sensing and biosensing devices have many advantages for detecting several analytes in complicated samples such as blood and clinical specimens. These systems exhibit exceptional sensitivity and selectivity. In this sensing platform, the electrochemical transducer's working electrode surface serves as the interface for bio/electrochemical events. The interaction between the analyte and the immobilized bioreceptor, which has selective binding affinity and catalytic activity, may be assessed by measuring the electrochemical signals produced. Various electrochemical methods, including potentiometry, amperometry, conductometry, impedance spectroscopy, and voltammetry, are commonly employed for the construction, calibration, and optimization of electrochemical sensors and biosensors. The following subsections will provide an overview of these electrochemical techniques [34].

6.4.1 Potentiometric Biosensors

Potentiometric biosensors use a biological sensing element and an electrochemical transducer, such as the working electrode, to produce a potential difference output signal [33]. Ion-selective electrodes (ISEs), which are made to respond to particular ions, or ion-sensitive field-effect transistors (FETs) can be used to convert biological activity into an electric potential signal. Versatile potentiometric sensing protocols can be designed to detect a variety of target analytes by employing a selective bioreceptor. Recently potentiometric nanobiosensors has been developed that use a three-dimensional MIP for quick COVID-19 diagnosis [35]. By differentiating the target coronavirus from other viruses including the H1N1 and

H3N2 influenza viruses and Middle East respiratory syndrome (MERS), COVID-19-potentiometric biosensors show good selectivity [36]. In addition, these virus biosensors show a very low limit of detection (LOD) for suspensions of pure viral particles and for spike proteins.

6.4.2 Amperometric Biosensors

A two-electrode or a three-electrode setup can be used to apply amperometric techniques. With this technique, the reference electrode—such as Ag/AgCl or calomel—and the sensor both run at a constant voltage. The redox reactions of the system produce the faradaic current at the working electrode surface. The amount of analyte in the supporting electrolyte determines how much redox current flows through it [37]. Nanomaterials, including graphite, graphene, CNTs, noble metals like Au, Pt, Cu, or metal oxides like ITO can be used to make the working electrode. Amperometric sensing has a large dynamic range, high accuracy, fast response, and excellent sensitivity. These features make them suitable for mass manufacturing and commercialization. However, amperometric methods have low selectivity, which can cause interference and cross-reactivity with other electroactive compounds [38]. Yaping Dong *et al.* developed an amperometric biosensor for creatinine clinical assessment in renal function evaluation. They found the sensor promising for on-site and at-home diagnostics [39].

6.4.3 Conductometric Biosensors

Conductometric biosensors can detect changes in a solution's electrical conductivity, allowing us to determine ionic composition changes from chemical and biological processes [40]. These biosensors have various advantages. (I) They eliminate the need for a reference electrode, simplifying sensor design; (II) they can be easily miniaturized using thin-film electrodes made of low-cost materials, allowing for large-scale production; and (III) conductometric transducers are light-insensitive and consume little power. Conductometric biosensors are made by coating electrodes with tiny layers of Au, chromium (Cr), Cu, and nickel (Ni). These modified electrodes operate as platforms for immobilizing biocatalytic enzymes, such as glucose oxidase and urease [41].

6.4.4 Impedimetric Biosensors

Impedimetric biosensors detect electrical impedance at the solid electrode surface contact while applying a modest alternating current potential, making them useful electrochemical biosensors [42]. The resistance changes are recorded over frequencies and shown in a Nyquist plot [43]. This technique enables the sensitive investigation of interfacial properties, particularly on functionalized surfaces, and facilitates monitoring diverse bio-recognition events [44]. Antibodies or aptamers may be used for developing impedimetric-immunosensors or aptasensors. The former forms an immune-complex on the sensor surface via particular antigen-antibody interactions, while the latter immobilizes aptamers, short single-stranded oligonucleotides, increasing electron transfer resistance. This allows label-free biomolecular detection [45]. Electrochemical impedance spectroscopy (EIS) devices are prevalent in biological and environmental studies due to their lab-on-a-chip compatibility, ease of manipulation, and on-site determination [46, 47]. Furthermore, EIS serves as a non-destructive technique applicable beyond biosensor development, providing valuable electrochemical information for characterizing various materials and processes such as fuel cell performance, corrosion analysis, battery charging/discharging, and other electrochemical phenomena [48]. Pathogen detection uses EIS to evaluate electrode/electrolyte interface impedance at a specific frequencies [49]. The detection probe changes impedance when

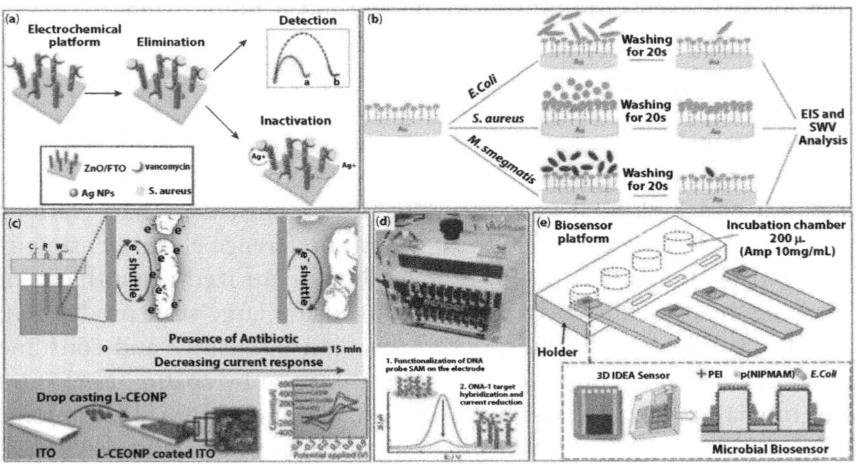

Figure 6.5 Illustrations of innovative biosensing platforms: (a) electrochemical sensor detecting, eliminating, and inactivating *S. aureus*. (b) Interaction of screen printed Au electrodes-van with various bacteria. (c) Step-by-step construction of L-CeONP/ITO working electrode for *E. coli* analysis. (d) Mini PCR machine for DNA amplification and measurement. (e) Microbial biosensor platform monitoring bacterial response to ampicillin. Reproduced (adapted) from Reference [51]. Copyright 2022, MDPI.

target bacteria bind, allowing measurement. The bio-recognition element might be antibodies, aptamers, RNA, or polymers [50]. Recently developed EIS biosensors use nanotubes and nanoparticles. For *Staphylococcus aureus* detection, a multifunctional sensor containing silver nanoparticles (AgNPs) coated zinc oxide nanorod arrays and vancomycin was designed (Figure 6.5) [51]. The synergistic germicidal effect of AgNPs and vancomycin may have made this sensor effective at eliminating bacteria [52].

6.4.5 Voltammetric Biosensors

Voltammetric biosensors operate by analyzing the concentration of a target through the measurement of faradaic current, which varies with the electric potential [53]. These sensors offer the advantage of simultaneous analysis of multiple analytes. Voltammetry detects electroactive analytes by detecting working electrode oxidation or reduction. Peak locations can be identified using potential values. Peak current strength, known as Faradic current, indicates target species concentration. Cyclic voltammetry (CV) provides quantitative and qualitative data and is commonly used. It details electrochemical processes such electron transfer kinetics, redox reaction thermodynamics, diffusion, and adsorption. CV is commonly used due to its versatility and ability to provide comprehensive information about the analyzed system. Voltammetric biosensors have been used for the determination of heavy metals [54, 55]. Blood glucose and urea levels can be monitored with disposable sensor chips. Voltammetry also able to evaluate plasma and dose forms of medicinal drugs. Using crown-ether macromolecules, voltammetric biosensors can detect biomarkers in microbial cultures.

6.5 Impact of Nanomaterials on Biosensor Performance

The utilization of nanosized materials in electrochemical biosensors has significantly improved their performance [56, 57]. An ideal nanomaterial should possess specific desirable properties, including easy preparation, availability, a plentiful supply of functional groups, and chemical stability, which make it suitable for constructing biosensors [58]. By incorporating nanomaterials onto the electrode surface, several advantages are achieved. These include facilitating rapid electron transfer kinetics, augmenting the electroactive surface area, and increasing the capacity to load biomolecules. Moreover, the large surface area provided by nanomaterials enhances the binding affinity of immobilized biomolecules, particularly through multivalent immobilization techniques [59]. In the past decade, remarkable advancements have been made in nanomaterials and sensing approaches, leading to improved specificity and sensitivity of biosensors. Noteworthy advancements include the use of graphene and CNTs, immunomagnetic nanoparticles or beads, SPEs, paper-based and inkjet-printed platforms, and interdigitated array microelectrodes [60–62]. These developments improve the performance and capabilities of biosensors. These innovations enhanced biosensor performance and have expanded the possibilities for precise and accurate sensing applications [63].

6.6 Noble Metal Nanomaterials

Noble metal nanoparticles, such as Au, Ag, Pt, Pd, and their alloys and core-shell structures, have significantly advanced electrochemical sensor and biosensor platforms in biomedical analyses. These nanoparticles possess exceptional physical, chemical, and electrochemical properties that depend on their size and shape [65]. Their use in electrochemical sensors holds great promise for enhancing sensitivity and selectivity through tailored signal amplification strategies (Figure 6.6) [64]. In biomedical applications, nano-engineering these materials with low toxicities and environmental impact helps selectively target cells and tissues and properly identify affected areas. Noble metal-based nanomaterials have found extensive application in clinical, pharmaceutical, and medical diagnostics, as well as in cancer therapeutics [66]. Synthesis of metal, bio-functionalized, and nanocomposites has been significant for sensor and biosensor development. This section describes how noble metal nanoparticles and nanocomposites offer extremely selective and sensitive biological and biomedical sensing platforms.

6.6.1 Gold Nanoparticles (AuNPs)

AuNPs are popular in electrochemical sensor for biomedical applications [67]. They can recover efficiently from biological redox processes, have adjustable physicochemical features, high surface area, and stability [68]. AuNPs and related nanocomposites have many benefits, including easy preparation, chemical stability, biocompatibility, wide electrochemical potential range, and catalytic activity. These characteristics make them promising for miniaturized sensing platforms in the future. In nanomedicine, AuNPs find extensive use due to their compatibility with a wide range of drugs and biomarkers [69]. AuNP-based sensor are able to precisely

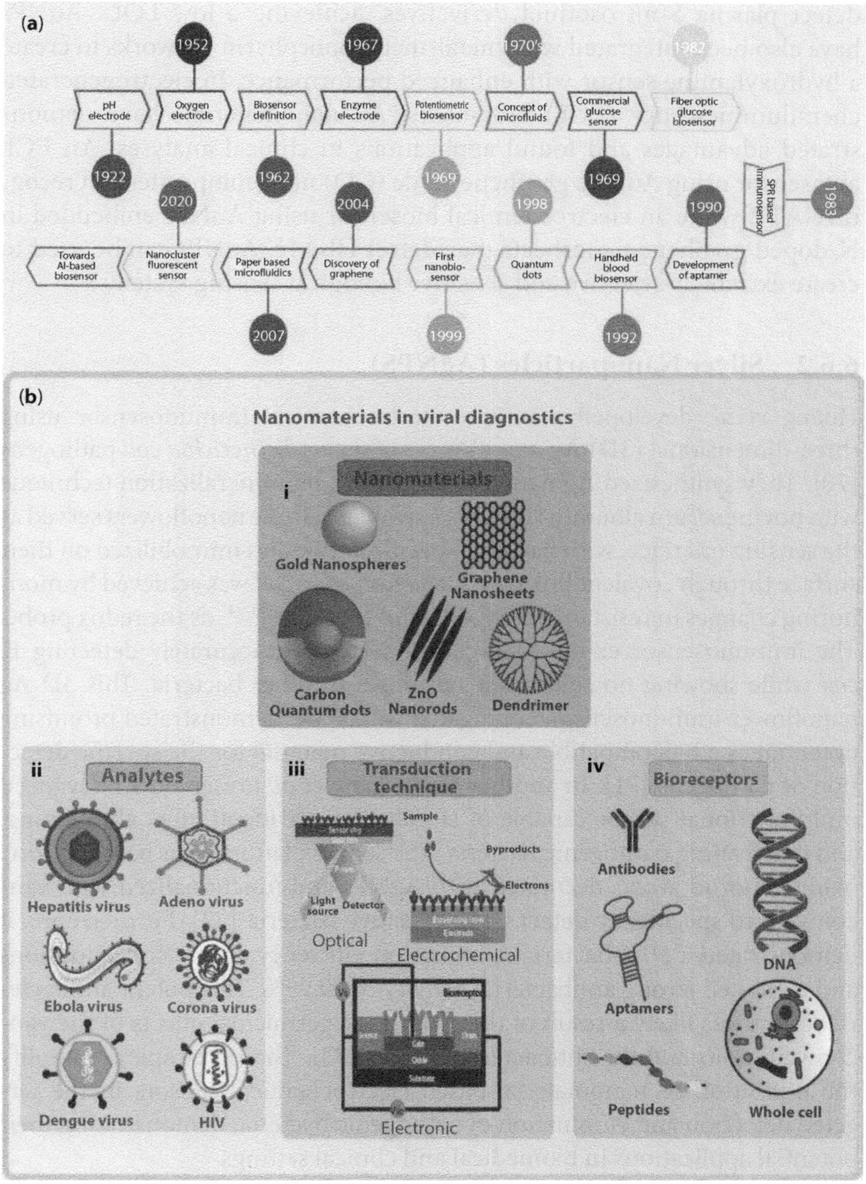

Figure 6.6 (a) Timeline for nanobiosensor development. (b) Schematic overview of nanomaterials, analytes, transduction techniques, and bioreceptors used in viral diagnostics. Reproduced (adapted) from Reference [64]. Copyright 2023, Springer Nature.

detect plasma S-nitrosothiol derivatives, achieving a low LOD. AuNPs have also been integrated with metal-metalloporphyrin networks to create a hydroxylamine sensor with enhanced performance. In electrogenerated chemiluminescence (ECL) biosensors, Au nanomaterials have demonstrated advantages and found applications in clinical analyses. An ECL aptasensor using AuNPs/graphene oxide (GO) nanocomposites can recognize ATP, while an electrochemical biosensor using AuNPs embedded in N-doped graphene nanosheets may detect HbA1c. AuNPs can be used to create extremely selective and sensitive biological sensing systems.

6.6.2 Silver Nanoparticles (AgNPS)

Huang *et al.* developed a unique electrochemical immunosensor using three-dimensional (3D) Ag nanoflowers to detect *Escherichia coli* pathogens [70]. They synthesized Ag nanoflowers using a biomineralization technique with bovine serum albumin (BSA) as a template. These nanoflowers served as the sensing interface, with bacteria-specific antibodies immobilized on their surface through covalent linkages. Detection of *E. coli* was achieved by monitoring changes in resistance using EIS and $[Fe(CN)_6]^{3-/4-}$ as the redox probe. The immunosensor exhibited excellent selectivity, accurately detecting *E. coli* while showing no significant responses to other bacteria. This 3D Ag nanoflower immunosensor, conjugated with BSA, demonstrated promising potential as a biocompatible and conductive material for the specific detection of pathogens [71]. In another study, Yang *et al.* presented an advanced multifunctional sensor capable of simultaneously identifying, eliminating, and inactivating pathogenic bacteria. The sensor platform was based on zinc oxide nanorod arrays decorated with AgNPs and functionalized with vancomycin to specifically detect Gram-positive bacteria [72]. Furthermore, it demonstrated a 50% bacterial elimination efficiency at low concentrations and exhibited strong antibacterial activity of 99.99%. This potent antibacterial effect was likely a result of the synergistic germicidal effects of the vancomycin units and the antibacterial AgNPs [73]. These examples exemplify the utilization of nanomaterial-based electrochemical sensors in the targeted detection and elimination of pathogenic bacteria, demonstrating their potential applications in biomedical and clinical settings.

6.6.3 Palladium Nanoparticles (PdNPs)

The exceptional catalytic and sensing activities of palladium nanoparticles (PdNPs) have made them highly sought after in the biomedical field. The ability to precisely control their size and shape is essential. Pd is cheaper than Au and Pt for electrochemical sensing and biosensing devices. For biomarker detection, Pd nanoparticles' unique electrical characteristics, increased catalytic activity, and selective sensing have been intensively studied. Rahi *et al.* developed a genosensor for *Brucella* detection by depositing PdNPs on an Au surface in an electrochemical setup. The immobilization and hybridization of a *Brucella*-specific probe were achieved using a voltammetric technique. The linear calibration range and a low LOD of the sensor demonstrates its potential for sensitive *Brucella* detection. Bimetallic Au-Pd NPs have also been widely used in catalytic and sensing studies. Au-Pd NPs were used to build electrochemical sensors for levodopa (LD) and uric acid (UA) detection in samples [74, 75]. Detecting LD and UA simultaneously was more sensitive and selective using the sensors. To enhance sensitivity and selectivity further, nanocomposites containing metal nanoparticles, graphene, and ionic liquids (IL) have been employed for sensing various analytes such as dopamine, nifedipine, hydroquinone, nitric oxide (NO), and DNA. Palladium nanoparticles on a graphene sheet and IL developed a non-enzymatic glucose sensor. Different ILs were optimized for detecting glucose and ascorbic acid, demonstrating the versatility of these nanocomposites in sensing applications.

6.6.4 Platinum Nanoparticles (PtNPs)

Electrochemical biosensors for biological applications have focused on Pt-based nanoparticles due to their unique electrical and electrocatalytic characteristics. In electron-transfer processes, platinum nanoparticles (PtNP) composition, surface environment, crystalline structure, and orientation influence their behavior. Electrodes made of Pt-derived nanocomposites have enabled the development of accurate bioanalytical techniques for detecting biomarkers and early-stage diseases. To measure glucose in human perspiration, a flexible electrochemical sensor using Pt-decorated graphite and glucose oxidase (GOx) has been reported [76]. The linear detection range and low LOD of this sensor made it suitable for non-invasive glucose measurement in human perspiration samples. Parrilla *et al.* developed a paper-based potentiometric sensor for biological fluid glucose sensing. Functional PtNPs offer a wide linear range and low LOD and improve biomolecule detection sensitivity and selectivity. They electrochemically detected nitric oxide (NO) using a nanocomposite of reduced graphene oxide (rGO), cobalt oxide (Co3O4) nanocubes, and PtNPs. The metal oxide nanocubes and PtNPs in rGO sheets together produced excellent catalytic activity. A nanocomposite of Pt-tungsten NPs, rGO, and IL was used to develop a highly sensitive electrochemical sensor for NO detection [77]. This sensor demonstrated a low LOD, high sensitivity, and excellent specificity against electroactive interferences. It successfully detected NO in human serum and urine samples.

6.7 Metal Oxide Nanomaterials

Metal oxide nanoparticles are used in electrochemistry, soft magnetism, catalysis, and sensing due to their unique features. Their reduced dimensions, large surface area to volume ratio, specialized facet exposure, and equivalent Debye length improve sensing sensitivity and selectivity. In particular, metal oxide nanomaterials have garnered significant attention in electrocatalysis and sensing. The strong electrocatalytic activity, affordability, and capability to capture organic compounds make metal oxide nanomaterials highly desirable in these applications. CeO_2, CuO, NiO, Fe_2O_3, Co_3O_4, MnO_2, ZnO, TiO_2, SnO_2, CdO, and others are used in electroanalysis. As electrocatalysts for sensing various analytes, these metal oxide nanoparticles promote biological and biomedical applications.

6.7.1 Cerium Oxide Nanomaterials

The enzyme or protein immobilization and catalytic activity of CeO_2 nanoparticles make them popular in biosensors. Using a nanocomposite of single-walled carbon nanohorns and CeO2 catalysts (CeO_2/SWCNH) produced an electrochemical sensor for H_2O_2 [78]. The CeO_2/SWCNH electrode performed well as an H_2O_2 sensor with a LOD of 0.1 mM. The sensor displays good stability over 2 weeks and outstanding H_2O_2 selectivity in complicated matrices and cleaning liquid samples. Sensitivity is improved by immobilizing primary antibodies (Ab1) in sandwich-type immunosensors. Synergistic sensing with CeO_2 nanocomposites and other semiconductor oxides has been reported. A CeO_2 nanocomposite (Co_3O_4@CeO_2-Au@Pt) was used to create an electrochemical SCCA immunosensor. This immunosensor detected SCCA with good sensitivity and specificity. These examples demonstrate the promise of CeO_2 nanoparticles in high-performance biosensors. Incorporating CeO_2 with other nanomaterials or catalysts enhances the sensing capabilities and provides opportunities for advanced sensing applications in biomedical research and diagnostics. CeO_2 nanomaterials offer promising prospects for designing and fabricating nanobiosensors with improved performance, contributing to advancing biomedical sciences and diagnostics.

6.7.2 Copper Oxide Nanomaterials

CuO nanomaterials are multifunctional sensing and biosensing materials. These include hierarchical nanostructures, high surface area, various valence states, and variable electron-transport performance. Yang *et al.* demonstrated a non-enzymatic glucose sensor using nanoneedle-like CuO on N-doped rGO (CuO/N-rGO) [78]. The CuO/N-rGO sensor responded quickly to glucose, with a linear range and a low LOD. Cu_2O@CeO_2-Au nanocomposites for prostate-specific antigen (PSA) sensing were invented by Li *et al.* Compared to pure Cu_2O, Au nanoparticles, and Cu_2O@CeO_2, the core-shell structure decorated with AuNPs (Cu_2O@CeO_2/Au NPs) reduced H_2O_2 electrocatalytically. The nanocomposites-based immunosensor detected PSA with a linear range of 0.1–100 ng/mL and a low LOD of 0.03 (S/N=3). Cu_2O@CeO_2-Au nanocomposites synergistically increased electrocatalytic activity, enabling PSA detection. The unique features of CuO nanoparticles and their combination with other materials boost their sensing capabilities, enabling numerous biosensor applications.

6.7.3 Magnetic Nanomaterials

Magnetic nanoparticles as NiO, Fe_2O_3, and Co_3O_4 are promising electrochemical biosensors [79]. The large accessible surface area and efficient electron-transfer behavior make them ideal for sensing. Numerous sectors have researched hierarchical porous metal oxide frameworks with controlled surface topologies and dimensions. A enzyme-free glucose sensor made on hierarchical porous Co_3O_4/graphene. The microsphere structure of Co_3O_4/graphene, characterized by scanning electron microscope (SEM), transmission electron microscopy (TEM), and energy-dispersive X-ray spectroscopy (EDS) analysis, exhibited aggregated and interconnected graphene sheets with dispersed Co_3O_4. Binary metal oxides are an effective way to improve individual metal oxide performance. Noble metal nanoparticles placed on semiconducting oxide substrates exhibit catalytic activity owing to interface polarization. As support materials for noble metal nanoparticles, metal oxide nanomaterials have higher catalytic activity than single-component nanomaterials. This increased activity is due to metal-oxide interface interplay. They enable electron transfer and enzyme immobilization in a biocompatible environment, improving electrode sensitivity. Fe_2O_3 nanoparticles and poly(3,4-ethylenedioxythiophene) (PEDOT)-rGO nanocomposite were used as acetylcholine detecting electrodes with a linear range and a low LOD. Genetically modified M13@MnO_2/GOx nanowires were used to prepare a glucose biosensor. These biosensing examples demonstrate sensitive and accurate analyte detection in numerous applications of metal oxide nanoparticles [80].

6.8 Carbon Nanomaterials

Carbon-based nanomaterials like single walled carbon nanotubes (SWNTs), multi walled carbon nanotubes (MWNTs), single-walled carbon nanohorns (SWCNHs), buckypaper, graphene, and fullerenes (like C60) have many advantages due to their high surface-to-volume ratio, electrical conductivity, chemical resilience, biocompatibility, and mechanical strength [81]. These properties have developed diverse carbon-based sensing electrodes capable of detecting various biological and biomedical analytes. The morphology is essential in enhancing their sensitivity, selectivity, and stability. Carbon nanomaterials have excellent electrocatalytic and sensing capabilities due to reactive surface functional groups, edge-plane-like domains, and impurities. Continuous advancements have been made in the fabrication techniques for integrating carbon nanomaterials into sensors, enabling the development of novel sensor platforms.

6.8.1 CNTs

Introducing nanomaterials, particularly CNTs, into pathogen biosensors has been pursued to enhance sensitivity and selectivity. CNTs offer advantages such as increased working electrode surfaces and improved electron transfer rates when combined with electrochemical devices. In a recent study by Yamada *et al.*, SWCNTs were incorporated into a multijunction sensor to detect bacteria simultaneously [82]. They developed a 2x2 junction array on disposable chip devices by introducing four SWCNT-coated wires. Each junction included bacteria-specific antibodies. Current changes were measured when pathogens from microbiological samples bonded to the sensor. Arvand *et al.* improved an *S. aureus* biosensor using MWCNTs [83]. CNTs may improve biosensor sensitivity and selectivity, making them appealing. However, useful sensing equipment must have easy manufacturing techniques. Pathogen detection using array systems has been investigated to identify numerous pathogens in physiological samples. Yamada *et al.* showed that a SWCNT-based multi-junction array device can detect bacteria simultaneously [82]. The sensor achieved a low LOD and offered a relatively short analysis time, highlighting the potential of such approaches in pathogen detection.

6.8.2 Graphene

Graphene has attracted significant interest in developing electrochemical immunosensors for pathogen detection. Minimizing nonspecific interactions and enhancing non-fouling properties are crucial to ensure effective graphene-based sensor design. A recent study focused on integrating rGO with a polyethyleneimine (PEI) thin film [84]. Adding bacterial ligands and antifouling units to the sensor resulted in it being very selective and sensitive to uropathogenic *E. coli* strains. A high surface area and good electrochemical characteristics were achieved by adding rGO, whereas amino groups permitted surface modification using *E. coli*-specific antibodies targeting fimbrial antigens. To reduce nonspecific adsorption, pyrene-PEG moieties were added to the sensor surface. In biological urine samples, the sensor can selectively recognize *E. coli* bacteria. CNTs and graphene are preferred biosensor nanomaterials owing to their electrical characteristics. *Pseudomonas aeruginosa* pyoverdine may be electrochemically detected using graphene/graphite-modified SPEs [85]. Figure 6.7 shows a sensor for electrochemical detection of *P. aeruginosa* pyoverdine. The sensor uses SPEs modified with graphene/graphite. Graphene is reduced electrochemically, then AuNPs are deposited. Pyoverdine may

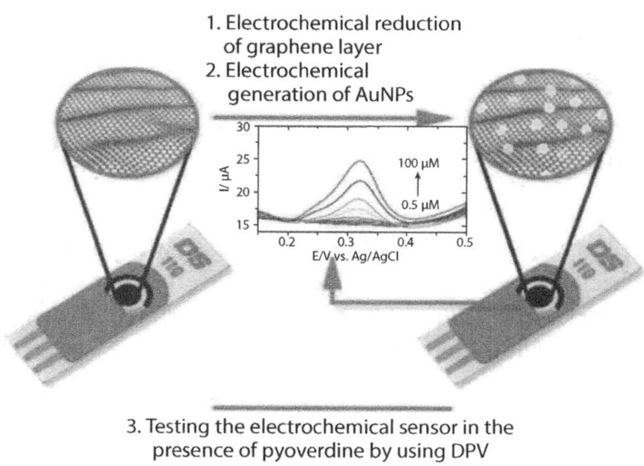

Figure 6.7 Illustration of the design of screen-printed electrodes (SPEs) modified with graphene/graphite for the electrochemical detection of pyoverdine from *Pseudomonas aeruginosa*. Reproduced (adapted) from Reference [85]. Copyright 2019, MDPI.

be detected using differential pulse voltammetry with modified electrodes. Their unique characteristics make them promising candidates for future biosensing applications, as they can enhance the sensitivity, selectivity, and performance of immunosensors for pathogen detection.

6.9 Polymer Nanomaterials

Polymer nanomaterials have become widely used in electrochemical sensors and biosensor platforms for biomolecule detection. These platforms rely on the interaction between biotic components (biomolecules) and abiotic components (transducers). The polymer-based sensors can be upgraded through factors such as bio-functionalization, durability, electrochemical signal transduction, and specificity. Bio-functionalization involves the attachment of biomolecules to the sensor surface, which enhances their durability. Efficient electrochemical signal transduction is achieved by ensuring fast kinetics of carriers and analytes, enabling efficient detection. High specificity is crucial for accurate detection and relies on the sensor's ability to recognize specific bio-analytes. DNA, enzymes, proteins, antigens, and metabolites may be detected using dendrimers, conducting polymers (CPs), and MIPs. In sensing applications, these materials provide great sensitivity, linearity, minimal hysteresis, and selectivity. Recently, polymer nanoparticles have been functionalized, integrated, and used in biomedical electrochemical sensors and biosensors. These developments contribute to the advancement of sensitive and specific detection methods in areas such as medical diagnostics, bioanalysis, and personalized medicine.

6.9.1 Dendrimers

Dendrimers have been widely employed as a basis for electrochemical biosensor platforms owing to their distinct structural features, which include structural consistency, accuracy, well-ordered composition, and biocompatibility [86]. These properties have prompted major attempts to construct dendrimer-based electrodes for electrochemical sensing. One emphasis has been on the direct electrochemistry of hemeproteins and enzymes. Dendrimers provide an efficient conducting interface that immobilizes active biomolecules without compromising their biological activity. These examples highlight the potential of dendrimers as effective platforms for electrochemical sensing, offering high sensitivity, detection limits, and selectivity for various bio-analytes in practical applications.

6.9.2 CPs

CPs including polypyrrole (PPy), polyaniline (PANI), polythiophene (PTP), and its derivatives have been widely studied for chemical sensing, biosensing, gas sensing, and supercapacitors because to their unique electrical characteristics. Conductivity makes nanostructured CPs and their functionalized derivatives promising for sensing and biosensing. AuNPs on PANI were used for fabricating an electrochemical neurotransmitter sensor [87]. Wang *et al.* designed a non-enzymatic blood glucose sensor using poly(o-phenylenediamine)/AgNPS (PoPD/AgNPs) as a composite. The sensor has a broad linear range, low LOD for glucose, and long-term stability, making it suitable for bioanalysis. Due to their well-defined polymeric structures without exposed surface bonds, CPs are compatible with living cells and cell components [88]. Bioelectronics and healthcare applications benefit from nanopolymers or functionalized polymers with tailored surface structures and morphologies to improve sensing device selectivity and biocompatibility. The ongoing research in CP-based sensors and biosensors aims to optimize their properties, selectivity, and biocompatibility, ultimately enhancing their performance and applicability in various biomedical fields.

6.9.3 MIPs

The application of MIPs in electrochemical biosensors for detecting bacteria and bacterial molecules provides several benefits, including exceptional selectivity, resistance to degradation, and tolerance to variations in temperature and pressure. The following key findings summarize the mentioned studies. Golabi *et al.* developed a cell-imprinted polymer (CIP) using 3-aminophenylboronic acid for the specific detection of bacterial cells [89]. Bacterial surface cis-diol boronic group molecules reacted preferentially and reversibly with polymer boronic acid groups. Chen *et al.* constructed an electrochemiluminescence biosensor using nitrogen-doped graphene quantum dots and polydopamine-imprinted polymers [90]. *E. coli* was tagged with bio-conjugated nitrogen-doped graphene quantum dots after electropolymerizing dopamine and target bacteria on the electrode surface. Jiang *et al.* designed magnetic nanoparticle-based MIPs on glassy carbon electrodes to collect Gram-negative bacteria's quorum sensing (QS) signaling molecules [91]. Diverse pulse voltammetry investigated bacterial QS molecule oxidative current responses. QS molecules were detectable at 0.8 nM and linearly across the 2.5–100 nM range. A SWCNTs coating on SPEs was used to imprint polymers to detect *S. aureus* protein A using EIS technique. With a recovery factor of 91.1 ± 6.6%, the LOD obtained was 16.83 nM. Polycyclic aromatic (PA) biomolecules were removed from polymeric layers to produce voids. Khan *et al.* reported *Proteus mirabilis* flagellar fragment detection using SWCNTs-SPEs and a handmade carbon-printed electrode [92]. The platform exhibited LODs of 0.7 ng/mL and 0.9 ng/mL, as determined by EIS and square wave voltammetry (SWV) methods, respectively, and demonstrated minimal interference from flagellar filaments or globular proteins of other bacteria. These studies underscore the potential of MIPs in electrochemical biosensors for the selective detection of bacteria with high selectivity, degradation resistance, and mimicking natural selectivity.

6.10 Bionanomaterials

Biomaterial-based nanocomposites have advanced in biosensors and other applications in recent years. These nanocomposites combine the benefits of proteins/enzymes, nucleic acids, and biopolymers with metals, semiconductors, carbon materials, and CPs. This integration of biotechnology and nanotechnology has given rise to innovative electrochemical biosensor platforms that offer remarkable selectivity, sensitivity, and biocompatibility. These bio-nanomaterial-based electrochemical biosensors have been successfully employed in various applications, including biosensing, cell targeting, bio-imaging, biomineralization, biocatalysis, and drug delivery. Combining biomaterials with nanocomposites offers efficient sensing performance, enabling sensitivity, selectivity, and rapid response in biological systems. Here are a few examples of biomaterial-based electrochemical sensor platforms for biomolecule detection:

6.10.1 Aptamers

Aptamers are increasingly used in electrochemical pathogen detection owing to their strong affinity and selectivity for target molecules. These aptamers can be immobilized on electrode surfaces through chemical modifications and undergo conformational changes upon binding to the target molecule, enabling electrochemical detection strategies. Recent studies have demonstrated the use of aptamers in electrochemical biosensors for detecting pathogens. Guo *et al.* designed a peroxidase-mimicking DNAzyme amplification electrochemical biosensor for sensitive and selective *E. coli* detection [71]. Their aptamer-prime probe technology elongated polymerase via rolling circle amplification (RCA) upon target binding. *E. coli* on the electrode surface formed G-quadruplex oligomers, which displayed high catalytic and electrochemical reactions to H_2O_2. Other pathogen detection researchers have used aptamers in sensor designs. Kaur *et al.* developed a 3D-hierarchical bridged rebar graphene (BRG) aptasensor to detect pathogenic *E. coli* [93]. The researchers used BRG nanostructures with anti-*E. coli* DNA aptamers to detect bacteria-DNA interactions on the aptasensor electrode. Brosel-Oliu *et al.* developed an impedimetric aptamer-based *E. coli* electrochemical sensor [94]. Tailoring 3D interdigitated $TaSi_2$ electrode arrays with *E. coli*-specific aptamers resulted in high selectivity (LOD = 290 CFU mL^{-1}). Regeneration for numerous usage and pathogen identification in actual samples were shown by the sensor. The impedimetric aptasensor by Reich *et al.* detects *S. aureus* [95]. A particular aptamer targeted *S. aureus* surface-bound virulence factor protein A. Detecting *S. aureus* with a LOD of 10 CFU mL^{-1} was possible by immobilizing the aptamer on Au electrode surfaces and binding target biomolecules to enhance impedance. These experiments show how electrochemical biosensors can specifically detect infections using aptamers. Aptamers have excellent specificity, easy synthesis, and the possibility for multiplexed array devices by functionalizing various aptamers on each electrode. They might build speedy, sensitive, and selective infectious disease detection technologies.

6.10.2 DNA Nanostructures

Pathogen detection with DNA sensors is popular. DNA is immobilized on a sensor surface and interacts with a pathogen gene fragment [96]. Voltammetric or impedimetric readouts assess interface DNA assembly, structure, and electrochemical properties. Guo *et al.* report a voltammetric sensor that detects bacteria quickly. Self-assembling methylene blue and ferrocene hairpin probes on Au electrodes was used. Primers hybridize with open-chain hairpin probes after hairpin probes hybridize with target DNA, triggering prolonged reactions assisted by polymerase and deoxyribonucleoside triphosphates. Redox labels and dissociated targets are released from hairpin probes during amplification, allowing them to attach to additional probes and activate new cycles. Enhancing current suppression facilitates quantitative detection of *E. coli* and *Salmonella typhimurium* DNA. Another work employs a DNA-antibody nanostructured sensor to detect *Streptococcus pneumoniae* [97]. Wang *et al.* attached a hollow DNA tetrahedron onto Au electrodes [88]. The potential of electrochemical DNA-based sensors as effective pathogen detection tools offers advantages such as specificity, stability, and cost-effectiveness.

6.11 Conclusions and Future Perspectives

Electrochemical nanobiosensors are essential for pathogen detection and might replace standard diagnostic procedures. Affordable, high-sensitivity, rapid analysis, excellent selectivity, and real-time readings without complicated pre-treatment processes are positive aspects of these biosensors. Electrochemical nanobiosensors for pathogen detection are still under development, and several challenges remain. The primary challenges include improving device sensitivity and selectivity, sample pre-treatment, analysis timeframes, and sensor portability and wearability. To overcome these challenges, further research is needed in various disciplines including chemistry, engineering, biology, medicine, and healthcare. Microfluidic devices, micro- and nano-electrode arrays, signal amplification, and aptamer design have contributed to selective and sensitive microbe detection. Electrochemical biosensors must be tested on many patient biological samples to ensure reliability and reproducible results. Additionally, reaction times and linear dynamic ranges must be characterized. Depending on the bacteria, disease, and patient, clinically relevant amounts of bacterial metabolites and colony-forming units should be considered during optimization. The electrochemical systems for bacteria detection in complicated biological settings including broths, urine, blood, and sputum for selectivity and performance need to be evaluated. The necessity for optimization in complicated growth medium is evident from the few sensors that have shown accurate identification with high replication and sensitivity. Developing wearable electrochemical sensors specific to infection sites also holds promise for future applications. In conclusion, ideal electrochemical biosensors for pathogen detection should possess selectivity, sensitivity, affordability, ease of assembly and use, and enable fast and accurate analysis. Overcoming challenges and advancing the field will facilitate the practical implementation of electrochemical sensors in clinical healthcare facilities, enabling pathogen detection, early diagnosis, and monitoring of infection stages. Continued research, interdisciplinary collaboration, and the development of intelligent nanomaterials are vital for the future progress of this field.

Acknowledgement

T.F. expresses gratitude to United International University, Bangladesh, for their support at the institutional level. A.B.I. acknowledges the assistance from the Committee for Advanced Studies and Research (CASR) at BUET.

References

1. Campuzano, S., Yáñez-Sedeño, P., Pingarrón, J.M., Molecular biosensors for electrochemical detection of infectious pathogens in liquid biopsies: current trends and challenges. *Sensors,* 17, 2533, 2017.
2. Kelley, S.O., What are clinically relevant levels of cellular and biomolecular analytes? *ACS Sensors,* 2, 2, 193–197, 2017.
3. Krejcova, L. *et al.*, Nanoscale virus biosensors: state of the art. *Nanobiosensors in Disease Diagnosis,* 4, 47–66, 2015.
4. Nagel, B., Dellweg, H., Gierasch, L., Glossary for chemists of terms used in biotechnology (IUPAC Recommendations 1992). *Pure Appl. Chem.*, 64, 143–168, 1992.
5. Grabowska, I., Malecka, K., Jarocka, U., Radecki, J., Radecka, H., Electrochemical biosensors for detection of avian influenza virus--current status and future trends. *Acta Biochim. Pol.*, 61, 3, 2014.
6. Nidzworski, D., Pranszke, P., Grudniewska, M., Król, E., Gromadzka, B., Universal biosensor for detection of influenza virus. *Biosens. Bioelectron.*, 59, 239–242, 2014.
7. Chowdhury, A.-N., Shapter, J., Imran, A.B., *Innovations in nanomaterials.* Nova Publishers, 2015.
8. Foyez, T. and Imran, A.B., Nanomaterials in Pharmaceuticals. *Emerg. Applications Nanomater.*, 141, 218–245, 2023.
9. de Eguilaz, M.R., Cumba, L.R., Forster, R.J., Electrochemical detection of viruses and antibodies: A mini review. *Electrochem. Commun.*, 116, 106762, 2020.
10. Zhou, J., Lai, W., Zhuang, J., Tang, J., Tang, D., Nanogold-functionalized DNAzyme concatamers with redox-active intercalators for quadruple signal amplification of electrochemical immunoassay. *ACS Appl. Mater. Interfaces,* 5, 2773–2781, 2013.
11. Nosrati, R. *et al.*, Helicobacter pylori point-of-care diagnosis: nano-scale biosensors and microfluidic systems. *TrAC, Trends Anal. Chem.*, 97, 428–444, 2017.
12. Altintas, Z., Akgun, M., Kokturk, G., Uludag, Y., A fully automated microfluidic-based electrochemical sensor for real-time bacteria detection. *Biosens. Bioelectron.*, 100, 541–548, 2018.

13. Barreda-García, S. *et al.*, Attomolar quantitation of Mycobacterium tuberculosis by asymmetric helicase-dependent isothermal DNA-amplification and electrochemical detection. *Biosens. Bioelectron.*, 68, 122–128, 2015.
14. Kaya, H.O., Cetin, A.E., Azimzadeh, M., Topkaya, S.N., Pathogen detection with electrochemical biosensors: Advantages, challenges and future perspectives. *JEAC*, 882, 114989, 2021.
15. Simoska, O. and Stevenson, K.J., Electrochemical sensors for rapid diagnosis of pathogens in real time. *Analyst*, 144, 22, 6461–6478, 2019.
16. Stotter, J. *et al.*, Comparison of the electrical, optical, and electrochemical properties of diamond and indium tin oxide thin-film electrodes. 17, 4880–4888, 2005.
17. Kimmel, D.W., LeBlanc, G., Meschievitz, M.E., Cliffel, D.E., Electrochemical sensors and biosensors. *Analy. Chem.*, 84, 685–707, 2012.
18. Lee-Montiel, F.T., Reynolds, K.A., Riley, M.R., Detection and quantification of poliovirus infection using FTIR spectroscopy and cell culture. *J. Bio. Eng.*, 5, 1–13, 2011.
19. Gabig-Ciminska, M., Developing nucleic acid-based electrical detection systems. *Microb. Cell Factories*, 5, 1, 1–8, 2006.
20. Padliya, N.D. and Cooper, B., Mass spectrometry-based proteomics for the detection of plant pathogens. *Proteomics*, 6, 4069–4075, 2006.
21. Alatraktchi, F.A.a., Johansen, H.K., Molin, S., Svendsen, W.E., Electrochemical sensing of biomarker for diagnostics of bacteria-specific infections. *Nanomedicine*, 11, 2185–2195, 2016.
22. Thevenot, D.R., Toth, K., Durst, R.A., Wilson, G.S., Electrochemical biosensors: recommended definitions and classification. *Pure Appl. Chem.*, 71, 2333–2348, 1999.
23. Hassan, R.Y., Advances in electrochemical nano-biosensors for biomedical and environmental applications: From current work to future perspectives. *Sensors*, 22, 7539, 2022.
24. Dyussembayev, K. *et al.*, Biosensor technologies for early detection and quantification of plant pathogens. *Front. Chem.*, 9, 636245, 2021.
25. Srimaneepong, V., Rokaya, D., Thunyakitpisal, P., Qin, J., Saengkiettiyut, K., Corrosion resistance of graphene oxide/silver coatings on Ni–Ti alloy and expression of IL-6 and IL-8 in human oral fibroblasts. *Sci. Rep.*, 10, 1–12, 2020.
26. Topkaya, S.N., Azimzadeh, M., Ozsoz, M., Electrochemical biosensors for cancer biomarkers detection: Recent advances and challenges. *Electroanalysis*, 28, 1402–1419, 2016.
27. Fortunati, S. *et al.*, Rapid quantification of SARS-CoV-2 spike protein enhanced with a machine learning technique integrated in a smart and portable immunosensor. *Biosensors*, 12, 426, 2022.
28. Bard, A.J., Faulkner, L.R., White, H.S., *Electrochemical methods: fundamentals and applications*, John Wiley & Sons, 2022.

29. Sande, M.G., Rodrigues, J.L., Ferreira, D., Silva, C.J., Rodrigues, L.R., Novel biorecognition elements against pathogens in the design of state-of-the-art diagnostics. *Biosensors,* 11, 418, 2021.
30. Altuntepe, A., Seyhan, A., Zan, R., Graphene for Si-based solar cells. *J. Mol. Struct.,* 1200, 2020.
31. Smutok, O. *et al.,* A reagentless bienzyme amperometric biosensor based on alcohol oxidase/peroxidase and an Os-complex modified electrodeposition paint. 113, 590–598, 2006.
32. Shi, J. and Porterfield, D.M., Surface modification approaches for electrochemical biosensors, in: *Biosensors-Emerging Materials and Applications, 1st ed.;* Serra, P.A. (Ed.), pp. 209–226, 2011.
33. Adetayo, A.E., Ahmed, T.N., Zakhidov, A., Beall, G.W., Improvements of Organic Light-Emitting Diodes Using Graphene as an Emerging and Efficient Transparent Conducting Electrode Material. 9, 2021.
34. Kipshidze, N., Yeo, N., Kipshidze, N., Photodynamic therapy for COVID-19, 14, 651–652, 2020.
35. Jeon, Y. *et al.,* A Review of Various Attempts on Multi-Functional Encapsulation Technologies for the Reliability of OLEDs. 13, 1478, 2022.
36. Harun-Ur-Rashid, M., Foyez, T., Jahan, I., Pal, K., Imran, A.B., Rapid diagnosis of COVID-19 via nano-biosensor-implemented biomedical utilization: a systematic review. *RSC Adv.,* 12, 9445–9465, 2022.
37. Polat, E.O. *et al.,* Transducer Technologies for Biosensors and Their Wearable Applications. *Biosensors,* 12, 385, 2022.
38. Eivazzadeh-Keihan, R. *et al.,* Applications of carbon-based conductive nanomaterials in biosensors. *Che. Eng. J.,* 136183, 2022.
39. Dong, Y. *et al.,* A disposable printed amperometric biosensor for clinical evaluation of creatinine in renal function detection. 248, 123592, 2022.
40. Dzyadevych, S. and Jaffrezic-Renault, N., Conductometric biosensors, in: *Biological Identification,* pp. 153–193, Elsevier, 2014.
41. Soldatkin, O. *et al.,* Development of conductometric biosensor array for simultaneous determination of maltose, lactose, sucrose and glucose. *Talanta,* 115, 200–207, 2013.
42. Magar, H.S., Hassan, R.Y., Mulchandani, A., Electrochemical impedance spectroscopy (EIS): Principles, construction, and biosensing applications. *Sensors,* 21, 6578, 2021.
43. Orazem, M.E. and Tribollet, B., Electrochemical impedance spectroscopy. *New Jersey,* 1, 383–389, 2008.
44. Leva-Bueno, J., Peyman, S.A., Millner, P., A review on impedimetric immunosensors for pathogen and biomarker detection. *Med. Microbiol. Immunol.,* 209, 343–362, 2020.
45. Bertok, T. *et al.,* Electrochemical impedance spectroscopy based biosensors: Mechanistic principles, analytical examples and challenges towards commercialization for assays of protein cancer biomarkers. *ChemElectroChem,* 6, 989–1003, 2019.

46. Mohammad-Razdari, A. *et al.*, An impedimetric aptasensor for ultrasensitive detection of Penicillin G based on the use of reduced graphene oxide and gold nanoparticles. *Microchim. Acta,* 186, 1–10, 2019.
47. Manickam, A., Chevalier, A., McDermott, M., Ellington, A.D., Hassibi, A., A CMOS electrochemical impedance spectroscopy (EIS) biosensor array. *IEEE Trans. Biomed. Circuits Syst.,* 4, 379–390, 2010.
48. Kim, D.C. and Kang, D.J., Molecular recognition and specific interactions for biosensing applications. *Sensors,* 8, 6605–6641, 2008.
49. Watanabe, H., Omoto, S., Hoshi, Y., Shitanda, I., Itagaki, M., Electrochemical impedance analysis on positive electrode in lithium-ion battery with galvanostatic control. *J. Power Sources,* 507, 230258, 2021.
50. Kuss, S., Amin, H.M., Compton, R.G., Electrochemical detection of pathogenic bacteria—recent strategies, advances and challenges. *Chemistry–Asian J.,* 13, 2758–2769, 2018.
51. Madhu, S., Ramasamy, S., Choi, J., Recent Developments in Electrochemical Sensors for the Detection of Antibiotic-Resistant Bacteria. *Pharmaceuticals,* 15, 1488, 2022.
52. Millner, P., Caygill, R., Conroy, D., Shahidan, M., Impedance interrogated affinity biosensors for medical applications: novel targets and mechanistic studies, in: *Biosensors for medical applications*, pp. 103–134, Elsevier, 2012.
53. Abd El-Raheem, H., Hassan, R.Y., Khaled, R., Farghali, A. & El-Sherbiny, I.M. Polyurethane-doped platinum nanoparticles modified carbon paste electrode for the sensitive and selective voltammetric determination of free copper ions in biological samples. *Microchem. J.,* 155, 104765, 2020.
54. Alfadaly, R.A. *et al.,* Microbial sensing and removal of heavy metals: Bioelectrochemical detection and removal of chromium (vi) and cadmium (ii). *Molecules,* 26, 2549, 2021.
55. Abd El-Raheem, H., Hassan, R.Y., Khaled, R., Farghali, A., El-Sherbiny, I.M., New sensing platform of poly (ester-urethane) urea doped with gold nanoparticles for rapid detection of mercury ions in fish tissue. *RSC Adv.,* 11, 31845–31854, 2021.
56. Sedki, M., Hassan, R.Y., Hefnawy, A., El-Sherbiny, I.M., Sensing of bacterial cell viability using nanostructured bioelectrochemical system: rGO-hyperbranched chitosan nanocomposite as a novel microbial sensor platform. *Sens. Actuators B, Chem.,* 252, 191–200, 2017.
57. Foyez, T. and Imran, A.B., Nanotechnology in Vaccine Development and Constraints, in: *Nanovaccinology as Targeted Therapeutics*, pp. 1–20, 2022.
58. Andronescu, C. and Schuhmann, W., Graphene-based field effect transistors as biosensors. *Curr. Opin. Electrochem.,* 3, 11–17, 2017.
59. Jayamohan, H. *et al.,* Highly sensitive bacteria quantification using immunomagnetic separation and electrochemical detection of guanine-labeled secondary beads. *Sensors,* 15, 12034–12052, 2015.
60. Bhardwaj, J., Devarakonda, S., Kumar, S., Jang, J., Development of a paper-based electrochemical immunosensor using an antibody-single walled

carbon nanotubes bio-conjugate modified electrode for label-free detection of foodborne pathogens. *Sens. Actuators B, Chem.*, 253, 115–123, 2017.
61. Hassan, R.Y., Sultan, M.A., ElAlamin, M.M.A., Atia, M.A., Aboul-Enein, H.Y., A Disposable carbon nanotubes-screen printed electrode (CNTs-SPE) for determination of the antifungal agent posaconazole in biological samples. *Electroanalysis,* 29, 843–849, 2017.
62. Moya, A., Gabriel, G., Villa, R., del Campo, F.J., Inkjet-printed electrochemical sensors. *Curr. Opin. Electrochem.,* 3, 29–39, 2017.
63. Killard, A.J., Disposable sensors. *Curr. Opin. Electrochem.*, 3, 57–62, 2017.
64. Ramalingam, M. *et al.*, Impact of nanotechnology on conventional and artificial intelligence-based biosensing strategies for the detection of viruses. *Discover Nano,* 18, 1–28, 2023.
65. Smith, B.R. and Gambhir, S.S., Nanomaterials for *in vivo* imaging. *Chem. Rev.,* 117, 901–986, 2017.
66. Chen, G., Roy, I., Yang, C., Prasad, P.N., Nanochemistry and nanomedicine for nanoparticle-based diagnostics and therapy. *Chem. Rev.,* 116, 28262885, 2016.
67. Rahman, A. and Bin Imran, A., Stimuli-responsive and Self-healing Multicomponent Hydrogels for Biomedical Applications, 2023.
68. Masitas, R.A., Allen, S.L., Zamborini, F.P., Size-dependent electrophoretic deposition of catalytic gold nanoparticles. *J. Am. Chem. Soc.,* 138, 15295–15298, 2016.
69. Baldim, V. *et al.,* Amperometric quantification of S-nitrosoglutathione using gold nanoparticles: a step toward determination of S-nitrosothiols in plasma. *Anal. Chem.,* 88, 3115–3120, 2016.
70. Wang, T. *et al.,* A Review on Graphene-Based Gas/Vapor Sensors with Unique Properties and Potential Applications. *Nanomicro Lett.,* 8, 95–119, 2016.
71. Guo, Y. *et al.*, Label-free and highly sensitive electrochemical detection of E. coli based on rolling circle amplifications coupled peroxidasemimicking DNAzyme amplification. 75, 315–319, 2016.
72. Yang, Z., Wang, Y., Zhang, D., A novel multifunctional electrochemical platform for simultaneous detection, elimination, and inactivation of pathogenic bacteria based on the Vancomycin-functionalised AgNPs/3D-ZnO nanorod arrays. *Biosens. Bioelectron.,* 98, 248–253, 2017.
73. Rahman, A., Solaiman, Foyez, T., Susan, M.A.B.H., Imran, A.B., Self-Healable and Conductive DoubleNetwork Hydrogels with Bioactive Properties. *Macromol. Chem. Phys.,* 221, 2000207, 2020.
74. Ranjbar, S. and Shahrokhian, S., Design and fabrication of an electrochemical aptasensor using Au nanoparticles/carbon nanoparticles/cellulose nanofibers nanocomposite for rapid and sensitive detection of Staphylococcus aureus. *Bioelectrochemistry,* 123, 70–76, 2018.
75. Rezaei, B., Shams-Ghahfarokhi, L., Havakeshian, E., Ensafi, A.A., An electrochemical biosensor based on nanoporous stainless steel modified by gold

and palladium nanoparticles for simultaneous determination of levodopa and uric acid. *Talanta,* 158, 42–50, 2016.

76. Jeerapan, I., A stretchable and screen-printed electrochemical sensor for glucose determination in human perspiration. *Biosens. Bioelectron.,* 91, 885–891, 2017.
77. Govindhan, M., Liu, Z., Chen, A., Design and electrochemical study of platinum-based nanomaterials for sensitive detection of nitric oxide in biomedical applications. *Nanomaterials,* 6, 211, 2016.
78. Bracamonte, M.V. *et al.,* H2O2 sensing enhancement by mutual integration of single walled carbon nanohorns with metal oxide catalysts: the CeO2 case. *Sens. Actuators B, Chem.,* 239, 923–932, 2017.
79. Pedrero, M., Campuzano, S., Pingarrón, J.M., Magnetic beads-based electrochemical sensors applied to the detection and quantification of bioterrorism/biohazard agents. *Electroanalysis,* 24, 470–482, 2012.
80. Jiang, H. *et al.,* Magnetic molecularly imprinted polymer nanoparticles based electrochemical sensor for the measurement of Gram-negative bacterial quorum signaling molecules (N-acyl-homoserine-lactones). 75, 411–419, 2016.
81. Okuno, J. *et al.,* Label-free immunosensor for prostate-specific antigen based on single-walled carbon nanotube array-modified microelectrodes. *Biosens. Bioelectron.,* 22, 2377–2381, 2007.
82. Yamada, K., Choi, W., Lee, I., Cho, B.-K., Jun, S., Rapid detection of multiple foodborne pathogens using a nanoparticle-functionalized multi-junction biosensor. *Biosens. Bioelectron.,* 77, 137–143, 2016.
83. Zhao, G., Zhan, X., Dou, W., A disposable immunosensor for Shigella flexneri based on multiwalled carbon nanotube/sodium alginate composite electrode. *Anal. Biochem.,* 408, 53–58, 2011.
84. Jijie, R. *et al.,* Reduced graphene oxide/polyethylenimine based immunosensor for the selective and sensitive electrochemical detection of uropathogenic Escherichia coli. *Sens. Actuators B, Chem.,* 260, 255–263, 2018.
85. Gandouzi, I. *et al.,* A nanocomposite based on reduced graphene and gold nanoparticles for highly sensitive electrochemical detection of Pseudomonas aeruginosa through its virulence factors. *Materials,* 12, 1180, 2019.
86. Miodek, A., Mejri, N., Gomgnimbou, M., Sola, C., Korri-Youssoufi, H., E-DNA sensor of Mycobacterium tuberculosis based on electrochemical assembly of nanomaterials (MWCNTs/PPy/PAMAM). *Anal. Chem.,* 87, 9257–9264, 2015.
87. Sadanandhan, N.K. and Devaki, S.J., Gold nanoparticle patterned on PANI nanowire modified transducer for the simultaneous determination of neurotransmitters in presence of ascorbic acid and uric acid. *J. Appl. Polym. Sci.,* 134, 2017.
88. Wang, W. *et al.,* DNA tetrahedral nanostructures for the biomedical application and spatial orientation of biomolecules. *Bioact. Mater.,* 33, 279–310, 2024.

89. Golabi, M., Kuralay, F., Jager, E.W., Beni, V., Turner, A.P., Electrochemical bacterial detection using poly (3-aminophenylboronic acid)-based imprinted polymer. *Biosens. Bioelectron.*, 93, 87–93, 2017.
90. Zhang, Y. *et al.*, A molybdenum disulfide/gold nanorod composite-based electrochemical immunosensor for sensitive and quantitative detection of microcystin-LR in environmental samples. 244, 606–615, 2017.
91. Li, J. *et al.*, Ag/N-doped reduced graphene oxide incorporated with molecularly imprinted polymer: An advanced electrochemical sensing platform for salbutamol determination. 90, 210–216, 2017.
92. Khan, M.A.R., Moreira, F.T., Riu, J., Sales, M.G.F., Plastic antibody for the electrochemical detection of bacterial surface proteins. *Sens. Actuators B, Chem.*, 233, 697–704, 2016.
93. Kaur, H., Shorie, M., Sharma, M., Ganguli, A.K., Sabherwal, P., Bridged Rebar Graphene functionalized aptasensor for pathogenic E. coli O78: K80: H11 detection. *Biosens. Bioelectron.*, 98, 486–493, 2017.
94. Brosel-Oliu, S. *et al.*, Novel impedimetric aptasensor for label-free detection of Escherichia coli O157: H7. *Sens. Actuators B, Chem.*, 255, 2988–2995, 2018.
95. Qaanei, M., Taheri, R.A., Eskandari, K., Electrochemical aptasensor for Escherichia coli O157: H7 bacteria detection using a nanocomposite of reduced graphene oxide, gold nanoparticles and polyvinyl alcohol. *Anal. Methods,* 13, 3101–3109, 2021.
96. Nordin, N., Yusof, N.A., Abdullah, J., Radu, S., Hushiarian, R. A simple, portable, electrochemical biosensor to screen shellfish for Vibrio parahaemolyticus. *Amb Express,* 7, 1–9, 2017.
97. Wang, J. *et al.*, Clinically relevant detection of Streptococcus pneumoniae with DNA-antibody nanostructures. 89, 12, 6900–6906, 2017.

7

Enzymatic Nanobiosensor Strategies to Contain the Spread of Infectious Diseases

Soumendu Patra[1], Harshita Shand[1], Swarnab Dutta[2], Rittick Mondal[3] and Suvankar Ghorai[1,4]*

[1]*Department of Microbiology, Raiganj University, North Dinajpur, West Bengal, India*
[2]*Department of Chemistry, Raiganj University, North Dinajpur, West Bengal, India*
[3]*Department of Sericulture, Raiganj University, North Dinajpur, West Bengal, India*
[4]*Department of Cellular and Molecular Medicine, Herbert Wertheim College of Medicine, Florida International University, Miami, FL, USA*

Abstract

Detection of disease at an early stage is the biggest challenge faced in the world of medicine. In the recent decade, nanobiosensors have become an integral part of disease diagnosis and detection. The use of nanomaterials in biosensor production have enhanced their sensitivity, specificity and selectivity. Nanobiosensor-based diagnostics offers diligent and targeted method for the detection of diseases and are crucial for molecular detection of disease associated biomarkers. Enzymes are biological catalyst that smoothens a variety of reactions in living system and help in sustaining of life. Enzymes can catalyze a large number of reactions with great precision, efficiency, and accuracy and easily manipulated which are important for biosensor designing. The use of enzymes in nanobiosensors needs modification and thorough knowledge of the type of enzyme that should be used. The technology of enzymatic nanobiosensors involves a technology that offers a fast and inexpensive solution to the recent perturb of disease detection. Several approaches have been made to create enzyme-based nanobiosensors for rapid detection of infectious diseases. In this book chapter, we will discuss about various enzymatic nanobiosensors developed for disease diagnosis.

***Keywords*:** Nanobiosensor, enzyme, diagnosis, infection, disease

**Corresponding author*: suvankarbt@gmail.com

Suvardhan Kanchi, Ayyappa Bathinapatla, Anitha Varghese and Phumlane Selby Mdluli (eds.) Nano-Biosensor Technologies for Diagnosis of Infectious Diseases, (231–240) © 2025 Scrivener Publishing LLC

7.1 Introduction

In the last decade, nanobiosensors have emerged as a major diagnostic technique in handling spread of infection by active diagnosis [1]. Hence, pharmaceutical industries have embodied nanoparticles in the manufacture of biosensors that offer precise sensitivity, accuracy, and particularity. The majorly used nanomaterials for building nanobiosensors are gold nanoparticles, carbon nanotubes, graphene, and photonic crystals which are known for their resourcefulness [2]. Nanoparticle-based biosensors have increased the precision and accuracy of disease detection system [3].

The precision of a biosensor can be boosted by enhancing its conductivity and through deposition of a sheet of nanoparticles (NPs) on the transducer surface, where a large variety of substances, such as biological materials, can be immobilized [4]. Materials with a high specific surface to volume ratio, like NPs, can increase the number of bioreceptor units in a smaller area while still serving as a transducing element [5]. Though, the disparity in shape, size, and constitution of NPs, accompanied by the general suspensions impermanence, can control performance of reaction and the time of response, potentially resulting lower reliability and negatively affecting mercenary interest [6].

Enzymes are biological catalysts which smoothens a huge array of reactions in living systems [7]. They are synthesized every life form including animals, plants, fungi, and microorganisms [8]. The enzymes are made up of amino acids that fold-up in a complex, forming a three-dimensional tertiary structure with hydrophobic core encompassed by hydrophilic amino acid chains [9]. The highly complex three-dimensional structure of an enzyme produced a chemical environment necessary to catalyze a particular mechanism of reaction and it has a specific region inside its structure known as the active site, which is the site of catalysis. When enzyme present in free state, it is water-soluble and hence become difficult to reuse and recover them, thus enzyme immobilization is very crucial step. Enzymatic immobilization technique can be classified into physical and chemical method. In physical method weak monovalent interactions like hydrogen bond, hydrophobic interactions, van der Waals interactions are present or mechanical containment of enzyme is accomplished with the support [10]. In chemistry, covalent bonds formation can occur via ether, thioester, amide, carbamide bonds between the enzymes, and the support [11]. The immobilization depends upon the specific process based on the desired specificity and sensitivity [12].

Adsorption is another physical technique for immobilization of enzyme with the support, it can be accomplished via physical, ionic or affinity-based methods. This technique allows easy removal of enzymes from the support ensuring reuse in followed immobilization cycles [13].

Enzyme incorporation in polymeric mesh can be done by covalent and non-covalent bonds in trapping method which allows substrate passage and enzyme retention [14]. This technique does not demand any attachment to support matrix. The advantage includes no requirement of extensive purification, it is rapid and fast and the reactions can also run in moderate conditions [15].

Glutaraldehyde is used as an enzyme crosslinking bifunctional agent for preparation of an immobilized enzyme without a target [16]. It involves high enzymatic activity and low-cost production. The non-essential functional group of enzymes are used for covalent bonding. Covalent bonds are stable bonds between enzyme and support which reduces enzyme percolation [17].

An enzyme-based biosensor consists of an important enzyme that catalyzes a biological reaction with a substrate molecule and a transducer that transforms the result of the reaction into a quantifiable signal [18]. Enzymatic biosensors have better sensing because of their extremely specific enzymatic interactions with tiny chemical molecules. This transducer was created using a variety of detecting methodologies, such as electrochemical, optical, or mass-based methods [19].

7.2 Components of Enzymatic Biosensor

Enzymatic biosensors are produced using metalloenzymes. As the metalloenzymes contains metal ions in the core which can perform enzymatic reactions with small molecules like antigen of bacteria and can be used to develop biosensors [20]. Enzymes are deposited on the surface of sensors and reacts with specific substrate to produce a recognizable signal. The enzyme first recognizes and binds with the target molecule through active site of the enzyme [21]. The enzyme then catalyzes chemical reaction and produces a quantifiable signal and the signal is quantified by using electrochemical or fluorescence technique. For example, H_2O_2 can be detected through an enzymatic reaction between iron containing metalloprotease and H_2O_2 resulting in iron ion of metalloenzyme being oxidized from Fe^{2+} to Fe^{3+} and producing H_2O. This redox reaction can be measured via electrochemical method. As the biomolecules are intrinsic it became difficult to detect them hence enzymatic nanobiosensors are developed.

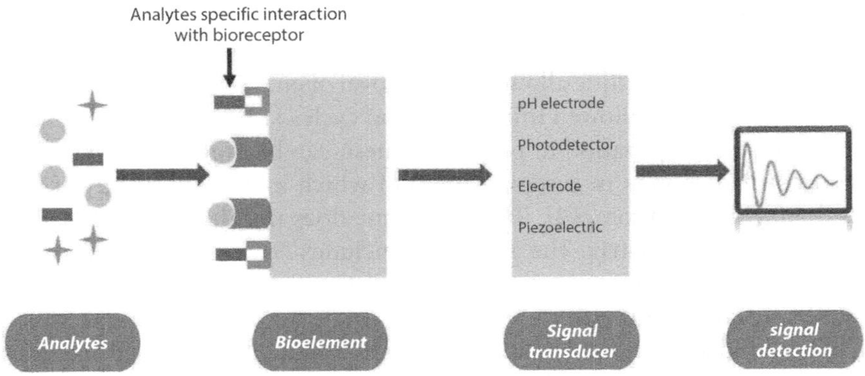

Figure 7.1 The schematic diagram of enzymatic nanobiosensor.

For boosting enzymatic processes, fostering electron transport, or creating useful conductive electrodes, electrochemical biosensors are being created that use nanomaterials [22]. Amperometric Tyr-based nanobiosensors are created using carbon nanomaterials and nanodiamonds (ND). The capacity to monitor analytes or targets in real time, as well as their great sensitivity and selectivity, have made fluorescent-based enzymatic nanobiosensors a hot topic in recent years. Enzymatic reactions enhance specificity, accuracy, and selective detection in fluorescence biosensors using nanomaterials. These sensors can be created by immobilizing fluorescent NPs on particular enzymes, which can then excite the fluorescent signal through an enzymatic reaction involving the enzymes and the targets. The schematic diagram of an enzymatic nanobiosensor is shown in Figure 7.1.

7.3 Enzymatic Nanobiosensors for Pathogen Detection

Researchers have developed an NP-enzyme sensor for rapid detection of bacteria with an olfactory output. These sensors consist of surface functionalized NP, pro-fragrance substance, enzyme for cleaving pro-fragrance molecule to produce olfactory output. The combination of these components collectively offers sensitive sensor system for detection of bacteria as low as 10^2 CFU/ml. These sensor uses NP for bacteria recognition and inhibiting fragrance generating enzyme. AuNP is used with benzyl head group at terminus as it strongly interacts with the anion cell surface of bacteria [23]. Lipase enzyme of *Candida Rugosa* was used as enzymatic amplifier. Gram positive and gram negative microorganisms that could be detected by this technique.

Table 7.1 Summarizing nanozyme-based detection of various pathogens and their LOD.

Pathogen	Detection technique	Limit of detection	Reference
Human norovirus	Colorimetry-based nanozyme strategy	30 virus/ml	[26]
Yersinia enterecolitica	GQD nanozyme biosensor	5 CFUL/ml	[27]
Gibberellazeae	AuNP-based nanozyme	10 ng/ml	[28]
Ebola virus	Lateral flow assay (LFA) along with magnetic NPs	240 PFU/ml	[31]
Dengue virus	LFA along with AgNP	150 ng/ml	[32]

Researchers have also developed a hybrid enzyme and colorimetry based nanobiosensor that amplifies enzyme and gives highly sensitive detection of bacterial pathogen in aqueous environment. The essential components of these sensors are an enzyme that provides signal amplification (β-galactosidase), a CPRG (chromogenic substrate, chlorophenol red β-D-galactopyranoside) that displays color, and a cationic NP that binds reversibly to β-Gal, to block the enzyme without denaturing it. The AuNPs used in this biosensor are carried out with quaternary ammonium ligands that gives steadiness, bio-compatibility, which makes a biosensor stable and sensitive. Interaction of anionic surface to analyte (bacteria) with cationic particle surface oust the β-Gal, resulting in restoration of enzymatic activity. The enzyme in activated gives a colorimetric readout by converting a pale-yellow substrate into red colored product [24]. Table 7.1 summarizes the nanozyme-based detection of various pathogens along with their limit of detection.

7.4 Nanozymes

Nanozymes are nanomaterials which act similarly to the enzymes and have the ability to bypass the impediments of biological enzymes as they are cost effective, generated exogenously to catalyze a wide range of biological processes and are highly stable for a long-time storage and have long shelf-life. Iron oxide is a nanoparticle that can imitate the traditional peroxidase enzymes [25]. Nanozymes are also know to detect antibiotics, heavy metal ions, pathogens and antioxidants. Many metals and metal oxide nanozymes are known to detect and identify infectious microorganisms. The infectious agents can be identified by performing magnetophoretic chromatography, magnetic nanozyme-linked immunosorbent assay, ELISA (enzyme-linked immunosorbent assay), immunochromatographic assay, colorimetric and electrochemical assay, colorimetric immunoassay, lateral flow assay and nanozyme strip assay.

A colorimetry-based nanozyme strategy has been developed for quick and extremely sensitive detection as well as identification of the food-borne human norovirus which causes gastroenteritis outbreaks globally. In response to this human norovirus, a blue color producing sensor has been developed; the intensity of the blue color is directly proportional to the virus concentration. This is due to the excellent binding specificity of the aptamer for the human norovirus combined with the nanozyme activity of AG3-aptamer coupled AuNPs. This assay provided a limit of detection (LOD) of 30 viruses/mL of sample [26].

A graphene quantum dot nanozyme immunosensor based on electrochemical sensor is developed which provides efficient and highly specific detection for virulent bacterial pathogen *Yersinia enterecolitica*, causative agent of yersiniosis. The immunosensor has an LOD of 5 CFU/mL for *Yersinia enterecolitica* in milk. The gold electrode is firstly treated with GQDs and then antibody is immobilized by 1-ethyl-3-(3-dimethylaminopropyl)-carbodimide, and finally treated with bovine serum albumin (BSA) and ethanolamine to inactivate carbonyl groups and elimination of non-specific analyte detection. The nanozyme redox current is suppressed by immobilized *Yersinia enterecolitica* and as a result, the detection signal is inversely proportional to the concentration of bacterial analyte in the sample [27].

A technique based on nanozymes can be used to detect the zearalenone (estrogenic metabolite) found in fungal extract produced by the fungal species *Gibberellazeae*. The AuNP nanozyme exhibits vivid blue coloration when 3,3′,5,5′-tetramethylbenzidine (TMB) is present. The nanozyme activity of AuNPs with strong zearalenone specificity of the zearalenone specific aptamer in that the aptamer capped with AuNP firstly loses its nanozyme activity, and recovers its activity as zearalenone replaces the aptamer layer on AuNP, and the zearalenone concentration is directly proportional to blue color intensity produced and can be seen by naked eyes. The LOD of this technique is 10 ng/mL [28].

Parasites can also be detected using nanozyme approach. *Leishmania* parasite has been detected by chronoamperometric method where nanogold was deposited on the electrode as the catalyst for converting hydrogen ions into hydrogen and monitored the faradic signal. *Trichinella spiralis* can be identified using a nanogold-based ELISA; in this technique, the horse radish peroxidase enzyme was swapped out for AuNPs as the nanozyme, and *Trichinella spiralis* was captured and immobilized using polyclonal antibody immunoglobulins [29].

7.5 Future Aspects

A combination of nanotechnology, biotechnology, and microbiology can make a breakthrough in diagnostics. Enzymatic nanobiosensors have profound application in the field of diagnosis, biorefineries, food industry, pharmaceuticals, etc. These applications demand the need of heat stable enzymes that can function well at robust environmental conditions.

Thermozyme-based nanobiosensors can be useful in detection of pathogen and prevention of spread of infection. Thermozymes are stable, have longer shelf life, and can perform catalytic functions in the presence of organic solvents at high temperature.

Researchers have developed an electrochemical-based thermozyme nanobiosensor using FAD-GDH (flavin adenine dinucleotide-dependent glucose dehydrogenase) thermozyme, derived from *Talaromycesemersonii*, a thermophilic fungus. The bio-anode was developed using single-walled carbon nanotubes attached on Au electrode surface. Glucose analyte was detected and had stability even at 70°C [30].

Alzheimer's disease detection has been done using acylpeptide hydrolase thermozyme extracted from *Sulfolobustokodaii* using gold nanorods. Thermozyme-based nanosensors can be used to detect infectious disease and more research is required in this aspect. Due to its high specificity with which it reacts with small molecule, enzymatic biosensors have proven to offer exceptionally high sensing capabilities. Apart from other biosensors, enzymatic biosensors have greater advantage which includes low detection range, a broad-spectrum range, high sensitivity, and specificity. The usage of non-enzymatic biosensor is not limited to medicine, it finds other such usage like monitoring of the environment and food safety. Having such a promising future, it is not devoid of limitation. One such problem is the concern about its stability and reproducibility, especially when the sample is biologically complex. Another such problem is the interference of other molecule present in the sample which affect our nanobiosensor's sensitivity and specificity. In addition, the incorporation of nanomaterials into enzymatic biosensors is an expensive procedure that demands specialized equipment and high human expertise.

References

1. Yang, W., Ratinac, K.R., Ringer, S.P., Thordarson, P., Gooding, J.J., Braet, F., Carbon nanomaterials in biosensors: should you use nanotubes or graphene? *Angew. Chem. Int. Ed.*, 49, 2114–2138, 2010.
2. Su, H., Li, S., Jin, Y., Xian, Z., Yang, D., Zhou, W., Kerman, K., Nanomaterial-based biosensors for biological detections. *Adv. Health Care Technol.*, 3, 19–29, 2017.
3. Kozitsina, A.N., Svalova, T.S., Malysheva, N.N., Okhokhonin, A.V., Vidrevich, M.B., Brainina, K.Z., Sensors based on bio and biomimetic receptors in medical diagnostic, environment, and food analysis. *Biosensors*, 8, 35, 2018.
4. López-Gallego, F., Jackson, E., Betancor, L., Heterogeneous systems biocatalysis: the path to the fabrication of self-sufficient artificial metabolic cells. *Chem. Eur. J.*, 23, 17841–17849, 2017.
5. Silva, T.A., Moraes, F.C., Janegitz, B.C., Fatibello-Filho, O., Electrochemical biosensors based on nanostructured carbon black: A review. *J. Nanomater.*, 2017, 1, 4571614, 2017.
6. Tian, K., Prestgard, M., Tiwari, A., A review of recent advances in nonenzymatic glucose sensors. *Mater. Sci. Eng., C*, 41, 100–118, 2014.
7. Singhania, R.R., Patel, A.K., Thomas, L., Goswami, M., Giri, B.S., Pandey, A., *Industrial Enzymes*, Elsevier, Amsterdam, The Netherlands, 2015.
8. Bhalla, T.C., Kumar, V., Kumar, V., Thakur, N., Savitri, Nitrile metabolizing enzymes in biocatalysis and biotransformation. *Appl. Biochem. Biotechnol.*, 185, 925–946, 2018.
9. Kuah, E., Toh, S., Yee, J., Ma, Q., Gao, Z., Enzyme mimics: advances and applications. *Chem. Eur. J.*, 22, 8404–8430, 2016.
10. Lima, G.V., da Silva, M.R., de Sousa Fonseca, T., de Lima, L.B., de Oliveira, M.D.C.F., de Lemos, T.L.G., de Mattos, M.C., Chemoenzymatic synthesis of (S)-Pindolol using lipases. *Appl. Catal. A-Gen.*, 546, 7–14, 2017.
11. Mohamad, N.R., Marzuki, N.H.C., Buang, N.A., Huyop, F., Wahab, R.A., An overview of technologies for immobilization of enzymes and surface analysis techniques for immobilized enzymes. *Biotechnol. Biotechnol. Equip.*, 29, 205–220, 2015.
12. Eş, I., Vieira, J.D.G., Amaral, A.C., Principles, techniques, and applications of biocatalyst immobilization for industrial application. *Appl. Microbiol. Biotechnol.*, 99, 2065–2082, 2015.
13. Lin, J., Wen, Q., Chen, S., Le, X., Zhou, X., Huang, L., Synthesis of amine-functionalized Fe3O4@ C nanoparticles for laccase immobilization. *Int. J. Biol. Macromol.*, 96, 377–383, 2017.
14. Sneha, H.P., Beulah, K.C., Murthy, P.S., Enzyme immobilization methods and applications in the food industry, in: *Enzymes in food biotechnology*, pp. 645–658, 2019.

15. Adnan, M., Li, K., Xu, L., Yan, Y., X-shaped ZIF-8 for immobilization rhizomucormiehei lipase via encapsulation and its application toward biodiesel production. *Catalysts*, *8*, 96, 2018.
16. Chung, Y., Christwardana, M., Tannia, D.C., Kim, K.J., Kwon, Y., Biocatalyst including porous enzyme cluster composite immobilized by two-step crosslinking and its utilization as enzymatic biofuel cell. *J. Power Sources*, 360, 172–179, 2017.
17. Mehrasbi, M.R., Mohammadi, J., Peyda, M., Mohammadi, M., Covalent immobilization of Candida antarctica lipase on core-shell magnetic nanoparticles for production of biodiesel from waste cooking oil. *Renew. Energy*, *101*, 593–602, 2017.
18. Cavalcante, F.T.T., de A. Falcão, I.R., da S. Souza, J.E., Rocha, T.G., de Sousa, I.G., Cavalcante, A.L.G., de Oliveira, A.L.B., de Sousa, M.C.M., dos Santos, J.C.S., Designing of Nanomaterials-Based Enzymatic Biosensors: Synthesis, Properties, and Applications. *Electrochem*, 2, 149–184, 2021.
19. Sohal, N., Maity, B., Shetti, N.P., Basu, S., Biosensors based on MnO2 nanostructures: A review. *ACS Appl. Nano Mater.*, *4*, 2285–2302, 2021.
20. Yang, Z., Liang, G., Xu, B., Enzymatic hydrogelation of small molecules. *Acc. Chem. Res.*, *41*, 315–326, 2008.
21. Rahmawati, I., Einaga, Y., Ivandini, T.A., Fiorani, A., Enzymatic biosensors with electrochemiluminescence transduction. *ChemElectroChem*, *9*, 202200175, 2022.
22. Eivazzadeh-Keihan, R., Noruzi, E. B., Chidar, E., Jafari, M., Davoodi, F., Kashtiaray, A., Mahdavi, M., Applications of carbon-based conductive nanomaterials in biosensors. *Chem. Eng. J.*, 442, 136183, 2022.
23. Li, X., Robinson, S.M., Gupta, A., Saha, K., Jiang, Z., Moyano, D.F., Rotello, V.M., Functional gold nanoparticles as potent antimicrobial agents against multi-drug-resistant bacteria. *ACS Nano*, *8*, 10682–10686, 2014.
24. Miranda, O.R., Li, X., Garcia-Gonzalez, L., Zhu, Z.J., Yan, B., Bunz, U.H., Rotello, V.M., Colorimetric bacteria sensing using a supramolecular enzyme–nanoparticle biosensor. *J. Am. Chem. Soc.*, *133*, 9650–9653, 2011.
25. Liang, M. and Yan, X., Nanozymes: from new concepts, mechanisms, and standards to applications. *Acc. Chem. Res.*, *52*, 2190–2200, 2019.
26. Weerathunge, P., Ramanathan, R., Torok, V.A., Hodgson, K., Xu, Y., Goodacre, R., Bansal, V., Ultrasensitive colorimetric detection of murine norovirus using NanoZymeaptasensor. *Anal. Chem.*, *91*, 3270–3276, 2019.
27. Savas, S. and Altintas, Z., Graphene quantum dots as nanozymes for electrochemical sensing of Yersinia enterocolitica in milk and human serum. *Materials*, *12*, 2189, 2019.
28. Sun, S., Zhao, R., Feng, S., Xie, Y., Colorimetric zearalenone assay based on the use of an aptamer and of gold nanoparticles with peroxidase-like activity. *Mikrochim. Acta*, *185*, 1–7, 2018.

29. Gomaa, M.M., Early diagnosis of experimental Trichinella spiralis infection by nano-based enzyme-linked immunosorbent assay (nano-based ELISA). *Exp. Parasitol.*, 212, 107867, 2020.
30. Iwasa, H., Hiratsuka, A., Yokoyama, K., Uzawa, H., Orihara, K., Muguruma, H., Thermophilic Talaromycesemersonii Flavin adenine dinucleotide-dependent glucose dehydrogenase bioanode for biosensor and biofuel cell applications. *ACS Omega, 2*, 1660–1665, 2017.
31. Duan, D., Fan, K., Zhang, D., Tan, S., Liang, M., Liu, Y., Zhang, J., Zhang, P., Liu, W., Qiu, X., Kobinger, G.P., Gao, G.F., Yan, X., Nanozyme-strip for rapid local diagnosis of Ebola. *Biosens. Bioelectron.*, 74, 134–141, 2015.
32. Yen, C.-W., de Puig, H., Tam, J.O., Gomez-Marquez, J., Bosch, I., Hamad-Schifferli, K., Gehrke, L., Multicolored silver nanoparticles for multiplexed disease diagnostics: distinguishing dengue, yellow fever, and Ebola viruses. *Lab Chip*, 15, 1638, 1641, 2015.

8

Development of Optical Nanosensors for the Detection of Infectious Diseases

Ndivhuwo Shumbula[1], Nosipho Moloto[2], Phumlane Mdluli[1,3] and Mbuso Mlambo[1,4,5*]

[1]Health Platform, Advanced Materials Division, Mintek, Randburg, South Africa
[2]Molecular Sciences Institute, School of Chemistry, University of the Witwatersrand, Johannesburg, South Africa
[3]Faculty of Applied Sciences, Durban University of Technology, Durban, South Africa
[4]Department of Physics, University of Pretoria, Pretoria, South Africa
[5]Department of Biotechnology and Chemistry, Vaal University of Technology, Vanderbijlpark, South Africa

Abstract

The difficulty in diagnosing infectious diseases when they are still in their early, curable stages is a direct contributor to society's high mortality rate. As a result, there is an increasing demand for the development of simple, sensitive, and selective point-of-care diagnostic devices. Numerous immunoassay diagnostic methods, including surface-enhanced Raman scattering (SERS), enzyme immunoassay, and fluoroimmunoassay, were developed and used to detect diseases early on when they are still asymptomatic. On the other hand, optical nanosensors have attracted significant attention for the early detection of infectious diseases. These biosensors can precisely measure and quantify several disease biomarkers (*e.g.*, deoxyribonucleic acid (DNA), proteins). This chapter will cover recent methods for the development of various optical biosensors based on the utilization of nanomaterial. This will include but is not limited to surface plasmon resonance (SPR), localized SPR, SERS, colorimetry, and fluorescence-based nanosensors.

***Keywords*:** Nanosensor, diagnosis, early detection, infectious diseases, point of care

**Corresponding author*: mbusom@mintek.co.za

Suvardhan Kanchi, Ayyappa Bathinapatla, Anitha Varghese and Phumlane Selby Mdluli (eds.) Nano-Biosensor Technologies for Diagnosis of Infectious Diseases, (241–278) © 2025 Scrivener Publishing LLC

8.1 Introduction

The prevalence of infectious diseases presents significant global health and socio-economic challenges, resulting in millions of fatalities each year, with a higher spread in developing countries. These diseases are caused by infectious pathogens, such as bacteria, viruses, fungi, and parasites and they can be easily transmitted between individuals through direct inoculation, airborne or waterborne routes. Pathogenic bacteria alone contribute to approximately one-third of infectious deaths, with notable examples including *Salmonella*, Listeria, *Escherichia coli* (*E. coli*), and *Staphylococcus aureus* (*S. aureus*). Viral infections also contribute significantly to disease burdens [1–6]. Over the past few decades, numerous severe viral infectious diseases have emerged, including human immunodeficiency virus acquired immunodeficiency syndrome (HIV AIDS), severe acute respiratory syndrome (SARS), novel influenza A (H1N1) flu, Ebola hemorrhagic fever, Middle East respiratory syndrome (MERS), and Zika virus disease (ZVD), which have profoundly impacted human health and the economy [7]. More recently, the global spread of coronavirus disease-2019 (COVID-19), caused by SARS-CoV-2, has become a pressing concern [8–10]. To date, the global impact of COVID-19 has been staggering, with over 750 million confirmed infections and an approximate death toll of 7 million worldwide [11]. In addition to the loss of lives, the COVID-19 pandemic has caused widespread mental distress and imposed significant economic burdens, marked by market crashes and recessions [12, 13].

The negative impacts of infectious diseases have prompted policymakers and researchers to focus on implementing sustainable mitigation measures. These measures encompass eliminating transmission routes, protecting individuals with chronic conditions, and prioritizing infection control. Among these strategies, infection control stands out as the most effective approach [14, 15]. Successful infection control relies on the development of efficient vaccines, early treatments, and accurate diagnostics. Remarkable achievements have been made in these areas, significantly impacting humanity [16–19, 23]. However, continuous research is still necessary to enhance existing vaccines, treatments, and diagnostics for the timely prevention and effective management of infectious disease outbreaks. The recent COVID-19 pandemic has emphasized the urgency of ongoing research, particularly in the field of point-of-care (POC) diagnostics.

Effective treatment of infectious diseases relies heavily on accurate detection, as it plays a crucial role in reducing transmission and preventing or mitigating outbreaks. Therefore, the development of diagnostic devices capable of early and highly accurate disease detection is of paramount importance. Traditional diagnostic techniques commonly employed in laboratories include typical immunoassays, nucleic acid amplification, cell culture, and microscopy. However, these methods have inherent limitations. Despite their high sensitivity and rapidity, immunoassays like enzyme-linked immunosorbent assay (ELISA) are prone to cross-reactivity, leading to potential false positive results. Additionally, ELISA necessitates trained personnel, and large samples, and can be costly. Sample preparation complexity in nucleic acid amplification tests such as polymerase chain reaction (PCR) can introduce contamination risks and result in false positives. Microscopic techniques exhibit reduced sensitivity with clinical specimens, while cell culture techniques are time-consuming and raise safety concerns. Another drawback of conventional diagnostic procedures is the multi-step process involved in sample collection, preparation, and result delivery. Samples obtained from POC locations, such as urine, blood, or swabs, are transported to designated laboratories, where they undergo preparation and analysis by specialized personnel. The data is then relayed back to clinicians for interpretation, and the results are communicated to patients after days of waiting [20–24]. This extended process hinders timely evidence-based treatment and can contribute to the overuse of antibiotics. Particularly in under-resourced healthcare systems, these challenges have detrimental effects on society, leading to deteriorating health. Overall, traditional diagnostic methods are time-consuming, and require sophisticated laboratories, specialized equipment, and trained personnel, making them cost-intensive.

8.2 Overview of Biosensor

Biosensors are highly advanced analytical tools that utilize a combination of a biological component or bioreceptor and a physicochemical transducer to effectively detect specific analytes. They achieve this by converting biochemical information into a detectable signal [21, 22]. Biosensors can be categorized based on the type of bioreceptor and/or signal transduction used. Bioreceptors are molecules that specifically recognize the analyte in the sample and can include enzymes, aptamers, antibodies, and cell-surface molecules. On the other hand, a transducer registers and converts the biological signal resulting from the interaction between the bioreceptor and the analyte into a detectable output [21, 25–28]. The type of transducer employed determines the form of the signal, which can be optical, electrical, electrochemical, magnetic, thermal, or piezoelectric [29–34]. For example, an electrochemical-based biosensor converts the biological signal from the analyte into an electrical signal, which is then detected and interpreted by a detector [27, 29]. Conversely, optical-based biosensors utilize the interaction of light or electromagnetic radiation with the biological sample to detect and quantify the analyte. The signal can manifest as absorbance, transmittance, emission, fluorescence, etc. [29, 34, 35]. Biosensors are characterized by their remarkable sensitivity and selectivity, rendering them suitable for a wide range of applications. Moreover, they are designed to be reusable, contributing to their cost-effectiveness and sustainability. Additionally, biosensors operate independently of physical parameters such as temperature and pH, enhancing their reliability and suitability for diverse environments.

Biosensors can also be classified based on the mode of interaction between the analyte and the bioreceptor, namely, catalytic and affinity. Catalytic biosensors utilize biocomponents capable of recognizing specific (bio)chemical species and catalyzing a chemical reaction to produce a detectable signal. This type of biosensor is represented mostly by enzymatic biosensors, which make use of specific enzymes or their combinations. Affinity biosensors leverage the specific binding capabilities between an analyte and a biorecognition element. Table 8.1 shows the comparison of catalytic and affinity biosensors ranging from their mechanisms to their types and applications. It is important to note that while these biosensors have distinct characteristics, there can also be an overlap in their applications and performance depending on the specific design and implementation [36–41].

Biosensors have garnered significant attention for their ability to detect infectious diseases in the biomedical field. The development of biosensors was facilitated by advancements in molecular amplification techniques, next-generation affinity probes, and the unique optical properties of nanoparticles such as quantum dots, graphene, palladium, silver, and gold [42–47]. The pursuit of high accuracy and sensitivity in a time-efficient and cost-effective manner has always been a priority in the development of robust and reliable biosensors. Figure 8.1 shows a schematic diagram generalizing how a biosensor is designed and operates.

A diverse range of biosensors was developed and gained considerable recognition for their effectiveness in detecting infectious diseases. Notable examples include lateral flow assays (LFA), electrochemical biosensors, and optical biosensors. Lateral flow assays are widely employed in the detection of tuberculosis (TB), HIV, and the more recent COVID-19 disease [48–50]. These user-friendly biosensors can be conveniently utilized in POCs settings such as clinics or homes, offering rapid results with turnaround times ranging from 5 to 15 min [50].

Table 8.1 Table comparing catalytic biosensors and affinity biosensors.

	Catalytic biosensor	Affinity biosensor
Principles	Measure the catalytic activity of enzymes or catalysts	Detect binding interactions between molecules
Sensing mechanisms	Change in reaction rate or product formation	Signal change upon specific binding
Selectivity	Achieve selectivity with specific enzymes/ catalysts	Selectivity based on recognition element specificity
Sensitivity	High sensitivity due to catalytic amplification	High sensitivity with specific recognition elements
Detection range	Wide range of analytes (small molecules, ions, biomolecules)	Primarily used for biomolecule detection (proteins, nucleic acids, antigens)
Response time	Rapid response due to efficient catalytic reactions	Response time can vary based on binding kinetics
Types	• Enzymatic biosensors • Polynucleotides or DNAzymes • Catalytic inhibitory • Whole-cell sensors	• Immunosensors • Nucleic acid biosensors • Receptor-ligand biosensors • Whole-cell biosensors
Applications	Medical diagnostics, Environmental monitoring, food safety, industrial process control	Medical diagnostics, pharmaceutical research, biomolecular interaction studies, bioprocess monitoring

Electrochemical biosensors provide specific and quantitative analytical information about target analytes by utilizing a working electrode as a transducer. The analyte interacts with a bioreceptor immobilized on the electrode's surface, generating a measurable signal in the form of a current (amperometric), voltage (potentiometric), or conductivity (conductometric), depending on the type of transducer employed [27]. While the electrochemical biosensor is commonly recognized for diabetes diagnosis, it has also been extensively applied in the detection of infectious diseases such as Ebola and hepatitis B [51, 52].

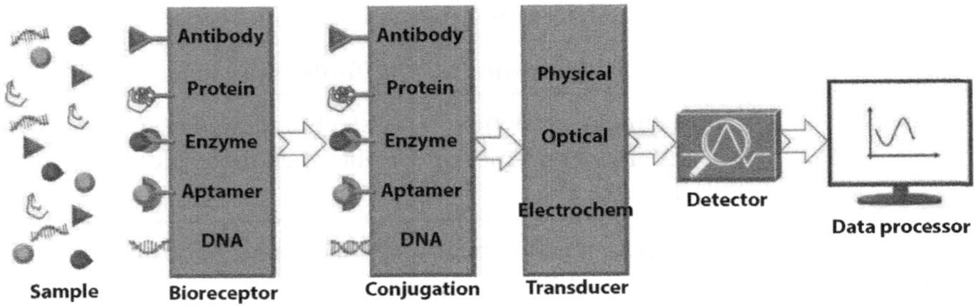

Figure 8.1 Basic schematic diagram of a biosensor.

Optical biosensors, renowned for their rapid detection, robustness, sensitivity, and selectivity, are widely employed in diagnosing infectious diseases. These biosensors employ optical transducers that generate an intensity-based optical signal proportionate to the concentration of the target analyte. Various transducer technologies are employed, including fluorescence, luminescence, internal reflection, surface plasmon resonance (SPR), absorption, and light scattering. Furthermore, the development of optical biosensors has significantly benefited from advancements in microelectronics, microelectromechanical systems (MEMS), molecular biology, biotechnology, chemistry, and nanotechnology [26, 34, 35, 53, 54].

Within the realm of optical biosensors, the integration of nanomaterials has given rise to optical-based nanosensors (optical nanosensors). These nanosensors have exhibited a remarkable impact in achieving accurate and early detection of infectious diseases. The rapid progress in nanoscience and nanotechnology has revolutionized the robustness and clinical significance of optical nanosensors for infectious disease diagnosis. This chapter highlights recent breakthroughs, challenges, and mechanisms of operation associated with optical nanosensor development. Additionally, we explore the integration of optical nanosensors with existing biosensor technologies, aiming to enhance early and accurate detection of infectious diseases.

8.3 Introduction to Optical Nanosensors

Optical nanosensors are devices or systems that utilize nanotechnology and optical principles to analyze and detect specific substances or properties at nanoscale levels. They combine the benefits of nanotechnology, which deals with materials and structures at the atomic and molecular level, with optical sensing techniques, which involve the interaction of light/electromagnetic radiation with matter. Optical nanosensors typically consist of nanoscale components such as nanoparticles, nanowires, nanotubes, or other nanostructured surfaces that can interact with the target analyte. These nanomaterials are engineered to exhibit specific optical properties, such as fluorescence, absorption, or scattering, which can be modified in the presence of the target substance [29, 34, 35, 54, 55].

The detection mechanism of optical nanosensors relies on the measurement of changes in the interaction of light with matter (nanomaterial or nanomaterials-bioreceptor cluster) caused by the presence of a target analyte. For example, when the target analyte binds to the surface of nanosensors, it can induce changes in the nanomaterial's optical properties, leading to alterations in the intensity, wavelength, or polarization of the emitted or transmitted light. These changes can then be quantified and correlated with the concentration or presence of the target analyte [53, 56].

Optical nanosensors have found a wide range of applications in various fields including diagnostics due to their ability to perform real-time and label-free detection with high speed and fast response time. For instance, they can be used for detecting biomarkers in biological samples, a characteristic vital to early detection of diseases. By integrating optical, biological, and nanotechnology principles, these sensors enable extremely sensitive and precise measurements of a wide range of biological analytes, including proteins, DNA, antibodies, enzymes, and small compounds such as drugs or toxins [26, 54, 56]. Moreover, nanosensors can directly be used to detect pathogens (*i.e.*, viruses, bacteria, etc.) [34, 35, 53]. They can be tailored to specifically recognize without any interference. This versatility allows researchers and medical practitioners to use optical nanosensors for the early detection of infectious diseases. Nanosensors based on SERS, SPR, and fluorescence have been commonly used for the detection of various infectious diseases through biomarkers or pathogen recognition mechanisms [47, 57, 58]. For instance, SERS has been used to simultaneously detect *E. Coli* and *S. Aureus.* Using vancomycin-modified Fe3O4@Au magnetic nanoparticles and aptamer-functionalized SERS tags, Zhang and co-workers managed to simultaneously detect *E. coli* and *S. aureus*. The limit of detection (LOD) of 20 and 50 cells mL^{-1} was achieved for *S. aureus* and *E. coli*, respectively [59]. Chung *et al.* developed an SPR biosensor for the detection of human hepatitis B virus (HBV) through the detection of anti-HBs antibodies, using several amplification methods such as sandwich assay of secondary antibodies, avidin-biotinylated antibodies, and peroxidase–anti-peroxidase (PAP) complex. The highest detection sensitivity was obtained using the PAP amplification method and the results comparable to that of commercial ELISA assay were obtained [60].

The ongoing advancements in nanotechnology, materials science, and biotechnology have further enhanced the performance and capabilities of optical biosensors. New materials with enhanced optical properties, such as plasmonic nanoparticles or fluorescent probes, are being integrated into sensor designs to improve sensitivity and selectivity. The next section discusses much more about the types of optical nanosensors and their specific application in the detection of infectious diseases.

8.3.1 Types of Optical Nanosensors—Towards Detection of Infectious Diseases

There are several types of optical biosensors, each using different principles for detection. This section will cover the nanosensors based on SERS, SPR, fluorescence, and colorimetry. It will discuss their principles of operation, application in the detection of infectious diseases, and their advantages, and disadvantages.

8.3.1.1 SERS-Based Nanosensors

Surface-enhanced Raman spectroscopy (SERS) is an advanced analytical technique that enhances the Raman signal of molecules adsorbed on a metallic surface, typically Au or Ag as shown in Figure 8.2. Developed in the early 1970s as an evolution of classic Raman spectroscopy, SERS is highly effective in detecting and identifying various chemicals and biological analytes. It offers exceptional sensitivity, allowing for the detection of single molecules, while also providing unique chemical fingerprints for characterization or identification. Furthermore, it enables the detection of small quantities of target molecules and provides detailed molecular information, making it a powerful analytical tool [61, 62].

By leveraging the plasmonic properties of metallic nanostructured surfaces, SERS amplifies the Raman scattering signal by several orders of magnitude (*i.e.*, over 10^{14}). This enhancement arises from the interaction between the incident light, the molecules of interest, and the surface plasmons on the surface of the metal. When the analyte molecules are near or adsorbed onto the metallic surface, the electromagnetic field at the surface is significantly enhanced, leading to a stronger Raman scattering signal [61]. Moreover, SERS provides a distinctive chemical fingerprint that is crucial for molecular identification. This is because each molecule has a unique Raman spectrum, consisting of characteristic peaks corresponding to its vibrational modes. These peaks can be used as spectral signatures for identifying specific molecules or differentiating between closely related compounds [63, 64]. These capabilities have yielded the development of SERS-based biosensors. SERS-based biosensors have been used in various applications, including diagnosis of infectious diseases diagnosis.

Surface-enhanced Raman scattering (SERS) biosensors have been designed to detect specific biomolecules, such as proteins, peptides, or nucleic acids that are present in bodily fluids or on the surface of pathogens. The detection of infectious diseases using SERS can either be in the form of a labeled or label-free detection mechanism [65]. Labeled detection involves the functionalization of SERS substrate (*e.g.*, Au or Ag NPs) with a molecular recognition moiety or a Raman reporter that captures the analytes or biomarkers, bringing them closer to the SERS substrate. The presence of the analyte is indirectly confirmed by monitoring the SERS signal of the reporter molecules. On the other hand, a label free SERS detection involves the adsorption of analyte onto or located near a SERS substrate, allowing for identification and quantification based on its unique SERS spectral fingerprints [61, 65, 66]. Table 8.2 summarizes the differences in the principles and characteristics of labelled and label-free SERS detection mechanisms.

Using either detection mechanism (*i.e.*, labeled or label-free), SERS biosensors have enabled the detection of various infectious diseases, either through biomarker or pathogen identification. SERS has found its widespread application in detecting diseases such as HIV, TB, hepatitis B and C, and malaria [67–74]. Moreover, SERS biosensors have the potential to detect multiple analytes simultaneously, which is important for the diagnosis of infectious diseases that may have multiple biomarkers [75]. In one study, a SERS biosensor was used to detect *Neisseria meningitidis*, *Streptococcus pneumoniae*, and *Haemophilus influenzae* in cerebrospinal fluid (CSF) using their unique cell membrane biomarkers such as lipids and polysaccharides. The SERS response was enhanced by simultaneous

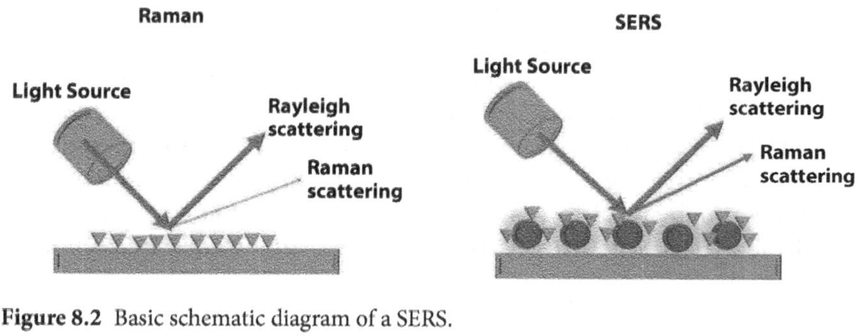

Figure 8.2 Basic schematic diagram of a SERS.

Table 8.2 Comparison of label and label-free detection.

Aspect	Labeled detection	Label-free detection
Principle	Indirect detection using a labeled Raman reporter molecule (*i.e.*, dyes, thiols, or other molecules with high Raman scattering cross-sections)	Direct detection based on intrinsic SERS signals
Detection method	Monitoring intense SERS signal of reporter molecule associated with the substrate	Analyte identification/quantification based on unique SERS spectral fingerprints
Sensitivity	High	High
Specificity	High	High
Interference from matrix	Minimal	Common issue
Molecular information	Provides information about the labeled reporter molecule	Provides information about the analyte itself
Recognition element	Required (antibodies, aptamers, enzymes, etc.)	Not required

filtration and immobilization of CSF on Au–Ag coated polycarbonate membranes allowing intrinsic identification of the biomarkers [76]. In another study by Owens and co-workers, tuberculosis antigen biomarkers were detected from clinical serum samples using a novel sandwich-based label-free SERS immunoassay. Their immunoassay was able to detect phospho-myo-inositol-capped lipoarabinomannan (PILAM), an antigenic biomarker presents in tuberculosis-infected patients [77]. However, to ensure accurate and specific detection, the method required sample pre-treatment. The pre-treatment step is crucial to eliminate any potential non-specific interactions between them and other naturally occurring constituents in a sample, serum in their case. This pretreatment step was essential in mitigating biological matrix interferences, which are vital for obtaining a high signal-to-noise ratio in the SERS signal [77]. Achieving a high analytical sensitivity is of paramount importance in accurately detecting PILAM and facilitating effective tuberculosis diagnosis. Ngo *et al.* developed an extremely sensitive SERS-based diagnostic assay for the detection of malaria ribosomal ribonucleic acid (rRNA) in blood lysates. Their lab-in-a-stick device uses ultrabright SERS-encoded nanorattles and a magnetic bead-based sandwich assay. This device was able to achieve an LOD of 200 fM without RNA extraction or PCR amplification, making it suitable for field transferable and POC SERS diagnostics [73]. Ngo *et al.* also fabricated Au and Ag nanowaves were SERS substrates to develop an extremely sensitive SERS biosensor for the detection of dengue viral DNA. This approach managed to detect the dengue DNA at extremely low amounts further achieving an LOD of 6 attomoles [78]. Using Au shell-isolated nanoparticles (NPs) as the SERS substrate, Camacho *et al.* demonstrated the use of SERS immunoassay to detect Zika virus. Their method was highly specific with no cross-reactivity and managed to detect ZIKV NS1 with an LOD of 10 ng/mL [79].

Surface-enhanced Raman scattering (SERS)-based biosensors have significantly advanced the detection of several diseases' biomarkers. These methods have undergone an evolution, leading to improved possibilities for POC diagnostics. This progress is attributed to the development of functional nanoparticles that enable precise biomarker identification and enhance SERS sensitivities. The effectiveness and accuracy of detecting target biomarkers are greatly influenced by the functionality of these nanoparticles [80–84].

8.3.1.2 SPR and LSPR-Based Nanosensors

8.3.1.2.1 Basic Principles of SPR and LSPR-Based Biosensors

Surface plasmon resonance (SPR) and localized SPR (LSPR) biosensors are powerful tools for label-free detection and analysis of biomolecular interactions [85]. Surface plasmon resonance (SPR) biosensors employ collective plasmon resonances in metallic films to monitor changes in the refractive index near the metal surface. By immobilizing biomolecules on the metal surface and detecting the resulting shifts in the SPR angle or wavelength, SPR biosensors enable real-time monitoring. They offer advantages such as label-free detection, high sensitivity, and the ability multiplex detection. Surface plasmon resonance (SPR) biosensors find applications in various fields including disease detection, drug discovery, and environmental monitoring [86].

On the other hand, LSPR biosensors focus on localized plasmon resonances in individual metallic NPs [85–87]. By tailoring the size, shape, and composition of NPs, LSPR biosensors can achieve enhanced sensitivity and tunability. Localized SPR biosensors offer versatile design options and can be used for DNA/protein detection, cellular imaging, and chemical/biological sensing [85, 87, 89]. Their localized nature allows for single-particle characterization and higher spatial resolution, enabling detection and analysis at nanoscale level [90]. The transition from SPR to LSPR biosensors represents a shift from macroscopic to nanoscale sensing, unlocking new possibilities for sensitive and tailored biosensing applications.

8.3.1.2.2 SPR-Based Nanosensors to Detect Infectious Diseases

The basic operation mechanism of SPR biosensors varies depending on the type of technique used (see Tables 8.3 and 8.4 for a summary). For example, in the case of a prism-based SPR biosensor, a thin metal film (commonly gold or silver) is deposited on a glass substrate, and a prism is placed on top of the metal film [91]. The metal film acts as a waveguide for surface plasmons in the metal. When light is incident on the prism at a specific angle, it undergoes total internal reflection at the prism-metal interface. The evanescent field of the reflected light extends into the metal film, interacting with the biomolecules immobilized on the metal surface. Binding events between the immobilized and analyte molecules in solution cause changes in the refractive index at the metal surface, leading to a shift in the angle at which the total internal reflection occurs (see Figure 8.3). By monitoring the angle of the reflected light, changes in the refractive index can be measured, providing information about the binding interactions between the immobilized molecules and the analyte molecules. This allows for the real-time monitoring of various biomolecular interactions, such as antigen-antibody binding, protein-protein interactions, DNA hybridization, and enzymatic reactions [91–94].

Surface plasmon resonance (SPR) biosensors offer several advantages over other biosensing technologies. They provide label-free detection, eliminating the need for fluorescent or radioactive labels, which simplifies the assay and reduces costs. Their ability to offer real-time monitoring allows for kinetic analysis of binding events [91]. Additionally, SPR biosensors are extremely sensitive, with LOD in the picomolar to femtomolar range, depending on the analyte. As a result, various SPR-based biosensors have been developed for POC detection of infectious diseases [20]. Surface plasmon resonance (SPR) biosensors have been applied to diagnose various stages of the Epstein–Barr virus by detecting antibodies against different viral antigens [95]. Additionally, SPR has been utilized to monitor the real-time binding kinetics between dengue IgM monoclonal antibodies and their corresponding antigens [96].

Table 8.3 Comparison of SPR and LSPR biosensors.

Technique	Principles	Advantages	Disadvantages	Applications
SPR	Based on the coherence oscillation of free or conduction electrons on metal surface.	• Label-free detection. • Real-time analysis. • High sensitivity. • Multiplexing analysis.	• Requires metal film or periodic arrays. • Limited spatial resolution. • Bulk refractive index changes.	• Biomolecular interaction analysis. • Protein-protein binding studies. • Drug discovery and development.
LSPR	Based on localized electric field enhancement of surface plasmons around metallic NPs.	• High sensitivity at nanoscale. • NPs size, and shape tuning. • Enhanced detection limits. • Tailored plasmon resonances. • Wide range of sensing targets.	• Limited compatibility with some samples. • Difficulty in uniformity of nanoparticles. • Aggregation-induced effects. • Limited spatial range.	• Environmental monitoring. • DNA/protein detection. • Cellular imaging. • Chemical and biological sensing.

Table 8.4 Types of SPR techniques and their mechanisms of operation.

SPR biosensor type	Mechanism
Prism-based SPR biosensors	Incident light passes through a prism and is reflected off a metal-coated surface, detecting changes in the resonance angle.
Grating-coupled SPR biosensors	A diffraction grating couples light into the surface plasmons. Changes in the diffraction angle or intensity indicate binding events.
Imaging SPR biosensors	The array of sensors captures reflected light at different positions, providing spatial information about binding events.
Localized SPR biosensors	Metallic nanoparticles or nanostructured surfaces generate localized surface plasmons, and changes in the plasmon resonance indicate binding events.
Fiber-optic SPR biosensors	Optical fibers coated with a thin metal film detect changes in the refractive index near the fiber surface, measuring shifts in transmitted or reflected light intensity.

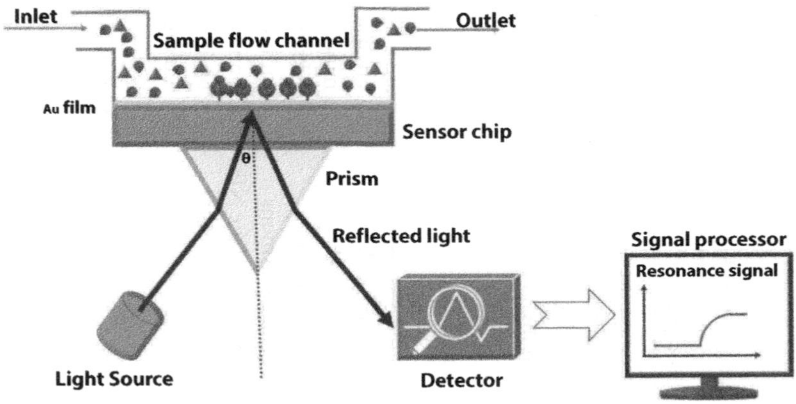

Figure 8.3 Basic schematic diagram of SPR biosensor.

Like any other technique, SPR-based biosensors incur some challenges including the requirement for relatively large liquid sample volumes, which can limit their practicality, especially in POC or portable applications. Additionally, the power consumption associated with conventional SPR-based biosensors can be significant, making them less suitable for remote or resource-limited settings [97]. To overcome these challenges, researchers have turned to nanomaterials among other multidisciplinary approaches to improve the applicability of SPR-based biosensors. By incorporating the sensor surface with nanomaterials, the signal amplification and LOD can further be enhanced. Nanomaterials enable the detection of lower analyte concentrations, thereby improving the sensitivity of the biosensor [89, 91, 94, 97, 98]. In their study, Prabowo *et al.* successfully demonstrated the development of a few-layer graphene (FLG)-based SPR nanosensor for the detection of *Mycobacterium tuberculosis* complex (MTBC) DNA. This nanosensor was achieved through the binding of either Au nanourchins (AuNu) or Au nanorods (AuNr) with the single-stranded DNA (ssDNA) probe of MTBC, after which it was deposited on the surface of the SPR sensor coated with FLG. The AuNu/AuNr-ssDNA probe was deposited and interacted with FLGO through π-π stacking between ssDNA and the graphene layer. Upon the injection of complimentary ssDNA (cssDNA), the SPR signal was significantly enhanced and the LOD in the femtomolar range was achieved [99]. Gür *et al.* developed an SPR nanosensor for diagnosis of urinary tract infections through real-time detection of *E. coli*. The bacteria were immobilized on a Au NPs imprinted SPR chip surface. Their SPR nanosensor system demonstrated ultrasensitive detection capabilities with a low LOD (1 CFU/mL) due to the synergistic effects of AuNPs and the surface imprinting technique. Selectivity experiments were conducted using *S. aureus*, *Klebsiella pneumoniae*, and *Pseudomonas aeruginosa*, but the highest response was observed for *E. coli*. Furthermore, the developed nanosensing system successfully achieved the recognition of *E. coli*, even in a complex medium such as an artificial urine sample [100]. Novel SPR nanosensors with the potential to be employed in the detection of infectious diseases have been designed using 2D nanostructured materials such as transition metal chalcogenides and graphene. Their sensitivity, figure of merit, and quality factor were investigated using the attenuated total reflection technique (ATR) [98]. In another study, a highly sensitive 2D MoS_2 nanosheets decorated fiber optic SPR immunosensor was developed for the detection of *E. coli*. *E. coli* monoclonal antibodies were successfully immobilized on the MoS_2 functionalized sensing platform through hydrophobic interactions. The immunosensor demonstrated rapid bacterial quantification and was evaluated for non-specificity and cross-reactivity with *Salmonella Typhimurium* and *S. aureus*. Compared to conventional fiber optic SPR sensors, this label-free immunosensor exhibited better performance with a low LOD (94 CFU/mL) and improved specificity [101].

8.3.1.2.3 LSPR-Based Nanosensors to Detect Infectious Diseases

The principles and mechanisms of operation of LSPR are summarized in Table 8.3 and Figure 8.4. The outstanding LSPR properties brought about plasmonic NPs has given rise to LSPR biosensing device for POC diagnostics. Like SPR biosensors, LSPR biosensors have been used in numerous times in the detection of diseases like HIV, influenza, TB, Ebola, Zika, pathogen toxins, etc. [102–110]. For instance, Lee and co-workers have designed an ultrasensitive label-free LSPR nanosensor for the detection of HIV using Au dots circular nanopatterns. This method was able to quantitatively detect various amounts of HIV-1 with an LOD of 200 fg/mL. These observations were attributed to the longitudinal wavelength shift in the ultraviolet-visible (UV-Vis) spectrum resulting from the change in the local refractive index caused by antigen-antibody interaction on the surface of the substrate [111]. Another study by Lee *et al.* demonstrated the fabrication of a label-free LSPR nanosensor for the detection of avian influenza H5N1 using multi-functional DNA three-way junction conjugated on the surface Au spike-like NP (hAuSN). The detection of H5N1 was conducted through the monitoring of hemagglutinin (HA) binding aptamer. This sensor was able to detect HA protein from the buffer solution and managed to achieve the LOD of 1 pM [112]. Sun and co-workers also developed LSPR nanosensor to detect MTB antibody using fusion protein CFP10-ESAT6 as an antigen. The protein was conjugated on the surface of Au nanorods (AuNRs) followed by subsequent incubation with serum of TB infected and non-infected human. The MTB was detected through monitoring the change or shift of LSPR band wavelength of AuNRs upon addition of the sample. This nanosensor achieved 79% sensitivity and 92% specificity [113]. Zandieh *et al.* developed LSPR sensor for the detection

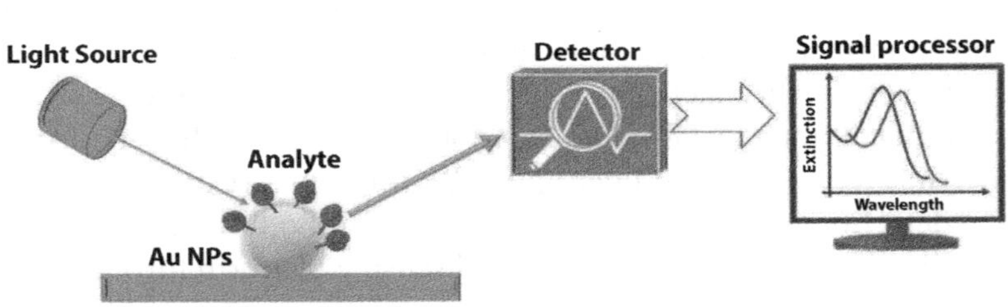

Figure 8.4 Basic schematic diagram of LSPR biosensor.

of *E. coli* endotoxin using Ag nanostructures capped with SAM of 3-mercaptopropionic acid (MPA). This biosensor was able to achieve LOD of 340 pg/mL and was selective to *E. coli* endotoxin over *Brucella abortus* (*B. abortus*) endotoxin [114].

8.3.1.3 Colorimetric Nanosensors

Colorimetric analysis is a method for analyzing colors, commonly used to determine the concentration or presence of an analyte by observing color changes. This technique has found diverse applications, including biosensing devices such as colorimetric nanosensors. Colorimetric nanosensors are nanoscale sensors that utilize colorimetric principles to detect and quantify specific analytes. They take advantage of the optical properties of plasmonic nanoparticles such as Au, Ag, and copper (Cu) NPs. These NPs have strong absorption in the visible range of the electromagnetic spectrum attributed to their unique SPR phenomenon. The SPR properties of these materials can be fully exploited by varying their size, shape, and interparticle distance. Colorimetric nanosensors are particularly useful in biosensing, where they can detect and measure biological or chemical analytes by inducing visible color changes. The presence of an analyte or the formation of a reaction can alter the absorption or reflection of light, resulting in a detectable color change [115–122]. Figure 8.5 shows the basic mechanism of operation of colorimetric biosensor.

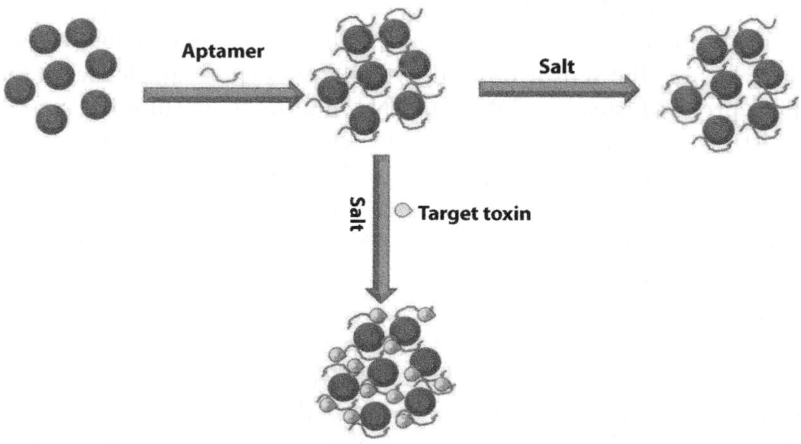

Figure 8.5 Schematic representation of colorimetric biosensor.

Colorimetric biosensors, including nanosensors, consist of a recognition element (*e.g.*, antibodies, enzymes, or DNA probes) and a transducer element. The recognition element selectively binds to the target analyte, initiating a specific reaction or signal generation. This reaction induces a change in color, intensity, or hue, which can be observed visually or measured using specialized instruments like spectrophotometers, colorimeters, or smartphone apps. Color change can arise from various mechanisms, such as enzyme-catalyzed reactions, pH changes, or the aggregation of nanomaterials [120–124]. These colorimetric biosensors have been developed to detect a wide range of target analytes, including pathogens, toxins, biomarkers, and other analytes of interest. They offer several advantages, such as simplicity, portability, cost-effectiveness, and ease of data interpretation, making them suitable for POC diagnostics and on-site monitoring applications [125–131].

Colorimetric nanosensors have been extensively researched for the detection of infectious diseases. They are capable of detecting biomolecules or pathogens by resulting in color changes when the target binds to receptor molecules on the sensor surface [125, 126]. For instance, gold nanoparticle-based colorimetric sensors can be functionalized with specific receptors that bind to the target biomolecule or pathogen, causing a change in the size and shape of the NPs and resulting in a color change [116]. Other types of colorimetric nanosensors utilize quantum dots, carbon nanotubes, or plasmonic nanoparticles [132–134]. These sensors have been successfully employed in the detection of various infectious diseases, including HIV, tuberculosis, and influenza [135–137].

Teengam *et al.* developed a POC paper-based colorimetric nanosensor for detecting Middle East respiratory syndrome coronavirus (MERS-CoV), Mycobacterium tuberculosis (MTB), and human papillomavirus (HPV) DNAs. The sensor utilized pyrrolidinyl peptide nucleic acid (acpcPNA) probes instead of DNA or RNA due to their superior chemical properties, ease of synthesis, and efficient hybridization with complementary DNA strands. In the absence of complementary DNA, the acpcPNA probe caused aggregation of citrate-capped silver nanoparticles (AgNPs). However, when the target DNA was present, the formation of anionic DNA-acpcPNA duplex resulted in the dispersion of AgNPs due to electrostatic repulsion, leading to a noticeable color change confirming the presence of the analyte. The technique achieved low detection limits of 1.53 nM (MERS-CoV), 1.27 nM (MTB), and 1.03 nM (HPV) [101]. Kim and co-workers have also showcased the development of a colorimetric assay for the detection of MERS-CoV. This nanosensor involved the formation of a self-assembled monolayer (SAM) of dsDNA on AuNPs under an electrolyte. The presence of MERS-CoV DNA was confirmed by a change in the color of AuNPs and a shift of LSPR in the UV-Vis range. This method was able to achieve an LOD of 1 pmol/μL of MERS-CoV [138]. A study by Choi *et al.* presented a combined paper-based biosensor and loop-mediated isothermal amplification (LAMP) system for detecting nucleic acids from samples, providing a complete solution from sample to result. The biosensor performs nucleic acid extraction, amplification, and detection functions. After amplification, the amplicons labelled with biotin bind to AuNP-detection probe conjugates, which are then immobilized at the test zone through the interaction of biotin and streptavidin (SA). This process generates a visible signal, enabling the detection of the *E. coli* from various samples at POC with the LOD range of 10–1000 CFU mL^{-1} [139]. A recent study developed a rapid and amplification-free colorimetric based nanosensor using AuNPs for detecting SARS-CoV-2 nucleic acids. Thiol-modified antisense oligonucleotides specific to the N-gene of SARS-CoV-2 were used to coat the AuNPs. In the presence of the target RNA sequence, the AuNPs agglomerate, causing a blue color change. Ribonuclease H (RNaseH) treatment further agglomerated the AuNPs, leading to a visible precipitate. The assay takes less than 10 min, with an LOD of 1.8×10^{-7} g/mL. The selectivity of this sensor was confirmed using MERS-CoV viral RNA [140]. A multiplexed colorimetric sensor for the detection of dengue, yellow fever, and Ebola viruses was designed using multi-colored AgNPs. Using this technique, the LODs which were 100–300X lower than the required detection concentration was achieved [123]. A novel and ultra-sensitive colorimetric biosensor for detecting the influenza virus (H5N1) was presented by Xu and co-workers. The biosensor utilized Au nanobi-pyramids and achieved an impressive limit of detection (LOD) of 0.086 mU/mL. It also showed a linear detection range spanning from 0.1 to 5 mU/mL, with the same LOD of 0.086 mU/mL [137].

Colorimetric nanosensors hold great promise in detecting infectious diseases with high sensitivity, rapid response times, simplicity, and cost-effectiveness. They have the potential for POC diagnostics and may significantly improve early and accurate disease detection and management.

8.3.1.4 Fluorescence Nanosensors

Fluorescence-based nanosensors have emerged as a highly promising approach for detecting and monitoring various infectious diseases. These types of biosensors utilize the phenomenon of fluorescence to detect and analyze specific biological targets. The principle behind these sensors lies in the emission of fluorescent light by certain molecules, known as fluorophores, when they absorb energy. Fluorophores can be organic dyes, nanomaterials, or fluorescent proteins. They possess unique optical properties, such as specific absorption and emission wavelengths, quantum yield, and fluorescence lifetime. By selecting the appropriate fluorophore, fluorescence biosensors can be tailored to detect specific biological targets, such as proteins, nucleic acids, small molecules, or even whole cells [141–144].

The detection mechanism of fluorescence biosensors typically involves the recognition element and binding of the target analyte by a recognition element, such as antibodies, aptamers, or molecular imprinted polymers. The recognition element is designed to specifically interact with the target analyte, leading to the formation of a complex. This binding event induces a change in the fluorescent properties of the fluorophore, such as fluorescence intensity, lifetime, or wavelength, which can then be measured and correlated to the presence or concentration of the target analyte [145–147].

One commonly used technique in fluorescence biosensing is the Förster resonance energy transfer (FRET). Förster resonance energy transfer (FRET) occurs when two fluorophores, a donor and an acceptor, are in proximity. The energy from the excited donor fluorophore is transferred to the acceptor fluorophore through non-radiative energy transfer. Förster resonance energy transfer (FRET)-based biosensors enable the detection of biomolecular interactions, conformational changes, or proximity assays, providing valuable information about the binding events and molecular dynamics. Fluorescence biosensors can be designed to operate in various formats, including homogeneous solution-based assays, solid-phase assays, or even imaging agents for cellular or tissue analysis. They can be implemented in simple and portable devices or integrated with advanced technologies, such as microfluidics, lab-on-a-chip systems, or biosensor arrays, to enhance their performance and multiplexing capabilities [141, 143, 145, 147–151].

One of the advantages of fluorescence biosensors is their high sensitivity, allowing for the detection of target analytes at low concentrations. Additionally, they offer real-time or near real-time detection, rapid response times, and the potential for miniaturization and POC applications. Fluorescence biosensors can be designed for quantitative measurements or qualitative screening, depending on the specific requirements of the application. However, like any other technique, there are also challenges associated with fluorescence biosensors. One challenge is the potential for interference from background fluorescence or auto-fluorescence from biological samples, which may affect the accuracy and specificity of the detection. Additionally, careful optimization of the assay conditions, including the choice of fluorophores, recognition elements, and assay buffers, is necessary to achieve optimal sensitivity and selectivity [143, 145, 147, 149, 152].

In recent years, advancements in nanotechnology have contributed to the development of nano-based fluorescence biosensors. Nanomaterials, such as quantum dots (QDs), carbon dots (CDs), or plasmonic NPs, offer unique properties, including high brightness, stability, tunable emission spectra, and increased surface-to-volume ratio, enhancing the performance of fluorescence biosensors [47, 153, 154]. These nanomaterials can be functionalized with recognition elements, amplification strategies, or surface modifications to improve the sensitivity and selectivity of the biosensors. For instance, QDs are semiconductor nanocrystals that emit light when excited by an external energy source, such as UV light. By attaching specific antibodies or antigens to the surface of QDs, they can be used to detect the presence of viruses or bacteria in biological samples [47, 151, 155–157]. When the QDs bind to the target biomolecule, they emit a specific color of light that can be detected and quantified. Another example of a fluorescence-based nanosensor is the use of AuNPs. AuNPs are plasmonic nanomaterials that can be functionalized with specific biomolecules, such as antibodies or DNA strands, which can then bind to target biomolecules in biological samples. When the AuNPs bind to the target biomolecule, they change their optical properties, which can be detected by measuring their fluorescence emission [158, 158].

Hu *et al.* developed a selective and sensitive immunosensor based on fluorescent multifunctional nanosphere (RNs@Au) containing QDs and Au NPs for the detection of Ebola virus (EBOV) glycoprotein. The RNs@Au was modified with antibody (Ab) and streptavidin (SA) to label the target and enhance the signal. This assay was able to detect 2 ng/mL of glycoprotein and achieve an LOD of 0.18 ng/mL in 20 min. Moreover, this assay was able to detect EBOV from spiked urine and tap water sample [160]. Another fluorescence based nanosensor for EBOV detection was developed by Zang and co-workers. In this study, an on-chip sensor for the detection of EBOV antigen was developed using a 3D plasmonic nanoantenna assay sensor. This nanosensor had successfully detects EBOV glycoprotein in human plasma at a concentration as low as 220 fg/mL [161]. Wen *et al.* also developed a graphene oxide (GO)-based fluorescence nanosensor for the detection of EBOV DNA. This sensor was found to be selective and highly sensitive with an LOD of 1.4 pM [162].

Takemura *et al.* reported the development of QDs-based immunofluorescence biosensors for the detection of non-structural protein 1 (NS1) of Zika virus (ZIKV) which operate using LSPR phenomenon from AuNPs.

The fluorescence intensity of QDs was amplified by the LSPR signal from AuNPs. This biosensor was able to detect a wide concentration range (10 fg/mL to 1 ng/mL) of NS1 with an LOD of 1.28 fg/mL [163]. Another ultrasensitive fluorescence-LFA-based biosensor was developed for the detection of thrombocytopenia syndrome virus (SFTSV). This biosensor used high quantum yield fluorescent CDs/SiO_2 nanospheres (CSNs) as reporters. The high sensitivity, specificity, and extended test lifetime of the biosensor with LOD as low as 10 pg mL^{-1} SFTSV nucleoprotein were made possible by the remarkable fluorescent properties of CSNs. Satisfactory results of this biosensor towards clinical samples of human serum of patients were obtained with ease, rendering it a potential POC device [164].

In a different study, Ji *et al.* developed a protein-Au nanocluster (NC)-based fluorescent nanosensor array for quick bacterial identification. This biosensor was able to distinguish between six different types of bacteria, including two different types of drug-resistant bacteria. By using these protein-AuNCs probes, 100% specificity accuracy was achieved [165]. Another study by Zhao *et al.* showcased the development of label-free fluorescence biosensors based on DNA template AgNCs probes for the detection of DNA of HIV type 1 (HIV-1). With an LOD of $2.9*10^{-10}$ mol/L of HIV-1 DNA, the proposed fluorescent sensor demonstrated excellent selectivity and high sensitivity against mismatched target DNA [166]. Li *et al.* demonstrated the fabrication of a user-friendly, highly sensitive fluorescence nanosensor for the detection of H1N1 through a fluorescent probe-based 'OFF–ON' detection protocol [167].

8.4 Remarks

Nanosensors have shown significant promise in the detection of infectious diseases, offering several advantages over traditional diagnostic methods. They often provide real-time detection capabilities of biological and chemical substances, enabling continuous or rapid measurements without the need for complex sample preparation. Additionally, optical nanosensors can be miniaturized and integrated into portable devices, enabling POC testing and on-site monitoring in various settings.

With impressive technological and technical growth in microbiology and gene sequencing, a wide range of nucleic acids (DNA, RNA, non-coding RNA) biomarkers for disease diagnosis are being reported more frequently than before. Having access to such rich genetic information is highly beneficial to the scientists working in POC diagnostics as it will allow for designing probes and assays that ensure high detection specificities and high-throughput detection protocols. This will immensely assist the performance of analysis in biological samples with no nucleic acid amplification carried out in the field-transferable adaptation and optimization of diagnostic assays.

Some remarks on the use of nanosensors for disease detection, including their advantages, drawbacks, recent breakthroughs, and gaps are highlighted below:

8.4.1 Advantages of Nanosensors for Infectious Disease Detection

- **Sensitivity:** nanosensors have the potential to detect diseases at very low concentrations, allowing for early and accurate diagnosis.
- **Speed:** nanosensors can provide rapid results, enabling quick identification of infectious agents and prompt initiation of appropriate treatment.
- **Portability:** nanosensors can be designed to be portable and handheld, making them suitable for use in resource-limited settings and point-of-care diagnostics.
- **Multiplexing:** nanosensors can be engineered to detect multiple pathogens simultaneously, allowing for efficient screening and identification of different infectious agents.
- **Cost-effectiveness:** nanosensors have the potential to be cost-effective, reducing the need for expensive laboratory equipment and personnel.

8.4.2 Drawbacks of Nanosensors for Infectious Disease Detection

- **Development challenges:** designing and fabricating nanosensors with high sensitivity and specificity can be technically challenging, requiring expertise in nanotechnology and materials science.
- **Standardization:** standardizing nanosensor platforms and assay protocols across different laboratories and settings can be difficult, hindering widespread adoption and comparability of results.

- **False Positives and false negatives:** nanosensors may face issues related to false-positive or false-negative results, which can impact the reliability of the diagnostic outcome.
- **Regulatory barriers:** the regulatory pathway for approving nanosensor-based diagnostic devices can be complex and time-consuming, leading to delays in commercialization and clinical implementation.
- **Long-term stability:** ensuring the long-term stability and durability of nanosensors can be a challenge, particularly in real-world settings where they may be exposed to various environmental conditions.

8.4.3 Recent Breakthroughs in Nanosensor-Based Disease Detection

- **Paper-based nanosensors:** researchers have developed paper-based nanosensors that can detect infectious diseases such as malaria, dengue, and Zika virus. These low-cost, portable devices show great potential for use in resource-limited settings.
- **Smartphone-based nanosensors:** integration of nanosensors with smartphones allows for on-site detection and data analysis, enabling real-time disease monitoring and surveillance.
- **Breath-based nanosensors:** nanosensors integrated into breath analysis platforms can detect volatile organic compounds (VOCs) released by infectious agents, offering non-invasive and rapid disease detection.
- **Multiplexed nanosensors:** advances in multiplexed nanosensors enable simultaneous detection of multiple pathogens, reducing the time and cost of diagnosis.

8.4.4 Gaps in Nanosensor-Based Disease Detection

- **Clinical validation:** many nanosensor-based diagnostic platforms are still in the research and development phase, requiring rigorous clinical validation to demonstrate their accuracy, sensitivity, and specificity in real-world patient populations.

- **Access and affordability:** ensuring widespread access to nanosensor-based diagnostics, particularly in low-resource settings, remains a challenge due to factors such as cost, infrastructure limitations, and regulatory barriers.
- **Integration with healthcare systems:** integrating nanosensor-based diagnostics into existing healthcare systems and workflows requires collaboration between researchers, clinicians, and policymakers to address implementation challenges and ensure seamless integration.
- **Privacy and ethical concerns:** the use of nanosensors for disease detection raises concerns about data privacy, security, and the ethical implications of collecting and analyzing personal health information.

While nanosensors hold great promise for infectious disease detection, continued research, technological advancements, and collaboration across disciplines are necessary to overcome the existing challenges and bridge the gaps to achieve their full potential in clinical practice.

References

1. Fauci, A.S., Infectious diseases: considerations for the 21st century. *Clin. Infect. Dis.*, *32*, 5, 675–685, 2001.
2. Bloom, D.E. and Cadarette, D., Infectious disease threats in the twenty-first century: strengthening the global response. *Front. Immunol.*, *10*, 549, 2019.
3. Rohr, J.R., Barrett, C.B., Civitello, D.J., Craft, M.E., Delius, B., DeLeo, G.A., Hudson, P.J., Jouanard, N., Nguyen, K.H., Ostfeld, R.S., Remais, J.V., Emerging human infectious diseases and the links to global food production. *Nat. Sustain.*, *2*, 6, 445–456, 2019.
4. 5 of the world's deadliest infectious diseases, https://www.weforum.org/agenda/2020/04/covid-19-infectious-diseases-tuberculosis-measles-malaria/ (accessed Jul. 17, 2023).
5. The World Health Report 2000: health systems: improving performance, https://apps.who.int/iris/handle/10665/79020 (accessed Jul. 17, 2023).
6. The world health report 1996: fighting disease, fostering development, https://apps.who.int/iris/handle/10665/178481 (accessed Jul. 17, 2023).
7. Ryu, W.S., New emerging viruses, in: *Molecular virology of human pathogenic viruses*, p. 289, 2017.
8. Azkur, A.K., Akdis, M., Azkur, D., Sokolowska, M., van de Veen, W., Brüggen, M.C., O'Mahony, L., Gao, Y., Nadeau, K., Akdis, C.A., Immune response to

SARS-CoV-2 and mechanisms of immunopathological changes in COVID-19. *Allergy, 75*, 7, 1564–1581, 2020.
9. D'Amore, F., Vinacci, G., Agosti, E., Cariddi, L.P., Terrana, A.V., Vizzari, F.A., Mauri, M., Giorgianni, A., Pressing issues in COVID-19: probable cause to seize SARS-CoV-2 for its preferential involvement of posterior circulation manifesting as severe posterior reversible encephalopathy syndrome and posterior strokes. *Am. J. Neuroradiol.*, 41, 10, 1800–1803, 2020.
10. Harapan, H., Itoh, N., Yufika, A., Winardi, W., Keam, S., Te, H., Megawati, D., Hayati, Z., Wagner, A.L., Mudatsir, M., Coronavirus disease 2019 (COVID-19): A literature review. *J. Infect. Public Health, 13*, 5, 667–673, 2020.
11. WHO Coronavirus (COVID-19) Dashboard, WHO Coronavirus (COVID-19) Dashboard With Vaccination Data, https://covid19.who.int/ (accessed Jul. 17, 2023).
12. Delardas, O., Kechagias, K.S., Pontikos, P.N., Giannos, P., Socio-economic impacts and challenges of the coronavirus pandemic (COVID-19): an updated review. *Sustainability, 14*, 15, 9699, 2022.
13. Nicola, M., Alsafi, Z., Sohrabi, C., Kerwan, A., Al-Jabir, A., Iosifidis, C., Agha, M., Agha, R., The socio-economic implications of the coronavirus pandemic (COVID-19): A review. *Int. J. Surg., 78*, 185–193, 2020.
14. Infection prevention and control, https://www.who.int/teams/integrated-health-services/infection-prevention-control (accessed Jul. 17, 2023).
15. Haque, M., McKimm, J., Sartelli, M., Dhingra, S., Labricciosa, F.M., Islam, S., Jahan, D., Nusrat, T., Chowdhury, T.S., Coccolini, F., Iskandar, K., Strategies to prevent healthcare-associated infections: a narrative overview. *Risk Manag. Healthc. Policy*, 13, 1765–1780, 2020.
16. Torres-Sangiao, E., Holban, A.M., Gestal, M.C., Advanced nanobiomaterials: vaccines, diagnosis and treatment of infectious diseases. *Molecules, 21*, 7, 867, 2016.
17. Arji, G., Ahmadi, H., Nilashi, M., Rashid, T.A., Ahmed, O.H., Aljojo, N., Zainol, A., Fuzzy logic approach for infectious disease diagnosis: A methodical evaluation, literature and classification. *Biocybern. Biomed. Eng., 39*, 4, 937–955, 2019.
18. Shih, H.I., Wu, C.J., Tu, Y.F., Chi, C.Y., Fighting COVID-19: A quick review of diagnoses, therapies, and vaccines. *Biomed. J., 43*, 4, 341–354, 2020.
19. Xu, K., Liang, Z.C., Ding, X., Hu, H., Liu, S., Nurmik, M., Bi, S., Hu, F., Ji, Z., Ren, J., Yang, S., Nanomaterials in the prevention, diagnosis, and treatment of mycobacterium tuberculosis infections. *Adv. Healthc. Mater., 7*, 1, 1700509, 2018.
20. Wang, C., Liu, M., Wang, Z., Li, S., Deng, Y., He, N., Point-of-care diagnostics for infectious diseases: From methods to devices. *Nano Today, 37*, 101092, 2021.
21. Sin, M.L., Mach, K.E., Wong, P.K., Liao, J.C., Advances and challenges in biosensor-based diagnosis of infectious diseases. *Expert Rev. Mol. Diagn., 14*, 2, 225–244, 2014.

22. Metkar, S.K. and Girigoswami, K., Diagnostic biosensors in medicine–a review. *Biocatal. Agric. Biotechnol.*, *17*, 271–283, 2019.
23. Chen, H., Liu, K., Li, Z., Wang, P., Point of care testing for infectious diseases. *Clin. Chim. Acta*, *493*, 138–147, 2019.
24. Caliendo, A.M., Gilbert, D.N., Ginocchio, C.C., Hanson, K.E., May, L., Quinn, T.C., Tenover, F.C., Alland, D., Blaschke, A.J., Bonomo, R.A., Carroll, K.C., Better tests, better care: improved diagnostics for infectious diseases. *Clin. Infect. Dis.*, *57*, suppl_3, S139–S170, 2013.
25. Qin, J., Wang, W., Gao, L., Yao, S.Q., Emerging biosensing and transducing techniques for potential applications in point-of-care diagnostics. *Chem. Sci.*, *13*, 10, 2857–2876, 2022.
26. Rich, R.L. and Myszka, D.G., Survey of the 2009 commercial optical biosensor literature. *J. Mol. Recognit.*, *24*, 6, 892–914, 2011.
27. Thévenot, D.R., Toth, K., Durst, R.A., Wilson, G.S., Electrochemical biosensors: recommended definitions and classification. *Biosens. Bioelectron.*, *16*, 1-2, 121–131, 2001.
28. Mehrotra, P., Biosensors and their applications–A review. *J. Oral Biol. Craniofac. Res.*, *6*, 2, 153–159, 2016.
29. Tyśkiewicz, R., Fedorowicz, M., Nakonieczna, A., Zielińska, P., Kwiatek, M., Mizak, L., Electrochemical, optical and mass-based immunosensors: A comprehensive review of Bacillus anthracis detection methods. *Anal. Biochem.*, *675*, 115215, 2023.
30. Gupta, G., Jha, K., Chaudhary, S., Biosensor Assays Types and Their Roles Toward Ligand–Receptor Interactions in Drug Discovery. *ASSAY Drug Dev. Technol.*, *21*, 5, 190–201, 2023.
31. Selvolini, G. and Marrazza, G., Sensor principles and basic designs, in: *Fundamentals of Sensor Technology*, pp. 17–43, Wood head Publishing, Sawston, UK, 2023.
32. Naresh, V. and Lee, N., A review on biosensors and recent development of nanostructured materials-enabled biosensors. *Sensors*, *21*, 4, 1109, 2021.
33. Dincer, C., Bruch, R., Costa-Rama, E., Fernández-Abedul, M.T., Merkoçi, A., Manz, A., Urban, G.A., Güder, F., Disposable sensors in diagnostics, food, and environmental monitoring. *Adv. Mater.*, *31*, 30, 1806739, 2019.
34. Mocan, T., Matea, C.T., Pop, T., Mosteanu, O., Buzoianu, A.D., Puia, C., Iancu, C., Mocan, L., Development of nanoparticle-based optical sensors for pathogenic bacterial detection. *J. Nanobiotechnol.*, *15*, 1–14, 2017.
35. Yoo, S.M. and Lee, S.Y., Optical biosensors for the detection of pathogenic microorganisms. *Trends Biotechnol.*, *34*, 1, 7–25, 2016.
36. Tu, J., Torrente-Rodríguez, R.M., Wang, M., Gao, W., The era of digital health: A review of portable and wearable affinity biosensors. *Adv. Funct. Mater.*, *30*, 29, 1906713, 2020.
37. Campuzano, S., Pedrero, M., Yáñez-Sedeño, P., Pingarrón, J.M., Nanozymes in electrochemical affinity biosensing. *Microchim. Acta*, *187*, 1–16, 2020.

38. Alhadrami, H.A., Biosensors: Classifications, medical applications, and future prospective. *Biotechnol. Appl. Biochem.*, *65*, 3, 497–508, 2018.
39. Mikkelsen, S.R. and Cortón, E., Biosensors, in: *Bioanalytical Chemistry*, pp. 131–149, 2004.
40. Bhattarai, P. and Hameed, S., Basics of biosensors and nanobiosensors, in: *Nanobiosensors: From Design to Applications*, pp. 1–22, 2020.
41. Şahin, S., Ünlü, C., Trabzon, L., Affinity biosensors developed with quantum dots in microfluidic systems. *Emergent Mater.*, *4*, 187–209, 2021.
42. Mitsuma, S.F., Mansour, M.K., Dekker, J.P., Kim, J., Rahman, M.Z., Tweed-Kent, A., Schuetz, P., Promising new assays and technologies for the diagnosis and management of infectious diseases. *Clin. Infect. Dis.*, *56*, 7, 996–1002, 2013.
43. Irkham, I., Ibrahim, A.U., Pwavodi, P.C., Al-Turjman, F., Hartati, Y.W., Smart graphene-based electrochemical nanobiosensor for clinical diagnosis. *Sensors*, *23*, 4, 2240, 2023.
44. Kerry, R.G., Ukhurebor, K.E., Kumari, S., Maurya, G.K., Patra, S., Panigrahi, B., Majhi, S., Rout, J.R., del Pilar Rodriguez-Torres, M., Das, G., Shin, H.S., A comprehensive review on the applications of nano-biosensor-based approaches for non-communicable and communicable disease detection. *Biomater. Sci.*, *9*, 10, 3576–3602, 2021.
45. Frasco, M.F. and Chaniotakis, N., Semiconductor quantum dots in chemical sensors and biosensors. *Sensors*, *9*, 9, 7266–7286, 2009.
46. Devi, P., Saini, S., Kim, K.H., The advanced role of carbon quantum dots in nanomedical applications. *Biosens. Bioelectron.*, *141*, 111158, 2019.
47. Su, S., Fan, J., Xue, B., Yuwen, L., Liu, X., Pan, D., Fan, C., Wang, L., DNA-conjugated quantum dot nanoprobe for high-sensitivity fluorescent detection of DNA and micro-RNA. *ACS Appl. Mater. Interfaces*, *6*, 2, 1152–1157, 2014.
48. Alhabbab, R.Y., Alfaleh, M.A., Alsulaiman, R.M., Alamri, S.S., Eyouni, M.S., ElAssouli, M.Z., Abuzenadah, A.M., Hashem, A.M., Amplifying lateral flow assay signals for rapid detection of COVID-19 specific antibodies. *Global Challenges*, *6*, 7, 2200008, 2022.
49. Andreani, M., Frola, C.E., Caceres, D.H., Canteros, C.E., Rolón, M.J., Chiller, T., Guelfand, L., Validation of a lateral flow assay for rapid diagnosis of histoplasmosis in advanced HIV disease, Buenos Aires, Argentina. *Appl. Microbiol.*, *2*, 4, 950–955, 2022.
50. Seele, P.P., Dyan, B., Skepu, A., Maserumule, C., Sibuyi, N.R.S., Development of gold-nanoparticle-based lateral flow immunoassays for rapid detection of TB ESAT-6 and CFP-10. *Biosensors*, *13*, 3, 354, 2023.
51. Ataide, V.N., Pradela-Filho, L.A., Ameku, W.A., Negahdary, M., Oliveira, T.G., Santos, B.G., Paixão, T.R., Angnes, L., based electrochemical biosensors for the diagnosis of viral diseases. *Microchim. Acta*, *190*, 7, 276, 2023.

52. Yan, J., Wang, J., Liu, H., Wang, L., Yu, K., Deng, L., Su, J., Chen, H., MiR-29b detection in serum using an electrochemical biosensor for the early diagnosis of gestational diabetes. *Anal. Biochem.*, *674*, 115209, 2023.
53. Petrovszki, D., Valkai, S., Gora, E., Tanner, M., Bányai, A., Fürjes, P., Dér, A., An integrated electro-optical biosensor system for rapid, low-cost detection of bacteria. *Microelectron. Eng.*, *239*, 111523, 2021.
54. Bassi, M.D.J., Araujo Todo Bom, M., Terribile Budel, M.L., Maltempi de Souza, E., Müller dos Santos, M., Roman, L.S., Optical Biosensor for the Detection of Infectious Diseases Using the Copolymer F8T2 with Application to COVID-19. *Sensors*, *22*, 15, 5673, 2022.
55. Antman-Passig, M., Wong, E., Frost, G.R., Cupo, C., Shah, J., Agustinus, A., Chen, Z., Mancinelli, C., Kamel, M., Li, T., Jonas, L.A., Optical nanosensor for intracellular and intracranial detection of amyloid-beta. *ACS Nano*, *16*, 5, 7269–7283, 2022.
56. Rabiee, N., Fatahi, Y., Ahmadi, S., Abbariki, N., Ojaghi, A., Rabiee, M., Radmanesh, F., Dinarvand, R., Bagherzadeh, M., Mostafavi, E., Ashrafizadeh, M., Bioactive hybrid metal-organic framework (MOF)-based nanosensors for optical detection of recombinant SARS-CoV-2 spike antigen. *Sci. Total Environ.*, *825*, 153902, 2022.
57. Kearns, H., Goodacre, R., Jamieson, L.E., Graham, D., Faulds, K., SERS detection of multiple antimicrobial-resistant pathogens using nanosensors. *Anal. Chem.*, *89*, 23, 12666–12673, 2017.
58. Singh, P., Surface plasmon resonance: A boon for viral diagnostics, in: *Reference Module in Life Sciences*, Elsevier, Amsterdam, Netherlands, 2017, B978-0-12-809633-8.12245-9.
59. Zhang, C., Wang, C., Xiao, R., Tang, L., Huang, J., Wu, D., Liu, S., Wang, Y., Zhang, D., Wang, S., Chen, X., Sensitive and specific detection of clinical bacteria via vancomycin-modified Fe_3O_4@ Au nanoparticles and aptamer-functionalized SERS tags. *J. Mater. Chem. B*, *6*, 22, 3751–3761, 2018.
60. Chung, J.W., Kim, S.D., Bernhardt, R., Pyun, J.C., Application of SPR biosensor for medical diagnostics of human hepatitis B virus (hHBV). *Sens. Actuators B: Chem.*, *111*, 416–422, 2005.
61. Le Ru, E. and Etchegoin, P., *Principles of Surface-Enhanced Raman Spectroscopy: and related plasmonic effects*, Elsevier, Amsterdam, Netherlands, 2008.
62. Huh, Y.S., Chung, A.J., Erickson, D., Surface enhanced Raman spectroscopy and its application to molecular and cellular analysis. *Microfluid. Nanofluid.*, *6*, 285–297, 2009.
63. Wang, Y., Zhao, C., Wang, J., Luo, X., Xie, L., Zhan, S., Kim, J., Wang, X., Liu, X., Ying, Y., Wearable plasmonic-metasurface sensor for noninvasive and universal molecular fingerprint detection on biointerfaces. *Sci. Adv.*, *7*, 4, eabe4553, 2021.

64. Qian, X.M. and Nie, S.M., Single-molecule and single-nanoparticle SERS: from fundamental mechanisms to biomedical applications. *Chem. Soc. Rev.*, *37*, 5, 912–920, 2008.
65. Liu, Y., Zhou, H., Hu, Z., Yu, G., Yang, D., Zhao, J., Label and label-free based surface-enhanced Raman scattering for pathogen bacteria detection: A review. *Biosens. Bioelectron.*, *94*, 131–140, 2017.
66. Keshavarz, M., Tan, B., Venkatakrishnan, K., Label-free SERS quantum semiconductor probe for molecular-level and *in vitro* cellular detection: a noble-metal-free methodology. *ACS Appl. Mater. Interfaces*, *10*, 41, 34886–34904, 2018.
67. Chen, H., Park, S.K., Joung, Y., Kang, T., Lee, M.K., Choo, J., SERS-based dual-mode DNA aptasensors for rapid classification of SARS-CoV-2 and influenza A/H1N1 infection. *Sens. Actuators B: Chem.*, *355*, 131324, 2022.
68. Yadav, S., Senapati, S., Desai, D., Gahlaut, S., Kulkarni, S., Singh, J.P., Portable and sensitive Ag nanorods based SERS platform for rapid HIV-1 detection and tropism determination. *Colloids Surf. B: Biointerfaces*, *198*, 111477, 2021.
69. Isola, N.R., Alarie, J.P., Griffin, G.D., Vo-Dinh, T., Development of a genosensor for Mycobacterium tuberculosis, in: *Biomedical Sensing, Imaging, and Tracking Technologies I*, 1996, April, vol. 2676, SPIE, pp. 228–240.
70. Botta, R., Chindaudom, P., Eiamchai, P., Horprathum, M., Limwichean, S., Chananonnawathorn, C., Patthanasettakul, V., Kaewseekhao, B., Faksri, K., Nuntawong, N., Tuberculosis determination using SERS and chemometric methods. *Tuberculosis*, *108*, 195–200, 2018.
71. Wang, L., Zhang, X.D., Tang, J.W., Ma, Z.W., Usman, M., Liu, Q.H., Wu, C.Y., Li, F., Zhu, Z.B., Gu, B., Machine learning analysis of SERS fingerprinting for the rapid determination of Mycobacterium tuberculosis infection and drug resistance. *Comput. Struct. Biotechnol. J.*, *20*, 5364–5377, 2022.
72. Kamińska, A., Witkowska, E., Winkler, K., Dzięcielewski, I., Weyher, J.L., Waluk, J., Detection of Hepatitis B virus antigen from human blood: SERS immunoassay in a microfluidic system. *Biosens. Bioelectron.*, *66*, 461–467, 2015.
73. Ngo, H.T., Gandra, N., Fales, A.M., Taylor, S.M., Vo-Dinh, T., Sensitive DNA detection and SNP discrimination using ultrabright SERS nanorattles and magnetic beads for malaria diagnostics. *Biosens. Bioelectron.*, *81*, 8–14, 2016.
74. Garrett, N.L., Sekine, R., Dixon, M.W., Tilley, L., Bambery, K.R., Wood, B.R., Bio-sensing with butterfly wings: naturally occurring nano-structures for SERS-based malaria parasite detection. *Phys. Chem. Chem. Phys.*, *17*, 33, 21164–21168, 2015.
75. Hassanain, W.A., Johnson, C.L., Faulds, K., Graham, D., Keegan, N., Recent advances in antibiotic resistance diagnosis using SERS: focus on the "Big 5" challenges. *Analyst*, *147*, 21, 4674–4700, 2022.
76. Kamińska, A., Witkowska, E., Kowalska, A., Skoczyńska, A., Ronkiewicz, P., Szymborski, T., Waluk, J., Rapid detection and identification of bacterial

meningitis pathogens in *ex vivo* clinical samples by SERS method and principal component analysis. *Anal. Methods*, *8*, 22, 4521–4529, 2016.
77. Owens, N.A., Laurentius, L.B., Porter, M.D., Li, Q., Wang, S., Chatterjee, D., Handheld Raman spectrometer instrumentation for quantitative tuberculosis biomarker detection: a performance assessment for point-of-need infectious disease diagnostics. *Appl. Spectrosc.*, *72*, 7, 1104–1115, 2018.
78. Ngo, H.T., Wang, H.N., Fales, A.M., Nicholson, B.P., Woods, C.W., Vo-Dinh, T., DNA bioassay-on-chip using SERS detection for dengue diagnosis. *Analyst*, *139*, 22, 5655–5659, 2014.
79. Camacho, S.A., Sobral-Filho, R.G., Aoki, P.H.B., Constantino, C.J.L., Brolo, A.G., Zika immunoassay based on surface-enhanced Raman scattering nanoprobes. *ACS Sens.*, *3*, 3, 587–594, 2018.
80. Hamm, L., Gee, A., De Silva Indrasekara, A.S., Recent advancement in the surface-enhanced Raman spectroscopy-based biosensors for infectious disease diagnosis. *Appl. Sci.*, *9*, 7, 1448, 2019.
81. Nie, S. and Emory, S.R., Probing single molecules and single nanoparticles by surface-enhanced Raman scattering. *Science*, *275*, 5303, 1102–1106, 1997.
82. Gehan, H., Fillaud, L., Chehimi, M.M., Aubard, J., Hohenau, A., Felidj, N., Mangeney, C., Thermo-induced electromagnetic coupling in gold/polymer hybrid plasmonic structures probed by surface-enhanced Raman scattering. *ACS Nano*, *4*, 11, 6491–6500, 2010.
83. Mlambo, M., Harris, R.A., Mashazi, P., Sabela, M., Kanchi, S., Madikizela, L.M., Shumbula, P.N., Moloto, N., Hlatshwayo, T.T., Mdluli, P.S., Computational and experimental evaluation of selective substitution of thiolated coumarin derivatives on gold nanoparticles: Surface enhancing Raman scattering and electrochemical studies. *Appl. Surf. Sci.*, *396*, 695–704, 2017.
84. Shumbula, N.P., Ndala, Z.B., Nkabinde, S.S., Kolokoto, T., Shumbula, P.M., Mdluli, P.S., Hlatshwayo, T., Mlambo, M., Moloto, N., Experimental and theoretical studies of 1-alkanethiols SAMs of various chain-lengths on gold nanoparticles and their effect on SERS, stability and surface properties. *J. Mol. Struct.*, *1221*, 128801, 2020.
85. Takemura, K., Surface plasmon resonance (SPR)-and localized SPR (LSPR)-based virus sensing systems: Optical vibration of nano-and micro-metallic materials for the development of next-generation virus detection technology. *Biosensors*, *11*, 8, 250, 2021.
86. Chiu, N.F., The current status and future promise of SPR biosensors. *Biosensors*, *12*, 11, 933, 2022.
87. Sagle, L.B., Ruvuna, L.K., Ruemmele, J.A., Van Duyne, R.P., Advances in localized surface plasmon resonance spectroscopy biosensing. *Nanomedicine*, *6*, 8, 1447–1462, 2011.
88. Balamurugan, S., Mayer, K.M., Lee, S., Soper, S.A., Hafner, J.H., Spivak, D.A., Nanostructure shape effects on response of plasmonic aptamer sensors. *J. Mol. Recognit.*, *26*, 9, 402–407, 2013.

89. Malhotra, B.D. and Ali, M.A., Nanomaterials in biosensors: Fundamentals and applications, in: *Nanomaterials for biosensors*, pp. 1–74, 2018.
90. Mayer, K.M. and Hafner, J.H., Localized surface plasmon resonance sensors. *Chem. Rev.*, *111*, 6, 3828–3857, 2011.
91. Singh, P., SPR biosensors: historical perspectives and current challenges. *Sens. Actuators B: Chem.*, *229*, 110–130, 2016.
92. Puiu, M. and Bala, C., Plasmonic Biosensors for Medical Applications, in: *Encyclopedia of Sensors and Biosensors*, vol. 1-4, 1, pp. 402–420, 2023.
93. Wang, Y., Zhou, J., Li, J., Construction of plasmonic nano-biosensor-based devices for point-of-care testing. *Small Methods*, *1*, 11, 1700197, 2017.
94. Homola, J., Present and future of surface plasmon resonance biosensors. *Anal. Bioanal. Chem.*, *377*, 528–539, 2003.
95. Riedel, T., Rodriguez-Emmenegger, C., de los Santos Pereira, A., Bědajánková, A., Jinoch, P., Boltovets, P.M., Brynda, E., Diagnosis of Epstein–Barr virus infection in clinical serum samples by an SPR biosensor assay. *Biosens. Bioelectron.*, *55*, 278–284, 2014.
96. Jahanshahi, P., Wei, Q., Jie, Z., Ghomeishi, M., Sekaran, S.D., Mahamd Adikan, F.R., Kinetic analysis of IgM monoclonal antibodies for determination of dengue sample concentration using SPR technique. *Bioengineered*, *8*, 3, 239–247, 2017.
97. Srinivasan, B. and Tung, S., Development and applications of portable biosensors. *J. Lab. Autom.*, *20*, 4, 365–389, 2015.
98. Rahman, M.M., Rana, M.M., Rahman, M.S., Anower, M.S., Mollah, M.A., Paul, A.K., Sensitivity enhancement of SPR biosensors employing heterostructure of PtSe2 and 2D materials. *Opt. Mater.*, *107*, 110123, 2020.
99. Prabowo, B.A., Purwidyantri, A., Liu, B., Lai, H.C., Liu, K.C., Gold nanoparticle-assisted plasmonic enhancement for DNA detection on a graphene-based portable surface plasmon resonance sensor. *Nanotechnology*, 32, 9, 095503, 2021.
100. Gür, S.D., Bakhshpour, M., Denizli, A., Selective detection of Escherichia coli caused UTIs with surface imprinted plasmonic nanoscale sensor. *Mater. Sci. Eng.: C*, *104*, 109869, 2019.
101. Kaushik, S., Tiwari, U.K., Pal, S.S., Sinha, R.K., Rapid detection of Escherichia coli using fiber optic surface plasmon resonance immunosensor based on biofunctionalized Molybdenum disulfide (MoS2) nanosheets. *Biosens. Bioelectron.*, *126*, 501–509, 2019.
102. Adegoke, O., Morita, M., Kato, T., Ito, M., Suzuki, T., Park, E.Y., Localized surface plasmon resonance-mediated fluorescence signals in plasmonic nanoparticle-quantum dot hybrids for ultrasensitive Zika virus RNA detection via hairpin hybridization assays. *Biosens. Bioelectron.*, *94*, 513–522, 2017.
103. Cao, J., Sun, T., Grattan, K.T., Gold nanorod-based localized surface plasmon resonance biosensors: A review. *Sens. Actuators B: Chem.*, *195*, 332–351, 2014.

104. Xu, T. and Geng, Z., Strategies to improve performances of LSPR biosensing: Structure, materials, and interface modification. *Biosens. Bioelectron., 174*, 112850, 2021.
105. Chowdhury, A.D., Nasrin, F., Gangopadhyay, R., Ganganboina, A.B., Takemura, K., Kozaki, I., Honda, H., Hara, T., Abe, F., Park, S., Suzuki, T., Controlling distance, size and concentration of nanoconjugates for optimized LSPR based biosensors. *Biosens. Bioelectron., 170*, 112657, 2020.
106. Hammond, J.L., Bhalla, N., Rafiee, S.D., Estrela, P., Localized surface plasmon resonance as a biosensing platform for developing countries. *Biosensors, 4*, 2, 172–188, 2014.
107. Zhao, J., Zhang, X., Yonzon, C.R., Haes, A.J., Van Duyne, R.P., Localized Surface Plasmon Resonance Biosensors. *Nanomedicine, 1*, 2, 219–228, 2006.
108. Hall, W.P., Ngatia, S.N., Van Duyne, R.P., LSPR biosensor signal enhancement using nanoparticle– antibody conjugates. *J. Phys. Chem. C, 115*, 5, 1410–1414, 2011.
109. Pellas, V., Hu, D., Mazouzi, Y., Mimoun, Y., Blanchard, J., Guibert, C., Salmain, M., Boujday, S., Gold nanorods for LSPR biosensing: synthesis, coating by silica, and bioanalytical applications. *Biosensors, 10*, 10, 146, 2020.
110. Damborský, P., Švitel, J., Katrlík, J., Optical biosensors. *Essays Biochem., 60*, 1, 91–100, 2016.
111. Lee, J.H., Kim, B.C., Oh, B.K., Choi, J.W., Highly sensitive localized surface plasmon resonance immunosensor for label-free detection of HIV-1. *Nanomed. Nanotechnol. Biol. Med., 9*, 7, 1018–1026, 2013.
112. Lee, T., Kim, G.H., Kim, S.M., Hong, K., Kim, Y., Park, C., Sohn, H., Min, J., Label-free localized surface plasmon resonance biosensor composed of multi-functional DNA 3 way junction on hollow Au spike-like nanoparticles (HAuSN) for avian influenza virus detection. *Colloids Surf. B: Biointerfaces, 182*, 110341, 2019.
113. Sun, W., Yuan, S., Huang, H., Liu, N., Tan, Y., A label-free biosensor based on localized surface plasmon resonance for diagnosis of tuberculosis. *J. Microbiol. Methods, 142*, 41–45, 2017.
114. Zandieh, M., Hosseini, S.N., Vossoughi, M., Khatami, M., Abbasian, S., Moshaii, A., Label-free and simple detection of endotoxins using a sensitive LSPR biosensor based on silver nanocolumns. *Anal. Biochem., 548*, 96–101, 2018.
115. dos Santos Pires, A.C., Soares, N.D.F.F., da Silva, L.H.M., da Silva, M.D.C.H., De Almeida, M.V., Le Hyaric, M., de Andrade, N.J., Soares, R.F., Mageste, A.B., Reis, S.G., A colorimetric biosensor for the detection of foodborne bacteria. *Sens. Actuators B: Chem., 153*, 1, pp.17–23, 2011.
116. Zhao, W., Brook, M.A., Li, Y., Design of gold nanoparticle-based colorimetric biosensing assays. *ChemBioChem, 9*, 15, 2363–2371, 2008.
117. Song, Y., Wei, W., Qu, X., Colorimetric biosensing using smart materials. *Adv. Mater., 23*, 37, 4215–4236, 2011.

118. Nayak, S.S. and Pattabiraman, T.N., A new colorimetric method for the estimation of glycosylated hemoglobin. *Clin. Chim. Acta*, *109*, 3, 267–274, 1981.
119. Muthwa, S.F., Zulu, N.S., Kistan, M., Onwubu, S.C., Shumbula, N.P., Moloto, N., Mpelane, S., Hlatshwayo, T., Mlambo, M., Mdluli, P.S., Unravelling mechanism for detecting chromium on functionalized gold nanoparticles via a smartphone and spectrophotometric-based systems supported by CIEL* a* b* colour space and molecular dynamics. *J. Mol. Struct.*, *1274*, 134394, 2023.
120. Muthwa, S.F., Makhanya, T.R., Mlambo, M., Shumbula, N.P., Shumbula, P.M., Onwubu, S.C., Moloto, N., Gengan, R.M., Mdluli, P.S., Synthesis and characterization of 3-(p-tolyl)-2, 3-dihydropyrazolo [3, 4-b] indole-1 (4H)-carbothioamide functionalized gold nanoparticles for the CIEL* a* b*/ Yxy colorimetric detection of Cr (VI). *J. Mol. Struct.*, *1209*, 127985, 2020.
121. Zarlaida, F. and Adlim, M., Gold and silver nanoparticles and indicator dyes as active agents in colorimetric spot and strip tests for mercury (II) ions: a review. *Microchim. Acta*, *184*, 45–58, 2017.
122. Ramesh, M., Janani, R., Deepa, C., Rajeshkumar, L., Nanotechnology-enabled biosensors: A review of fundamentals, design principles, materials, and applications. *Biosensors*, *13*, 1, 40, 2022.
123. Yen, C.W., de Puig, H., Tam, J.O., Gómez-Márquez, J., Bosch, I., Hamad-Schifferli, K., Gehrke, L., Multicolored silver nanoparticles for multiplexed disease diagnostics: distinguishing dengue, yellow fever, and Ebola viruses. *Lab Chip*, *15*, 7, 1638–1641, 2015.
124. Azab, M.Y., Hameed, M.F.O., Obayya, S.S., Overview of optical biosensors for early cancer detection: fundamentals, applications and future perspectives. *Biology*, *12*, 2, 232, 2023.
125. Li, D., Cheng, W., Yan, Y., Zhang, Y., Yin, Y., Ju, H., Ding, S., A colorimetric biosensor for detection of attomolar microRNA with a functional nucleic acid-based amplification machine. *Talanta*, *146*, 470–476, 2016.
126. Li, Y., Zhu, J., Zhang, H., Liu, W., Ge, J., Wu, J., Wang, P., High sensitivity gram-negative bacteria biosensor based on a small-molecule modified surface plasmon resonance chip studied using a laser scanning confocal imaging-surface plasmon resonance system. *Sens. Actuators B: Chem.*, *259*, 492–497, 2018.
127. Yu, T., Xu, H., Zhao, Y., Han, Y., Zhang, Y., Zhang, J., Xu, C., Wang, W., Guo, Q., Ge, J., Aptamer based high throughput colorimetric biosensor for detection of staphylococcus aureus. *Sci. Rep.*, *10*, 1, 9190, 2020.
128. Guo, J., Wu, S., Wang, Y., Zhao, M., A label-free fluorescence biosensor based on a bifunctional MIL-101 (Fe) nanozyme for sensitive detection of choline and acetylcholine at nanomolar level. *Sens. Actuators B: Chem.*, *312*, 128021, 2020.
129. Wu, Y., Boonloed, A., Sleszynski, N., Koesdjojo, M., Armstrong, C., Bracha, S., Remcho, V.T., Clinical chemistry measurements with commercially available test slides on a smartphone platform: Colorimetric determination of glucose and urea. *Clin. Chim. Acta*, *448*, 133–138, 2015.

130. Choi, C.K., Shaban, S.M., Moon, B.S., Pyun, D.G., Kim, D.H., Smartphone-assisted point-of-care colorimetric biosensor for the detection of urea via pH-mediated AgNPs growth. *Anal. Chim. Acta, 1170*, 338630, 2021.
131. Hao, N., Lu, J., Zhou, Z., Hua, R., Wang, K., A pH-resolved colorimetric biosensor for simultaneous multiple target detection. *ACS Sens., 3*, 10, 2159–2165, 2018.
132. Zheng, J., Song, D., Chen, H., Xu, J., Alharbi, N.S., Hayat, T., Zhang, M., Enhanced peroxidase-like activity of hierarchical MoS2-decorated N-doped carbon nanotubes with synergetic effect for colorimetric detection of H2O2 and ascorbic acid. *Chin. Chem. Lett., 31*, 5, 1109–1113, 2020.
133. Wang, B., Liu, F., Wu, Y., Chen, Y., Weng, B., Li, C.M., Synthesis of catalytically active multielement-doped carbon dots and application for colorimetric detection of glucose. *Sens. Actuators B: Chem., 255*, 2601–2607, 2018.
134. Zhong, Q., Chen, Y., Su, A., Wang, Y., Synthesis of catalytically active carbon quantum dots and its application for colorimetric detection of glutathione. *Sens. Actuators B: Chem., 273*, 1098–1102, 2018.
135. Ma, X. and Miao, P., Silver nanoparticle@ DNA tetrahedron-based colorimetric detection of HIV-related DNA with cascade strand displacement amplification. *J. Mater. Chem. B, 7*, 16, 2608–2612, 2019.
136. Sandlund, J., Lim, S., Queralto, N., Huang, R., Yun, J., Taba, B., Song, R., Odero, R., Ouma, G., Sitati, R., Murithi, W., Development of colorimetric sensor array for diagnosis of tuberculosis through detection of urinary volatile organic compounds. *Diagn. Microbiol. Infect. Dis., 92*, 4, 299–304, 2018.
137. Xu, S., Ouyang, W., Xie, P., Lin, Y., Qiu, B., Lin, Z., Chen, G., Guo, L., Highly uniform gold nanobipyramids for ultrasensitive colorimetric detection of influenza virus. *Anal. Chem., 89*, 3, 1617–1623, 2017.
138. Kim, H., Park, M., Hwang, J., Kim, J.H., Chung, D.R., Lee, K.S., Kang, M., Development of label-free colorimetric assay for MERS-CoV using gold nanoparticles. *ACS Sens., 4*, 5, 1306–1312, 2019.
139. Choi, J.R., Hu, J., Tang, R., Gong, Y., Feng, S., Ren, H., Wen, T., Li, X., Abas, W.A.B.W., Pingguan-Murphy, B., Xu, F., An integrated paper-based sample-to-answer biosensor for nucleic acid testing at the point of care. *Lab Chip, 16*, 3, 611–621, 2016.
140. Moitra, P., Alafeef, M., Dighe, K., Frieman, M.B., Pan, D., Selective naked-eye detection of SARS-CoV-2 mediated by N gene targeted antisense oligonucleotide capped plasmonic nanoparticles. *ACS Nano, 14*, 6, 7617–7627, 2020.
141. Kim, S.J., Choi, S.J., Jang, J.S., Cho, H.J., Kim, I.D., Innovative nanosensor for disease diagnosis. *Acc. Chem. Res., 50*, 7, 1587–1596, 2017.
142. Saylan, Y. and Denizli, A., Virus detection using nanosensors, in: *Nanosensors for smart cities*, pp. 501–511, Elsevier, Amsterdam, Netherlands, 2020.
143. Deng, J., Zhao, S., Liu, Y., Liu, C., Sun, J., Nanosensors for diagnosis of infectious diseases. *ACS Appl. Bio Mater., 4*, 5, 3863–3879, 2020.

144. Liu, M., Qiu, J.G., Ma, F., Zhang, C.Y., Advances in single-molecule fluorescent nanosensors. *Wiley Interdiscip. Rev. Nanomed. Nanobiotechnol.*, *13*, 5, e1716, 2021.
145. Tainaka, K., Sakaguchi, R., Hayashi, H., Nakano, S., Liew, F.F., Morii, T., Design strategies of fluorescent biosensors based on biological macromolecular receptors. *Sensors*, *10*, 2, 1355–1376, 2010.
146. VanEngelenburg, S.B. and Palmer, A.E., Fluorescent biosensors of protein function. *Curr. Opin. Chem. Biol.*, *12*, 1, 60–65, 2008.
147. De Lorimier, R.M., Smith, J.J., Dwyer, M.A., Looger, L.L., Sali, K.M., Paavola, C.D., Rizk, S.S., Sadigov, S., Conrad, D.W., Loew, L., Hellinga, H.W., Construction of a fluorescent biosensor family. *Protein Sci.*, *11*, 11, 2655–2675, 2002.
148. Chen, G., Song, F., Xiong, X., Peng, X., Fluorescent nanosensors based on fluorescence resonance energy transfer (FRET). *Ind. Eng. Chem. Res.*, *52*, 33, 11228–11245, 2013.
149. He, H., Sun, D.W., Wu, Z., Pu, H., Wei, Q., On-off-on fluorescent nanosensing: Materials, detection strategies and recent food applications. *Trends Food Sci. Technol.*, *119*, 243–256, 2022.
150. Chen, M., Yan, Z., Han, L., Zhou, D., Wang, Y., Pan, L., Tu, K., Upconversion Fluorescence Nanoprobe-Based FRET for the Sensitive Determination of Shigella. *Biosensors*, *12*, 10, 795, 2022.
151. Chini, M.K., Kumar, V., Javed, A., Satapathi, S., Graphene quantum dots and carbon nano dots for the FRET based detection of heavy metal ions. *Nano-Struct. Nano-Objects*, *19*, 100347, 2019.
152. Liu, R., Ali, S., Haruna, S.A., Ouyang, Q., Li, H., Chen, Q., Development of a fluorescence sensing platform for specific and sensitive detection of pathogenic bacteria in food samples. *Food Control*, *131*, 108419, 2022.
153. Zhang, Z., Shi, Y., Pan, Y., Cheng, X., Zhang, L., Chen, J., Li, M.J., Yi, C., Quinoline derivative-functionalized carbon dots as a fluorescent nanosensor for sensing and intracellular imaging of Zn 2+. *J. Mater. Chem. B*, *2*, 31, 5020–5027, 2014.
154. Liang, N., Hu, X., Li, W., Mwakosya, A.W., Guo, Z., Xu, Y., Huang, X., Li, Z., Zhang, X., Zou, X., Shi, J., Fluorescence and colorimetric dual-mode sensor for visual detection of malathion in cabbage based on carbon quantum dots and gold nanoparticles. *Food Chem.*, *343*, 128494, 2021.
155. Tong, X., Shi, S., Tong, C., Iftikhar, A., Long, R., Zhu, Y., Quantum/carbon dots-based fluorescent assays for enzyme activity. *TrAC, Trends Anal. Chem.*, *131*, 116008, 2020.
156. He, J., Zhang, H., Zou, J., Liu, Y., Zhuang, J., Xiao, Y., Lei, B., Carbon dots-based fluorescent probe for "off-on" sensing of Hg (II) and I–. *Biosens. Bioelectron.*, *79*, 531–535, 2016.
157. Khan, Z.G. and Patil, P.O., A comprehensive review on carbon dots and graphene quantum dots based fluorescent sensor for biothiols. *Microchem. J.*, *157*, 105011, 2020.

158. Ding, C. and Tian, Y., Gold nanocluster-based fluorescence biosensor for targeted imaging in cancer cells and ratiometric determination of intracellular pH. *Biosens. Bioelectron.*, *65*, 183–190, 2015.
159. Gaviria-Arroyave, M.I., Cano, J.B., Peñuela, G.A., Nanomaterial-based fluorescent biosensors for monitoring environmental pollutants: A critical review. *Talanta Open*, *2*, 100006, 2020.
160. Hu, J., Jiang, Y.Z., Wu, L.L., Wu, Z., Bi, Y., Wong, G., Qiu, X., Chen, J., Pang, D.W., Zhang, Z.L., Dual-signal readout nanospheres for rapid point-of-care detection of ebola virus glycoprotein. *Anal. Chem.*, *89*, 24, 13105–13111, 2017.
161. Zang, F., Su, Z., Zhou, L., Konduru, K., Kaplan, G., Chou, S.Y., Ultrasensitive Ebola virus antigen sensing via 3D nanoantenna arrays. *Adv. Mater.*, *31*, 30, 1902331, 2019.
162. Wen, J., Li, W., Li, J., Tao, B., Xu, Y., Li, H., Lu, A., Sun, S., Study on rolling circle amplification of Ebola virus and fluorescence detection based on graphene oxide. *Sens. Actuators B: Chem.*, *227*, 655–659, 2016.
163. Takemura, K., Adegoke, O., Suzuki, T., Park, E.Y., A localized surface plasmon resonance-amplified immunofluorescence biosensor for ultrasensitive and rapid detection of nonstructural protein 1 of Zika virus. *PLoS One*, *14*, 1, e0211517, 2019.
164. Xu, L.D., Zhang, Q., Ding, S.N., Xu, J.J., Chen, H.Y., Ultrasensitive detection of severe fever with thrombocytopenia syndrome virus based on immunofluorescent carbon dots/SiO2 nanosphere-based lateral flow assay. *ACS Omega*, *4*, 25, 21431–21438, 2019.
165. Ji, H., Wu, L., Pu, F., Ren, J., Qu, X., Point-of-care identification of bacteria using protein-encapsulated gold nanoclusters. *Adv. Healthc. Mater.*, *7*, 13, 1701370, 2018.
166. Zhao, H., Wang, L., Zhu, J., Wei, H., Jiang, W., Label-free nucleic acids detection based on DNA templated silver nanoclusters fluorescent probe. *Talanta*, *138*, 163–168, 2015.
167. Li, X., Lu, D., Sheng, Z., Chen, K., Guo, X., Jin, M., Han, H., A fast and sensitive immunoassay of avian influenza virus based on label-free quantum dot probe and lateral flow test strip. *Talanta*, *100*, 1–6, 2012.

9

Aptasensors: Selective and Powerful Tools for Infectious Diseases Diagnosis

Seele, P. P.[1,2*], Van der Walt, H.[1,2], Sibuyi, N.R.S.[1,3] and Maserumule, M.C.[1,2]

[1]*Health Platform, Advanced Materials Division, Mintek, Randburg, Johannesburg, South Africa*
[2]*Department of Science and Innovation/Mintek Nanotechnology Innovation Centre, Biolabels Development Node, Advanced Materials Division, Mintek, Randburg, Johannesburg, South Africa*
[3]*Department of Science and Innovation/Mintek Nanotechnology Innovation Centre, Biolabels Research Node, Department of Biotechnology, University of the Western Cape, Bellville, South Africa*

Abstract

A timely and accurate diagnosis is a critical aspect of disease prevention, management and treatment; which requires rapid, cost-effective, easy-to-use and robust tools at point of care (PoC). The conventional rapid diagnostic systems are based on antibody-antigen interactions. In recent years, aptamers have been shown to display similar characteristics to antibodies, and unlike antibodies, they are stable and possess high selectivity. Aptamers fold into three-dimensional (3D) structures that recognize and target various biomolecules such as metal ions, proteins, and whole cells. Aptamer-based detection systems (aptasensors) are gaining popularity in disease diagnosis, in this case, the aptamers interact with the target analyte with high specificity like antibody-antigen reactions, and a signal is generated which can be detectable by various types of biosensors including optical and electrochemical-based techniques. The colorimetric method affords visualization of the test results through the naked eye, without a need for analytical instrumentation, making PoC diagnosis deliverable. This chapter delves into the use of aptasensors for the diagnosis of infectious diseases. Aptasensors were extensively studied for the detection of pathogenic microorganisms such as severe acute respiratory syndrome coronavirus 2 (SARS CoV-2), *Mycobacterium tuberculosis*, *Brucella*

**Corresponding author*: palesas@mintek.co.za

Suvardhan Kanchi, Ayyappa Bathinapatla, Anitha Varghese and Phumlane Selby Mdluli (eds.) Nano-Biosensor Technologies for Diagnosis of Infectious Diseases, (279–314) © 2025 Scrivener Publishing LLC

melitensis, human immunodeficiency virus-1 (HIV-1), *Listeria monocytogenes*, and avian influenza virus, among others. A detailed overview will be given on the potential use of aptamers in diagnostics, especially in the diagnosis of infectious diseases, and the status of their commercialization to date.

Keywords: Aptamers, aptasensors, biomarkers, diagnostics, infectious diseases, point of care

9.1 Introduction

By definition, biosensors are analytical tools that can provide qualitative and quantitative data from the interaction of a biological probe with a target analyte producing a chemical or physical signal that is converted by a transducer into a signal that is measurable [1–3]. The very first sensor that was developed was the pH sensor, in 1909 by Søren Sørensen, however, the grand work on biosensors was laid by Leland C. Clark, who was considered as the 'father of biosensors,' the first one to demonstrate the concept of a glucose test in 1962 [2, 4]. Antibodies have been the 'gold standard' biomolecular capture probe for clinical diagnostic systems. The introduction of aptamers, which have been termed 'chemical or artificial antibodies' [1] might offer relief and improve the performance of these systems. The growing interest in aptamers over antibodies stems from their flexible structure, small size, ease and robustness in their chemical production, ease of modification, cost-effectiveness, high stability and lack of immunogenicity [5, 6].

In aptasensors, the biosensor employs aptamers as the biomolecular capture probe. Aptasensors were first described in 1996 by Drolet and colleagues [7], following the emergence of the enzyme-linked oligonucleotide assay (ELONA) [7, 8]. Since then, different types of aptasensors have been developed and categorized into optical, electrochemical, and acoustic, depending on the type of biomolecular recognition probe method of analysis or transducer [9]. Aptasensors can, in addition, be classified as either being labelled or label-free [3, 10]. Usually, optical and electrochemical types of aptasensors are well suited for point of care (PoC), hence, the focus in the following sections.

Infectious diseases causing endemics or pandemics have prompted the need to innovate in the field of diagnostics, especially with devices that cater at PoC. Diagnostic tools that are robust and timely are critical in combating the spread of diseases. Aptasensors have been extensively researched for the detection of various pathogenic microorganisms such as *Mycobacterium tuberculosis* [11–14], *Brucella melitensis* [15], severe acute respiratory syndrome coronavirus 2 (SARS CoV-2) [16–20], human immunodeficiency virus-1 (HIV-1) [21–24], *Listeria monocytogenes* [25, 26], *Plasmodium falciparum* [27–29], and avian influenza virus, among others [9]. However, the most recent advances in electrochemical aptasensors have been limited to non-infectious diseases mostly related to cancer and cardiac disease; with glucose electrochemical aptasensors being the most patented [9]. This can be attributed to various factors including the type of biomarker or the test sample of the infectious disease. Biomarkers are an important aspect of consideration when designing appropriate aptasensors for infectious diseases, with some targets thought to be more 'aptagenic' than others [30]. Biomarkers can be classified according to their properties and/or their applications; where in the first group they can either be imaging biomarkers or biomarkers that are sub-grouped into body fluid, biopsy or volatile organic compounds from breath. Secondly, the diagnostic purpose determines their classification where they could either be for diagnostics, prognosis, disease staging, for susceptibility and for pharmacodynamics/response [9]. Thus, aptasensors are not only limited to just diagnostics but their use can be implemented for different types of tests. This chapter will expand on the characteristics of aptasensors that influence their diagnostic application, their commercial progress, and clinical relevance.

9.2 Aptamers as Selective and Powerful Tools for Diagnostics

Aptamers are single stranded-DNA (ss-DNA) or single stranded RNA (ss-RNA) oligonucleotides, that can fold into three-dimensional (3D) structures that recognize and target various biological molecules from simple metal ions to more complex macromolecules such as proteins, and whole cells [2, 8, 9]. The word 'aptamer' originates from Latin and Greek words 'aptus,' meaning to fit and 'meros,' which means particle [10]. A broad spectra of aptamer targets contrasts that of antibodies, which makes them versatile as probes. Several structural motifs are adopted by the nucleic acid chains, which are between 10 and 100 bases long; hairpins, loops, G-quadruplex structures, kissing complexes, purine rich bulges, stems, and pseudoknots [2, 11]. Structural conformation is a determinant and dictating factor of specificity and function; this is a principle that is well-known.

Adaptive conformation, also known as induced conformational change, allows aptamers to bind to a variety of targets, which undergo their own conformational changes to bind with high affinity and specificity to the aptamer [11, 12]. The dissociation constant for these molecules can be as low as in the picomolar range, and impressively so, aptamers are able to differentiate structural variations as subtle a change as a methyl group and between enantiomers [2, 13, 14]. The interactions between aptamer and target are mainly through the van der Waals forces, hydrophobic interactions, hydrogen bonding and electrostatic interactions [2, 11, 15].

It is well-accepted that the function of molecules is often determined by their structure and, in some instances, inferred from knowledge accumulated from molecules with similar morphologies. Aptamers as biorecognition molecules of aptasensors, are an inherent component to their functionality, making it essential to understand how their structure influences their proper functioning. Although aptamers can be either DNA or RNA, the instability of RNA (especially in the selection process) has been one of the drawbacks toward commercial use of aptasensors [9], hence, most of the discovered aptamers are DNA rather than RNA [30]. This counters their advantageous flexibility of being able to form various structural morphologies as opposed to the more rigid and stable double-stranded helical DNA structure [31, 32]. Proteins are the main targets of aptamers, followed by small molecules, cells, nucleic acids, and viruses [30]. The selectivity and specificity of aptasensors is very dependent on the structure of aptamers, which fold into 3D structures that mimic the conventionally used antibodies.

Aptamers have stronger binding affinity to their targets, comparable and at times higher than that of antibodies. Their dissociation constant's (K_d value) is usually in the low- to sub-nanomolar range for protein targets, and in the low- to sub-micromolar range for small molecule targets [30, 32]. The successful application of aptamers is equally dependent on their specificity, which is reflected by the high K_d of the respective target versus K_d of the non-target. Interactions and binding forces between aptamers and their targets influence these parameters which include the van der Waals forces, stacking of aromatic rings, hydrogen bonding, electrostatic forces, salt bridges, and shape complementarity [1, 8, 9].

Both RNA and DNA aptamers can be highly selective and discriminate against their target analyte. An interesting study by Jenison and colleagues successfully showed selectivity in their study wherein an oligonucleotide from a library was able to differentiate caffeine from theophylline by the steric hindrance emanating from the methyl group situated at R3 of the N-7 atom, with an impressive 10^4 fold higher K_d for caffeine [33]. Moreover, selectivity is determined by the intermolecular hydrogen bonding between the 1-2-3 stacking and S-turn motifs, and the interlocking base zipper [34]. Another interesting example where the selectivity of aptamers was displayed, was with the L- and D-enantiomers of the oligopeptide arginine-vasopressin. The DNA aptamer was able to discriminate between the two enantiomers with a K_d of 1.1 μM^{-1} for the L-enantiomer and a reported non-significant binding to the D-enantiomers [35, 36]. Given that shape is a determinant of molecular recognition and binding, conformational adaptations that occur at the binding site together with the target analyte (especially occurs with peptide targets) [37, 38] are important to characterize and comprehend so that relevant surface modifications can be made to enhance stability and the binding affinity. These interactions can be complex and multi-layered, due to the diverse architectural patterns of the target molecules, the design and behavior of the aptamers will be impacted; Patel and colleagues have reviewed this phenomenon [32]. As previously mentioned, certain aptamer targets are considered more 'aptagenic' than others, these are targets that have been successfully used to produce aptamers *in vitro* [30]. Moreover, conformational modifications of aptamers can be ascribed to environmental changes such as pH, temperature, ionic strength, charge, and due to immobilization of biomolecules on the surface of the electrode [39].

9.2.1 Architecture of RNA and DNA Aptamer Molecules

Nucleic acid–protein interactions are not new to biology, they serve in intricate biological processes such as in replication, transcription, translation, DNA repair, splicing, modifications, packaging, and RNA silencing [40, 41]. Therefore, it is not surprising that the aptamer design is such that binding of molecules is efficient in selectivity towards its target. Nucleic acids act as scaffolds that provide various binding sites such as the wider but shallow major groove, the narrower but deeper minor groove formed by the double-stranded helix, the electron rich purine rings and through intercalation of molecules between adjacent base pairs [41–43]. Tags can also be introduced by modifications at the C5 position [42]. Purine bases or adjacent base pairs are able to partition moderately, allowing the binding of small molecules with planar, aromatic groups such as those in adenosine monophosphate (AMP) and flavin mononucleotide (FMN) [32]. Stacking between the aromatic rings culminates in an increase in hydrophobic interactions as well as the van der Waals interactions, which are cumulative thereby increasing the stability of the established interaction. This can be observed in the FMN-RNA aptamer complex, wherein the isoalloxazine ring and the adjacent triple base displayed extensive overlap through the nuclear magnetic resonance (NMR) and molecular dynamics [44]. The nucleic acid grooves are ideal binding sites for proteins as they provide potential for hydrophobic interactions, hydrogen bonding and electrostatic interactions. The negatively charged backbone of the aptamers has a propensity for binding to the positively charged protein moieties, such as those in thrombin, nuclear factor-κB, tumor necrosis factor-α, and MS2 capsid of the bacteriophage [45–47]. It is this well-defined distribution of aptamer charges that enhances the affinity to targets, better than with antibodies [48]; it is postulated that the initial interactions between aptamer and ligand occurs via electrostatic forces [49].

Nucleic acid backbone and/or base replacements can be employed through some specific sequence chemistries to improve the binding affinity and stability of aptamers [50]. This malleability of aptamers offers a great advantage over antibodies which would require more complex methods to modify them. Peptide nucleic acids have been widely used as replacement due to their amide rich backbone which renders them neutral and able to escape nuclease degradation [50]. Selectivity is dependent on the potential intermolecular hydrogen bonding pattern between the nucleic acid bases with the target molecule, an interaction that is able to discriminate between the acceptor-donor hydrogen partners. The molecular recognition between a DNA aptamer and its target, argininamide was elegantly described by Lin and Patel using NMR and molecular dynamics for resolving the structural complex [51]. The binding pocket was formed by the DNA stem-loop fold; encapsulating the argininamide moiety which resulted in hydrogen bonding and stacking with nucleic acid bases thereby stabilizing the interactions. For the binding to occur, the 24-base long DNA aptamer undergoes conformational adaptation that creates an architectural landscape where the argininamide can fit in and bind [51]. Adaptive conformational changes will be dealt with in the section to follow, describing structural changes that the interacting molecules undergo, be it folding and/or unfolding to accommodate particular shapes and binding specificities.

9.2.2 Adaptive Conformational Change

In an aqueous environment, biological molecules including nucleic acids and proteins, are not rigid but exist dynamically, which allows them to transform into their optimum conformation upon recognition and binding of their ligand. Upon this transition, optimum functionality can thus be assumed and sometimes, requiring that the ligand also undergo adaptive conformational changes. Adaptive or induced conformational changes are most often a requisite for complex formation; wherein complementarity and recognition elements are formulated by structural adjustment of partners [34, 35, 52]. In general, ss-RNA aptamers are flexible, partially structured and capable of various tertiary structural configurations that ultimately create a pocket to fit the ligand [32, 45]. Otherwise aptamers can be engineered to undergo adaptive conformational changes which increases binding affinity [1].

Once binding has occurred, the labile and unstructured segments of the aptamer are stabilized becoming more rigid and structured [53]. Figure 9.1 demonstrates this phenomenon.

There are numerous examples demonstrating this phenomenon; a study by Battiste and colleagues described the adaptive conformational changes of an aptamer that selectively binds to the human immunodeficiency virus type 1 (HIV-1) Rev peptide. The phosphate backbone adjacent to the G48.G71 purine base pairs structurally adopts an unusual S-shaped conformation that widens the surface of the major groove, crafting a pocket which accommodates a deeper penetration of the HIV-1 Rev α-helix [38]. It is the enclosure of the larger surface area of the protein by the nucleic acid which is reported to be fundamental for recognition and specificity between the molecules, this is a well-known phenomenon. In this case, mutational studies show that four amino acids (Arg35, Arg39, Arg44, and Asn40) are indispensable for the hydrogen bonding interaction formed with the U45, G46, and G47 bases [38]. These residues were reported to be crucial for the recognition and specificity required for the hydrogen bonding. This further emphasizes the importance of characterizing the interactions between aptamer and target, so that the aptamer can be tailored to achieve optimum binding affinity and selectivity.

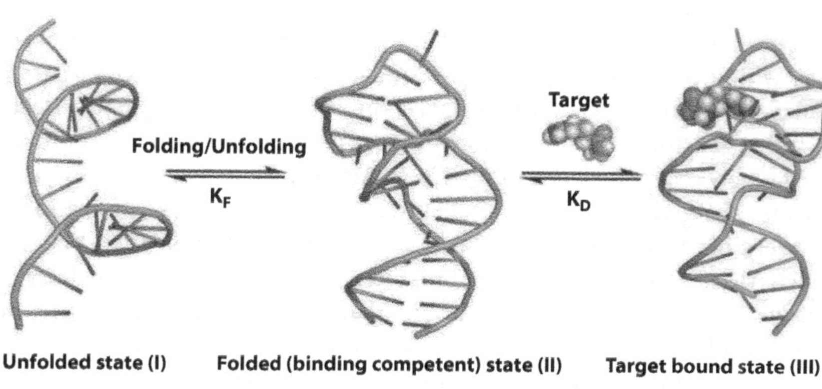

Figure 9.1 Illustration of adaptive conformational change.
The K_F represents the equilibrium constant between the unfolded (I) and the binding competent state (II) of the aptamer; the K_D, is the dissociation constant between the binding competent state (II) and the bond state (III). Reproduced with permissions from MDPI [53].

9.2.3 Impact of the Electrode Surface on Aptamer Conformation

Immobilization of biological molecules such as enzymes and aptamers may be detrimental for their intended function as shape complementarity is of utmost importance for ligand recognition, specificity and binding. It is therefore important that we comprehend how the electrode surface affects their conformation. Tello *et al.* reviewed this aspect of aptasensors—the charge of the surface electrode and environmental conditions were discussed in relation to the biosensor sensitivity, selectivity, reproducibility, stability and shelflife of the aptasensor [39]. The electrodes used are usually gold and/or carbon-based material such as graphene, however, gold is more popular because of its inert nature and ability to form well-structured crystals [54]. There are different methods that are used in immobilizing aptamers on electrodes including covalent conjugation, physical immobilization, supramolecular interactions, and entrapment within polymers, with the ideal method being one that induces minimum conformational changes, rendering the availability and accessibility of the ligand [39]. For a thrombin aptasensor, immobilization by avidin provided maximum sensitivity, compared to dendrimer layers and chemisorption, which gave moderate and low sensitivity, respectively [55]. This was attributed to the flexibility of the aptamers on the avidin, which was in turn related to the flexibility of the linker, whereas the other two methods administered a more rigid attachment. Charge is an essential element dictating the fold of the aptamer, for instance, when the negative charge of the phosphate backbone of ss-DNA is dominating well enough to cause repulsion, then the aptamer will exist in the unfolded state. In contrast to a neutralized backbone, which permits folding into secondary structural elements that are stabilized by intramolecular hydrogen bonding [39]. Consequently, negatively charged aptamer molecules will assume linearity when immobilized in negatively charged electrodes, such as AuNPs, whereas the folded more neutrally charged ones evade the electrostatic repulsion [39]. Interestingly, the extent of the folding affects the ligand binding and intensity of the signal, whereas in the unfolded, more preferred state, the aptamer undergoes adaptive conformational changes upon ligand binding which enhances the specific binding of the ligand compared to the less flexible folded aptamers that tend to produce reduced signals [56]. Density of the aptamer on the electrode is an essential element to consider because of its direct impact on molecular crowding and steric hindrance which affect conformational folding and ligand binding [1, 57]. Establishing a balance between optimum aptamer concentrations while minimizing molecular crowding and steric hindrance is crucial, although, Ilgu and Nilsen-Hamilton suggested that it was the molecular crowding that aids the folding of ss-DNA aptamer perhaps increasing the binding affinity of the ligand [58]. Orientation of the aptamer can dictate the way in which it folds, usually binding to the electrode via its 5′ or 3′end affecting its functionality. All of these factors have proved to be important building blocks to a functional aptasensor.

9.2.4 Impact of Nucleotide Sequence, Ionic Strength, Temperature, and pH on Functionality of Aptasensors

Not only do intramolecular forces drive aptamer folding, the nucleotide sequence, length and environmental factors dictate the physicochemical properties of the final 3D structure. Consequently, the conformational stability of the aptamers and the type of intermolecular interactions that it may have with the target is impacted; with direct implications on the binding affinity. Variations in the ionic strength, temperature and pH of an aqueous environment can directly affect the sensitivity and selectivity of the aptasensor [59].

The functionality of a thrombin aptasensor designed by Hianik and colleagues, was investigated against a variety of ionic strength and pH conditions. Increasing the concentration of NaCl resulted in decreased sensitivity. The suggested underlying mechanisms were either that increasing the concentration of NaCl caused a shielding effect on the thrombin surface and/or DNA backbone by the Na^+ ions such that methylene blue, which was the positively charged detector molecule used, would be repelled and hindered from binding. Alternatively, the increased salt concentration induced conformational changes modifying the binding site [55]. Further studies led the authors to conclude that the Na^+ ions shielded the binding of the positively charged detector molecule, methylene blue, from binding to the DNA aptamer and binding to the protein thus limiting sensing of the electrochemical shifts detected by the methylene blue. At pH 7.5, sensitivity was optimum whereas at lower or higher pH, sensitivity decreased with a similar pattern, this excluded protonation/deprotonation as a contributing factor, but rather, induced conformational changes were deemed a more fitting explanation [55]. The effect of ionic strength has also been shown to affect aptamer binding to immunoglobulins, which are exploited as aptamer targets for diagnosing infections [57]. Electrostatic interactions between the positively charged Fc domain of IgE and the D17.4 DNA aptamer were found to be important for the binding affinity, wherein increasing the salt concentration from 50 to 300 mM NaCl resulted in an increase in K_d [60]. This was attributable to the transformed conformational state of the IgE loop. Given that natural nucleic acids are endowed with a high negative density and lack hydrophobic patches, it is difficult to recognize and bind immunoglobulins which have hydrophobic moieties and negatively charged groups [57]. Therefore, when developing an aptasensor it is absolutely critical not to consider the aptamer-target analyte interaction in vacuum, rather the environmental conditions that will allow accessibility of the binding site as well as improve the compatibility between the interacting molecules.

9.3 Synthesis of Aptamers

In order to raise aptamers for specific targets, an *in vitro* systematic evolution of ligands by exponential enrichment (SELEX) technique is used [61, 62]. SELEX was first described in 1990 by Tuerk and Gold [63], since then it has been widely and successfully used for generating aptamers. The steps involved in basic SELEX are outlined in Figure 9.2: it begins with the binding of a random library of primed oligonucleotides with a specific biomarker or target. This target could be anything, from small molecules, toxins [62, 64], antibodies [65], proteins [66] to cells [67]. After sufficient time is allowed for binding, the unbound oligonucleotides are removed. The oligonucleotide-target complex is then isolated, followed by dissociation of the oligonucleotide from the target [62, 64, 65].

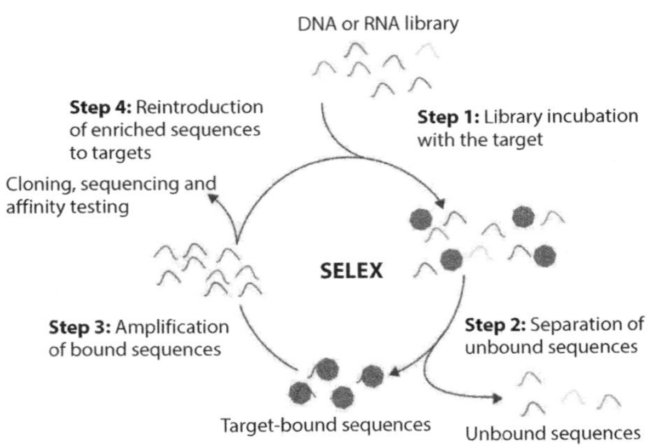

Figure 9.2 Basic steps of isolating aptamers using SELEX. Reprinted with permission from MDPI [53].

The collected oligonucleotides are subjected to a round of polymeric chain reaction (PCR) amplification, resulting in an enriched oligonucleotide library. This enriched library is then subjected to the same steps, repeated for 6 to 20 cycles, to obtain oligonucleotides high-affinity and -specificity for the target [61, 62, 64, 68]. The enriched library is cloned, sequenced and truncated to obtain the final set of target-specific aptamers. Since then, variations were introduced in the steps to reduce the time of synthesis, giving rise to a variety of SELEX techniques [61, 62, 64, 68].

9.3.1 Types of SELEX Methodologies

9.3.1.1 Isolation-Based Techniques

The first set of variations stem from the removal or isolation of the unbound oligonucleotides from the target bound oligonucleotides. With a large enough variation in size between the unbound and bound oligonucleotide, the target complex can be trapped on a nitrocellulose membrane, with pore size smaller than the complex [68]. The unbound oligonucleotides are then simply washed off. The trapped oligonucleotides can be eluted with the addition of a base or through heating and precipitated with ethanol, for recovery [68].

Target-specific aptamers can also be bound to resin microspheres of agarose or dextran and added into a column as stationary phase, to purify through an affinity chromatography [64, 68]. The oligonucleotide library, as a stationary phase, is incubated on the column, binding to the immobilized target molecules. The unbound oligonucleotides are washed off the column and the bound oligonucleotides are recovered by elution from the column [62, 68].

The specific oligonucleotides can also be attached to magnetic beads (MBs), to perform MBs-SELEX [68, 69] or capture-SELEX [70]. Here, the target is bound to the MBs and the oligonucleotide library incubated with the beads-target. After affinity binding, a magnetic field is applied and the unbound oligonucleotides are washed away. The target bound oligonucleotides are then removed and processed further [62, 68, 70].

In addition, graphene oxide (GO) can be added to an oligonucleotide-target mixture, after conjugation, to bind the free, unreacted oligonucleotides. This technique, known as GO-SELEX [71], eliminates the need for target fixing and retains the target conformation. This results in higher affinity aptamers, with reports showing 24-fold increase in binding affinity from MB-SELEX to GO-SELEX [62, 72].

9.3.1.2 PCR Modification-Based Techniques

This is based on modifications related to changes in PCR steps. Fluorescent MB-SELEX uses fluorophore-labelled primers for quantitative amplification during PCR [68]. The aptamer quantity bound to the target molecule can be measured in real time, by following the fluorescent intensity. This increases the screening efficiency and reduces the number of SELEX cycles required [68, 73].

Tailing SELEX uses a modified oligonucleotide library, with 4 and 6 bases, fixed sequences at the end of each molecule [68]. During amplification, the target-bound oligonucleotides, links to the primer fragment, using the bridging sequence, while the same bridging sequence is used as the primer for reverse transcription. Alkaline lysis removes the primer and this secondary library is then used in consecutive screenings, still containing the fixed sequences at each end. This allows for the direct removal of oligonucleotides without the primer binding sites and makes further aptamer truncation unnecessary [68].

Similarly, sequence-constructive SELEX [74] elongates first-round aptamers with seven randomized bases at both the 5' and 3' ends. The aptamers selected after the second SELEX round is again elongated with seven bases and subjected to another round of SELEX. The final aptamers are modified to remove hairpin structures and exhibit 30-fold enhancement in binding affinity from the first round of SELEX products [74].

9.3.1.3 SELEX Process Modification Techniques

Other SELEX modification techniques are used to improve the process or to conduct the process on other target or library types. In order to remove potentially cross-reactive aptamers, the oligonucleotide library can be subjected to binding with either unrelated or closely related targets, prior to target binding. Negative SELEX allows for the removal of non-specific oligonucleotides, by eliminating them from the library before target binding is commenced [68, 73, 75].

In order to identify an aptamer that has multiple targets, usually with a common domain structure, switching SELEX is employed. Here, screening is switched between the various targets in consecutive sets to raise an aptamer to detect them all [68].

Mirror SELEX uses the chiral molecules optical properties in mirror screening, where the *D*-nucleic acid library is screened with the target's optical isomer. The selected sequences are expressed as target aptamers in

L-nucleic acids. This results in highly stable aptamers, not easily identified by the ribonucleases [68].

The use of the full genome of an organism, as oligonucleotide library, is known as genome SELEX. This library is used for screening against polysaccharides, proteins and antibodies as targets, allowing for the recognition of natural sequences to these targets [62, 68].

The use of intact cells as target, cell SELEX, sees specific cells incubated with a library [68]. The library is bound to not only the target cells, but closely related cells; a normal cell line *vs.* a cancerous cell line for example, when raising aptamers for a specific cancer type [66–68, 76–79]. This removes oligonucleotides that bind to both the normal and cancerous cell lines, allowing greater specificity. The cell suspension is centrifuged and the free nucleic acids in the supernatant are discarded (forward screening method). Following that, the cell-attached nucleic acids are denatured and again separated via centrifugation. The nucleic acids, now in the supernatant, are amplified and used in the next round of screening. The unbound nucleic acids, after cell incubation and centrifugation, can be used after amplification, as the secondary library (reverse screening method). The final denatured and collected nucleic acids, are aptamers to the cells or the cells' membrane proteins. This technique is often used to identify membrane biomarkers specific to a disease without bias or distinguish between closely related subspecies of a disease [66–68, 76–79]. *In situ* tissue-based SELEX uses a similar technique to cell-SELEX, but exposes the library to specific diseased tissues *vs.* the healthy tissues [77].

Incorporating the SELEX process onto a microchip, allows for single step detection and selection of aptamers. Microfluidic-SELEX combines automation and miniaturization to obtain enriched aptamer pools in a single step [73]. The whole SELEX process, from target binding, to separation, to PCR, is combined on a chip. Although this speeds up the reaction, the microfluidic device requires a large amount of optimization before use [73, 77, 80–82].

Quartz crystal microbalance-SELEX (QCM-SELEX) measures the mass change of the target, with and without bound oligonucleotides, using a QCM [68]. A change in mass, when a nucleotide is bound to the target, causes a change in the resonance frequency, allowing for aptamer affinity characterization. Although this technique allows for real-time observation and eliminates free nucleic acid separation, it decreases efficiency, by increasing the amount of screening rounds required to obtain the aptamer of choice [62, 68].

The high selectivity and separation efficiency of capillary electrophoresis (CE) improves screening efficacy dramatically, reducing 8–15 rounds of screening to only 2–4 rounds [68]. The CE-SELEX allows binding of the target to the library in solution and requires no elution step, utilizing the mobility differences of target-bound macromolecular aptamers vs the free library [61, 68, 80, 83]. The CE-SELEX process can be tailored based on application; typically, the library is incubated with the target until equilibrium is reached, prior to being injetced into the capillary. This reduces the selection process to 1–3 rounds. Equilibrium capillary electrophoresis of equilibrium mixture (ECEEM) [68, 83] allows for target-library equilibrium, but maintains dynamic equilibrium during the CE process.

Micro free flow electrophoresis (μFFE) requires a special μFFE device [68]. It allows for single round screening, with small sample sizes, simple, large volume collection and reduction in non-specific amplification and contamination. It applies the electric field perpendicular to the flow of buffer, with lateral mobility separation [68, 80, 83]. Separation of molecules is by charge and size typical of electrophoretic techniques. Lastly, fracture capillary electrophoresis (FCE)-SELEX follows nonequilibrium capillary electrophoresis of equilibrium mixtures (NECEEM). The target and library peaks are broken into various fractions, that are amplified separately by PCR [68, 83].

In silico SELEX uses computer software to predict target binding sites and allows for the optimization, modification and truncation of aptamers [62]. It can calculate aptamer affinity and simulate specific aptamer position points after mutation. It allows for the optimization of the selection process and a significant reduction in experimental costs [62].

9.3.1.4 Aptamer Modifications-Based Techniques

Aptamers can also be modified, post selection. Various techniques exist for modifications of each component of the oligonucleotide [53, 84, 85]. These modifications can be included, either from the start of the *in vitro* selection process or toward the end. In the case of RNA aptamers, 2' position modifications of the sugar (ribose) can confer nuclease resistance, but can affect the activity of the aptamer. The most common is 2'-aminopyrimidine modifications, to prevent the A-mediated degradation of RNase [53, 84]. Modification of the phosphodiester linkage can also confer nuclear resistance, but has the potential to increase the non-specific binding of these aptamers to plasma proteins. These include phosphorothioate or boranophosphate linkages, rather than the traditional phosphodiesters [84, 86].

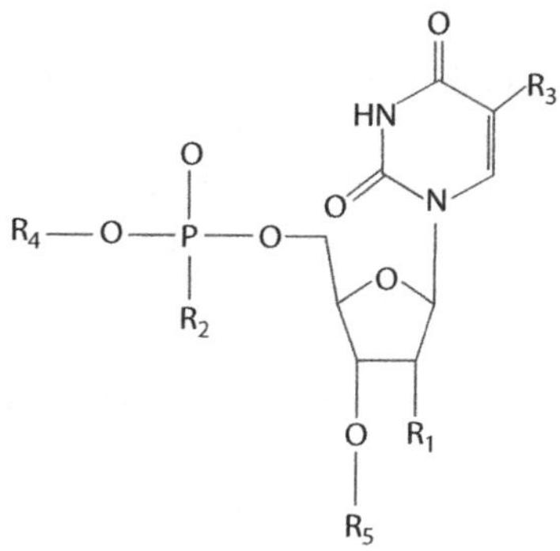

R1: OMe, NH2, F, LNA etc.
R2: S, S2, CH3, etc.
R3: a variety of functional side chains
R4: 5'-end modification, PEG, Choleserol, etc,
R5: 3'-end modification, PEG, Cholesterol, inverted dT, etc.

Figure 9.3 Possible modification of short chain oligonucleotides.
The figure was reproduced from a review by Adachi *et al.* with permissions from MDPI publishers [30].

Shortening of the aptamers' length, truncation, minimizes cross reactivity and reduces the production cost of these materials. PEGylation [53], or the addition of polyethylene glycol to the aptamer decreases its glomerular filtration rate *in vivo*, improving its plasma half-life by protecting the aptamer from enzymatic degradation. Modification through the inclusion of artificial or unnatural nucleotide bases or sugar phosphates can change the aptamers target preference, expand its activity or improve its affinity [84]. Examples of the possible modifications are illustrated in Figure 9.3.

The widely used, modified base, aminoallyl uridine, is T7 RNA polymerase acceptable. Additionally, amino-reactive agent modification, can incorporate hydrophobic aromatic side chains, said to improve acquisition probability and affinity. The increased aromatic chains does however, also increase the plasma clearance rate of the aptamer [53, 84].

9.4 Application of Aptasensors in PoC Diagnostics

Aptamers are now adopted as biorecognition or capture probes in various bioanalytical (molecular and immuno-based) assays for detection of disease biomarkers [87–89]. Most of the clinically relevant diagnostic tests use antibody-based assays [89], however, antibodies and their derivatives (antigen-binding fragments, single-chain variable fragments, or single-chain antibodies) are faced with many limitations that might affect the performance of these systems [1]. Antibody production takes a long time which might not be conducive when faced with pandemics, they are irreproducible, are very costly and used for immunogenic targets [87, 89]. Aptamers, also called chemical or artificial antibodies due to their high specificity and strong affinity for their targets, are now explored as alternatives to antibodies [89, 90]. Aptamers serve as biorecognition elements for various targets including non-immunogenic targets [91], whole cells [92], drugs, hormones [1], toxins, and heavy metals [93]. Unlike antibodies, aptamers have high thermal stability and do not require any special storage conditions or handling, making them mostly suited for PoC.

Aptasensors follow a similar detection principle as traditional biosensors [94]. They are made up of a target recognition moiety (aptamers), transducer (optical [88], electrochemical or mass-based [91]), and/or signal processor. The single stranded aptamers fold into a 3D structure (aptatope) that will recognize and bind to their target receptors in the test sample (Figure 9.4) [93]. Ultimately, the formation of aptamer-receptor complex will generate a signal that can be visually detected and/or quantifiable through optical and electrochemical assays.

Optical and electrochemical aptasensors are conducive for PoC [93], and will be discussed in the context of *in vitro* diagnostics. They are by far the easiest to operate, rapid, cheap, user friendly and generate visual or digital output that is easy to interpret [88, 92, 95]. Incorporating aptamers in existing PoC techniques such as lateral flow, electrochemical and optical sensing systems improved sensitivity and test performance [91, 96]. The emergence of COVID-19 demonstrated the importance of PoC

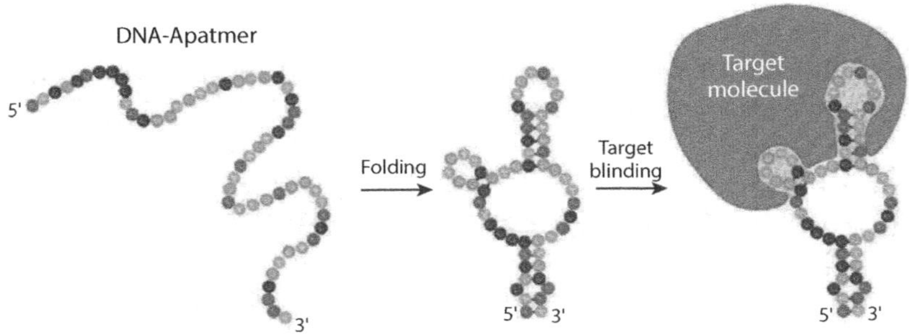

Figure 9.4 Principle of aptasensors. Aptamers fold into various tertiary structures that allow them to bind to their associate biomarkers). Reprinted with permission from Wiley Analytical Science [1].

systems, which will in turn enable containment of infectious diseases and prevent their spread. Aptasensors have many benefits in this regard, and even more beneficial for the resource limited areas, which have been shown to be the most vulnerable. These systems meet the World Health Organization (WHO) ASSURED (Affordable, Sensitive, Specific, User-friendly, Rapid and robust, Equipment-free, and Deliverable to end-users) criteria for PoC diagnostic tests [87].

9.4.1 Colorimetric Aptasensors

Colorimetric assays are the most commonly explored *in vitro* aptasensors for detection of disease biomarkers. Several in-solution and lateral flow colorimetric aptasensors were fabricated and tested for various targets. Gold nanoparticles (AuNPs) aggregation-based in-solution assay is one of the in-solution colorimetric assays adopted into the aptamer technology [88, 95, 97]. AuNPs have unique optical properties that are exploited for the in-solution colorimetric aptasensors. They have a distinct ruby red to grape colors that change to purple or blue when they are destabilized by changes in pH or addition of salts. AuNPs-based salt induced aggregation aptasensors are based on the aptamer absorption-desorption mechanisms, where AuNPs aggregate in the presence of the cognate target resulting in color change after addition of salt. Two of the mechanisms for this assay are illustrated in Figure 9.5; aptamer-AuNP conjugates are protected

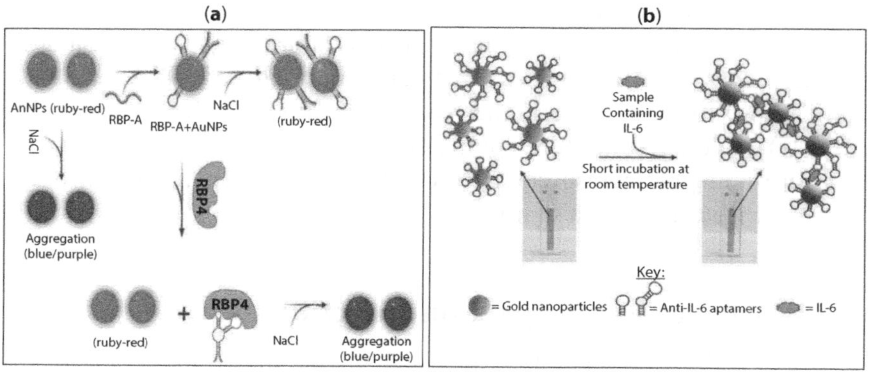

Figure 9.5 Mechanisms of AuNPs-based aggregation aptasensors for the detection of disease biomarkers. The assays can be designed to show salt-induced aggregation in the presence of target analyte following aptamer absorption-desorption mechanism (a). Reprinted with permission from MDPI [88], or AuNP-aptamer-sandwich test format (b). Reprinted with permission from Springer Nature [95].

against salt aggregation and only destabilized in the presence of the target. The example in Figure 9.5 (a) shows a competitive aptasensor, in a positive test, the aptamer will bind to the analyte detaching from the AuNPs. Displacement of aptamers from the AuNPs will cause flocculation (AuNPs color change) upon addition of NaCl. In a negative test, the aptamer-AuNP conjugates and the AuNPs retains their color as they are protected from salt-induced aggregation [88]. Conversely, salt-induced AuNPs aggregation was used to represent a positive test, as shown in Figure 9.5 (b). In this assay, complementary aptamers were conjugated to the AuNPs and upon binding of the analyte to the aptamer-AuNP conjugates, the electrostatic repulsive forces between the AuNPs was weakened leading to AuNPs aggregation and color change. Whereas in the absence of the target, the color remains the same [95].

Aptamers are versatile and compatible with various technologies, besides AuNPs as detection probes, enzymatic colorimetric aptasensors were also used for the detection of circulating tumor-related materials (CTRMs) in blood samples. This strategy was combined with the target enrichment step by using magnetic separation of aptamer-bound biomolecules, then followed by visual detection of alkaline phosphatase catalyzed colored visual signal [92]. Aptamer-based LFAs require less effort, no assay set-up, using less amount of test samples and no instruments for assessing the results. These assays were also effective in the detection of infectious agents and had a short turnaround time compared to the commonly used molecular tests. The SARS-CoV-2 spike protein was detected in phosphate-buffered saline (PBS) in less than 30 min (sample preparation inclusive), contrary to the afore mentioned aptasensors, binding of the analyte stabilized the AuNPs [99]. The sensitivity of this assay was commended for clinical diagnostics, where it showed a sensitivity and specificity of >97% for SARS-CoV-2 spike protein in fresh and frozen human saliva. The AptameXTM COVID-19 test results were comparable to the tests done using reverse transcription polymerase chain reaction (RT-PCR) on nasopharyngeal, oropharyngeal and saliva samples [97].

9.5 Aptasensors Impact on Infectious Disease Diagnosis

So far, aptamers against protein targets that are mainly associated with mammalian cells, infectious agents and others have been isolated. Dunn and colleagues reported that 378 aptamers are for mammalian protein targets, while 149 were for infectious agents; and 58 pathogenic microbes *vs* 53 cancer and 30 non-cancer mammalian cells [30]. This trend is encouraging, showing that efforts in research and development of aptamers in infectious disease is prominent. To diagnose infectious diseases, biomarkers have become popular targets, more especially proteins/antigens that are secreted into host cells. Biomarkers have been very evolutionary to the diagnostic field; helping to diagnose and differentiate between early and late disease, active and latent disease, risk of disease, and pharmacodynamics [9].

9.5.1 Aptasensors for PoC (Vs. Other Clinical Methods of Diagnosis)

The current gold standard for diagnosing infectious diseases, including emerging and dangerous pathogens (EDPs) is RT-PCR, which detects the pathogen's genomic material with high sensitivity and specificity [100]. Other tests include those that detect either viral antigens or host antibodies such as enzyme linked immunosorbent assay (ELISA) and immunofluorescence-based methods [101, 102]. The latter methods can however yield variable results depending on pathogenesis and disease progression specific to the pathogen, as well as the time window within which the test samples are collected [102]. Moreover, much like PCR, these assays require laboratories, reliable infrastructure and power supply, and skilled personnel [102–104]. Therefore, aptasensors may offer more convenient and efficient methods for diagnosing infectious diseases, particularly in resource-limited settings.

Effective infectious disease control requires reliable tests that meet the WHO ASSURED guidelines for PoC diagnostics [105]. The need and potential impact for such diagnostics were further underscored by the recent COVID-19 pandemic, which necessitated speedy diagnosis and isolation of positive cases to curb disease spread and promptly initiate treatment. These PoC diagnostics can be particularly impactful in low-middle income countries (LMICs) where resources are scarce, distribution channels are in some, if not most cases inefficient, and power supply is often unreliable.

Rapid diagnostics such as lateral flow devices (LFDs) can be used at the PoC to aid infectious disease control measures. Examples of these include those deployed during the height of the COVID-19 pandemic for the detection of SARS-Cov-2 antigens and antibodies [20, 106]. These tests made moderate impact in certain settings, particularly in ruling out potential cases and determining the number of previously infected cases in epidemiological studies. However, these types of LFDs are prone to sensitivity and specificity issues among other challenges, which are associated with using antibodies and proteins as recognition elements. Aptasensors have a number of key advantages over antibody-and-protein-based diagnostics, and are poised to address some of the challenges outlined above.

9.5.2 Aptasensors Versus Immunosensors

The more commonly used non-molecular tests for the diagnosis of infectious diseases are ELISAs. These assays are reportedly more sensitive when using aptamers as target-recognition elements [107]. Aptamers can be isolated against diverse targets (immunogenic and non-immunogenic), while antibodies can only be selected against immunogenic targets [107], which is a significant advantage. In biological assays, it is estimated that reagent-related variations contribute approximately 36% of the total irreproducibility [108]; and antibodies are the most widely used reagents (biomarkers/target-recognition/detection probes) in these assays [108]. However, antibodies are prone to cross-reactivity (even monoclonal antibodies) [109] with batch-to-batch variability, which further exacerbates the reproducibility issue [110].

While antibodies have effectively served the industry for decades and continue to be useful in some aspects, superior molecules are required to significantly improve the current detection systems, advance the field, and pave the way for next-generation diagnostics. Aptamers, which have key advantages over antibodies (as outlined above), can potentially replace and/or complement antibodies in various assays including biosensors, lateral flow and microfluidics [11, 14, 108, 111, 112].

In previous studies comparing aptamers and antibodies in various detection assays including ELISA [13, 113–115], lateral flow, immunohistochemistry [108], and immunophenotyping [116–118], aptamers were found to be superior [108]. An interesting example of an aptamer-based LFD, which highlighted the high sensitivity and specificity or target selectivity of aptamers, was one developed by Le and colleagues, which distinguished between strains of the influenza virus [98]. This LFD used an antibody-aptamer pair, which could confirm the presence of the virus, with the aptamer in this case being able to distinguish a particular influenza strain with great precision [98]. Hybridization of aptamers and antibodies is a rational that is derived from the successful use of complementary primary and secondary antibodies in sandwich assays; and due to the costly, unstable and limited spectra of antibody targets [1, 119].

This approach has been explored in the detection of a few biomarkers such as thrombin, platelet-derived growth factor-BB, porcine reproductive and respiratory syndrome virus type II, adipokines, β-amyloid, C-reactive protein, α-1 antitrypsin, epidermal growth factor receptor (EGFR), MUCIN-1, prostate-specific antigen, glycated hemoglobin, Factor IX, IgE, (TNF- α), and Tau protein for different types of aptasensors [119, 120]. An enhanced limit of detection (LOD) was evident in some of these studies, decreasing to more than 5000-fold, and detecting as low as in the pico Molar range when compared to conventional aptasensors [1, 119]. Selectivity was reported to be enhanced by this hybrid system; accuracy being an indispensable element in developing a successful diagnostic instrument where human health is concerned. To date, the only infectious disease in which this hybrid method has been reported is the influenza virus [120]. A more focused research is required for the application of the aptamer-antibody hybrid system in the diagnosis of infectious diseases. Next generation aptasensors incorporating mineral-based probes (*e.g.*, AuNP, AgNPs), other fluorescent reporter systems such as nanocellulose, as well as the latest molecular target-detecting technologies such as the CRISPR-Cas13 SHERLOCK (specific high-sensitivity enzymatic reporter unlocking) [121], have the potential to achieve even more superior PoC diagnostic performance.

9.5.3 Aptasensors Used in Infectious Disease Diagnosis: The Research Versus Commercial Picture

There is currently a limited number of aptasensors in the commercial diagnostic development pipelines for the detection of infectious disease biomarkers. Numerous efforts towards the development of aptamer-based infectious disease diagnostics have been reported in literature, including viruses on the WHO's emerging and dangerous pathogens [122]. Examples of these include aptamer-based qPCR signal amplification and detection of SARS-CoV-2 [123], aptamer-trapping and PCR fluorescence probe-based detection for the influenza virus, ELONA-based sensor for the Zika virus [124], magnetic-based detection chip for Ebola virus [125], as well as the fluorescence probe or ELONA-based diagnostics for tuberculosis [12, 14, 97, 126, 127].

The advancement of aptamer-based systems was previously delayed by the SELEX process patents, which spanned two decades [128]. This may have been one of the main stumbling blocks in the commercial development of aptasensors. However, these patents have recently expired, and this has presented more avenues for the application of aptamers in diagnostics, therapeutics, and theranostics, and the enterprise thereof. In anticipation of this shift, a 2015 Markets and Markets report projected that the global aptamer market would reach $244.93 million by 2020 [129]. This trend was reportedly driven by the SELEX patents expiration, greater investments in Biotech and Pharma R&D, and the notably high efficiency and low cost of aptamers relative to antibodies [129]. The latest reports indicate that the global aptamer market was valued at USD 1.6 billion in 2021, with an expected CAGR of 23.5% from 2022 to 2030 [130]. Aptamer-based diagnostic products are expected to drive market growth, and potentially become the preferred detection probes in diagnostic and pathology labs (due to their high specificity, selectivity, and efficacy, and low manufacturing cost as some of the key advantages over antibodies) [130].

9.5.4 Aptasensors in the Commercialization Pipeline

In 2018, it was reported that over 40 companies were actively working on aptamer-based diagnostics and therapeutics [131]. However, there are limited reports on aptasensors that are currently on the market. In terms of aptasensors either in or closer to the market, there are a few companies who are players in this space. For instance, UK-based companies Aptamer Group and Mologic co-developed the aptamer-based SARS-Cov-2 Antigen tests able to detect the main SARS-CoV-2 variants that were set for clinical evaluation and Conformité Européenne subsequent marking. To our knowledge, there are currently no reports on the clinical and commercial progress thereof. Another UK-based company, Hutano Diagnostics, is currently developing a lateral flow device to diagnose sepsis, which is caused by dysregulation in infection-driven host immune response [132]. Another product offering is a universal aptamer-based LFD kit that can be customized to the client's needs, which may aid other rapid diagnostic tests (RDTs) developers and manufacturers including those who are developing tests for infectious disease markers [133]. Much closer to the market, a Switzerland-based company, Achiko AG, developed and launched the AptameX COVID-19 rapid test whose readout uses a UV Spectrophotometer, which is reported to have obtained Conformité Européenne marking [134].

9.5.5 Potential Impact

From a product development perspective, aptasensors may offer relative ease and minimal complexities. With the potential of simpler multiplexing (particularly for LFDs) relative to antibody-based assays which are affected by less than optimal binding kinetics between the antibody and target analyte [135], aptasensors with high affinity can be developed to detect multiple biomarkers in a single step and inform prompt and appropriate intervention. Also, due to their particularly amenable chemistries, aptamers can be readily conjugated to fluorescent reporters and linkers for various detection needs without laborious post-synthesis modifications [14, 108, 111, 112].

In terms of finished products, aptasensors can potentially offer overall robustness, stability and ease of distribution. For instance, antibody-based LFDs typically have a 2-year lifespan, which restricts stockpiling prospects for epidemic preparedness. Antibody-based LFDs also require specific environmental conditions, and in some cases, cold chain transport and refrigerated storage of the entire test kit or certain components thereof. Temperatures during storage and use of LFD RDTs are known to affect test performance—both sensitivity and specificity [136]. Recommended temperatures are typically in the 2–30°C range, which are not always practical in the global South countries with average temperatures of over 30°C [136]. Moreover, these environmental conditions requirements may have low product penetration implications in certain regions such as Africa, where most clinics have unreliable or inconsistent power supply. However, since aptamers are both thermostable and have a much longer shelf life (estimated to be over three times that of antibodies [103, 137, 138], aptasensors (LFD or other suitable PoC formats) will have comparatively less complex distribution channels and enable stockpiling in areas of need.

On the economic front, the use of aptasensors may also help reduce the inflated financial burdens on healthcare systems, through cost-saving resulting from using PCR tests primarily for confirmatory purposes rather than as first-line tests. Additionally, since the cost of aptamer production (pre-and post-isolation/generation) is significantly lower than that of antibodies, aptasensor prices may ultimately be lower than antibody-based tests or immunosensors. For instance, producing 1 mg of biotinylated aptamer is estimated to cost around $100 compared with $1,000 for the same antibody amount [139]. Moreover, aptamers can be generated within weeks, while antibody production takes several months [139]. Aptamer production is both precisely controlled and easily scalable, and does not require the use of animals (which may be beneficial for the environment). Once selected and characterized, aptamer sequences can be stored digitally, and batches of specific aptamers can subsequently be efficiently synthesized at a required scale.

Aptasensors will also be particularly useful for deployment to regions prone to infectious disease outbreaks affecting humans and animals.

The potential impact in agricultural settings can be enormous—for instance, where diseases such as Bovine tuberculosis and Brucellosis affect life stock and infected herds have to be quarantined [140, 141].

On the patients' health and economic perspective, particularly in highly unequal societies, readily deployable and more affordable technologies such as aptasensors will be of great benefit to historically marginalized groups. These types of solutions can help close health access gaps in cases where individuals do not have private health insurance or the funds to readily afford lab-based tests, and accordingly curb delays in diagnosis and interventions.

9.6 Drawbacks and Potential Future Work

Some older studies highlighted concerns about some aptamers generated against proteins expressed in prokaryotic cells not interacting efficiently or stably with the same proteins when expressed in eukaryotic models and linked this phenomenon to post-translational modifications within the eukaryotic cells [142]. However, proper understanding of specific targets prior to initiation of the SELEX process can help eliminate these challenges. Other potential challenges include the complexities that may come with selecting aptamers against whole viral or bacterial cells versus purified targets such as enzymes or proteins due to structural differences thereof, although some technical advances have been made over the years in this regard. There are also reports of aptamers having weaker interactions with the natural targets after being optimized on recombinant targets (reviewed by Wan *et al.*, 2021) [122].

Many of the challenges associated with aptamers in clinical use apply more to therapeutics rather than diagnostics. These include the rapid degradation of RNA aptamers in particular, by nucleases in the blood and their swift renal clearance, which can be corrected by chemical modifications such as the inclusion of PEG ligands or 3'- and 5'-end modifications to improve stability and bioavailability [143].

Aptasensors may pose some challenges during the product development process, which simultaneously present additional avenues for innovation. For instance, while proteins and antibodies can be readily captured on the LFD nitrocellulose membranes through both hydrophobic and hydrophilic interactions [25, 144], this sort of 'passive adsorption' approach is less effective for smaller (<1000 Da) molecules such as aptamers and peptides, which typically require some protein-based anchoring such as the Biotin-Streptavidin interactions [144]. Covalently attaching these smaller molecules to the nitrocellulose membranes would offer high binding strength, test stability and reproducibility [145, 146]; and removing any protein-based substrate anchors and relying on specific chemical interactions would eliminate the need for special storage conditions to maintain test functionality. This offers several opportunities for innovation—from using various chemical linkers as membrane anchors for aptamers in LFDs, to generating specific nucleic acid-based complex molecules that can be immobilized on various substrates as needed, and creating new biosensor design formats that both exploit and support the desirable features of aptamers.

9.7 Conclusions

Through research and development, pandemic readiness is a possibility, and a necessity since the outbreak of infectious diseases has been unpredictable with outcomes that are highly detrimental, and at times irreparable. In this regard, PoC devices have been at the epicentre of research for their ease of use, cost-effectiveness and accessibility especially for resource-limited communities, improving the control and management of infectious diseases. Aptasensors have the potential to improve PoC test performance (sensitivity and specificity), robustness, field-readiness, shelf life and the cost of production/manufacturing (by at least 10-fold, thus driving the price down), as well as substantially decrease complexities in scalability, distribution, storage, and use.

References

1. Arshavsky-Graham, S., Heuer, C., Jiang, X., Segal, E., Aptasensors versus Immunosensors—Which Will Prevail? *Eng. Life Sci.*, *22*, 319–333, **2022**, doi: 10.1002/elsc.202100148.
2. Kaur, H. and Shorie, M., Nanomaterial Based Aptasensors for Clinical and Environmental Diagnostic Applications. *Nanoscale Adv.*, *1*, 2123–2138, **2019**, doi: 10.1039/c9na00153k.
3. Lim, Y.C., Kouzani, A.Z., Duan, W., Aptasensors: A Review. *J. Biomed. Nanotechnol.*, *6*, 93–105, **2010**, doi: 10.1166/jbn.2010.1103.
4. Yoo, E.H. and Lee, S.Y., Glucose Biosensors: An Overview of Use in Clinical Practice. *Sensors*, *10*, 4558–4576, **2010**, doi: 10.3390/s100504558.
5. Zhou, J., Rossi, J., Biology, C., Rd, E.D., Aptamers as Targeted Therapeutics: Current Potential and Challenges. *Nat. Rev. Drug Discov.*, *8*, 181–202, **2017**, doi: 10.1038/nrd.2016.199.Aptamers.

6. Amin, A.G., De, P., Graham, B., Calderon, R.I., Franke, M.F., Chatterjee, D., Urine Lipoarabinomannan in HIV Uninfected, Smear Negative, Symptomatic TB Patients: Effective Sample Pretreatment for a Sensitive Immunoassay and Mass Spectrometry. *Sci. Rep.*, *11*, 1–13, **2021**, doi: 10.1038/s41598-021-82445-4.
7. Drolet, D.W., Moon-McDermott, L., Romig, T.S., An Enzyme-Linked Oligonucleotide Assay. *Nat. Biotechnol.*, *14*, 1021–1025, **1996**, doi: 10.1038/nbt0896-1021.
8. Moreno, M., Aptasensors, in: *Encycl. Anal. Sci.*, pp. 150–153, **2019**.
9. Majdinasab, M., Marty, J.L., Recent Advances in Electrochemical Aptasensors for Detection of Biomarkers. *Pharmaceuticals (Basel)*, 15, 8, 995, **2022 Aug 12**, doi: 10.3390/ph15080995.
10. Shaban, S.M. and Kim, D.H., Recent Advances in Aptamer Sensors. *Sensors (Switzerland)*, *21*, 1–31, **2021**, doi: 10.3390/s21030979.
11. Tang, X.-L., Zhou, Y.-X., Wu, S.-M., Pan, Q., Xia, B., Zhang, X.-L., CFP10 and ESAT6 Aptamers as Effective Mycobacterial Antigen Diagnostic Reagents. *J. Infect.*, *69*, 569–580, **2014**, doi: 10.1016/j.jinf.2014.05.015.
12. Ansari, N., Ghazvini, K., Ramezani, M., Shahdordizadeh, M., Yazdian-Robati, R., Abnous, K., Taghdisi, S.M., Selection of DNA Aptamers against Mycobacterium Tuberculosis Ag85A, and Its Application in a Graphene Oxide-Based Fluorometric Assay. *Microchim. Acta*, *185*, 21, **2018**, doi: 10.1007/s00604-017-2550-3.
13. Sypabekova, M., Bekmurzayeva, A., Wang, R., Li, Y., Nogues, C., Kanayeva, D., Selection, Characterization, and Application of DNA Aptamers for Detection of Mycobacterium Tuberculosis Secreted Protein MPT64. *Tuberculosis*, *104*, 70–78, **2017**, doi: 10.1016/j.tube.2017.03.004.
14. Rotherham, L.S., Maserumule, C., Dheda, K., Theron, J., Khati, M., Selection and Application of SsDNA Aptamers to Detect Active TB from Sputum Samples. *PLoS One*, *7*, 1–11, **2012**, doi: 10.1371/journal.pone.0046862.
15. Bayramoglu, G., Ozalp, V.C., Oztekin, M., Arica, M.Y., Rapid and Label-Free Detection of Brucella Melitensis in Milk and Milk Products Using an Aptasensor. *Talanta*, *200*, 263–271, **2019**, doi: 10.1016/j.talanta.2019.03.048.
16. Mandal, M., Dutta, N., Dutta, G., Aptamer-Based Biosensors and Their Implications in COVID-19 Diagnosis. *Anal. Methods*, *13*, 5400–5417, **2021**, doi: 10.1039/D1AY01519B.
17. Svobodova, M., Skouridou, V., Jauset-Rubio, M., Viéitez, I., Fernández-Villar, A., Cabrera Alvargonzalez, J.J., Poveda, E., Bofill, C.B., Sans, T., Bashammakh, A. *et al.*, Aptamer Sandwich Assay for the Detection of SARS-CoV-2 Spike Protein Antigen. *ACS Omega*, *6*, 35657–35666, **2021**, doi: 10.1021/acsomega.1c05521.
18. Yang, G., Li, Z., Mohammed, I., Zhao, L., Wei, W., Xiao, H., Guo, W., Zhao, Y., Qu, F., Huang, Y., Identification of SARS-CoV-2-against Aptamer with High Neutralization Activity by Blocking the RBD Domain of Spike

Protein 1. *Signal Transduct. Target. Ther.*, *6*, 227–2010, **2021**, doi: 10.1038/s41392-021-00649-6.

19. Dias, S.A., Freire, J.M., Pérez-Peinado, C., Domingues, M.M., Gaspar, D., Vale, N., Gomes, P., Andreu, D., Henriques, S.T., Castanho, M.A.R.B. *et al.*, New Potent Membrane-Targeting Antibacterial Peptides from Viral Capsid Proteins. *Front. Microbiol.*, *8*, 1–9, **2017**, doi: 10.3389/fmicb.2017.00775.
20. Koteswara Rao, V., Point of Care Diagnostic Devices for Rapid Detection of Novel Coronavirus (SARS-NCoV19) Pandemic: A Review. *Front. Nanotechnol.*, *2*, 1–9, **2021**, doi: 10.3389/fnano.2020.593619.
21. Bala, J., Chinnapaiyan, S., Dutta, R.K., Unwalla, H., Aptamers in HIV Research Diagnosis and Therapy. *RNA Biol.*, *15*, 327–337, **2018**, doi: 10.1080/15476286.2017.1414131.
22. Malatji, K., Fru, P.N., Mufhandu, H., Alexandre, K., Synthesis of Fluorescence Labelled Aptamers for Use as Low-Cost Reagents in HIV/AIDS Research and Diagnostics. *Biomed. Rep.*, *16*, 1–9, **2022**, doi: 10.3892/br.2021.1491.
23. Tombelli, S., Minunni, M., Luzi, E., Mascini, M., Aptamer-Based Biosensors for the Detection of HIV-1 Tat Protein. *Bioelectrochemistry*, *67*, 135–141, **2005**, doi: 10.1016/j.bioelechem.2004.04.011.
24. Rahim Ruslinda, A., Tanabe, K., Ibori, S., Wang, X., Kawarada, H., Effects of Diamond-FET-Based RNA Aptamer Sensing for Detection of Real Sample of HIV-1 Tat Protein. *Biosens. Bioelectron.*, *40*, 277–282, **2013**, doi: 10.1016/j.bios.2012.07.048.
25. Bruno, J.G., Application of DNA Aptamers and Quantum Dots to Lateral Flow Test Strips for Detection of Foodborne Pathogens with Improved Sensitivity versus Colloidal Gold. *Pathogens*, *3*, 341–355, **2014**, doi: 10.3390/pathogens3020341.
26. Chen, W., Cui, L., Song, Y., Chen, W., Su, Y., Chang, W., Xu, W., Detection of Listeria Monocytogenes Using Luminol-Functionalized AuNF-Labeled Aptamer Recognition and Magnetic Separation. *ACS Omega*, *6*, 26338–26344, **2021**, doi: 10.1021/acsomega.1c03527.
27. Miranda, G.F., *Development of Electrochemical Aptasensors for the Highly Sensitive, Selective, and Discriminatory Detection of Malaria Biomarkers*, Bioelectronics (IBI-3) Institute; German, Jülich. vol. 75, ISBN 9783958065895.
28. Chakma, B., Jain, P., Singh, N.K., Goswami, P., Development of Electrochemical Impedance Spectroscopy Based Malaria Aptasensor Using HRP-II as Target Biomarker. *Electroanalysis*, *30*, 1839–1846, **2018**, doi: 10.1002/elan.201800142.
29. Singh, N.K., Arya, S.K., Estrela, P., Goswami, P., Capacitive Malaria Aptasensor Using Plasmodium Falciparum Glutamate Dehydrogenase as Target Antigen in Undiluted Human Serum. *Biosens. Bioelectron.*, *117*, 246–252, **2018**, doi: 10.1016/j.bios.2018.06.022.
30. Dunn, M.R., Jimenez, R.M., Chaput, J.C., Analysis of Aptamer Discovery and Technology. *Nat. Rev. Chem.*, *1*, 1–16, **2017**, doi: 10.1038/S41570-017-0076.

31. Edwards, B.A.L., Batey, R.T., Sam-i, S.A.M., Sam-ii, S.A.M., Sam-iii, S.A.M., Riboswitches: A Common RNA Regulatory Element Riboregulation in Bacteria How Are Riboswitches Categorized? *Cold Spring Harb. Perspect. Biol., 2008*, 1–6, **2010**.
32. Patel, D.J., Suri, A.K., Jiang, F., Jiang, L., Fan, P., Kumar, R.A., Nonin, S., Structure, recognition and adaptive binding in RNA aptamer complexes. *J. Mol. Biol.*, 272, 5, 645–64, **1997 Oct 10**, doi: 10.1006/jmbi.1997.1281.
33. Jenison, R.D., Gill, S.C., Pardi, A., Polisky, B., High-Resolution Molecular Discrimination by RNA. *Science (80-.), 263*, 1425–1429, **1994**, doi: https://doi.org/10.1126/science.7510417.
34. Hermann, T. and Patel, D.J., Adaptive Recognition by Nucleic Acid Aptamers. *Science (80-.), 287*, 820–825, **2000**, doi: 10.1126/science.287.5454.820.
35. Michaud, M., Jourdan, E., Villet, A., Ravel, A., Grosset, C., Peyrin, E.A., DNA Aptamer as a New Target-Specific Chiral Selector for HPLC. *J. Am. Chem. Soc., 125*, 8672–8679, **2003**, doi: 10.1021/ja034483t.
36. Williams, K.P., Liu, X.-H., Schumacher, T.N.M., Lin, H.Y., Ausiello, D.A., Kim, P.S., D.P.B., Bioactive and Nuclease-Resistant L-DNA Ligand of Vasopressin. *Proc. Natl. Acad. Sci. USA, 94*, 11285–11290, **1997**.
37. Ye, X., Gorin, A., Ellington, A.D., Patel, D.J., Deep Penetration of an α-Helix into a Widened RNA Major Groove in the HIV-1 Rev Peptide-RNA Aptamer Complex. *Nat. Struct. Biol., 3*, 1026–1033, **1996**, doi: 10.1038/nsb1296-1026.
38. Battiste, J.L., Mao, H., Rao, N.S., Tan, R., Muhandiram, D.R., Kay, L.E., Frankel, A.D., Williamson, J.R., α Helix-RNA Major Groove Recognition in an HIV-1 Rev Peptide-RRE RNA Complex. *Science (80-.), 273*, 1547–1551, **1996**, doi: 10.1126/science.273.5281.1547.
39. Tello, A., Cao, R., Marchant, M.J., Gomez, H., Conformational Changes of Enzymes and Aptamers Immobilized on Electrodes. *Bioconjug. Chem., 27*, 2581–2591, 2016.
40. Ferraz, R.A.C., Lopes, A.L.G., da Silva, J.A.F., Moreira, D.F.V., Ferreira, M.J.N., de Almeida Coimbra, S.V., DNA–Protein Interaction Studies: A Historical and Comparative Analysis. *Plant Methods, 17*, 1–21, **2021**, doi: 10.1186/s13007-021-00780-z.
41. Cozzolino, F., Iacobucci, I., Monaco, V., Monti, M., Protein-DNA/RNA Interactions: An Overview of Investigation Methods in the -Omics Era. *J. Proteome Res., 20*, 3018–3030, **2021**, doi: 10.1021/acs.jproteome.1c00074.
42. Swayze, E.E., Griffey, R.H., Bennett, C.F., Nucleic Acids (Deoxyribonucleic Acid and Ribonucleic Acid). *Pharmaceuticals*, I. 2, 26, 1037–1052, **2007**.
43. Biebricher, A.S., Heller, I., Roijmans, R.F.H., Hoekstra, T.P., Peterman, E.J.G., Wuite, G.J.L., The Impact of DNA Intercalators on DNA and DNA-Processing Enzymes Elucidated through Force-Dependent Binding Kinetics. *Nat. Commun., 6*, 7304–7319, **2015**, doi: 10.1038/ncomms8304.
44. Fan, P., Suri, A.K., Fiala, R., Live, D., Patel, D.J., Molecular Recognition in the FMN-RNA Aptamer Complex. *J. Mol. Biol., 258*, 480–500, **1996**, doi: 10.1006/jmbi.1996.0263.

45. Adachi, T. and Nakamura, Y., Aptamers: A Review of Their Chemical Properties. *Molecules*, *24*, 4229, **2019**.
46. Asadzadeh, H., Moosavi, A., Alexandrakis, G., Mofrad, M.R.K., Atomic Scale Interactions between RNA and DNA Aptamers with the TNF- α Protein. *Biomed Res. Int.*, *2021*, **2021**, doi: 10.1155/2021/9926128.
47. Ma, X., Gosai, A., Balasubramanian, G., Shrotriya, P., Aptamer Based Electrostatic-Stimuli Responsive Surfaces for on-Demand Binding/Unbinding of a Specific Ligand. *J. Mater. Chem. B*, *5*, 3675–3685, **2017**, doi: 10.1039/c6tb02386j.
48. Ren, R., Wang, X., Cai, S., Zhang, Y., Korchev, Y., Ivanov, A.P., Edel, J.B., Selective Sensing of Proteins Using Aptamer Functionalized Nanopore Extended Field-Effect Transistors. *Small Methods*, *4*, 2000356–2000363, **2020**, doi: 10.1002/smtd.202000356.
49. Shraim, A.S., Abdel Majeed, B.A., Al-Binni, M.A., Hunaiti, A., Therapeutic Potential of Aptamer–Protein Interactions. *ACS Pharmacol. Transl. Sci.*, *5*, 1211–1227, 2022, doi: 10.1021/acsptsci.2c00156.
50. Yoo, H., Jo, H., Oh, S.S., Detection and beyond: Challenges and Advances in Aptamer-Based Biosensors. *Mater. Adv.*, *1*, 2663–2687, **2020**, doi: 10.1039/d0ma00639d.
51. Lin, C.H. and Patel, D.J., Encapsulating an Amino Acid in a DNA Fold. *Nat. Struct. Biol.*, *3*, 1046–1050, **1996**, doi: 10.1038/nsb1296-1046.
52. Patel, D.J., Structural Analysis of Nucleic Acid Aptamers. *Curr. Opin. Chem. Biol.*, *1*, 32–46, **1997**, doi: 10.1016/s1367-5931(97)80106-8.
53. Elskens, J.P., Elskens, J.M., Madder, A., Chemical Modification of Aptamers for Increased Binding Affinity in Diagnostic Applications: Current Status and Future Prospects. *Int. J. Mol. Sci.*, *21*, 1–31, **2020**, doi: 10.3390/ijms21124522.
54. Oberhaus, F.V., Frense, D., Beckmann, D., Immobilization Techniques for Aptamers on Gold Electrodes for the Electrochemical Detection of Proteins: A Review. *Biosensors*, *10*, 45–90, **2020**, doi: 10.3390/BIOS10050045.
55. Hianik, T., Ostatná, V., Sonlajtnerova, M., Grman, I., Influence of Ionic Strength, PH and Aptamer Configuration for Binding Affinity to Thrombin. *Bioelectrochemistry*, *70*, 127–133, **2007**, doi: 10.1016/j.bioelechem.2006.03.012.
56. Xiao, Y., Uzawa, T., White, R.J., DeMartini, D., Plaxco, K.W., On the Signaling of Electrochemical Aptamer-Based Sensors: Collision- and Folding-Based Mechanisms. *Electroanalysis*, *21*, 1267–1271, **2009**, doi: 10.1002/elan.200804564.
57. Bognár, Z. and Gyurcsányi, R.E., Aptamers against Immunoglobulins: Design, Selection and Bioanalytical Applications. *Int. J. Mol. Sci.*, *21*, 1–40, **2020**, doi: 10.3390/ijms21165748.
58. Ilgu, M. and Nilsen-Hamilton, M., Aptamers in Analytics. *Analyst*, *141*, 1551–1558, 2016.
59. Kawasaki, M. and Oda, M., Effects of Chain Length, Temperature, and Ionic Strength on Association and Dissociation Thermodynamics of

DNA. *Chem. Thermodyn. Therm. Anal., 3–4*, 100015, **2021**, doi: 10.1016/j.ctta.2021.100015.

60. Poongavanam, M.V., Kisley, L., Kourentzi, K., Landes, C.F., Willson, R.C., Ensemble and Single-Molecule Biophysical Characterization of D17.4 DNA Aptamer-IgE Interactions. *Biochim. Biophys. Acta, Proteins Proteomics, 1864*, 154–164, **2016**, doi: 10.1016/j.bbapap.2015.08.008.
61. Zhu, C., Li, L., Yang, G., Irfan, M., Wang, Z., Fang, S., Qu, F., High-Efficiency Selection of Aptamers for Bovine Lactoferrin by Capillary Electrophoresis and Its Aptasensor Application in Milk Powder. *Talanta, 205*, 120088–120095, **2019**, doi: 10.1016/j.talanta.2019.06.088.
62. Wang, Z.J., Chen, E.N., Yang, G., Zhao, X.Y., Qu, F., Research Advances of Aptamers Selection for Small Molecule Targets. *Chin. J. Anal. Chem., 48*, 573–582, 2020.
63. Tuerk, C. and Gold, L., Systematic Evolution of Ligands by Exponential Enrichment: RNA Ligands to Bacteriophage T4 DNA Polymerase. *Science (80-.), 249*, 505–510, **1990.**
64. Yang, K.A., Pei, R., Stojanovic, M.N., *In Vitro* Selection and Amplification Protocols for Isolation of Aptameric Sensors for Small Molecules. *Methods, 106*, 58–65, 2016.
65. Donia, T., Jyoti, B., Suizu, F., Hirata, N., Tanaka, T., Ishigaki, S., F, P.T.J., Nio-Kobayashi, J., Iwanaga, T., Chiorini, J.A. *et al.*, Identification of RNA Aptamer Which Specifically Interacts with PtdIns(3)P. *Biochem. Biophys. Res. Commun., 517*, 146–154, **2019**, doi: 10.1016/j.bbrc.2019.07.034.
66. Xiong, H., Yan, J., Cai, S., He, Q., Peng, D., Liu, Z., Liu, Y., Cancer Protein Biomarker Discovery Based on Nucleic Acid Aptamers. *Int. J. Biol. Macromol., 132*, 190–202, 2019.
67. Bayat, P., Taghdisi, S.M., Rafatpanah, H., Abnous, K., Ramezani, M., *In Vitro* Selection of CD70 Binding Aptamer and Its Application in a Biosensor Design for Sensitive Detection of SKOV-3 Ovarian Cells. *Talanta, 194*, 399–405, **2019**, doi: 10.1016/j.talanta.2018.10.063.
68. ZOU, X.M., ZHOU, J.W., SONG, S.H., CHEN, G.H., Screening of Oligonucleotide Aptamers and Application in Detection of Pesticide and Veterinary Drug Residues. *Chin. J. Anal. Chem., 47*, 488–499, 2019.
69. Shen, H., Deng, W., He, Y., Li, X., Song, J., Liu, R., liu, H., Yang, G., Li, L., Ultrasensitive Aptasensor for Isolation and Detection of Circulating Tumor Cells Based on CeO2@Ir Nanorods and DNA Walker. *Biosens. Bioelectron., 168*, 112516–112523, **2020**, doi: 10.1016/j.bios.2020.112516.
70. Tian, H., Duan, N., Wu, S., Wang, Z., Selection and Application of SsDNA Aptamers against Spermine Based on Capture-SELEX. *Anal. Chim. Acta, 1081*, 168–175, **2019**, doi: 10.1016/j.aca.2019.07.031.
71. Wu, S., Duan, N., Zhang, W., Zhao, S., Wang, Z., Screening and Development of DNA Aptamers as Capture Probes for Colorimetric Detection of Patulin. *Anal. Biochem., 508*, 58–64, **2016**, doi: 10.1016/j.ab.2016.05.024.

72. Park, J.-W., Tatavarty, R., Kim, D.W., Jung, H.-T., Gu, M.B., Immobilization-Free Screening of Aptamers Assisted by Graphene Oxide. *Chem. Commun.*, *48*, 2071–2073, **2012**, doi: DOI https://doi.org/10.1039/C2CC16473F.
73. Ning, Y., Hu, J., Lu, F., Aptamers Used for Biosensors and Targeted Therapy. *Biomed. Pharmacother.*, *132*, 110902–110923, 2020.
74. Wang, C.Y., Wu, C.Y., Hung, T.C., Wong, C.H., Chen, C.H., Sequence-Constructive SELEX: A New Strategy for Screening DNA Aptamer Binding to Globo H. *Biochem. Biophys. Res. Commun.*, *452*, 484–489, **2014**, doi: 10.1016/j.bbrc.2014.08.086.
75. Wang, J., Wang, Q., Luo, Y., Gao, T., Zhao, Y., Pei, R., *In Vitro* Selection of SsDNA Aptamers That Can Specifically Recognize and Differentiate Riboflavin and Its Derivative FAD. *Talanta*, *204*, 424–430, **2019**, doi: 10.1016/j.talanta.2019.06.039.
76. Li, W., Wang, S., Zhou, L., Cheng, Y., Fang, J., An SsDNA Aptamer Selected by Cell-SELEX for the Targeted Imaging of Poorly Differentiated Gastric Cancer Tissue. *Talanta*, *199*, 634–642, **2019**, doi: 10.1016/j.talanta.2019.03.016.
77. Wu, L., Zhu, L., Huang, M., Song, J., Zhang, H., Song, Y., Wang, W., Yang, C., Aptamer-Based Microfluidics for Isolation, Release and Analysis of Circulating Tumor Cells. *TrAC, Trends Anal. Chem.*, *117*, 69–77, 2019.
78. Wang, L., Li, P., Xiao, X., Li, J., Li, J., Yang, H.H., Tan, W., Generating Lung-Metastatic Osteosarcoma Targeting Aptamers for *in Vivo* and Clinical Tissue Imaging. *Talanta*, *188*, 66–73, **2018**, doi: 10.1016/j.talanta.2018.05.011.
79. Zhang, W.Y., Chen, H.L., Chen, Q.C., *In Vitro* Selection of Aptamer S1 against MCF-7 Human Breast Cancer Cells. *Bioorg. Med. Chem. Lett.*, *29*, 2393–2397, **2019**, doi: 10.1016/j.bmcl.2019.06.002.
80. Blind, M. and Blank, M., Aptamer Selection Technology and Recent Advances. *Mol. Ther. Nucleic Acids*, *4*, e223, **2015**, doi: 10.1038/mtna.2014.74.
81. Davydova, A., Vorobyeva, M., Bashmakova, E., Vorobjev, P., Krasheninina, O., Tupikin, A., Kabilov, M., Krasitskaya, V., Frank, L., Venyaminova, A., Development and Characterization of Novel 2′-F-RNA Aptamers Specific to Human Total and Glycated Hemoglobins. *Anal. Biochem.*, *570*, 43–50, **2019**, doi: 10.1016/j.ab.2019.02.004.
82. Ahmadyousefi, Y., Malih, S., Mirzaee, Y., Saidijam, M., Nucleic Acid Aptamers in Diagnosis of Colorectal Cancer. *Biochimie*, *156*, 1–11, 2019.
83. Zhu, C., Yang, G., Ghulam, M., Li, L., Qu, F., Evolution of Multi-Functional Capillary Electrophoresis for High-Efficiency Selection of Aptamers. *Biotechnol. Adv.*, *37*, 107432–107447, 2019.
84. Adachi, T. and Nakamura, Y., Aptamers: A Review of Their Chemical Properties and Modifications for Therapeutic Application. *Molecules*, *24*, 4229–4242, 2019.
85. Ni, S., Zhuo, Z., Pan, Y., Yu, Y., Li, F., Liu, J., Wang, L., Wu, X., Li, D., Wan, Y. *et al.*, Recent Progress in Aptamer Discoveries and Modifications for Therapeutic Applications. *ACS Appl. Mater. Interfaces*, *13*, 9500–9519, **2021**, doi: 10.1021/acsami.0c05750.

86. Amero, P., Amero, P., Lokesh, G.L.R., Chaudhari, R.R., Cardenas-Zuniga, R., Schubert, T., Attia, Y.M., Montalvo-Gonzalez, E., Elsayed, Ivan, C., Wang, Z., Cristini, V., de Franciscis, V. *et al.*, Conversion of RNA Aptamer into Modified DNA Aptamers Provides for Prolonged Stability and Enhanced Antitumor Activity. *J. Am. Chem. Soc.*, *143*, 7655–7670, **2021**, doi: https://doi.org/10.1021/jacs.9b10460.
87. Martin, D.R., Sibuyi, N.R., Dube, P., Fadaka, A.O., Cloete, R., Onani, M., Madiehe, A.M., Meyer, M., Aptamer-Based Diagnostic Systems for the Rapid Screening of Tb at the Point-of-Care. *Diagnostics*, *11*, 1–24, **2021**, doi: 10.3390/diagnostics11081352.
88. Moabelo, K.L., Lerga, T.M., Jauset-Rubio, M., Sibuyi, N.R.S., O'Sullivan, C.K., Meyer, M., Madiehe, A.M. A Label-Free Gold Nanoparticles-Based Optical Aptasensor for the Detection of Retinol Binding Protein 4. *Biosensors*, 12, 1061, 2022. https://doi.org/10.3390/bios12121061
89. Jayasena, S.D., Aptamers: An Emerging Class of Molecules That Rival Antibodies in Diagnostics. *Clin. Chem.*, *45*, 1628–1650, **1999**, doi: 10.1093/CLINCHEM/45.9.1628.
90. Zhou, G., Wilson, G., Hebbard, L., Duan, W., Liddle, C., George, J., Qiao, L., Zhou, G., Wilson, G., Hebbard, L. *et al.*, Aptamers: A Promising Chemical Antibody for Cancer Therapy. *Oncotarget*, *7*, 13446–13463, **2016**, doi: 10.18632/ONCOTARGET.7178.
91. Aslan, Y., Atabay, M., Chowdhury, H.K., Göktürk, I., Saylan, Y., Inci, F., Aptamer-Based Point-of-Care Devices: Emerging Technologies and Integration of Computational Methods. *Biosens.*, *13*, 569, **2023**, doi: 10.3390/BIOS13050569.
92. Shen, L., Jia, K., Bing, T., Zhang, Z., Zhen, X., Liu, X., Zhang, N., Shangguan, D., Detection of Circulating Tumor-Related Materials by Aptamer Capturing and Endogenous Enzyme-Signal Amplification. *Anal. Chem.*, *92*, 5370–5378, **2020**, doi: 10.1021/ACS.ANALCHEM.0C00051/ASSET/IMAGES/LARGE/AC0C00051_0004.JPEG.
93. Arghya, S., Suradip, D., Pragya, S., Utpal, B., Aptasensors in Health, Environment and Food Safety Monitoring. *Open J. Appl. Biosens.*, *2012*, 21880, **2012**, doi: 10.4236/OJAB.2012.12002.
94. Fu, Z., Lu, Y.-C., Lai, J.J., Recent Advances in Biosensors for Nucleic Acid and Exosome Detection. *Chonnam Med. J.*, *55*, 98, **2019**, doi: 10.4068/CMJ.2019.55.2.86.
95. Giorgi-Coll, S., Marín, M.J., Sule, O., Hutchinson, P.J., Carpenter, K.L.H., Aptamer-Modified Gold Nanoparticles for Rapid Aggregation-Based Detection of Inflammation: An Optical Assay for Interleukin-6. *Microchim. Acta*, *187*, 1–11, **2020**, doi: 10.1007/S00604-019-3975-7/FIGURES/5.
96. Lavania, S., Das, R., Dhiman, A., Myneedu, V.P., Verma, A., Singh, N., Sharma, T.K., Tyagi, J.S., Aptamer-Based TB Antigen Tests for the Rapid Diagnosis of Pulmonary Tuberculosis: Potential Utility in Screening for Tuberculosis. *ACS Infect. Dis.*, *4*, 1718–1726, **2018**, doi: 10.1021/acsinfecdis.8b00201.

97. Chotimah, S.N., Putra, Y.E., Ng, J.H., Tijptaningrum, A., Sahara, N.S., Arfianti, Amaliah, R., Edel, M.J., Rodriguez, M.O., Llerga, T.M. *et al.*, DNA Aptamer Gold Nanoparticle Colorimetric Diagnostic Test Kit of Saliva Samples for SARS_Cov2 Virus Linked to Mobile Phone Application (Aptamex^{TM}), medRxiv, **2022**, 2022.02.09.22269224, doi:10.1101/2022.02.09.22269224.
98. Le, T.T., Chang, P., Benton, D.J., McCauley, J.W., Iqbal, M., Cass, A.E.G., Dual Recognition Element Lateral Flow Assay Toward Multiplex Strain Specific Influenza Virus Detection. *Anal. Chem.*, *89*, 6781–6786, **2017**, doi: 10.1021/acs.analchem.7b01149.
99. Aithal, S., Mishriki, S., Gupta, R., Sahu, R.P., Botos, G., Tanvir, S., Hanson, R.W., Puri, I.K., SARS-CoV-2 Detection with Aptamer-Functionalized Gold Nanoparticles. *Talanta*, *236*, 122841, **2022**, doi: 10.1016/J.TALANTA.2021.122841.
100. Fenollar, F. and Mediannikov, O., Emerging Infectious Diseases in Africa in the 21st Century. *New Microbes New Infect.*, *26*, S10–S18, **2018**, doi: 10.1016/J.NMNI.2018.09.004.
101. Perkins, M.D. and Kessel, M., What Ebola Tells Us about Outbreak Diagnostic Readiness. *Nat. Biotechnol. - Comment.*, *33*, 464–469, **2015**.
102. Goudarzi, M., Fazeli, M., Azad, M., Seyedjavadi, S.S., Survey of Clinical Features, Pathogenesis and Therapeutic Options for Ebola Haemorrhagic Fever. *Arch. Adv. Biosci.*, *6*, 145–152, **2015**.
103. Ngom, B., Guo, Y., Wang, X., Bi, D., Development and Application of Lateral Flow Test Strip Technology for Detection of Infectious Agents and Chemical Contaminants: A Review. *Anal. Bioanal. Chem.*, *397*, 1113–1135, **2010**, doi: 10.1007/s00216-010-3661-4.
104. Kiszewski, A., Johns, B., Schapira, A., Delacollette, C., Crowell, V., Tan-Torres, T., Ameneshewa, B., Teklehaimanot, A., Nafo-Traoré, F., Estimated Global Resources Needed to Attain International Malaria Control Goals. *Bull. World Health Organ.*, *85*, 623–630, **2007**, doi: 10.2471/BLT.06.039529.
105. John, A.S. and Price, C.P., Existing and Emerging Technologies for Point-of-Care Testing. *Clin. Biochem. Rev.*, *35*, 155–167, **2014**.
106. Majdinasab, M., Badea, M., Marty, J.L., Aptamer-Based Lateral Flow Assays: Current Trends in Clinical Diagnostic Rapid Tests. *Pharmaceuticals*, *15*, 90–116, **2022**, doi: 10.3390/ph15010090.
107. Kulabhusan, P.K., Hussain, B., Yüce, M., Current Perspectives on Aptamers as Diagnostic Tools and Therapeutic Agents. *Pharmaceutics*, *12*, 1–23, **2020**, doi: 10.3390/pharmaceutics12070646.
108. Bauer, M., Strom, M., Hammond, D.S., Shigdar, S., Anything You Can Do, i Can Do Better: Can Aptamers Replace Antibodies in Clinical Diagnostic Applications? *Molecules*, *24*, 1–13, **2019**, doi: 10.3390/molecules24234377.
109. Pauly, D. and Hanack, K., How to Avoid Pitfalls in Antibody Use. *F1000Research*, *4*, 1–7, **2015**, doi: 10.12688/f1000research.6894.1.

110. Bordeaux, J., Welsh, A.W., Agarwal, S., Killiam, E., Baquero, M.T., Hanna, J.A., Anagnostou, V.K., Rimm, D.L., Antibody Validation. *Biotechniques*, *48*, 197–209, **2010**, doi: 10.2144/000113382.
111. Ciancio, D.R., Vargas, M.R., Thiel, W.H., Bruno, M.A., Giangrande, P.H., Mestre, M.B., Aptamers as Diagnostic Tools in Cancer. *Pharmaceuticals*, *11*, 1–23, **2018**, doi: 10.3390/ph11030086.
112. Moutsiopoulou, A., Broyles, D., Dikici, E., Daunert, S., Deo, S.K., Molecular Aptamer Beacons and Their Applications in Imaging and Diagnostics. *Small*, *35*, 1–35, **2019**, doi: 10.1002/smll.201902248.Molecular.
113. Toh, S.Y., Citartan, M., Gopinath, S.C.B., Tang, T.H., Aptamers as a Replacement for Antibodies in Enzyme-Linked Immunosorbent Assay. *Biosens. Bioelectron.*, *64*, 392–403, **2015**, doi: 10.1016/J.BIOS.2014.09.026.
114. Song, S., Wang, L., Li, J., Fan, C., Zhao, J., Aptamer-Based Biosensors. *TrAC, Trends Anal. Chem.*, *27*, 108–117, **2008**, doi: 10.1016/J.TRAC.2007.12.004.
115. Navani, N.K. and Li, Y., Nucleic Acid Aptamers and Enzymes as Sensors. *Curr. Opin. Chem. Biol.*, *10*, 272–281, **2006**, doi: 10.1016/J.CBPA.2006.04.003.
116. Shigdar, S., Qian, C., Lv, L., Pu, C., Li, Y., Li, L., Marappan, M., Lin, J., Wang, L., Duan, W., The Use of Sensitive Chemical Antibodies for Diagnosis: Detection of Low Levels of Epcam in Breast Cancer. *PLoS One*, *8*, e57613–e57620, **2013**, doi: 10.1371/journal.pone.0057613.
117. Henri, J., Bayat, N., Macdonald, J., Shigdar, S.A., Guide to Using Nucleic Acid Aptamers in Cell Based Assays. *Aptamers*, *3*, 4–9, **2019**.
118. Macdonald, J., Houghton, P., Xiang, D., Duan, W., Shigdar, S., Truncation and Mutation of a Transferrin Receptor Aptamer Enhances Binding Affinity. *Nucleic Acid Ther.*, *26*, 348–354, **2016**, doi: 10.1089/nat.2015.0585.
119. Seo, H.B. and Gu, M.B., Aptamer-Based Sandwich-Type Biosensors. *J. Biol. Eng.*, *11*, 1–7, **2017**, doi: 10.1186/s13036-017-0054-7.
120. Jarczewska, M. and Malinowska, E., The Application of Antibody-Aptamer Hybrid Biosensors in Clinical Diagnostics and Environmental Analysis. *Anal. Methods*, *12*, 3183–3199, **2020**, doi: 10.1039/d0ay00678e.
121. Kaminski, M.M., Alcantar, M.A., Lape, I.T., Greensmith, R., Huske, A.C., Valeri, J.A., Marty, F.M., Klämbt, V., Azzi, J., Akalin, E. *et al.*, A CRISPR-Based Assay for the Detection of Opportunistic Infections Post-Transplantation and for the Monitoring of Transplant Rejection. *Nat. Biomed. Eng.*, *4*, 601–609, **2020**, doi: 10.1038/s41551-020-0546-5.
122. Wan, Q., Liu, X., Zu, Y., Oligonucleotide Aptamers for Pathogen Detection and Infectious Disease Control. *Theranostics*, *11*, 9133–9161, **2021**, doi: 10.7150/thno.61804.
123. Shen, J., Liu, G., Ji, Y., Liu, Q., Cheng, L., Guan, K., Zhang, M., Liu, G., Xiong, J., Yang, J. *et al.*, 2D MXene Nanofilms with Tunable Gas Transport Channels. *Adv. Funct. Mater.*, *28*, 1–13, **2018**, doi: 10.1002/adfm.201801511.
124. Lee, K.H. and Zeng, H., Aptamer-Based ELISA Assay for Highly Specific and Sensitive Detection of Zika NS1 Protein. *Anal. Chem.*, *89*, 12743–12748, **2017**, doi: 10.1021/acs.analchem.7b02862.

125. Hong, S.-L., Xiang, M.-Q., Tang, M., Pang, D.-W., Zhang, Z.-L., Ebola Virus Aptamers: From Highly Efficient Selection to Application on Magnetism-Controlled Chips. *Anal. Chem.*, *91*, 3367–3373, **2019**, doi: 10.1021/acs.analchem.8b04623.
126. Dhiman, A., Haldar, S., Mishra, S.K., Sharma, N., Bansal, A., Ahmad, Y., Kumar, A., Sharma, T.K., Tyagi, J.S., Generation and Application of DNA Aptamers against HspX for Accurate Diagnosis of Tuberculous Meningitis. *Tuberculosis*, *112*, 27–36, **2018**.
127. Kumari, P., Lavania, S., Tyagi, S., Dhiman, A., Rath, D., Anthwal, D., Gupta, R.K., Sharma, N., Gadpayle, A.K., Taneja, R.S. *et al.*, A Novel Aptamer-Based Test for the Rapid and Accurate Diagnosis of Pleural Tuberculosis. *Anal. Biochem.*, *564–565*, 80–87, **2019**, doi: 10.1016/J.AB.2018.10.019.
128. Yoon, S. and Rossi, J.J., Future Strategies for the Discovery of Therapeutic Aptamers. *Expert Opin. Drug Discov.*, *12*, 317–319, **2017**, doi: 10.1080/17460441.2017.1290077.
129. Markets, Markets Aptamers Market. Mark. Mark. Report Code: BT 3550, 2015.
130. Report, M.A., Global Aptamers Market Size, Share & Growth Report, 2030. Mark. Anal. Reports, 2018.
131. Kaur, H., Bruno, J.G., Kumar, A., Sharma, T.K., Aptamers in the Therapeutics and Diagnostics Pipelines. *Theranostics*, *8*, 4016–4032, **2018**, doi: 10.7150/thno.25958.
132. Diagnostics, H., Sepsis LFD, Hutano Diagnostics, 2022.
133. Diagnostics, H., Universal Kit Service, 2022.
134. International, L., First-of-Its-Kind Saliva-Based Rapid Test Uses DNA Aptamer to Detect COVID-19, LabMedica, 2022.
135. Cytiva How to Guide Multiplexing Your Lateral Flow Assay, GE Life Sci., 2019.
136. Haage, V., de Oliveira-Filho, E.F., Moreira-Soto, A., Kühne, A., Fischer, C., Sacks, J.A., Corman, V.M., Müller, M.A., Drosten, C., Drexler, J.F., Impaired Performance of SARS-CoV-2 Antigen-Detecting Rapid Diagnostic Tests at Elevated and Low Temperatures. *J. Clin. Virol.*, *138*, 104796, **2021**, doi: 10.1016/j.jcv.2021.104796.
137. Wild, D., Sheehan, C., Binder, S., Chapter 7.1 - Introduction to Immunoassay Product Technology in Clinical Diagnostic Testing, in: *Immunoass. Handb.*, Fourth Ed., pp. 509–515, 2013.
138. Thiviyanathan, V. and Gorenstein, D.G., Aptamers and the next Generation of Diagnostic Reagents. *Proteomics Clin. Appl.*, *6*, 563–573, **2012**, doi: 10.1002/prca.201200042.
139. Raybiotech Custom Aptamer Services, Raybiotech, 2021.
140. Arnot, L.F. and Michel, A., Challenges for Controlling Bovine Tuberculosis in South Africa. *Onderstepoort J. Vet. Res.*, *87*, 1–8, 2020.

141. Tulu, D., Bovine Brucellosis: Epidemiology, Public Health Implications, and Status of Brucellosis in Ethiopia. *Vet. Med. Res. Rep.*, *13*, 21–30, **2022**, doi: 10.2147/vmrr.s347337.
142. Lakhin, A.V., Tarantul, V.Z., Gening, L.V., Aptamers: Problems, Solutions and Prospects. *Acta Naturae*, 5, 34–43, 2013.
143. Aljohani, M.M., Cialla-may, D., Popp, J., Chinnappan, R., Al-kattan, K., Zourob, M., Aptamers: Potential Diagnostic and Therapeutic Agents for Blood Diseases. *Molecules*, 27, pp 383–403, 1–21, **2022**.
144. Mueggler, P.A., Buck, R.J., Wang, H., H.T.P., Covalent Bonding of Molecules to an Activated Solid Phase Material, 2001.
145. Corporation, E.M.D.M., Rapid Lateral Flow Test Strips Considerations for Product Development, EMD Millipore, 2013.
146. Ellington, A.A., Kullo, I.J., Bailey, K.R., Klee, G.G., Antibody-Based Protein Multiplex Platforms: Technical and Operational Challenges. *Clin. Chem.*, *56*, 186–193, **2010**, doi: 10.1373/clinchem.2009.127514.

10

Nanobiosensors: A Platform for the Diagnosis of Microbial Pathogens

Ranjita Misra* and Naomi Sanjana Sharath

Department of Biotechnology, School of Sciences, Jain University, Bangalore, Karnataka, India

Abstract
Microbes are present all over the earth. They can be found in plants, animals as well as human beings. Some are harmless, whereas some can be harmful and are hence considered to be pathogens. The early detection of microbial pathogens is very crucial for effective disease management and prevention. In recent years, nanobiosensors have emerged as a promising tool for the rapid and sensitive detection of microbial pathogen. This book chapter provides an overview of the types of pathogens, the disease caused by them, and the nanomaterials and nanoparticles that can be used for efficient development of nanobiosensors with the prime focus on microbial pathogen detection. There are different types of biosensors such as electrochemical, optical, and thermal biosensors, which are discussed in detail. Recent research on the use of nanobiosensors for detecting viral and bacterial pathogens such as *E. coli* and *Salmonella*, are of great potential for the development of future nanobiosensors for better and more rapid detection. Some of the challenges that the nanobiosensors can face include sample preparation, stability, reproducibility, and so on, are also discussed. This chapter concludes by highlighting the potential of nanobiosensors as a platform for the diagnosis of microbial pathogens and outlining the possible future directions of research in this particular area.

Keywords: Nanobiosensors, pathogens, nanomaterials, diagnostics, diseases, stability, drug delivery, biocompatibility

**Corresponding author*: ranjita.biotech@gmail.com; ranjita.misra@jainuniversity.ac.in

Suvardhan Kanchi, Ayyappa Bathinapatla, Anitha Varghese and Phumlane Selby Mdluli (eds.) Nano-Biosensor Technologies for Diagnosis of Infectious Diseases, (315–340) © 2025 Scrivener Publishing LLC

10.1 Introduction

Microbes have evolved along with diversification of land, species of plants, and animals for millions of years. They inhabit every part of the earth, the soil, air, plants, and animals, and are even found in the gut of human beings. Human beings have 31 trillion cells and they have approximately the same number of bacterial cells in their gut [1]. Over 1400 human pathogens have been discovered so far. There are 1 trillion microbial species on the earth, which still remain unknown and uncharacterized [3]. Most of the microbes are usually harmless, but other harmful microbes which are thereby considered pathogens can sometimes cause health issues to the human body or sometimes, can even be potentially fatal to the host which it tries to invade itself, thereby affecting rural and urban areas [2]. Pathogens are of two types, obligative and facultative pathogens. Facultative pathogens cause infections occasionally. For example, *Candida*, which is a facultative parasite that can resort on its parasitic activity, but it does not rely on its host for a full completion of its life cycle. Obligate pathogens on the other hand, require a host/multiple hosts in order to complete their life cycle. Zika virus, for example, requires a multiple host for its life cycle completion [1]. Infections can be caused by any type of pathogen, be it bacteria, fungi, parasites, or viruses .Therefore, it is essential for us to take general precautions such as following basic hygiene and consuming healthy and nutritious foods, which are the basic keys on promoting and sustaining human life. Infectious diseases are not just caused by pathogens from unclean food or air. Situations such as wars, famines, and pandemics can cause the pathogens to thrive in massive amounts and cause a huge number of death tolls across the globe [1, 2]. An example is the influenza virus that causes respiratory illness which has the utmost ability to spread at an extremely fast rate, and which caused the most severe influenza pandemic in 1918, leading to more than 20–50 million deaths worldwide [4]. Pathogens can cause diseases in a variety of ways. One of the main methods is through production of toxins which damages the tissues/cells of the hosts. Whereas another way for the pathogen to spread the infection is to spread itself from an infected host to an uninfected host. The conventional method of diagnosis of these pathogens are not precise. Diagnosis of these pathogens at an early stage is very important. Incorrect diagnosis can often lead to consequences of the patient's health. Some of the common conventional methods include culturing of the microbes on the plates, colony counting, immunological techniques such as antigen–antibody interactions and polymerase chain reaction (PCR). If pathogens are detected at low concentrations, can result in improper diagnosis. Also, if the pathogen is unknown, there is no knowledge about the physical and biochemical properties of the bacterial species, making it a great disadvantage for the application of these detection techniques.

So, early and rapid detection methods is a must in order to reduce mortality and to make them cost effective in order to provide improvement for better clinical diagnostic purposes in the future.

10.2 Microbial Pathogens

A pathogen is an organism that invades its host causing their specific diseases and it can lead to virulence (the severity of the disease) depending upon the degree of damage caused by the pathogens to its fellow host. The main aim of the pathogen is to invade the host and thrive in it in a successful manner. Pathogens are very taxonomically diverse in because of their multiple strains that coexist, and widely exist in the biology systems. There are many different kinds of pathogens. Some examples are:

10.2.1 Bacteria

Bacteria are microscopic pathogens that reproduce at a rapid rate and are very diverse. Bacteria ranges from different shapes and sizes. The main five classes of bacteria are rod (bacilli), corkscrew (spirochaetes), comma (vibrios), spiral (spirilla), and spherical (cocci) [6]. Bacteria are also classified as to how they respond to oxygen in various environments. So, they are classified as aerobes, anaerobes, and facultative anaerobes. One unique feature about bacteria is that they can survive under extreme conditions. This is because of they are surrounded by a tough protective coating. When they attack the immune system, they show resistance to white blood cells (WBC) of the human body, hence, causing successful invasion of the host. They can be treated with antibiotics, but some bacteria can develop resistance to these antibiotics making the treatment process a huge challenge. Most of them are not harmful, they make up in the microtome of the human gut. Whereas, around 5% of the bacteria are harmful, hence causing infections [6]. Sometimes, the microbial inhabitants in our body cause trouble only if the immune system becomes weak or if they get a chance to access sterile parts of the body. For example, when a bowel perforation begins to enable the gut flora to enter the peritoneal cavity of the abdomen, it causes peritonitis (a condition of membrane inflammation lining in the abdominal wall which covers the abdominal organs, causing bursting of the appendix and can be life threatening). If the immune system is compromised by a virus, the human body can be more prone to immune infections and it can further increase the severity of the disease. In order for a bacterial pathogen to successfully survive the host, they need to colonize the host, find a nutritional niche (a sheltered or a private like space) which is compatible in the human body, replicate then its host through its sources, and exit the host and spread to another host [5]. After entering the host, they release their toxins that cause the required damage to the cells of the host.

10.2.2 Fungi

Fungi is eukaryotic in nature and there are more than 6 million species of fungi in the world. Approximately 300 fungal species are known to be pathogenic to humans. One of the most prominent fungal species, *Candida*, can cause infections in humans with weak immune systems [6]. Other common fungal species include *Aspergillus fumigatus* that causes allergies and *Aspergillus flavus* that causes fungal infection through food contamination. Other fungal diseases caused in humans include meningitis, pneumonia, skin and nail fungal infections, and asthma. The fungal cell wall is a thick structure with immensely great plasticity which is used to maintain cell integrity and viability [7]. Hence, making it a great defense mechanism to prevent itself from getting destroyed by any treatment. Some fungal pathogens can use signals to anticipate any possible threats to their survival, which they can then prepare themselves to defense.

10.2.3 Parasites

Parasites live in a host and rely on their host's food for their survival. The three classes of parasites that can cause human diseases are protozoa, helminths, and ectoparasites. Protozoa are single-celled microorganisms that can be either free living or parasitic in nature. They can spread through contaminated food or water, hence causing food borne diseases, insect bites or person to person transmission. Malaria, for example, is caused by *Plasmodium malariae*, which causes malarian infection. Helminths are parasitic worms that are larger multi-celled organisms that root themselves in the human digestive tract. These parasites cannot multiply or divide within a human body, so they then they are eliminated through stools. Ectoparasites are small organisms that live on the outside of the body that

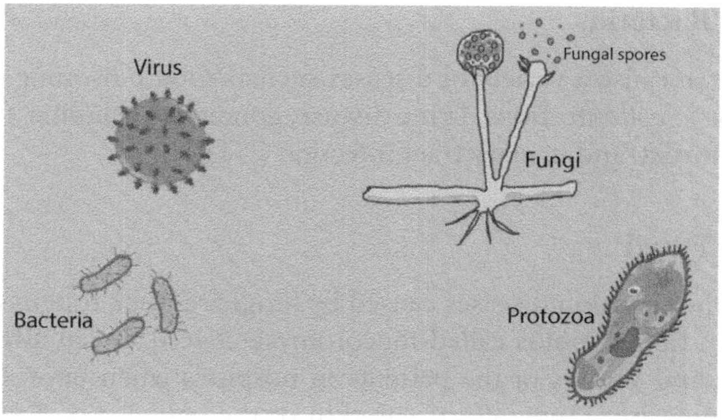

Figure 10.1 Types of microbes.

feed off the skin of the host for survival. These include ticks, fleas, mosquitoes, and lice. Figure 10.1 shows the different types of microbes.

10.3 Diseases Caused by Pathogens

Pathogen can cause diseases that range in terms of their severity and how they are transmitted. Some of the diseases caused by pathogens are:

10.3.1 Viruses

Viruses are infectious microbes that consist of a segment of nucleic acids deoxyribonucleic acid (DNA) or ribonucleic acid (RNA) that is enclosed by coating protein called the capsid. They cause a number of infections such as human immunodeficiency virus (HIV), chickenpox/shingles, measles, viral gastroenteritis, skin warts, flu, and meningitis. They have the unique ability to invade the host, replicate itself using the host's components, producing more viruses. After the completion of the replication cycle, the newly formed viruses are exported out of the host cell and they cause further damage to other cells of the host. Sometimes, when a person appears to have recovered from a viral infection, but falls sick again, it is because of the dormant nature of some of the viruses which remain dormant for some time before multiplying itself again. The use of antibiotics is of no use when it comes to treatment of the viruses. However, different vaccines and antiviral drug therapies can be used for treating them.

10.3.2 Bacteria

Bacteria can cause a variety of diseases such as food poisoning caused by *S. typhi* or *E. coli* infections, Lyme disease, gonorrhea cellulitis, tuberculosis, strep throat, and urinary tract infection (UTI).

10.3.3 Fungi

Some of the common diseases caused by fungi are black fungus caused by a group of fungal molds called mucoromycetes [63]. It can affect mainly the lungs and sinuses of the patients in hospitals when upon inhalation of the spores, but rarely affects the skin of the patient [7]. Vaginal yeast infections are also another major fungal infection in women that is caused by *Candida*. Other infections include ringworm, jock itch, and fungal nail infections (onychomycosis).

10.3.4 Parasites

Some of the diseases caused by parasites are public lice that are found in the genital area that can extensively feed on human blood, malaria, and intestinal worms that feed off the human body such as tapeworms, hookworms, pinworms, and *Trichinella*, which causes trichinosis (a disease caused by eating raw or undercooked meat). Other common parasites are *Giardia lamblia*, which causes beaver fever, and *Entamoeba histolytica*, which causes amoebic dysentery.

10.4 Importance of Pathogen Detection or Disease Diagnosis

The use of the traditional conventional methods such as the classical cell culture techniques or biochemical detection methods which are time consuming and selective. This can sometimes generate improper or false results such as improper colony counting on the agar culture plates. This makes the detection of the disease very challenging and difficult. Therefore, the need for developing rapid, sophisticated, cost effective, and more advanced systems for microbial detection causing infections is a must in order to prevent further spreading of the disease and to provide better management and for other applications such as better food and environmental monitoring, drug quality testing, bioresearch, and so on. Food pathogens arise from contaminated food sources due to unhygienic preparation such as street food, undercooked food, ready to eat processed foods, or foods prepared using contaminated water and so on. This causes various foodborne disease that are a major threat to public health. Many kinds of food pathogens cause various food borne diseases, ranging from mild cases to chronic conditions which may lead to hospitalization and even death [13]. For example, *S. aureus* is known to cause nosocomial infections (additional infections that are health care/hospital associated that arise when the patient is admitted in the hospital with diseases such as pneumonia) can also a cause of toxin-related food poisoning [8]. Technological advancements and various strategies for better identification and detection of microbes are on the rise with various scientific research conducted. So, the requirement of developing rapid, robust, and highly reliable detection tests for better and faster analysis and for the safety of human welfare is a major requirement. The various multi-disciplinary research fields have been conducted in order to develop more commercially feasible products and are on clinical trials. Whereas some of them have reached the market. Clinical laboratories to this day still use conventional methods for bacterial identification such as 16S rRNA sequencing and PCR for foodborne bacterial pathogen detection. Matrix-assisted laser desorption/ionization time-of-flight mass spectrometry (MALDI-TOF MS), which is one of the latest methods used in hospital microbial laboratories and is also used bacterial typing (discriminate various bacterial strains from one another), could help in detecting nosocomial outbreaks and have a beneficial impact on the patient's safety [9]. Matrix-assisted laser desorption/ionization time-of-flight mass spectrometry (MALDI TOF MS) also has improved in processing samples and expanded databases have facilitated rapid identification of microorganisms from some clinical samples thereby making it fast and efficient. Another rapid, highly specific, and reliable method is the loop-mediated isothermal amplification (LAMP) method. The reason LAMP is so rapid is because it uses four primers to target six specific regions of the template and it provides rapid amplification, which generated greater yield of amplification products than PCR assay. The final DNA products are formed in a cauliflower or stem loop like DNAs of different sizes. For example, the Loopamp detection kit (Eiken Chemical) is commercially available for foodborne pathogen detection such as *Shigella*. These kinds of desirable methods can be made more efficient and should take some parameters into consideration which would then lead to a cost-effective, more optimal and portable, point of source-detection system such as being able to detect multiple pathogens in just one single run. These types of systems can be developed in the near future.

10.5 Biosensors

Many microbiological tests are centralized in labs due to the requirement of complex instruments and highly qualified technical staff. In the recent years, lot of research has been conducted in order to find a better and simpler solution for these tests so that they can be designed in such a way that they can be performed virtually anywhere and can be more portable and accessible. Hence, the biosensor technology is well suited for this purpose due to its ability to give 'on- spot' interpretation of results [16]. Due to many situations such as the recent pandemic (COVID-19), the requirements of generating rapid and accurate diagnostic tests results and to treat various infections are on demand nowadays. Biosensors are analytical devices which convert biological responses into signals which can be electrical or optical signals. It was first discovered in 1962 by Clark known as the 'Father of Biosensors' where he developed an oxidase enzyme electrode required for glucose detection [17]. It consists of two main components, a biorecognition element (bioreceptor) which can be a cell, organelle or a tissue which recognizes the target analyte and a transducer, which converts it into an electrical signal that can be measurable. The promising areas for biosensors are food analysis, environmental monitoring, clinical diagnostics, and environmental monitoring. Biosensors are highly specific and sensitive, which can also allow detection of a wide range of samples such as blood, urine, and serum, thereby generating the required results. Biosensors are classified into four basic groups: optical, electrochemical, thermal sensors and mass-based biosensors, based upon on their signal transduction method.

10.5.1 Optical-Based Biosensors

Optical-based biosensors are very beneficial for the detection of bacterial pathogens due to their high selectivity and sensitivity. They offer a large number of subclasses based upon absorption, reflection, refraction, Raman, or infrared spectroscopy (IR). These subclasses, however, require a suitable spectrometer to record the properties of the analyte. The most common technique for optical detection used is surface plasmon resonance.

10.5.2 Electrochemical-Based Biosensors

Electrochemical-based biosensors are used for the quantification and identification of foodborne pathogens. They are further classified as potentiometric-, amperometric-, impedimetric-, and conductometric-based biosensors based upon parameters such as potential, current, impedance, and conductance. They have certain advantages such as the ability to work with turbid samples, low cost, and can be combined with other biosensing techniques for improved detection. However, they have their limitations such as limited sensitivity and selectivity.

10.5.3 Thermal Biosensors

Thermal biosensors measure the heat released or absorbed in a biochemical reaction [15]. It can be used to detect organic contaminants in water using the chemical oxygen demand (COD) parameter, home monitoring purposes such as checking the glucose levels for diabetic control, and so on. It has long term stability, accuracy, and low cost. But it has limitations such as complicated thermostating and weak sensitivity.

10.5.4 Mass-Based Biosensors

Mass-based biosensors are also called piezoelectric biosensors that produce electrical signals when mechanical force is applied. It consists of a piezoelectric crystal which is the sensor part working on the principle of change in oscillations due to the mass that is bound on the piezoelectrical surface. An example is the quartz crystal microbalance that is used in electronic devices. They have high stability and sensitivity but cannot be used for truly static measurements.

10.6 Nanobiosensors as Diagnostic Platform

Biosensors consist of bioreceptors as one of their components which can be glycoproteins, glycopeptides, carbohydrates, lipids, and lipoproteins that can facilitate multiple functions such as adhesion onto cell surfaces, involvement in infection processes, and intake of nutrients. These biosensors serve as biorecognition elements which is one common feature of all the biosensors [2]. Over the years, nanotechnology-based materials have been studied and used for the development of better diagnostic tests for the prevention and control of infectious diseases. The various chemical and physical properties of nanoparticles allow rapid, accurate and cost-efficient diagnostics. For example, fluorescent silica nanoparticles have been used in an antibody-based diagnosis to develop *Mycobacterium tuberculosis* complex (MTB) within a span of 4 h [17]. Even the incorporation of europium (Eu III) polymeric nanoparticles has been used for the detection of anthrax antibodies by using a fluorescence enzyme linked immunosorbent assay (ELISA) which proved to be successful. Work done by Wang *et al.* showed a recently developed detection-based method for *S. typhimurium* based on a combination of silica fluorescence nanoparticles and single-stranded DNA aptamers [18]. The use of liposomes is a great benefit for toxin detection. In order to detect the toxins, they are labelled with fluorescent markers like rhodamine dyes which can be incorporated into a sandwich fluoro-immunoassay on an antibody-coated microtiter plate [19]. This method was demonstrated by Ahn Yoon *et al.* where cholera toxins were detected using the limit of detection as 10 (fg/mL). This proved to be a successful reaction which was complete in just 20 min [20]. Similarly, tri-sialo-ganglioside GT1b-liposomes which is a botulism (BT) receptor was used to detect botulinum toxin (BT) on a nitrocellulose membrane strip [21]. The development of nanobiosensors can also be used with quantum dots. Quantum dots (QDs) are inorganic fluorophores that have, to some extent, replaced conventional fluorescent dyes, which exhibit unique light emitting properties which show strong resistance to chemical degradation and photobleaching and have high quantum yield and can detect several biomarkers. Quantum dots are nanocrystalline semiconductors composed of materials such as zinc sulfide (ZnS), zinc selenium (ZnSe), cadmium sulfide (CdS), cadmium selenium (CdSe), cadmium telluride (CdTe), and indium phosphide (InP). Their limitations are unstable fluorescence generation and cytotoxicity. A strategy developed for *E. coli* detection used QDs which were coated with mannose and the nanostructures in the device were able to identify a mannose-specific lectin called FimH from *E. coli* surfaces in their suspensions with less than 10^4 cells mL^{-1}. *E. coli* stands out as the most studied targeted bacteria using nanobiosensors [2]. The application of QDs in nanomedicine are of various kinds such as biosensors for use in therapeutics agents and can also be applied in florescent probes. Quantum dots have also shown to be a boon for the rapid detection of bacterial and viral proteins. Work done by Zhang and Hu [24] developed a multiplex assay using QD-based nanosensor for detecting HIV1 and HIV2. This generated results with high sensitivity, short analysis time, and had low sample consumption. Because of this, they can be used for a variety of applications such as development of point care tests, high throughput screening, and gene expression studies. Later in 2016, Zhang *et al.* demonstrated an immunosensor-based technique to screen and isolate *Salmonella sp.* with 10 cell/mL as the detection limit [26]. This proved to be a very efficient technique [25].

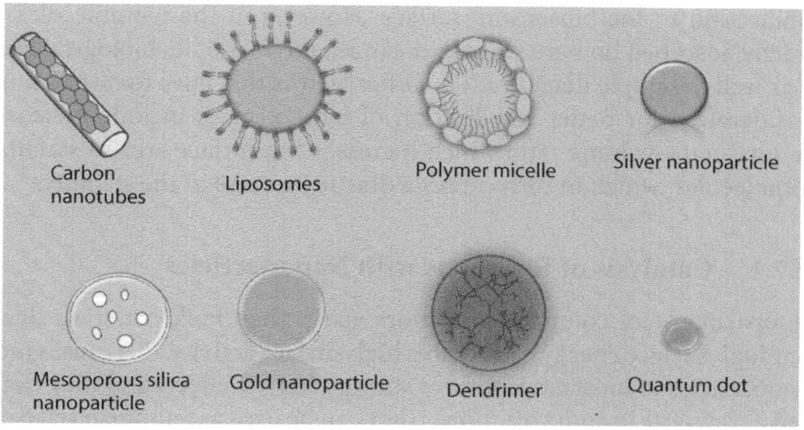

Figure 10.2 Types of nanoparticles.

Materials with sizes less than 100 nm are considered as nanobiosensors [2]. The use of nanostructures such as nanowires, nanotubes, nanopores, nanocomposites, and self-adhesive monolayers, can be combined with DNA or proteins that can lead to the development of a biomaterial-nanostructure interface with the required properties and a desirable performance. The use of self-adhesive monolayers, which are being used to design nanostructured interfaces have attracted great amount of attention. For example, self-adhesive monolayers on gold are used to make sensor joints (devices used to measure any high speed motion, used mainly for athletes). The thickness of the monolayer's surface can be controlled by selecting the appropriate molecules which is an advantage of using this material [27]. Other advantages include specific interaction between the layer and the desired analyte, and proper selectivity in the field of electrochemical biosensors [28]. Nanoparticles also have a variety of applications in the fabrication of sensors for the purposes of improved electron transfer and any kind of detection reactions (Figure 10.2).

10.7 Stabilization of Biomolecules with Nanoparticles

Certain chemical, physical, and electronic properties of nanoparticles help in distinguishing themselves from large scale materials. Due to their high amount of free surface energy and surface area, the nanoparticles strongly adsorb the biomolecules to their surface. This helps in giving proper stabilization to the biosensors surface. However, if the biomolecules are directly adsorbed on bare surface, it can lead to loss of its biological activity as well as sample denaturation. Other factors that have to be taken into consideration for better stabilization of biomolecules include increasing the surface-to-volume ratio which increases the surface area of stabilized biomolecules, which in turn causes a drastic increase in the stability.

10.7.1 Catalysis of Reactions with Nanoparticles

Nanoparticles are comparatively more active than bulk materials due to their high surface energy. Due of this high surface activity, this causes metal nanoparticles in most cases to have strong catalytic effects and they can be used as catalysts in some reactions. They can be used as nanomaterial based catalysts that can be easily separated and recycled [29]. For example, catalytic metal nanoparticles exhibit various desired catalytic reactions [30]. When they are present in electrochemical sensors and biosensors, they can reduce the excess potential of some reversible redox reactions, other than the usage of conventional electrodes which are unmodified.

10.7.2 Improving Electron Transfer with Nanoparticles

The use of redox proteins (proteins that are involved in a wide range of biochemical processes like DNA repair, etc.) such as enzymes, when electrically attached to the electrode surface, is significant in the fabrication of biosensors. The active sites of the redox proteins are surrounded by a thick layer protein shell, which is non-conducting in nature [31]. This is the reason there is no electron transfer between the electrode and the active site [30]. For example, transition metal nanomaterials can be used and are important to non-enzymatic glucose sensing mainly due to their electrocatalytic ability and immensely good selectivity [33].

10.7.3 Nanoparticles Used in Labelling of Biomolecules

Nanoparticles that label biomolecules such as DNA, antigens and antibodies have a significant role in the rising demand of electrochemical biosensors. Once the nanoparticles label the biomolecules, their biological activity is retained [34]. In order to measure the effects of metals, which can make it a possibility to measure the analytes present in the amounts of traces, the nanoparticles labels (metal or semiconductor nanoparticles) need to undergo proper dissolution and the ions can be measured by voltammetry stripping (a voltametric method for quantitative determination of specific ionic species that can determine certain trace elements). Work described by Cao *et al.* [24] demonstrated a series of platinum copper (PtCu) nanochains with different modified compositions. $Pt_{88}Cu_{12}$, $Pt_{75}Cu_{25}$, and $Pt_{50}Cu_{50}$ had been used in creating modified electrodes for glucose detection. Among them, the modified electrode made out of $Pt_{75}Cu_{25}$ showed the best results.

10.7.4 Nanoparticles as Reactants

Nanoparticles are considered to be more active than bulk materials and they show various chemical properties in certain cases, depending upon what type of reaction they undergo. This advantage shows some new development in electrochemical decomposition systems.

10.7.5 Nanomaterials for the Development of Biosensors

Nanomaterials have shown a great impact on the development of biosensors due to their unique physical and chemical properties such as high sensitivity, high surface area, and easy functionalization (Figure 10.3). Some commonly used nanomaterials include:

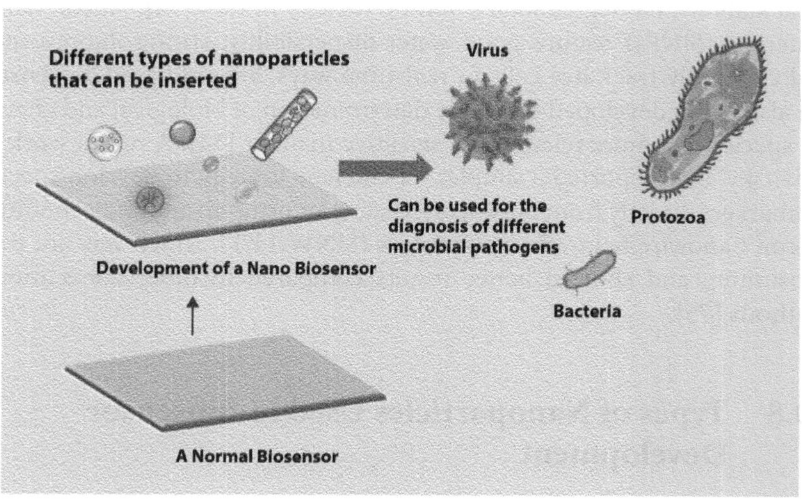

Figure 10.3 Development of nanobiosensors for diagnosis of microbial pathogens.

10.7.5.1 Quantum Dots

Quantum dots are semiconductor nanocrystals that emit light when excited. They can be used as fluorescent tags for biological detection of molecules.

10.7.5.2 Carbon Nanotubes

Carbon nanotubes (CNT) are cylindrical carbon molecules with high surface area and excellent electrical conductivity. They can be used in transducers in electrochemical biosensors. Recently, CNT field-effect transistor-type biosensors have been successfully used in the rapid detection of COVID-19 [73].

10.7.5.3 Graphene

Graphene is a two-dimensional (2D) carbon material with high electrical conductivity, large surface area, and excellent biocompatibility for electrochemical and optical detection.

10.7.5.4 Silicon-Based Nanomaterials

Silicon-based nanomaterials have attracted great attention because of their great stability, biodegradability, and no toxicity *in vivo* [74]. Silicon nanomaterials (SiNPs) feature good water-dispersibility, strong fluorescence, and can easily track live cells in real-time ways. Silicon-based biosensors are also being developed to enable determination of biological and chemical species in a more reliable and sensitive manner [74]. A recent work by Krivitsky *et al.* reported a simple, yet efficient strategy to develop specific biomarkers directly from unprocessed samples using an antibody-modified silicon nanowire field-effect transistors (SiNW-FET), which was not time consuming and showed better results compared to other conventional methods [75].

10.8 Types of Nanoparticles Used in Biosensor Development

10.8.1 Metal Nanoparticles

Among other metals, silver, gold, and platinum have received utmost attention due to their ease of making nanoparticles. Some of them are:

10.8.2 Gold Nanoparticles

Gold nanoparticles are the most widely used nanoparticles mainly in protein stabilization because of their unique ability to stabilize the proteins through covalent bonding. The smaller the nanoparticle, the more beneficial it is for enzyme stabilization. The gold nanoparticles have great biocompatibility that helps in preserving the biological activity of stabilized biomolecules. In order to design a gold nanoparticle based upon the desired application required by the biosensor, if one has to create a 2D structure of the gold nanoparticles, the process of self-assembly is used, *i.e.*, the process where nanoscale building blocks spontaneously organize themselves into ordered structures without any human intervention. Sometimes, they are even used to create 3D structures, for instance, the sol-gel method is used that has a greater number of binding sites for gold nanoparticles than the 2D method. An example of a natural substance called chitosan, which is derived from the shells of crustacean species like shellfish, can be used on the electrode surface for better adsorption of gold nanoparticles in order to generate higher sensor sensitivity. Work done by Wen *et al.* used dopamine-induced (3D) assembly of gold nanoparticles for the formation of gold hydrogel networks. After supercritical drying, they got converted to aerogels and exhibited high surface porosity which in turn, showed better catalytic progress in electrochemical reactions [37]. These electrodes can be used in manufacturing propulsion or enzymatic sensors. According to work done be Nunna *et al.*, an electrochemical immunosensor was developed using gold nanoparticles when fixed with a cancer antigen CA125, was used for detection of cancer. The study demonstrated that the experimental results provided a strong evidence of disease specific detection of CA125 at different concentrations with an increase in the capacitance signal and it proved that gold nanoparticles showed an enhancement in sensor signal response. Measuring the capacitance can provide advantages like minimizing the sensor size, etc. [38]. Another work done by Park *et al.* developed an electrochemical biosensor composed of a multifunctional DNA four-way junction (4WJ), which can specifically bind to haemagglutinin and carboxyl molybdenum disulfide (carboxyl-MoS2) to detect the presence of influenza virus H1N1. This biosensor showed specific and selective results.

The determination of glucose in a rapid and accurate manner is very crucial in terms of detecting diabetes as it is a rapid and a disease which can silently take one's life away. For this particular reason, the development of a proper glucose sensor is very significant. So, for the techniques used in designing an enzymatic electrochemical biosensor, one of the most prominent issues faced is enzyme stabilization, that has to be taken into consideration. Materials which are being selected, have to provide suitable conditions for effective loading while maintaining the enzyme's biological activity. Kang *et al.* demonstrated the development of the surface of a biosensor which was modified on a nanovoltaic electrode. The surface of the electrodes was electrochemically precipitated using gold nanoparticles. It was them immersed in a bath mixture containing chitosan, par benzoquinone glucose oxidase and ionic liquid on order to complete the preparation of enzymatic electrode. A comparative study was done using another set of normal gold electrodes without gold nanoparticles, the gold electrodes incorporated with gold nanoparticles showed that the response time was lesser and the sensitivity was 2.8 times more than a normal electrode without gold nanoparticles [40].

10.8.3 Silver Nanoparticles

They have great biocompatibility and is more stable in water and air, which helps in preserving the biological activity of the biomolecules and are also used for surface modification of electrodes. The application of silver nanoparticles in the fields of biosensor and imaging are widely researched and prominent in the recent days. Work done by Tang *et al.* used silver nanoparticles as a substrate for fixing hepatitis B antibodies on platinum (Pt) electrodes which were modified with silver colloids and polyvinyl butyral (PVB) as matrixes for immunoassays in order to proper detect the hepatitis B surface antigen (HBsAg) [42]. Some electrochemical reactions can be catalyzed by silver nanoparticles. Different kinds of enzymatic sensors of glucose and hydrogen peroxide can be developed using silver nanoparticles [34, 43]. Silver nanoparticles have high electrical conductivity like gold nanoparticles. Silver nanoparticles are also used in a variety of applications such as drug delivery, environmental and electronic fields of study and even as an antimicrobial agent [43].

10.8.4 Other Metal Nanoparticles

There are other nanoparticles that can be used in the field of biosensors other than gold or silver nanoparticles. One of them is platinum nanoparticles, which is of great relevance and has promising potentiality, and they have a vast range of applications such as drug delivery, biomedical applications such as for nanodiagnostic purposes, radiotherapy, implants, and photothermal therapy. They possess antimicrobial and anticancer properties [44]. Additionally, another work demonstrated the deposition of platinum iron Pt-Fe (III) nanoparticles on the surfaces of the electrodes and this made successful development of sensors to detect of nitic oxide (NO), nitrite (NO_2), and uric acid ($C_5H_4N_4O_3$) [45]. Copper nanoparticles are also of great significance because copper is inexpensive and abundant in nature, and they also play a vital role in the field of catalysis [46]. Copper nanoparticles are used in modifying electrodes in electrochemical sensor in order to accelerate their catalytic activity. Palladium nanoparticles (PdNPs) are also of great choice because of their intrinsic features such as their brilliant electronic, catalytic and optical properties, as well as their diversity in sizes and shapes. They can be used for different applications such as photothermal therapy because of its optical property and thermal stability. Because of their optical property, PdNPs have also been used in photoacoustic imaging (PAI) as an upcoming technique for early tumor diagnosis and cancer management [47]. Palladium nanoparticles have also recently shown to be a promising tool for gene or drug delivery systems. Since they have a high surface area which can load a large amount of the drug, stability and non-toxicity of these nanoparticles makes them work as efficient nanocarriers for drug development [48, 49]. Because of their photocatalytic properties which are outstanding, PdNPs have been discovered as prodrug activators which has been well studies in the standard drugs, 5-fluorouracil (5FU) and gemcitabine [50, 51]. Creation of hybrid nanoparticles such as gold-silver nanoparticles have been used in designing electrochemical sensors [2]. Also, work done by Yi *et al.* had developed a 3D nanoporous electrochemical biosensor which consisted of PdNPs electrodeposited onto self-supporting nanoporous gold (NPG) wires, which can detect traced dopamine (DA) at early stages, which is crucial for the diagnosis of Alzheimer's, Parkinson's, and Schizophrenia [52]. The PdNPs possess excellent superiority in detecting DA. It showed good stability, selectivity and is reusable as well. The facet of a palladium nanocrystal has the unique ability to disrupt microorganism membranes [53]. They can generate reactive oxygen species (ROS), which leads to DNA damage, protein denaturation, and cell membrane disruption, which eventually causes bacterial death. This proves that PdNPs have anti-pathogenic activity and can have a high potential for antimicrobial applications [56].

10.9 Challenges and Future Prospects

Nanotechnology on the whole is growing by leaps and bounds with the application of different kinds of nanomaterials that are used mainly for diagnostic purposes with modified biosensors on different kinds of material substrates. A few nanomaterials such as carbon nanotubes, graphene, nanocomposites, and quantum dots are used for the development of nanobiosensors [60]. Nanobiosensors have emerged as a promising technology tool to detect microbial pathogens. The main reasons are because of their high stability, rapid response time, and sensitivity. However, there are some challenges that are to be faced. Sample preparation is one crucial step that needs to be optimized to ensure that the target pathogen is properly isolated. If not monitored and prepared properly, it may lead to generation of improper results. The nanobiosensors should also have proper sensitivity and selectivity, which can target the pathogen in the presence of other microorganisms, environmental contaminants, or any interfering substances that may have the chance of affecting the signal [58]. These devices need considerable developments in order to address the requirements in various fields such as pharmaceutical, drug delivery, biomedical diagnostics performed *in vitro*, and pathogen detection fields [57]. The nanobiosensor should also maintain proper stability and reproducibility throughout, in order to show results over repeated usage. Another challenge is the cost effectiveness of the nanobiosensor. One of the first applications of the nanobiosensor was developed for glucose sensing [58]. Regulatory concerns and ethical concerns are also very important and crucial when it comes to nanobiosensor development. Lack of international guidelines, which is required for the development, for the safety and usage of nanobiosensors with ethical needs, have to be addressed carefully before these products can be commercially feasible, and has become a great concern to the public [60]. However, when it comes to commercialization of the nanobiosensors as a product, there were several challenges that had to be faced in the post-hype of field [59, 60]. Nanobiosensors, as a whole, holds great promise for microbial pathogen detection due to their high sensitivity, selectivity, and rapid response time, which can detect viruses, bacteria, and fungi, making them an immensely valuable tool for diagnosis of infectious disease in the future. The future prospects for microbial pathogen detection are bright, as there is a huge demand in the requirement of rapid and accurate diagnostic tools for future diseases as well [61, 62]. The idea of developing more advanced and miniaturized biosensors that are easily portable and accessible in hospitals, clinics, or even remote areas for the early diagnosis of infections is gaining significant attention.

Some of the key areas where nanobiosensors can make a significant impact include environmental monitoring wherein nanobiosensors can be used to monitor the water, air, or soil for the presence of microbial pathogens that can help spreading of infections. Food safety, where nanobiosensors can be developed in point of care devices that can detect harmful pathogens rapidly, thereby preventing the spreading of foodborne illnesses. Another future prospect is the development of nanobiosensors in the field of biodefense, where they can be used to detect biological agents in terms of any bioterrorism attack, enabling rapid response and containment. The development of nanobiosensors for microbial pathogen detection has the potential to revolutionize the fields of disease diagnosis and treatment, which can lead to better outcomes for the future.

Conclusion

Microbes are found everywhere and some are harmful to humans. These harmful microbes or pathogens can cause infections or diseases depending upon the type of pathogen and the severity of the disease be it bacteria, fungi, viruses, or parasites. The use of nanobiosensor research aims to develop more innovating technologies to provide better diagnosis and treatment for these diseases. It also offers a promising approach to detect microbial pathogens accurately and quickly. These sensors use nanotechnology to create highly specific probes that can even detect small amount of pathogen biomolecules in a sample such as food pathogen detection. The use of nanobiosensors can lead to a faster, accurate diagnosis, enabling earlier treatment and prevent spreading of the disease. Many different kinds of biosensors which can be optical, thermal, or electrochemical, and using various nanoparticles such as gold, silver, and palladium, have the potential to revolutionize the fields of pathogen detection. Each of them will face challenges that need to be addressed carefully so that it can be widely used in clinical settings. These also include the demand for better integration with the existing diagnostic tools and protocols as well as more extensive testing to ensure better safety and efficacy of these nanobiosensors. Overall, nanobiosensors have the potential to make an immensely significant impact in the field of microbial pathogen, providing faster and more accurate results, and improving the outcome of the patient, which can lead to a better and safer world to live in.

References

1. Balloux, F. and van Dorp, L., Q&A: What are pathogens, and what have they done to and for us? *BMC Biol.*, 15, 1, 91, 2017 Oct 19, doi: 10.1186/s12915-017-0433-z. PMID: 29052511.
2. Beltran-Pineda, M., Pena, D., Sierra, C., Nanobiosensors for Pathogenic Agents Detection. *J. Braz. Chem. Soc.*, 32, 1687–1710, 2021, doi:10.21577/0103-5053.20210081.
3. Microbiology by numbers. *Nat. Rev. Microbiol.*, 9, 9, 628, 2011 Sep, doi: 10.1038/nrmicro2644. PMID: 21961177.
4. Salzberger, B., Mohr, A., Hitzenbichler, F., Die Influenza 1918. *DMW - Dtsch. Med. Wochenschr.*, 143, 25, 1858–1863, 2018, doi: 10.1055/a-0666-1962.
5. Hauck, C.R., Introduction: bacterial pathogens. *Methods Mol. Biol.*, 470, 3–4, 2009, doi: 10.1007/978-1-59745-204-5_1 PMID: 19089371.
6. Stanford, K., Introduction to the Special Issue "Molecular Basis and the Pathogenesis of Enterohemorrhagic Escherichia coli Infections". *Toxins (Basel)*, 12, 12, 763, 2020 Dec 3, doi: 10.3390/toxins12120763. PMID: 33287118.
7. Garcia-Rubio, R., de Oliveira, H.C., Rivera, J., Trevijano-Contador, N., The Fungal Cell Wall: *Candida, Cryptococcus*, and *Aspergillus* Species. *Front. Microbiol.*, 10, 2993, 2020 Jan 9, doi: 10.3389/fmicb.2019.02993. PMID: 31993032.
8. Crago, B., Ferrato, C., Drews, S.J., Svenson, L.W., Tyrrell, G., Louie, M., Prevalence of Staphylococcus aureus and methicillin-resistant S. aureus (MRSA) in food samples associated with foodborne illness in Alberta, Canada from 2007 to 2010. *Food Microbiol.*, 32, 1, 202–5, 2012 Oct, doi: 10.1016/j.fm.2012.04.012. Epub 2012 May 9. PMID: 22850394.
9. Rodríguez-Sánchez, B., Cercenado, E., Coste, A.T., Greub, G., Review of the impact of MALDI-TOF MS in public health and hospital hygiene, 2018. *Euro Surveill.*, 24, 4, 1800193, 2019 Jan, doi: 10.2807/1560-7917.ES.2019.24.4.1800193. PMID: 30696525.
10. Ohtsuka, K., Yanagawa, K., Takatori, K., Hara-Kudo, Y., Detection of Salmonella enterica in naturally contaminated liquid eggs by loop-mediated isothermal amplification, and characterization of Salmonella isolates. *Appl. Environ. Microbiol.*, 71, 11, 6730–5, 2005 Nov, doi: 10.1128/AEM.71.11.6730-6735.2005. PMID: 16269703.
11. Akbulut, D., Grant, K.A., McLauchlin, J., Development and application of Real-Time PCR assays to detect fragments of the Clostridium botulinum types A, B, and E neurotoxin genes for investigation of human foodborne and infant botulism. *Foodborne Pathog. Dis.*, 1, 4, 247–57, 2004 Winter, doi: 10.1089/fpd.2004.1.247. PMID: 15992287.
12. Bang, J., Beuchat, L.R., Song, H., Gu, M.B., Chang, H.I., Kim, H.S., Ryu, J.H., Development of a random genomic DNA microarray for the detection and identification of Listeria monocytogenes in milk. *Int. J. Food Microbiol.*, 161,

2, 134–41, 2013 Feb 1, doi: 10.1016/j.ijfoodmicro.2012.11.023. Epub 2012 Dec 6. PMID: 23313851.

13. Law, J.W., Ab Mutalib, N.S., Chan, K.G., Lee, L.H., Rapid methods for the detection of foodborne bacterial pathogens: principles, applications, advantages and limitations. *Front. Microbiol.*, 5, 770, 2015 Jan 12, doi: 10.3389/fmicb.2014.00770. PMID: 25628612.
14. Velusamy, V., Arshak, K., Korostynska, O., Oliwa, K., Adley, C., An overview of foodborne pathogen detection: in the perspective of biosensors. *Biotechnol. Adv.*, 28, 2, 232–54, 2010 Mar-Apr, doi: 10.1016/j.biotechadv.2009.12.004. Epub 2009 Dec 16. PMID: 20006978.
15. Zheng, Y.H., Hua, T.C., Xu, F., A thermal biosensor based on enzyme reaction. *Conf. Proc. IEEE Eng. Med. Biol. Soc.*, 2005, 1909–12, 2005, doi: 10.1109/IEMBS.2005.1616824. PMID: 17282593.
16. Tetyana, P., Shumbula, P.M., Njengele-Tetyana, Z., Biosensors: Design, Development and Applications, in: *Nanopores*, Ameen, S., Shaheer Akhtar, M., Shin, H.-S., (Eds.), IntechOpen, 319, 2021. https://doi.org/10.5772/intechopen.97576.
17. Qin, D., He, X., Wang, K., Zhao, X.J., Tan, W., Chen, J., Fluorescent nanoparticle-based indirect immunofluorescence microscopy for detection of Mycobacterium tuberculosis. *J. Biomed. Biotechnol.*, 2007, 7, 89364, 2007, doi: 10.1155/2007/89364. PMID: 18273415.
18. Wang, Q.Y. and Kang, Y.J., Bioprobes Based on Aptamer and Silica Fluorescent Nanoparticles for Bacteria Salmonella typhimurium Detection. *Nanoscale Res. Lett.*, 11, 1, 150, 2016 Dec, doi: 10.1186/s11671-016-1359-z. Epub 2016 Mar 16. PMID: 26983430.
19. Pöhlmann, C. and Elßner, T., Multiplex Immunoassay Techniques for On-Site Detection of Security Sensitive Toxins. *Toxins (Basel)*, 12, 11, 727, 2020 Nov 20, doi: 10.3390/toxins12110727. PMID: 33233770.
20. Ahn-Yoon, S., DeCory, T.R., Baeumner, A.J., Durst, R.A., Ganglioside-liposome immunoassay for the ultrasensitive detection of cholera toxin. *Anal. Chem.*, 75, 10, 2256–61, 2003 May 15, doi: 10.1021/ac026428t. PMID: 12918964.
21. Hatheway, C.L. and Ferreira, J.L., Detection and identification of Clostridium botulinum neurotoxins. *Adv. Exp. Med. Biol.*, 391, 481–98, 1996, doi: 10.1007/978-1-4613-0361-9_39. PMID: 8726084.
22. Balasaria, S., Mukhopadhyay, B., Chemical Synthesis of the Pentasaccharide Repeating Unit of the O-Antigen from Escherichia Coli Strain SDLZB008 in the Form of Its 2-Aminoethyl Glycoside. *Carbohydr Res.*, 523, 108734, 2023. https://doi.org/10.1016/j.carres.2022.108734.
23. Abdellatif, A.A.H., Younis, M.A., Alsharidah, M., Al Rugaie, O., Tawfeek, H.M., Biomedical Applications of Quantum Dots: Overview, Challenges, and Clinical Potential. *Int. J. Nanomed.*, 17, 1951–1970, 2022 May 2, doi: 10.2147/IJN.S357980. PMID: 35530976.

24. Zhang, C.Y. and Hu, J., Single quantum dot-based nanosensor for multiple DNA detection. *Anal. Chem.*, 82, 5, 1921–7, 2010 Mar 1, doi: 10.1021/ac9026675. PMID: 20121246.
25. Xiang, F., Pang, D.W., Chen, Z., Shao, J.B., Xiong, L.H., Xiang, Y.P., Xiong, Y., Wu, K., Ai, H.W., Zhang, H., Zheng, X.L., Lv, J.R., Liu, W.Y., Hu, H.B., Mei, H., Zhang, Z., Sun, H., Xiang, Y., Sun, Z.Y., Visual and efficient immunosensor technique for advancing biomedical applications of quantum dots on Salmonella detection and isolation. *Nanoscale*, 8, 8, 4688–98, 2016 Feb 28, doi: 10.1039/c5nr07424j. PMID: 26853517.
26. Yufenyuy, E.L. and Parekh, B.S., Development of a Multiplex Assay for Concurrent Diagnoses and Detection of HIV-1, HIV-2, and Recent HIV-1 Infection in a Single Test. *AIDS Res. Hum. Retroviruses*, 34, 12, 1017–1027, 2018 Dec, doi: 10.1089/AID.2017.0279. Epub 2018 Sep 22. PMID: 30056751.
27. Huang, X., Zhu, Y., Kianfar, E., Nano Biosensors: Properties, applications and electrochemical techniques. *J. Mater. Res. Technol.*, *12*, 1649–1672, 2021, doi: 10.1016/j.jmrt.2021.03.048.
28. Chen, X., Wu, G., Cai, Z., Oyama, M., Chen, X., Advances in enzyme-free electrochemical sensors for hydrogen peroxide, glucose, and uric acid. *Microchim. Acta*, 181, 689–705, 2013, 10.1007/s00604-013-1098-0.
29. Zhu, Y. and Zhang, S., Antibacterial Activity and Mechanism of Lacidophilin From *Lactobacillus pentosus* Against *Staphylococcus aureus* and *Escherichia coli*. *Front. Microbiol.*, 11, 582349, 2020 Oct 29, doi: 10.3389/fmicb.2020.582349. PMID: 33193219.
30. Xu, D., Lv, H., Liu, B., Encapsulation of Metal Nanoparticle Catalysts Within Mesoporous Zeolites and Their Enhanced Catalytic Performances: A Review. *Front. Chem.*, 6, 550, 2018. https://doi.org/10.3389/fchem.2018.00550.
31. Dai, Z., Yang, A., Bao, X., Yang, R., Facile Non-Enzymatic Electrochemical Sensing for Glucose Based on Cu_2O-BSA Nanoparticles Modified GCE. *Sens. (Basel)*, 19, 12, 2824, 2019 Jun 24, doi: 10.3390/s19122824. PMID: 31238594.
32. Lin, Y., Yu, P., Hao, J., Wang, Y., Ohsaka, T., Mao, L., Continuous and simultaneous electrochemical measurements of glucose, lactate, and ascorbate in rat brain following brain ischemia. *Anal. Chem.*, 86, 8, 3895–901, 2014 Apr 15, doi: 10.1021/ac4042087. Epub 2014 Mar 24. PMID: 24621127.
33. Freeman, M.H., Hall, J.R., Leopold, M.C., Monolayer-protected nanoparticle doped xerogels as functional components of amperometric glucose biosensors. *Anal. Chem.*, 85, 8, 4057–65, 2013 Apr 16, doi: 10.1021/ac3037188. Epub 2013 Mar 26. PMID: 23472762.
34. Cao, X., Wang, N., Jia, S., Shao, Y., Detection of glucose based on bimetallic PtCu nanochains modified electrodes. *Anal. Chem.*, 85, 10, 5040–6, 2013 May 21, doi: 10.1021/ac400292n. Epub 2013 May 3. PMID: 23641874.
35. Dhara, K., Stanley, J., T, R., Nair, B.G., T.G., S.B., Pt-CuO nanoparticles decorated reduced graphene oxide for the fabrication of highly sensitive non-enzymatic disposable glucose sensor. *Sens. Actuators, B*, *195*, 197–205, 2014, doi: 10.1016/j.snb.2014.01.044.

36. Zhang, J., Sun, J., Chen, Q., Zuo, C., Resolution Analysis in a Lens-Free On-Chip Digital Holographic Microscope. *IEEE Trans. Comput. Imaging*, 6, 697–710, 2020, https://doi.org/10.1109/TCI.2020.2964247.
37. Wen, D., Liu, W., Haubold, D., Zhu, C., Oschatz, M., Holzschuh, M., Wolf, A., Simon, F., Kaskel, S., Eychmüller, A., Gold Aerogels: Three-Dimensional Assembly of Nanoparticles and Their Use as Electrocatalytic Interfaces. *ACS Nano*, 10, 2, 2559–67, 2016 Feb 23, doi: 10.1021/acsnano.5b07505. Epub 2016 Jan 25. PMID: 26751502.
38. Nunna, B.B., Mandal, D., Lee, J.U., Singh, H., Zhuang, S., Misra, D., Bhuyian, M.N.U., Lee, E.S., Detection of cancer antigens (CA-125) using gold nano particles on interdigitated electrode-based microfluidic biosensor. *Nano Convergence*, 6, 1, 3, 2019 Jan 17, doi: 10.1186/s40580-019-0173-6. PMID: 30652204.
39. Park, J.A., Kim, J., Kim, S.M., Sohn, H., Park, C., Kim, T.H., Lee, J.H., Lee, M.H., Lee, T., Fabrication of Electrochemical Influenza Virus (H1N1) Biosensor Composed of Multifunctional DNA Four-Way Junction and Molybdenum Disulfide Hybrid Material. *Materials (Basel)*, 14, 2, 343, 2021 Jan 12, doi: 10.3390/ma14020343. PMID: 33445498.
40. Kang, X., Wang, J., Wu, H., Aksay, I.A., Liu, J., Lin, Y., Glucose oxidase-graphene-chitosan modified electrode for direct electrochemistry and glucose sensing. *Biosens. Bioelectron.*, 25, 4, 901–5, 2009 Dec 15, doi: 10.1016/j.bios.2009.09.004. Epub 2009 Sep 6. PMID: 19800781.
41. Lu, L., Sun, R.W., Chen, R., Hui, C.K., Ho, C.M., Luk, J.M., Lau, G.K., Che, C.M., Silver nanoparticles inhibit hepatitis B virus replication. *Antivir. Ther.*, 13, 2, 253–62, 2008. PMID: 18505176.
42. Qian, J., Feng, S., Li, Y., Tao, T., Han, J., Chen, Q., Zuo, C., Single-shot absolute 3D shape measurement with deep-learning-based color fringe projection profilometry. *Opt. Lett.*, 45, 7, 1842–1845, 2020 Apr 1, doi: 10.1364/OL.388994. PMID: 32236013.
43. Tan, P., Li, H., Wang, J., Gopinath, S.C.B., Silver nanoparticle in biosensor and bioimaging: Clinical perspectives. *Biotechnol. Appl. Biochem.*, 68, 6, 1236–1242, 2021 Dec, doi: 10.1002/bab.2045. Epub 2020 Oct 11. PMID: 33043496.
44. Bloch, K., Pardesi, K., Satriano, C., Ghosh, S., Bacteriogenic Platinum Nanoparticles for Application in Nanomedicine. *Front. Chem.*, 9, 624344, 2021 Mar 5, doi: 10.3389/fchem.2021.624344. PMID: 33763405.
45. Menendez, N., Mazario, E., Herrasti, P., Fe3O4-Nanoparticle-Modified Sensor for the Detection of Dopamine, Uric Acid and Ascorbic Acid. *Chemosensors (Basel)*, 11, 2, 2023, doi: 10.3390/chemosensors11020079.
46. Gawande, M.B., Goswami, A., Felpin, F.X., Asefa, T., Huang, X., Silva, R., Zou, X., Zboril, R., Varma, R.S., Cu and Cu-Based Nanoparticles: Synthesis and Applications in Catalysis. *Chem. Rev.*, 116, 6, 3722–811, 2016 Mar 23, doi: 10.1021/acs.chemrev.5b00482. Epub 2016 Mar 3. PMID: 26935812.

47. Beard, P., Biomedical photoacoustic imaging. *Interface Focus*, 1, 4, 602–31, 2011 Aug 6, doi: 10.1098/rsfs.2011.0028. Epub 2011 Jun 22. PMID: 22866233.
48. Wu, D., Huang, L., Jiang, M.S., Jiang, H., Contrast agents for photoacoustic and thermoacoustic imaging: a review. *Int. J. Mol. Sci.*, 15, 12, 23616–39, 2014 Dec 18, doi: 10.3390/ijms151223616. PMID: 25530615.
49. O'Neal, D.P., Hirsch, L.R., Halas, N.J., Payne, J.D., West, J.L., Photo-thermal tumor ablation in mice using near infrared-absorbing nanoparticles. *Cancer Lett.*, 209, 2, 171–6, 2004 Jun 25, doi: 10.1016/j.canlet.2004.02.004. PMID: 15159019.
50. Weiss, J.T., Dawson, J.C., Macleod, K.G., Rybski, W., Fraser, C., Torres-Sánchez, C., Patton, E.E., Bradley, M., Carragher, N.O., Unciti-Broceta, A., Extracellular palladium-catalysed dealkylation of 5-fluoro-1-propargyl-uracil as a bioorthogonally activated prodrug approach. *Nat. Commun.*, 5, 3277, 2014, doi: 10.1038/ncomms4277. PMID: 24522696.
51. Weiss, J.T., Dawson, J.C., Fraser, C., Rybski, W., Torres-Sánchez, C., Bradley, M., Patton, E.E., Carragher, N.O., Unciti-Broceta, A., Development and bio-orthogonal activation of palladium-labile prodrugs of gemcitabine. *J. Med. Chem.*, 57, 12, 5395–404, 2014 Jun 26, doi: 10.1021/jm500531z. Epub 2014 Jun 12. PMID: 24867590.
52. Yi, X., Wu, Y., Tan, G., Yu, P., Zhou, L., Zhou, Z., Chen, J., Wang, Z., Pang, J., Ning, C., Palladium nanoparticles entrapped in a self-supporting nanoporous gold wire as sensitive dopamine biosensor. *Sci. Rep.*, 7, 1, 7941, 2017 Aug 11, doi: 10.1038/s41598-017-07909-y. PMID: 28801614.
53. Adams, C.P., Walker, K.A., Obare, S.O., Docherty, K.M., Size-dependent antimicrobial effects of novel palladium nanoparticles. *PLoS One*, 9, 1, e85981, 2014 Jan 20, doi: 10.1371/journal.pone.0085981. PMID: 24465824.
54. Fang, G., Li, W., Shen, X., Perez-Aguilar, J.M., Chong, Y., Gao, X., Chai, Z., Chen, C., Ge, C., Zhou, R., Differential Pd-nanocrystal facets demonstrate distinct antibacterial activity against Gram-positive and Gram-negative bacteria. *Nat. Commun.*, 9, 1, 129, 2018 Jan 9, doi: 10.1038/s41467-017-02502-3. PMID: 29317632.
55. Anand, K., Tiloke, C., Phulukdaree, A., Ranjan, B., Chuturgoon, A., Singh, S., Gengan, R.M., Biosynthesis of palladium nanoparticles by using Moringa oleifera flower extract and their catalytic and biological properties. *J. Photochem. Photobiol., B*, 165, 87–95, 2016 Dec, doi: 10.1016/j.jphotobiol.2016.09.039. Epub 2016 Sep 29. PMID: 27776261.
56. Azizi, S., Mahdavi Shahri, M., Rahman, H.S., Rahim, R.A., Rasedee, A., Mohamad, R., Green synthesis palladium nanoparticles mediated by white tea (*Camellia sinensis*) extract with antioxidant, antibacterial, and anti-proliferative activities toward the human leukemia (MOLT-4) cell line. *Int. J. Nanomed.*, 12, 8841–8853, 2017 Dec 14, doi: 10.2147/IJN.S149371. Retraction in: Int J Nanomedicine. 2022 Mar 18;17:1227-1228. PMID: 29276385.

57. Kant, K., Shahbazi, M.A., Dave, V.P., Ngo, T.A., Chidambara, V.A., Than, L.Q., Bang, D.D., Wolff, A., Microfluidic devices for sample preparation and rapid detection of foodborne pathogens. *Biotechnol. Adv.*, 36, 4, 1003–1024, 2018 Jul-Aug, doi: 10.1016/j.biotechadv.2018.03.002. Epub 2018 Mar 10. PMID: 29534915.
58. Shetti, N.P., Bukkitgar, S.D., Reddy, K.R., Reddy, C.V., Aminabhavi, T.M., ZnO-based nanostructured electrodes for electrochemical sensors and biosensors in biomedical applications. *Biosens. Bioelectron.*, 141, 111417, 2019 Sep 15, doi: 10.1016/j.bios.2019.111417. Epub 2019 Jun 8. PMID: 31202187.
59. Banigo, A., Azeez, T., Ejeta, K., Lateef, A., Ajuogu, E., Nanobiosensors: Applications in Biomedical Technology. *IOP Conf. Ser.: Mater. Sci. Eng.*, 805, 1, 012028, 2020. https://doi.org/10.1088/1757-899X/805/1/012028.
60. Kulkarni, M.B., Ayachit, N.H., Aminabhavi, T.M., Recent Advancements in Nanobiosensors: Current Trends, Challenges, Applications, and Future Scope. *Biosensors (Basel)*, 12, 10, 892, 2022 Oct 18, doi: 10.3390/bios12100892. PMID: 36291028.
61. Soy, S., Sharma, S. R., Nigam, V. K., Bio-Fabrication of Thermozyme-Based Nano-Biosensors: Their Components and Present Scenario. *J. Mater Sci. Mater. Electron.*, 33, 8, 5523–5533, 2022. https://doi.org/10.1007/s10854-022-07741-9.
62. Kulkarni, M.B., Yashas, E.P.K., Amreen, K., Goel, S., Internet of Things enabled portable thermal management system with microfluidic platform to synthesize MnO_2 nanoparticles for electrochemical sensing. *Nanotechnology*, 31, 42, 425504, 2020 Aug 4, doi: 10.1088/1361-6528/ab9ed8. Epub ahead of print. PMID: 32748804.
63. Ravel, J., Gajer, P., Abdo, Z., Schneider, G.M., Koenig, S.S., McCulle, S.L., Karlebach, S., Gorle, R., Russell, J., Tacket, C.O., Brotman, R.M., Davis, C.C., Ault, K., Peralta, L., Forney, L.J., Vaginal microbiome of reproductive-age women. *Proc. Natl. Acad. Sci. U.S.A.*, 108, Suppl 1, 4680–7, 2011 Mar 15, doi: 10.1073/pnas.1002611107. Epub 2010 Jun 3. PMID: 20534435.
64. Dan, M., Leshem, Y., Yeshaya, A., Performance of a rapid yeast test in detecting Candida spp. in the vagina. *Diagn. Microbiol. Infect. Dis.*, *67*, 1, 52–55, 2010, doi: 10.1016/j.diagmicrobio.2009.1. PMID: 20226617.
65. Li, W., Kuai, L., Qin, Q., Geng, B., Ag–Au bimetallic nanostructures: co-reduction synthesis and their component-dependent performance for enzyme-free H2O2 sensing. *J. Mater. Chem. A*, *1*, 24, 7111, 2013, doi: 10.1039/c3ta00106g.
66. Stephenson, T., Li, Z., Olsen, B., Mitlin, D., Lithium ion battery applications of molybdenum disulfide (MoS 2) nanocomposites. *Energy Environ. Sci.*, 7, 209–231, 2014, 10.1039/C3EE42591F.
67. Cipolatti, E.P., Silva, M.J.A., Klein, M., Feddern, V., Feltes, M.M.C., Oliveira, J.V., de Oliveira, D., Current status and trends in enzymatic nanoimmobilization. *J. Mol. Catal. B: Enzym.*, *99*, 56–67, 2014, doi: 10.1016/j.molcatb.2013.10.019.

68. Matta, L.L. and Alocilja, E.C., Carbohydrate Ligands on Magnetic Nanoparticles for Centrifuge-Free Extraction of Pathogenic Contaminants in Pasteurized Milk. *J. Food Prot.*, 81, 12, 1941–1949, 2018 Dec, doi: 10.4315/0362-028X.JFP-18-040. PMID: 30452292.
69. Ercole, C., Del Gallo, M., Mosiello, L., Baccella, S., Lepidi, A., Escherichia coli detection in vegetable food by a potentiometric biosensor. *Sens. Actuators, B*, 91, 163–168, 2003, 10.1016/S0925-4005(03)00083-2.
70. Ertl, P., Wagner, M., Corton, E., Mikkelsen, S.R., Rapid identification of viable Escherichia coli subspecies with an electrochemical screen-printed biosensor array. *Biosens. Bioelectron.*, 18, 7, 907–16, 2003 Jul, doi: 10.1016/s0956-5663(02)00206-3. PMID: 12713914.
71. Dubois, E., Hennechart, C., Merle, G., Burger, C., Hmila, N., Ruelle, S., Perelle, S., Ferré, V., Detection and quantification by real-time RT-PCR of hepatitis A virus from inoculated tap waters, salad vegetables, and soft fruits: characterization of the method performances. *Int. J. Food Microbiol.*, 117, 2, 141–9, 2007 Jun 30, doi: 10.1016/j.ijfoodmicro.2007.02.026. Epub 2007 Mar 30. PMID: 17490771.
72. Malhotra, B.D. and Ali, M.A., Nanomaterials in Biosensors: Fundamentals and Applications, in: *Nanomaterials for Biosensors*, pp. 1–74, 2018, doi: 10.1016/B978-0-323-44923-6.00001-7. Epub 2017 Oct 21. PMCID: PMC7152100.
73. Meskher, H., Mustansar, H.C., Thakur, A.K., Sathyamurthy, R., Lynch, I., Singh, P., Han, T.K., Saidur, R., Recent trends in carbon nanotube (CNT)-based biosensors for the fast and sensitive detection of human viruses: A critical review. *Nanoscale Adv.*, 5, 4, 992–1010, 2022 Nov 9, doi: 10.1039/d2na00236a. PMID: 36798507.
74. Ji, X., Wang, H., Song, B., Chu, B., He, Y., Silicon Nanomaterials for Biosensing and Bioimaging Analysis. *Front. Chem.*, 6, 38, 2018 Feb 28, doi: 10.3389/fchem.2018.00038. PMID: 29541633.
75. Krivitsky, V., Zverzhinetsky, M., Patolsky, F., Antigen-Dissociation from Antibody-Modified Nanotransistor Sensor Arrays as a Direct Biomarker Detection Method in Unprocessed Biosamples. *Nano Lett.*, 16, 10, 6272–6281, 2016 Oct 12, doi: 10.1021/acs.nanolett.6b02584. Epub 2016 Sep 6. PMID: 27579528.

11

Micro/Nanofluidics-Integrated Biosensors for Respiratory Viral Diseases Diagnosis

Aiswarya Chandrasekaran and G.H.R. Eranga Karunaratne*

Faculty of Science, Horizon Campus, Malabe, Sri Lanka

Abstract

Respiratory viral diseases like the corona virus disease and human influenza H1N1 pose major threats to mankind as they spread worldwide, causing threats to human life as well as to the economy. The outbreaks of these viral diseases in recent years have raised major concerns for the development of rapid, reliable, and sensitive detection methods which would help in curtailing the spread of disease and support effective treatment measures. Several modified diagnostic methods including nucleic acid-based and immunological assay-based point-of-care testing (POCT) incorporating various technologies have been used in recent past for the fast diagnosis. Recently, biosensor-based diagnostic technologies have been gaining more attention as they have proven to overcome the limitations of the conventional diagnostic methods. Furthermore, the integration of micro- and nanofluidic technologies into biosensor systems have shown promising scope of very quick and sensitive detection especially in times of critical pandemics and high demand settings as they possess high sensitivity, advanced analysis capabilities and wide range of possible applications. Biosensors integrated with micro/nanofluidic chips have been used for the detection of influenza virus H1N1, severe acute respiratory syndrome coronavirus (SARS-CoV), zika, hepatitis A virus and norovirus showing satisfactory results. The advent of new fluidic phenomena has led to the advancement of microfluidic to nanofluidic systems. However the use of micro/nanofluidics has some limitations related to multiplexed detection, selectivity and ability to be used for field analysis. By overcoming these limitations with extensive research, intense application and industrialization the biosensor detection systems in micro- and nanofluidic chips could replace the traditional apparatus and analysis methods used in chemical laboratories.

**Corresponding author*: ekarunarathne@horizoncampus.edu.lk

Suvardhan Kanchi, Ayyappa Bathinapatla, Anitha Varghese and Phumlane Selby Mdluli (eds.) Nano-Biosensor Technologies for Diagnosis of Infectious Diseases, (341–364) © 2025 Scrivener Publishing LLC

Keywords: Microfluidics, nanofluidics, biosensors, respiratory virus, diagnosis, SARS-CoV, influenza, polydimethylsiloxane

11.1 Introduction

Recently the world witnessed the outbreak of the COVID-19 pandemic, caused by the SARS-CoV-2. Over the recent years several similar respiratory viruses including the SARS-CoV, Middle East respiratory syndrome coronavirus (MERS-CoV), H1N1 influenza, H5N1 influenza, H7N9 influenza and enterovirus D68 have posed threats to mankind [1]. Early diagnosis of people infected by these viruses is essential as the absence of timely and accurate diagnosis might worsen the patient condition and cause treatment failure while promoting the spread of disease. In many cases cell culture-based tests that evaluate the structural changes in the host cell and hemadsorption caused by the virus has been used as the gold standard tests to diagnose respiratory viral infections [2]. In addition, several nucleic acid-based and antibody-based techniques are commonly in use for the detection of these viruses. Meanwhile highly sensitive, reliable, rapid and specific diagnostic methods are essential for the proper diagnosis. Biosensors come with these benefits and play a vital role in detecting these infections even under asymptomatic conditions. Biosensors have good selectivity, high automation, fast analysis, miniaturization, integration and low cost. Furthermore the combination of fluidic technologies such as micro- and nanofluidics together with biosensors has shown promising scope especially in diagnostic applications. Micro/nanofluidic systems are based on the manipulation of small volumes of fluids, which lead to optical detection methods. To facilitate the detection of viral components micro/nanofluidic technology has been used together with common viral ribonucleic acid (RNA) detection techniques such as reverse transcriptase–polymerase chain reaction (RT-PCR), nested PCR, fluorescence-based assays and enzyme-linked immunosorbent assay (ELISA), which aid in rapid diagnosis. Similarly, micro/nanotechnologies have been used in immunoassay-based diagnosis platforms to instantaneously detect IgG and IgM antibodies at different stages of infection. In recent years, a wide range of POCT devices, based on nucleic acid detection techniques and immunological assays, have been developed for the quick diagnosis of respiratory viral infections. The recent advances in micro/nanofluidic devices have made it possible to improvise these POCT platforms which facilitates the rapid detection of antibodies or infectious pathogens even by self-test that could be easily performed by individuals at the comfort of their home. These miniaturized devices are designed with all the vital steps integrated via fluidic manipulation and have the potential to carry out complex real-time detection [3]. To meet the increased demand of the infectious pathogen detection during conditions like a severe epidemic, specifically in resource-limited settings, there is a critical need for portable and integrated micro/nanofluidic platforms that can produce rapid, precise, and even multiplex diagnosis at the POCT [4]. This chapter provides a comprehensive overview of the micro/nanofluidic technology and the novel and latest micro/nanofluidics-integrated biosensors which have been designed to be used for the diagnosis of some major respiratory viral diseases. This may provide insights to future researchers to further develop more efficient POCT devices based on micro- and nano-based technologies integrated with biosensors.

11.2 Common Respiratory Viruses and Their Detection Components

Most of the viruses that cause diseases such as influenza, severe acute respiratory syndrome (SARS) and other respiratory diseases contain RNA as their genetic material [5]. Influenza A virus is an infectious virus transmitted by birds and mammals and capable of causing human influenza [6]. It belongs to the *Orthomyxoviridae* family and contains single-stranded RNA [1]. There are three subtypes of influenza virus: H1N1, H5N1, and H7N9 viruses [7]. This infection usually occurs via inhalation of virus-borne particles and leads to acute respiratory infection with cough and fever [8]. Three other major respiratory viral outbreaks witnessed by the world have been caused by the viruses belonging to *Coronaviridae* family [1]. This includes SARS-CoV in 2003 [9], MERS-CoV in 2012 [10], and SARS-CoV-2 in 2019 [1]. Coronaviruses are RNA viruses that originate from multiple species [11] and these viruses' transmission occurs from animals to humans and easily spreads through respiratory droplet causing high rates of infection [1]. Some strains cause mild cold-like symptoms including cough, fever and respiratory difficulties while some other strains can lead to fatal respiratory complications and eventually death. However some infections caused by these viruses, especially MERS can be asymptomatic [10]. The incubation period of the MERS-CoV ranges between 2 to 14 days and might result in symptoms like severe respiratory syndrome, pneumonia, and multi-organ damage. Pregnant women, children, and young adults are more susceptible to it [12]. The respiratory syncytial virus is another virus of interest which belongs to the *Paramyxoviridae* family [13]. This single-stranded RNA virus has been reported to cause lethality especially in children less than 5 years of age by causing pneumonia and bronchiolitis [14]. In addition to these viruses, some other lower virulent viruses such as human adenovirus (AdV), human bocavirus (HBoV), human parainfluenza viruses (HPIV), human metapneumovirus (HMPV), and human rhinovirus (HRV) also cause infections to human with similar mild symptoms like the other respiratory viruses [15].

Generally the whole viral particle, gene sequences, structural proteins and antibodies could be used as targets for virus detection [16]. Thus studying the genome sequence, proteins structure, transmission, pathogenicity, and host response will aid in developing novel and more specific detection methods for rapid diagnosis of the pathogen with high reliability and accuracy [1]. For detection of influenza virus, the whole virus and the structural proteins such as M1 protein, neuraminidase (NA), and hemagglutinin (HA) are used as targets [16]. In addition, gene sequences including the RNA transcripts and deoxyribonucleic acid (DNA) oligonucleotides of HA gene can be used as targets for influenza virus detection [17]. The immunological molecule-based detection methods for influenza virus use immunoglobulin M, G and antibodies for HA and NA present in blood of the infected patients [16]. MERS-CoV has four types of structural proteins which includes the envelop protein (E), spike protein(S), nucleocapsid protein (N), and membrane protein (M). The spike protein plays a major role in binding of the virus to the host cell receptors and it is most commonly used as the antigen for detection [18]. MERS-CoV genome has 30,119 nucleotides with 11 open reading frames (ORF), out of them the first open reading frames which is thought to code for nonstructural proteins serves as the detection target [19]. The World Health Organization (WHO) has also recommended the RNA upstream of the E gene for the detection of MERS-CoV [20]. Furthermore, antibodies against S and N proteins can be used, but that method is not ideal for early detection as those antibodies are found only in convalescent phase [21]. For SARS-CoV-2 detection, the whole virus and all four of their structural proteins including S, M, N, and E proteins could be used as targets. However S and N proteins have served as the most common biomarkers for early diagnosis of COVID-19 [22]. In addition to these proteins, the receptor binding domain protein can also be used as target of SARS-CoV-2 detection because the S1 subunit of S protein exhibits the receptor binding domain with high affinity for the host angiotensin-converting enzyme II receptors [23]. ORF 1a, ORF 1b, non-structural RNA-dependent RNA polymerase, S gene, and N gene of SARS-CoV are used for SARS-CoV detection [24].

IgG and IgM antibodies found in blood also can be used for SARS-CoV 2 detection after a specific number of days of getting infected [16].

11.3 Biosensors

Biosensors are devices used for analyzing various molecules or biologic components by converting biological reactions into quantifiable signals [1]. They are composed of bio-recognition element, transducer and signal output structures [25]. The bio-recognition elements like enzymes, antibodies, cell receptors, microorganisms, and tissues immobilized on transducers interact with the analyte forming a biochemical reaction. This reaction is then converted to a measurable signal by the transducer, which is then measured by digital detector module, as demonstrated in Figure 11.1. Antibodies, peptides, and aptamer are the most widely used bio-recognition elements in the respiratory virus detection [16]. Most biosensors use optical, electrochemical, and piezoelectric transducer-based detection systems [1]. In electrochemical biosensors, the electrodes act as both transduction and immobilization platforms and this biosensor monitor changes in charge distribution. Target analyte combine with secondary biorecognition probe and produce a redox molecule which diffuses to the surface of the electrode, thus producing a quantifiable signal. The biochemical reaction is measured via this digital electronic signal that it converted by a transducer [26]. The transducers of the electrochemical biosensors include gold electrodes, carbon paste electrodes, glassy carbon electrodes, and screen-printed electrode [16]. A large number of electrochemical biosensors with techniques such as differential pulse voltammetry, cyclic voltammetry, and electrochemical impedance spectroscopy are used in respiratory virus detection, but the most commonly used electrochemical biosensors for respiratory virus detection is voltammetric biosensors [27]. Optical biosensors measure changes in optical characteristics on the surface of the transducer. Piezoelectric biosensors use transducers which resonate in the

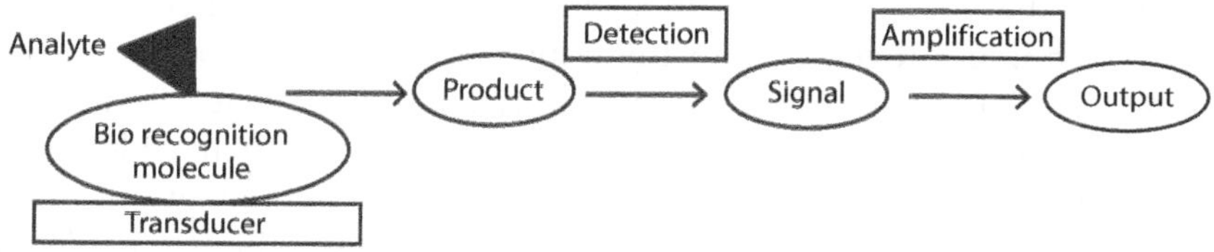

Figure 11.1 Fundamental mechanism of biosensor.

presence of an external current and then measure the change in resonance frequency that occurs due to the interaction between the analyte and the immobilized molecule [28].

Based on the type of analyte and biochemical reaction, biosensors can be further classified as enzymatic biosensors, whole-cells biosensors, immunosensors and DNA/nucleic acid-based biosensors [1]. Immunosensors work based on the antibodies antigens interactions [29] while enzymatic biosensors make use of enzymes which have specificity to target substrates [1]. In immunosensors, the most important antibodies used for respiratory virus detection includes polyclonal antibodies, monoclonal antibodies, and antibody single-chain Fv fragments. Based on the use of labels, immunosensors can be classified as label-based and label-free immunosensors. In label-free immunosensors, virus is captured by antibodies on modified electrodes, which cause a change in properties of the sensing surface. Label-free biosensors are the simplest and quickest biosensors with non-cross reactivity and largely used for respiratory virus detection. Several antibody-based electrochemical biosensors have been designed for the detection of H1NI, H3N2, MERS-CoV, HCoV, H5N1, AIV H7, and influenza virus using label-free detection methods and for H9N2, H7N9, and H5N1 using label-based detection methods [16]. DNA biosensors are capable of detecting specific nucleic acid sequences, thus differentiate between different organisms and detect pathogens. They have nucleic acids immobilized on the surface of transducers. They hybridize with complementary nucleic acid sequence and this reaction is detected with the use of suitable hybridization indicators [30]. The most common bio-recognition element used in nucleic acid-based electrochemical biosensors is single-stranded DNA [16]. For DNA-based analysis using biosensors, the selection of model DNA molecules, fluidics, and data analysis are very vital [31]. Whole-cell biosensors use cell surface antigens like proteins, peptidoglycan, and lipopolysaccharides as targets for bio-recognition [32].

Conventional biosensors were based on microarray systems containing cantilever or field-effect devices in which the target analytes are adsorbed onto sensing elements. The modern biosensors use microfluidic and nanofluidic sensors along with optical detection method [33]. The advent of modern technologies such as nanotechnology and advancement of fabrication techniques together with the improvements in micro- and nanofluidic techniques have enhanced miniaturization, design, sensitivity, and detection limits of biosensors with promising scope of their widespread use in future. For instance, the integration of nanotechnology into biosensors has enhanced detection sensitivity by increasing biocompatibility and resistance towards nonspecific adsorption [34]. The microfabrication technology has allowed the integration of multiple biochemical sensors into a microfluidic chip enabling multiplexed detection. However when carrying out multiplexed sensing, it is essential to distinguish signals without mutual interference [26].

11.4 Fluidic Technology

Micro/nanofluidic systems are miniaturized platforms used for the detection of different biological molecules with all steps integrated into the same platform with various elements to manipulate fluidic transport. Steps such as nucleic acid extraction, amplification, and detection are integrated on miniaturized devices which aid real-time diagnosis and these systems are characterized by their high surface-area-to-volume ratio. The fluid flow within the channels of these devices is based on principles such as pressure-driven flow and electrokinetics. However, when decreasing the channel dimension achieving the pressure-driven flow might be difficult, thus electric-field driven or electrokinetic flows are most commonly employed in fluidic devices [33]. Electrokinetics can be defined as the motion induced by electricity and is classified into two categories: electrokinetics over insulating surfaces and electrokinetics over conducting surfaces. Electrokinetics over insulating surfaces induces fluid velocity that is linearly proportional to the strength of supplied electric field whereas electrokinetics over conducting surfaces induces fluid velocity that is linearly proportional to the square of supplied electric field strength [35]. Electroosmosis is another popular method to move a fluid in channels by supplying a voltage [36]. Most of the microfluidic and nanofluidic devices are composed of materials like silicon, glass, and different polymers as the substrate. Glass comes with advantages like optical transparency, inertness, thermal stability, and hydrophilic features that facilitate fluid movement [37]. In most microfluidics and nanofluidics systems, the particles interact with the channel walls. The channels composed of glass gains a negative charge when placed in contact with an aqueous solution at neutral pH. This allows the oppositely charged ions to arrange along the glass channel walls. However, owing to the high cost of glass and expensive fabrication methods recently, the use of hydrophobic polymers like poly(dimethyl siloxane) (PDMS) for fluidic channel fabrication has become popular [33]. This allows accurate transfer of target analytes, provides good optical transparency, sealing properties, and is compatible with different organic solvents [38], however, achieving dimensional stability remains challenging [37]. Other polymers such as poly(methyl methacrylate) (PMMA) and poly(phenylene vinylene) are also used in some cases [39]. The undesirable deformations that may occur during replication of micro/nanochannel structures is a drawback associated with the use of polymer-based materials [37]. The two common fabrication methods used in micro/nanofluidic technology are the bottom-up method and top-down method. The bottom-up method is done by manipulation of individual blocks one at a time and the top-down method includes series of steps such as definition of channels, lithography, and etching [40].

11.4.1 Microfluidics

Microfluidics is defined as "devices and methods for controlling and manipulating fluid flows with length scale less than a millimeter" [41]. The first microfluidic device was the inkjet printer invented in 1965 [42] which was followed by the development of a miniaturized silicon wafer gas chromatograph [43]. Since then, the field of microfluidics has been widely studied and explored. The microfluidic chips are a type of micro-total analysis systems (μTAS) which has supported the portable and high-throughput biosensor technology [26]. It is a technique which works by controlling microfluidic fluids in a micrometer-sized channel. Microfluidics systems are made of network of wells and channels embedded on polymer or glass chips (Figure 11.2), which allows activities including sample handling, dilution, mixing, separation, and detection [44]. These devices have dimension ranging between 10 μm and a few 100 μm and provide variety of benefits like small amount of reagents and sample, high automatization, portability, quick analysis, short reaction times compared to standard laboratory apparatus, and possibility to high throughput applications. Microfluidic technology has been used so far in new biocatalysts characterization, sample purification and separation, and determination of different parameters like cell viability and enzymatic productivity [45]. Recently, the use of microfluidics in genomic applications, drug delivery, and point-of-care diagnosis

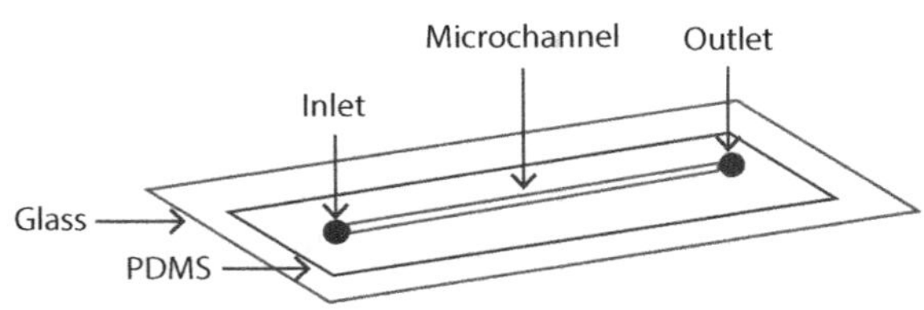

Figure 11.2 Schematic diagram of a microfluidic chip.

has been implemented [46]. Furthermore, microfluidics has opened avenues for novel concepts like organs on chip [47] and digital PCR [48].

Microfluidic platforms have supported many diagnostic tests such as nested and RT-PCR, reverse transcription loop-mediated isothermal amplification (RT-LAMP), clustered regularly interspaced short palindromic repeats (CRISPR), ELISA, rolling circle amplification, porous silicon nanowire forest (pSiNW), and silica sol-gel coating/bonding [11]. Microfluidic systems combined with electrochemical detection offer advantages as they allow integration of multiple steps into a monolithic and disposable device [49]. For the successful use of microfluidic technology, reliable methods to measure the flow rates and functional components like sensors and actuators are crucial. In majority of the microfluidic systems, flow rate measurement is made with the help of active sensor systems like the thermal and Coriolis flowmeters, while some others use passive sensor systems [46]. Other sensor technologies such as field-effect transistors, nanopores, and digital chips enhance the potential of microfluidic integrated sensors [50].

11.4.1.1 Types of Microfluidic Devices

Microfluidic devices can be classified into different types: paper-based, channel-based droplet-based, and digital microfluidics [51]. The paper-based system is composed of a sequence of nitrocellulose fibers or hydrophilic cellulose that allows liquid flow in paper via capillary movement [11]. Paper-based microfluidic sensors have also gained interest due to their unique characteristics such as passive liquid transport, compatibility, simplicity, portability, and ability of easy disposal. Recently, a sliding-paper device that combines LAMP with dopamine to detect SARS-COV-2 DNA in 25 min was developed [52]. Channel-based microfluidic devices have channels that allow the flow of reagents and are fabricated by methods such as lamination, three-dimensional printing (3D), nanofabrication, and molding [53]. Laminating involves the fusion of layers cut using laser to form the channel. However, this method will not provide sub-micrometer levels [54]. Molding incorporates steps such as shaping the mold, selecting suitable polymer, curing polymer and then releasing it from the mold [55]. Three-dimensional printing fabrication is done layer by layer which involves computer aided designing and building up of the two-dimensional (2D) layers using a 3D printer [56]. However, 3D printing has several limitations with regard to the material to be used, biocompatibility, and resolution requirements. Nanofabrication uses three methods: extreme ultraviolet lithography, nanoimprint lithography, and electron beam lithography.

Out of the three methods, nanoimprint lithography has wide applications in microfluidic fabrication due to its affordable cost [57].

A channel-based microfluidic electrochemical quantitative loop-mediated thermal amplification (MEQ-LAMP) system was developed for the detection of pathogenic DNA samples. The system uses LAMP which can be used to amplify non-denatured genomic DNA samples using six different primers that are highly specific and enable real-time MEQ-LAMP to detect pathogenic DNA. The system monitors the incorporation of DNA-binding methylene blue redox reporter molecules into newly made amplicons with a set of integrated electrodes. This device contains a single microfluidic chamber that functions as the LAMP reaction vessel as well as the electrochemical measurement cell. The chip is made of glass and PDMS layers using standard microfabrication techniques. The microfluidic chamber contains reference electrodes, platinum counter, and gold working electrode. Initially, the free MB molecules encounter with the gold working electrode and produce a measurable current. But as the LAMP reaction proceeds, the MB molecules intercalate into the amplicons. This causes a reduction in the number of free MB molecules which then causes the reduction of the redox current and enable monitoring of the reaction [49].

Digital microfluidics is a developing technology for liquid manipulation which consists of an open array of electrodes providing control over picolitre to microliter-sized individual droplets. This allows the droplets to be controlled as individual units, thus prevents cross-reactions among reagents and samples [50]. The digital microfluidics platform has been successfully employed for real-time assay of *bla*(CTX-M-15) gene along with isothermal DNA amplification which resulted in a detection limit of a single copy within 15 min [58]. During recent years, microfluidic technologies have been developed by generating monodispersed droplets in immiscible fluids [59]. Most studies have designed systems in which water droplets are dispersed in oil phase. Sampling of human blood glucose concentration and DNA analysis are some of the applications of droplet-based microfluidic technology. A study has been designed to use microfluidic droplet biosensors for DNA sensing. The study reports the use of molecular beacon in picoliter droplets for breast cancer gene (BRCA1) detection, which was fast, simple, very selective, and reproducible, and utilized very small amounts of sample [60].

11.4.2 Nanofluidics

Nanofluidics is the study of fluid behavior, especially ionic fluids within nanostructures like nanochannels [61]. The fluid dynamics of fluids at nanoscale considerably differs compared to the behavior and dynamics of fluids in large dimensions. Nanofluidic devices are defined as the devices with at least one critical dimension in the range of 1–100 nm [62]. Out-of-plane nanopores, in-plane nanochannels, and nanopipets are the typical types of nanofluidic structures. Nanochannels are ultrasmall closed space systems which function as the core components of most nanofluidic devices [37]. Three types of nanochannels used in nanofluidic devices are the planar nanochannel, square nanochannel, and deep nanochannel. Planar nanochannels are fabricated using standard photolithography and are most commonly used due to their high flow rate which offers easy detection and operation. The square nanochannels have an aspect ratio closer to 1 and they have also been widely used in recent years. Deep nanochannels are not very common as they require highly complicated nanofabrication methods. Nanofluidic devices could be made with a single nanochannel or an array of nanochannels that are arranged parallel to each other [63]. The fluid flow through the nanochannels occur via external driving forces such as electrical potential gradient, colloidal forces such as hydrophobic interactions and electrostatic forces, and friction forces between the molecules and the wall [64]. Nanopores can be formed in lipid bilayers, created in 2D materials like graphene and nanofabricated on thin solid membranes [65]. Nanopores are typically out-of-plane structures that are formed perpendicular to the plane of a substrate and can be symmetrical or asymmetrical. Symmetric nanopores are either cylindrically in shape or hourglass-shaped while asymmetric nanopores come with a wide base and a narrow tip. Nanopipets are special type of modified nanopores fabricated using fused silica capillaries or pulled glass and are conical in shape [66]. The development of a range of nanofabrication methods has made it possible for fabrication of well-defined nanofluidic devices [67]. Common fabrication methods such as electron-beam lithography, conventional photolithography, nanoimprint lithography, polymer nanochannels, and interference lithography are used in developing nanofluidic-based biosensors [63].

Nanofluidic biochips can be used for DNA analysis including DNA stretching and separation with high sensitivity. Nanofluidic biochips based on nanochannels have also been explored to detect viral RNA. These chips allow biomolecule transport, bioseparation, and biodetection [63]. Such nanofluidic devices are capable of probing the dynamic, entropic, and configuration characteristics of DNA molecules and are able to categorize DNA and identify the location of genetic information distributed in lengthy DNA molecules [68]. Moreover, the unique features of nanostructures, such as biocompatibility, high surface-to-volume ratio, high tensile strength, tunable pore size at nanoscale, and lightweight are beneficial for detection and analysis of virus [69]. Recently, carbon nanotubes have gained more interest as they can be directly used as nanofluidic channels without undergoing lithography. In addition to that, the use of 2D materials like graphene and graphene oxide in nanofluidics are also explored widely [70].

11.5 Applications of Micro/Nanofluidic-Based Biosensors in Respiratory Virus Detection

In recent years, a number of microfluidic devices capable of multiplexed detection of different viral targets with matching primer sets placed in each channel have been developed [3]. A microfluidic device integrated with automated processes like sample treatment, one step RT-PCR, and direct fluorescence detection was made to detect influenza-A viruses [71]. The device was capable of carrying out automatic sample lysis using glycan-coated magnetic beads. Clinical studies show that this device with pre-loaded reagents could be improvised for the detection of SARS-CoV-2 [3].

A multi-detectable nanoflow immunosensor based on ZnO nanorods for detecting H1N1, H5N1, and H7N9 influenza viruses was studied. The device used electrochemical detection method in which nanostructured ZnO nanorods having high isoelectric point interacted electrostatically with proteins having lower isoelectric point. This immunosensor utilized lower sample volumes and employed PDMS microfluidic chamber integrated with thin-film three-gold electrodes which had nanoliter-volumetric space for reaction. ZnO nanorods were grown on the inner wall of the PDMS channel using hydrothermal method and capture antibodies of H1N1, H5N1, and H7N9 were immobilized on the ZnO nanorods. Sandwich ELISA was used to detect the three influenza viruses by color change of 3,3',5,5'-tetramethylbenzidine containing solution and electrochemical redox current. This immunosensor was identified as an ideal device to be used as a POCT with low limit detection [7].

A reduced graphene oxide-based electrochemical immunosensor integrated with microfluidic chip was designed for the label-free detection of H1N1 influenza virus. Three microelectrodes were fabricated on a glass substrate and encapsulated with PDMS microchannel. Monoclonal antibodies specific to the H1N1 virus were immobilized on the working electrode with the help of reduced graphene oxide. The antibody immobilization was facilitated by the formation of covalent bonds between the amine group of the antibody and carboxyl group of the reduced graphene oxide.

The large surface area provided by reduced graphene oxide holds many electroactive sites, thus offered increased sensitivity. This immunosensor showed enhanced limit of detection of 0.5PFU mL^{-1} [72].

A 3D bead-based microfluidic chip was designed for the detection of influenza HA. This chip carries out isolation of HA based on paramagnetic beads and detection via electrochemical methods. Streptavidin-modified paramagnetic particles conjugated with biotinylated glycan and HA bound to glycan was used to facilitate isolation. Following isolation, the indirect detection was done using cadmium sulfide quantum dots label. This microfluidic device was fabricated via 3D printing and differential pulse voltammetry was used for the electrochemical detection [73].

An electrochemical DNA (E-DNA) sensor for multiplexed detection of influenza virus from throat swabs was developed by integrating on-chip genetic amplification and sample preparation by immunomagnetic separation into the sequence-specific E-DNA. This platform carries out label-free detection in single-step. The E-DNA sensors were integrated into a microfluidic chip and on-chip genetic amplification techniques were incorporated into it. Electrode-bound, redox-reporter-modified DNA probe generates a current change after undergoing conformational change that is caused by hybridization. The on-chip genetic amplification techniques used here were PCR and LAMP [74]. A microfluidic device was designed with pSiNW forest for size-based, label-free detection of H5N2 influenza virus. The device was able to capture virus inside the interwire spacing of pSiNW forest and also it had the unique feature of releasing the virus by inducing self-degradation of the pSiNW in the presence of aqueous surrounding [69].

Since the emergence of COVID-19, a larger number of studies have focused on designing and commercializing POCT devices for detection of SARS-CoV-2. Gene SoC®, a device built on microfluidic principle is a high-speed reciprocal flow reverse transcription-quantitative polymerase chain reaction (RT-qPCR) system that amplifies and detects specific genes within 15 min. This device demonstrated its ability to be employed for low-throughput detection of SARS-CoV-2 [75]. As an approach for cost cutting, paper-based microfluidic devices based on isothermal amplification have been proposed for COVID-19 diagnosis [76]. These paper-based microfluidic devices are very much ideal for infectious pathogen detection as they can be discarded soon after use and thus prevents cross infection especially among healthcare staff [3].

A highly sensitive SARS-CoV-2 detection kit was developed with nanoporous anodic aluminum oxide membrane (AAO). The AAO membrane surface was coated with 3-aminopropyltrimethoxysilane and glutaraldehyde as linking reagents. Complementary oligonucleotides with fluorescent modifier were coated onto the nanofluidic channels. This sensor was capable of detecting SARS-CoV-2 target DNA and RNA via hybridization method. The fabrication of AAO is simple and cost-effective compared to the other nanofabrication techniques [77].

A fast responding and quantitative capacitive aptasensor with a specific aptamer for detecting N-protein of SARS-CoV-2 was designed based on a low cost microelectrode array chip. The device uses solid–liquid interface capacitance as sensitive indicator. This sensor is capable of detecting the virus even at an ultratrace N-protein concentration of 10^{-1} ng/mL or lower. This sensor functions as a disposable device which carries out low-cost, label-free, real-time diagnosis of SARS-CoV-2 [78]. A microfluidic platform which collected serum by a pulling-force spinning top and paper-based microfluidic ELISA was designed to quantitatively detect IgA/IgM/IgG antibodies for screening of SARS-CoV-2. The special characteristic of this device is the use of pulling-force spinning top which enables human blood–serum separation thus allows people to isolate the serum without any clinical apparatus [79].

Another study focused on designing a nanofluidic-based ultra-sensitive, and multi-functional plasmonic biosensor for COVID-19 detection. The nanofluidic channel is formed by depositing metal nanostructure on infrared-transmitting magnesium fluoride substrate. The detection is carried out based on surface-enhanced infrared absorption. The infrared light emitted is passed through the MgF_2 top layer, through the nanostructure, nanofluidic channel and bottom layer of MgF_2. This is then collected by the infrared detector and analyzed using a computer. The detector uses the whole viral particle as the target and when the virus particles cover the metal structure's surface, the transmission spectrum changes significantly [80].

Apart from the direct detection, fluid technology-based biosensors have also been used for indirect detection approaches as well. Positive test results for Epstein-Barr virus (EBV) infection can be used as an index for diagnosing nasopharyngeal carcinoma. It was recently found that in patients with both EBV and COVID-19 infection, systemic inflammatory responses could be triggered and lead to respiratory failure quickly. Thus early detection of EBV and COVID-19 with sensitivity devices is necessary. A nanofluidic preconcentrator integrated with a nanoslit Fano resonance biosensor was designed for the detection of the membrane protein of EBV virus with a limit of detection of 100 pg/ml. The nanostructure was nanoimprinted on a cyclo-olefin polymer and then aluminum was deposited on the nanostructure to form the nanofluidic preconcentrator [81]. With the

Table 11.1 Microfluidic devices designed for the detection of SARS-CoV-2.

Detection target	Method	Mode of detection	Type of microfluidic device	Reference
IgG/IgM/ antigens	Immunoassay	Fluorescent	Channel-based MF	[82]
IgG antibodies	Immunoassay	Fluorescent	Channel-based MF	[83]
Anti-SARS-CoV-2 spike protein antibodies	Immunoassay	Localized surface plasmon resonance	Channel-based MF	[84]
Antigen; N-protein	Immuno-biosensing	Electrochemical	Channel-based MF	[85]
	Aptamer-based biosensor	Capacitance	Digital microfluidics	[86]
Nucleic acid	Reverse transcription-recombinase polymerase amplification (RT-RPA) and CRISPR-based assay	Fluorescent	Channel-based MF	[87]
	Nested-RPA and CRISPR-based assay	Fluorescent	Channel-based MF	[88]
	RT-LAMP	Fluorescent	Paper-based MF	[89]
	RT-PCR	Fluorescent	Digital microfluidics	[90]
	Magnetic bead, RT-RPA, CRISPR-based assay	Fluorescent	Digital microfluidics	[91]
	Droplet digital PCR	Fluorescent	Droplet-based microfluidics	[92]
	Lab-in-fiber detection	Fluorescent	Droplet-based microfluidics	[93]
Viral particles	Size-based capture	Surface enhanced Raman spectroscopy	Channel-based MF	[94]
IgG/IgM antibodies	Immunoassay	Electrochemical	Paper-based MF	[95]

emergence of COVID–19, a large number of studies focused on developing POCT devices to facilitate diagnosis. Some other fluidic-based biosensors used for SARS-CoV-2 detection, their mode of detection, and method of analysis are summarized in Table 11.1.

11.6 Advantages of Micro/Nanofluidic Diagnosis Tools Over the Other Diagnostic Methods

The incorporation of micro- and nanofluidics into biosensors has offered several advantages like the ability to process small sample volumes, quick turnaround times, easy integration and multiplexing, and high compactness and portability [44]. These devices also provide high-throughput detection along with the ability to integrate multiple complex processes into a single platform and enable automated detection [51]. Furthermore, the commonly used nucleic acid diagnostic methods like the PCR are not ideal to be used for POCT and are expensive [97]. Compared to these diagnostic methods, the micro/nano devices provide better advantages due to their reduced size which makes it portable and ideal to be used as POCT devices, high heat transfer efficiency, and requirement of less reagents [98]. Moreover, most biosensors require the handling of biomolecules in an aqueous electrolyte solution that is similar to their natural environment. Thus, micro- and nanofluidic systems allow for the development and function of such sensors [33]. In addition to that, at times of critical viral outbreaks like the COVID-19 pandemic, these novel techniques will facilitate quicker diagnosis for large number of samples to meet the demand and thus support the rapid curtailing of the spread of the disease. Also, some biosensors like the paper-based microfluidic devices are relatively environmentally friendly and available at affordable prices.

11.7 Conclusion and Future Perspectives

Diseases caused by respiratory viruses are of major concern due to their high rates of spreading and high rates of mortality. Moreover, these viruses are capable of undergoing mutations and evolving with time probably with increased virulence and thus making humans more susceptible to the diseases caused by them. Timely diagnosis and treatment is essential to curtail the spread of such viral outbreaks. Generally, analytical methods based on isolation of viruses followed by RT-PCR or immunological assays are used. However, most of these diagnostic methods are time consuming, expensive, and may produce false-positives and false-negatives results. Micro/nanofluidic biosensor-based detection devices are gaining interest due to the many advantages that they offer. Micro/nanofluidics play a vital role in improving biosensor performance with regard to automation, sustainability, suitability to be used for POCT, and on-field analysis. In fact, the future of biosensors depends on the combination of all these technologies in a single integrated device capable to carter the wide range of diagnostic necessities associated with human wellbeing. However the use of micro/nanofluidics has some limitations related to multiplexed detection, selectivity, and ability to be used for field analysis. In addition, microfluidic- and nanofluidic-based biosensors should improve on factors concerning reproducibility, stability, sample preparation, cost, and environmental friendliness. Despite the immense progress, the existing fluidic technology-based detection devices are yet to improve their performance to meet the gold standard methods, thus preventing their wide spread use in detection of pathogen. Noteworthy future advances in sensor technologies will likely concentrate on prompt detection, improving sensitivity, high throughput capacity, and portability. Getting a deeper understanding about the movement of solutes, solvents, ions, and biomolecules through micro- and nanochannels and gaining more control over new fabrication techniques would help these fluidic technologies to overcome the current challenges and evolve with enhanced sensing ability and efficiency, thus revolutionizing the use of biosensors for viral detection in future. By overcoming these limitations with extensive research, profound application and industrialization the biosensor detection systems in micro- and nanofluidic chips could replace the traditional apparatus and analysis methods used in chemical laboratories.

References

1. Ribeiroa, B.V., Cordeiro, T.A.R., Oliveira e Freitas, G.R., Ferreira, L.F., Biosensors for the detection of respiratory viruses: A review. *Talanta Open*, 2, 100007, 2020.
2. Das, S., Dunbar, S., Tang, Y., Laboratory diagnosis of respiratory tract infections in children – the state of the art. *Front. Microbiol.*, 9, 2478, 2018.
3. Wang, Y., Xu, H., Dong, Z., Wang, Z., Yang, Z., Yu, X., Chang, L., Micro/nano biomedical devices for point-of-care diagnosis of infectious respiratory diseases. *Med. Novel Technol. Devices*, 14, 100116, 2022.

4. Dong, X., Liu, L., Tu, Y., Zhang, J., Miao, G., Zhang, L., Ge, S., Xia, N., Yu, D., Qiu, X., Rapid PCR powered by microfluidics: a quick review under the background of COVID-19 pandemic. *Trends Anal. Chem.*, 143, 116377, 2021.
5. Weiss, S.R. and Navas-Martin, S., Coronavirus pathogenesis and the emerging pathogen severe acute respiratory syndrome coronavirus. *Microbiol. Mol. Biol. Rev.*, 69, 4, 635–664, 2005.
6. Boon, C.M., French, A.M.F., Fleming, D.M., Zambon, M.C., Detection of influenza a subtypes in community-based surveillance. *J. Med. Virol.*, 61, 163–170, 2001.
7. Ji-Hoon, H., Dongyoung, L., Charleson, H.C.C., Taeheon, K., James, J.P., A multi-virus detectable microfluidic electrochemical immunosensor for simultaneous detection of H1N1, H5N1, and H7N9 virus using ZnO nanords for sensitivity enhancement. *Sens. Actuators, B*, 228, 36–42, 2015.
8. Choi, Y.K., Goyal, S.M., Kang, S.W., Farnham, M.W., Joo, H.S., Detection and subtyping of swine influenza H1N1, H1N2 and H3N2 viruses in clinical samples using two multiplex RT-PCR assays. *J. Virol. Methods*, 106, 53–59, 2002.
9. Gong, S. and Bao, L., The battle against SARS and MERS coronaviruses: reservoirs and Animal models. *Anim. Models Exp. Med.*, 1, 125–133, 2018.
10. World Health Organization (WHO), Middle East respiratory syndrome coronavirus (MERS-CoV), 2019, https://www.who.int/emergencies/mers-cov/en/.
11. Basiri, A., Heidari, A., Nadi, M.F., Fallahy, M.T., Nezamabadi, S.S., Sedighi, M., Saghazadeh, A., Rezaei, N., Microfluidic devices for detection of RNA viruses. *Rev. Med. Virol.*, 31, 1–11, 2020.
12. Memish, Z.A., Al-Tawfiq, J.A., Assiri, A., AlRabiah, F.A., Al-Hajjar, S., Albarrak, A., Flemban, H., Alhakeem, R.F., Makhdoom, H.Q., Alsubaie, S., Al-Rabeeah, A.A., Middle East respiratory syndrome coronavirus disease in children. *Pediatr. Infect. Dis. J.*, 33, 904–906, 2014.
13. Paramyxoviridae and Pneumoviridae, in: *Fenner's veterinary virology*, 5th Editio, E.D.N.J. MacLachlan (Ed.), pp. 327–356, Academic Press, New York, 2017.
14. Collins, P.L. and Graham, B.S., Viral and host factors in human respiratory syncytial virus pathogenesis. *J. Virol.*, 82, 2040–2055, 2008.
15. Schildgen, O., Human bocavirus: lessons learned to date. *Pathogens*, 2, 1–12, 2013.
16. Zhao, Z., Huang, C., Huang, Z., Lin, F., He, Q., Jaffrezic-Renault, N., Guo, Z., Advancements in electrochemical biosensing for respiratory virus detection: A review. *Trends Anal. Chem.*, 139, 116253, 2021.
17. Bhalla, N., Pan, Y., Yang, Z., Payam, A.F., Opportunities and challenges for biosensors and nanoscale Analytical tools for pandemics: COVID-19. *ACS Nano*, 14, 7783–7807, 2020.
18. Malik, Y.A., Properties of coronavirus and SARS-CoV 2. *Malays. J. Pathol.*, 42, 1, 3–11, 2020.

19. Yan, Y., Chang, L., Wang, L., Laboratory testing of SARS-CoV, MERS-CoV, and SARS-CoV-2 (2019-nCoV): current status, challenges, and countermeasures. *Rev. Med. Virol.*, 30, e2106, 2020.
20. Zumla, D.H., Middle East respiratory syndrome. *Lancet*, 386, 995–1007, 2015.
21. Woo, P.C.Y., Lau, S.K.P., Wong, B.H.L., Tsoi, H.W., Fung, A.M.Y., Kao, R.Y.T., Chan, K.H., Peiris, J.S.M., Yuen, K.Y., Differential sensitivities of severe acute respiratory syndrome (SARS) coronavirus spike polypeptide enzyme-linked immunosorbent assay (ELISA) and SARS coronavirus nucleocapsid protein ELISA for serodiagnosis of SARS coronavirus pneumonia. *J. Clin. Microbiol.*, 43, 3054–3058, 2005.
22. Udugama, P.K., Kadhiresan, P., Kozlowski, H.N., Malekjahani, A., Osborne, M., Li, V.Y.C., Chen, H., Mubareka, S., Gubbay, J.B., Chan, W.C.C., Diagnosing COVID 19: the disease and tools for detection. *ACS Nano*, 14, 3822–3835, 2020.
23. Pokhrel, P., Hu, C., Mao, H., Detecting the coronavirus (CoVID-19). *ACS Sens.*, 5, 2283–2297, 2020.
24. Corman, V.M., Landt, O., Kaiser, M., Molenkamp, R., Meijer, A., Chu, D.K.W., Bleicker, T., Brünink, S., Schneider, J., Schmidt, M.L., Mulders, D.G.J.C., Haagmans, B.L., van der Veer, B., van den Brink, S., Wijsman, L., Goderski, G., Romette, J.L., Ellis, J., Zambon, M., Detection of 2019 novel coronavirus (2019-nCoV) by real-time RT-PCR. *Euro Surveill.*, 25, 2000045, 2020.
25. Valizadeh, N.S., Electrochemical detection of HIV-1 by nanomaterials, artificial cells. *Nanomed. Biotechnol.*, 45, 1467–1477, 2017.
26. Liao, Z., Wang, J., Zhang, P., Zhang, Y., Miao, Y., Gao, S., Deng, Y., Geng, L., Recent advances in microfluidic chip integrated electronic biosensors for multiplexed detection. *Biosens. Bioelectron.*, 121, 272–280, 2018.
27. Xiao, T., Huang, J., Wang, D., Meng, T., Yang, X., Au and Au-Based nanomaterials: synthesis and recent progress in electrochemical sensor applications. *Talanta*, 206, 120210, 2020.
28. Chorsi, M.T., Curry, E.J., Chorsi, H.T., Das, R., Baroody, J., Purohit, P.K., Ilies, H., Nguyen, T.D., Piezoelectric biomaterials for sensors and actuators. *Adv. Mater.*, 31, e1802084, 2019.
29. Abbas, A.K., Lichtmn, A.H., Pillai, S., *Cellular and molecular immunolog*, Elsevier, Philadelphia, 2015.
30. Rashid, J.I.A. and Yusof, N.A., The strategies of DNA immobilization and hybridization detection mechanism in the construction of electrochemical DNA sensor: A review. *Sens. Bio-Sens. Res.*, 16, 19–31, 2017.
31. Persson, F. and Tegenfeldt, J.O., DNA in nanochannels - directly visualizing genomic information. *Chem. Soc. Rev.*, 39, 985–372, 2010.
32. Xu, Y., Xie, X., Duan, Y., Wang, L., Cheng, Z., Cheng, J., A review of impedance measurements of whole cells. *Biosens. Bioelectron.*, 77, 824–836, 2016.

33. Prakash, S., Pinti, M., Bhushan, B., Theory, fabrication and applications of microfluidic and nanofluidic biosensors. *Philos. Trans. R. Soc.*, 370, 2269–2303, 2015.
34. Lan, W.C., Huang, T.S., Cho, Y.C., Huang, Y.T., Walinski, C.J., Chiang, P.C., Rusilin, M., Pai, F.T., Huang, C.C., Huang, M.S., The potential of a nanostructured titanium oxide layer with self-assembled monolayers for biomedical applications: surface properties and biomechanical behaviours. *Appl. Sci.*, 10, 590–595, 2020.
35. Yang, C. and Zhao, C., Advances in electrokinetics and their applications in micro/nano fluidics. *Microfluid. Nanofluid.*, 13, 179–203, 2012.
36. Plecis, A., Schoch, R.B., Renaud, P., Ionic transport phenomena in nanofluidics: experimental and theoretical study of the exclusion-enrichment effect on a chip. *Nano Lett.*, 5, 1147–1155, 2005.
37. Xu, Y., Nanofluidics: A new arena for materials science. *Adv. Mater.*, 30, 1702419, 2017.
38. Gervais, L.D., Microfluidic chips for point-of-care immunodiagnostics. *Adv. Mater. (Deerfield Beach, Fla.)*, 23, 151–176, 2011.
39. Chikkaveeraiah, B.V., Liu, H., Mani, V., Papadimitrakopoulos, F., A microfluidic electrochemical device for high sensitivity biosensing: detection of nanomolar hydrogen peroxide. *Electrochem. Commun.*, 11, 819–822, 2009.
40. Madou, M., *Fundamentals of microfabrication*, CRC Press LLC, FL, 1997.
41. Stone, H.A., Stroock, A.D., Ajdari, A., Engineering flows in small devices. *Annu. Rev. Fluid Mech.*, 36, 1, 381–411, 2004.
42. Sweet, R.G., High frequency recording with electrostatically deflected ink jets. *Rev. Sci. Instrum.*, 36, 2, 131–136, 1965.
43. Terry, S.C., A gas chromatography system fabricated on a silicon wafer using integrated circuit technology, Stanford University, 1975.
44. Arduini, F.C., Cinti, S., Scognamiglio, V., Moscone, D., Palleschi, G., How cutting-edge technologies impact the design of electrochemical (bio)sensors for environmental analysis. A review. *Anal. Chim. Acta*, 959, 15–42, 2017.
45. Viefhues, M., Analytics in microfluidic systems. *Adv. Biochem. Eng./Biotechnol.*, 179, 191–209, 2022.
46. Cavaniol, C., Cesar, W., Descroix, S., Viovy, J.L., Flowmetering for microfluidics. *Lab Chip*, 22, 3603–3617, 2022.
47. Ingber, D.E., Human organs-on-chips for disease modelling, drug development and personalized medicine. *Nat. Rev. Genet.*, 23, 467–491, 2022.
48. Hindson, M., Chevillet, J.R., Briggs, H.A., Gallichotte, E.N., Ruf, I.K., Hindson, B.J., Absolute quantification by droplet digital PCR versus analog real time PCR. *Nat. Methods*, 10, 10, 1003–1005, 2013.
49. Kuangwen, H., Adriana, S., Patterson, B., Ferguson, S., Plaxco, K.W., Soh, H.T., Rapid, sensitive, and quantitative detection of pathogenic DNA at the point of care through microfluidic electrochemical quantitative loop-mediated isothermal amplification. *Angew. Chem. Int.*, 51, 1–6, 2012.

50. Arce, C.L., Silicon photonic sensors incorporated in a digital microfluidic system. *Anal. Bioanal. Chem.*, 404, 2887–2894, 2012.
51. Shi, J., Zhang, Y., Yang, M., Recent development of microfluidics-based platforms for respiratory virus detection. *Biomicrofluidcs, 17*, 2, 024104, 2023.
52. Nguyen, H.A. and Lee, N.Y., Polydopamine aggregation: a novel strategy for power-free readout of loop-mediated isothermal amplification integrated into a paper device for multiplex pathogens detection. *Biosens. Bioelectron.*, 189, 113353, 2021.
53. Xia, Y., Si, J., Li, Z., Fabrication techniques for microfluidic paper-based analytical devices and their applications for biological testing: a review. *Biosens. Bioelectron.*, 77, 774–789, 2016.
54. Mahmud, M.A., Blondeel, E.J.M., Kaddoura, M., MacDonald, B.D., Features in microfluidic paper-based devices made by laser cutting: how small can they be? *Micromachines*, 220, 9, 220–228, 2018.
55. Li, D., *Encyclopedia of microfluidics and nanofluidics*, Springer Science & Business Media, NewYork, 2008.
56. Bhattacharjee, N., Urrios, A., Kang, S., Folch, A., The upcoming 3D-printing revolution in microfluidics. *Lab Chip*, 16, 1720–1742, 2016.
57. Gale, B.K., Jafek, A.R., Lambert, C.J., Goenner, B., Moghimifam, H., Nze, U.C., Kamarapu, S.K., A review of current methods in microfluidic device fabrication and future commercialization prospects. *Inventions*, 60, 3, 60–67, 2018.
58. Kalsi, S., Valiadi, M., Tsaloglou, M.N., Parryjones, L., Jacobs, A., Watson, R., Turner, C., Amos, R., Hadwen, B., Buse, J., Rapid and sensitive detection of antibiotic resistance on a programmable digital microfluidic platform. *Lab Chip*, 15, 14, 3065–3075, 2015.
59. Cristini, V. and Tan, Y.C., Theory and numberical simulation of droplet dynamics in complex flows - a review. *Lab Chip*, 4, 4, 257–264, 2004.
60. Lee, A.P., Yung-Chieh, T., Collins, J., Lung-Hsin, H., Tsung-His, H., Micro and nano fluidic chips for biosensors. *Sensors*, Irvine, CA, USA, 2005.
61. Ul Haque, A. and A.K., Nanochannels for nanofluidics: Fabrication aspects, in: *Encyclopedia of Nanotechnology*, B. Bhushan (Ed.), Springer, Dordrecht, 2012.
62. Prakash, S., Karacor, M.B., Banerjee, S., Surface modification in microsystems and nanosystems. *Surf. Sci. Rep.*, 64, 233–254, 2009.
63. Xia, D., Yan, J., Hou, S., Fabrication of Nanofluidic Biochips with Nanochannels for Applications in DNA Analysis. *Small*, 8, 2787–2801, 2012.
64. Sparreboom, W., van den Berg, A., Eijkel, J.C.T., Transport in nanofluidic systems: a review of theory and applications. *New J. Phys.*, 12, 015004, 2010.
65. Haque, F., Li, J., Wu, H., Liang, X., Guo, P., Solid-state and biological nanopore for real-time sensing of single chemical and sequencing of DNA. *Nano Today*, 56, 8, 56–74, 2013.

66. Haywood, D.G., Saha-Shah, A., Baker, L.A., Jacobson, S.C., Fundamental studies of nanofluidics: nanopores, nanochannels, and nanopipets. *Anal. Chem.*, 87, 172–187, 2015.
67. Duan, C., Wang, W., Xie, Q., Review article: Fabrication of nanofluidic devices. *Biomicrofluidics*, 7, 2, 26501, 2013.
68. Levy, L.S. and Craighead, H.G., DNA manipulation, sorting, and mapping in nanofluidic systems. *Chem. Soc. Rev.*, 39, 1133–1152, 2010.
69. Xia, Y., Tang, Y., Yu, X., Wan, Y., Chen, Y., Lu, H., Si-Yang, Z., Label-free virus capture and release by a microfluidic device integrated with porous silicon nanowire forest. *Small*, 13, 1–11, 2017.
70. Koltonow, A.R. and Huang, J., Ionic transport, two-dimensional nanofluidics. *Science*, 351, 1395–1396, 2016.
71. Shen, K., Sabbavarapu, N., Fu, C., Jan, J., Wang, J., Hung, S., Lee, G., An integrated microfluidic system for rapid detection and multiple subtyping of influenza A viruses by using glycan-coated magnetic beads and RT-PCR. Viruses by using glycan-coated magnetic beads and RT-PCR. *Lab on a Chip*, 19, 1277–1286, 2019.
72. Singh, R., Hong, A., Jang, J., Viruses using a Reduced Graphene Oxide-based Electrochemical Immunosensor Integrated with a Microfluidic Platform. *Sci. Rep.*, 7, 42771, 2017.
73. Krejcova, L., Nejdl, L., Rodrigo, M.A.M., Zurek, M., Matousek, M., Hynek, D., Zitka, O., Kopel, P., Adam, V., Kizek, R., 3D printed chip for electrochemical detection of influenza virus labeled with CdS quantum dots. *Biosens. Bioelectron.*, 54, 421–427, 2014.
74. Hsieh, K., Ferguson, B.S., Eisenstein, M., Plaxco, K.W., Soh, H.T., Integrated electrochemical microsystems for genetic detection of pathogens at the point of care. *Acc. Chem. Res.*, 48, 4, 911–920, 2015.
75. Sakai, J., Tarumoto, N., Orihara, Y., Kawamura, R., Kodana, M., Matsuzaki, N., Matsumura, R., Okane, K., Kawamura, T., Takeuchi, S., Evaluation of a high-speed but low-throughput RT-qPCR system for SARS-CoV-2 detection. *J. Hosp. Infect.*, 105, 615–618, 2020.
76. Yang, Z., Xu, G., Reboud, J., Ali, S.A., Kaur, G., McGiven, J., Boby, N., Gupta, P.K., Chaudhuri, P., Cooper, J.M., Rapid veterinary diagnosis of bovine reproductive infectious diseases from semen using paper-origami DNA microfluidics. *ACS Sens.*, 3, 403–409, 2018.
77. Makela, M., Lin, Z., Lin, P.T., Surface functionalized anodic aluminum oxide membrane for opto-nanofluidic SARS-CoV-2 genomic target detection. *IEEE Sens. J.*, 21, 22645–22650, 2021.
78. Qi, H., Hu, Z., Yang, Z., Zhang, J., Wu, J.J., Cheng, C., Wang, C., Zheng, L., Capacitive aptasensor coupled with microfluidic enrichment for real-time detection of trace SARS-CoV-2 nucleocapsid protein. *Anal. Chem.*, 94, 2812–2819, 2022.
79. Gong, F., Wei, H., Qi, J., Ma, H., Liu, L., Weng, J., Zheng, X., Li, Q., Zhao, D., Fang, H., Liu, L., He, H., Ma, C., Han, J., Sun, A., Wang, B., Jin, T., Li,

B., Li, B., Pulling -force spinning top for serum separation combined with paper-based microfluidic devices in COVID-19 ELISA diagnosis. *ACS Sens.*, 6, 2709–2719, 2021.

80. Li, D., Zhou, H., Hui, X., He, X., Mu, X., Plasmonic biosensor augmented by a genetic algorithm for ultra-rapid, label-free, and multi-functional detection of COVID-19. *Anal. Chem.*, 93, 9437–9444, 2021.
81. Hsieh, H., Luo, J., Shen, Y., Lo, S., Hsu, Y., Tahara, H., Fan, Y., Wei, P., Sheen, H., A nanofluidic preconcentrator integrated with an aluminum-based nanoplasmonic sensor for Epstein-Barr virus detection. *Sens. Actuators, B*, 335, 131327, 2022.
82. Lin, Q., Wen, D., Wu, J., Liu, L., Wu, W., Fang, X., Kong, J., Microfluidic immunoassays for sensitive and simultaneous detection of IgG/IgM/antigen of SARS-CoV-2 within 15 min. *Anal. Chem.*, 92, 9454–9458, 2020.
83. Swank, Z., Michielin, G., Yip, H.M., Cohen, P., Andrey, D.O., Vuilleumier, N., Kaiser, L., Eckerle, I., Meyer, B., Maerkl, S.J., A high-throughput microfluidic nanoimmunoassay for detecting anti-SARS-CoV-2 antibodies in serum or ultralow-volume blood samples. *Proc. Natl. Acad. Sci. U.S.A.*, 118, 2021.
84. Funari, R., Chu, K.Y., Shen, A.Q., Detection of antibodies against SARS-CoV-2 spike protein by gold nanospikes in an opto-microfluidic chip. *Biosens. Bioelectron.*, 169, 112578–112586, 2020.
85. Li, J. and Lillehoj, P.B., Microfluidic magneto immunosensor for rapid, high sensitivity measurements of SARS-CoV-2 nucleocapsid protein in serum. *ACS Sens.*, 6, 1270–1278, 2021.
86. Qi, H., Hu, Z., Yang, Z., Zhang, J., Wu, J.J., Cheng, C., Wang, C., Zheng, L., Capacitive aptasensor coupled with microfluidic enrichment for real-time detection of trace SARS-CoV-2 nucleocapsid protein. *Anal. Chem.*, 94, 2812–2819, 2022.
87. DePuig, H., Lee, R.A., Najjar, D., Tan, X., Soekensen, L.R., Angenent-Mari, N.M., Donghia, N.M., Weckman, N.E., Ory, A., Ng, C.F., Nguyen, P.Q., Mao, A.S., Ferrante, T.C., Lansberry, G., Sallum, H., Niemi, J., Collins, J.J., Minimally instrumented SHERLOCK (miSHERLOCK) for CRISPR-based point-of-care diagnosis of SARS-CoV-2 and emerging variants. *Sci. Adv.*, 7, 2944, 2021.
88. Liu, J., Wang, H., Zhang, L., Lu, Y., Wang, X., Shen, M., Li, N., Feng, L., Jing, J., Cao, B., Zou, X., Cheng, J., Xu, Y., Sensitive and Rapid Diagnosis of Respiratory Virus Coinfection Using a Microfluidic Chip-Powered CRISPR/Cas12a System. *Small*, 18, 2200854, 2022.
89. Garneret, P., Coz, E., Martin, E., Manuguerra, J.C., Brient-Litzler, E., Enouf, V., González Obando, D.F., Olivo-Marin, J.C., Monti, F., van der Werf, S., Vanhomwegen, J., Tabeling, P., Performing point-of-care molecular testing for SARS-CoV-2 with RNA extraction and isothermal amplification. *PLoS One*, 16, e0243712, 2021.

90. Huang, H., Huang, K., Sun, Y., Luo, D., Wang, M., Chen, T., Li, M., Duan, J., Huang, L., Dong, C., A Digital microfluidic RT-qPCR platform for multiple detections of respiratory pathogens. *Micromachines*, 13, 1650, 2022.
91. Chen, F.E., Lee, P.W., Trick, A.Y., Park, J.S., Chen, L., Shah, K., Mostafa, H., Carroll, K.C., Hsieh, K., Wang, T.H., Point-of-care CRISPR-Cas-assisted SARS-CoV-2 detection in an automated and portable droplet magnetofluidic device. *Biosens. Bioelectron.*, 190, 113390, 2021.
92. Shen, K.M., Sabbavarapu, N.M., Fu, C.Y., Jan, J.T., Wang, J.R., Hung, S.C., Lee, G.B., An integrated microfluidic system for rapid detection and multiple subtyping of influenza A viruses by using glycan-coated magnetic beads and RT-PCR. *Lab Chip*, 19, 1277–1286, 2019.
93. Parker, H.E., Sengupta, S., Harish, A.V., Soares, R.G., Joensson, H.N., Margulis, W., Russom, A., Laurell, F., A Lab-in-a-fiber optofluidic device using droplet microfluidics and laser-induced fluorescence for virus detection. *Sci. Rep.*, 12, 1–10, 2022.
94. Jadhav, S.A., Biji, P., Panthalingal, M.K., Murali Krishna, C., Rajkumar, S., Joshi, D.S., Sundaram, N., Development of integrated microfluidic platform coupled with Surface-enhanced Raman Spectroscopy for diagnosis of COVID-19. *Med. Hypotheses*, 146, 110356, 2021.
95. Yakoh, A., Pimpitak, U., Rengpipat, S., Hirankarn, N., Chailapakul, O., Chaiyo, S., Paper-based electrochemical biosensor for diagnosing COVID-19: Detection of SARS-CoV-2 antibodies and antigen. *Biosens. Bioelectron.*, 176, 112912, 2021.
96. Shi, B., He, G., Wu, W., A PCR microreactor machinery with passive micropump and battery-powered heater for thermo-cycled amplifications of clinical-level and multiplexed DNA targets. *Microchim. Acta*, 185, 467, 2018.
97. Song, Q., Sun, X., Dai, Z., Gao, Y., Gong, X., Zhou, B., Wu, J., Wen, W., Point of care testing detection methods for COVID 19. *Lab Chip*, 21, 1634–1660, 2021.
98. Wang. C., Liu, M., Wang, Z., Li, S., Deng, Y., He, N., Point-of-care diagnostics for infectious diseases: From methods to devices. *Nanotoday*, 37, 101092, 2021.

12

Nanobiosensor System: A Robust Analytical Tool for Pandemics

Mohammad Harun-Ur-Rashid[1], Israt Jahan[2] and Abu Bin Imran[3]*

[1]Department of Chemistry, International University of Business Agriculture and Technology, Dhaka, Bangladesh
[2]Department of Cell Physiology, Graduate School of Medicine, Nagoya University, Nagoya, Japan
[3]Department of Chemistry, Bangladesh University of Engineering and Technology, Dhaka, Bangladesh

Abstract

The world has experienced five major pandemics over the past two decades, including severe acute respiratory syndrome (SARS) (2002), swine flu (2009), Middle East respiratory syndrome (MERS) (2012), Ebola (2014), and coronavirus disease (COVID-19) (2019). The latter two are still active, with COVID-19 exhibiting unprecedented infection rates and complexity. Over 200 countries have been severely impacted by the pandemic, which has overwhelmed health-care systems and caused economic turmoil. As a result, global efforts have been made to comprehend the evolution of pathogens, develop nanobiosensors, and create effective treatments and vaccines. Modern nanobiosensors are based on electrical, chemical, electrochemical, optical, magnetic, aptameric, and plasmonic techniques. These sensors allow for the early detection of infectious microorganisms, aiding in managing pandemics. Estimating pathogen loads requires accurate, rapid, and cost-effective analytical instruments for determining infection stages and recovery durations. For point-of-care (POC) diagnostics, nanobiosensors provide a more efficient alternative to non-portable and labor-intensive conventional methods such as computed tomography-assisted polymerase chain reaction (CT-PCR). They employ bioreceptors with diverse transduction methods for the detection of specific biomarkers. Numerous organic-inorganic nanoparticles have been utilized to create analytical instruments with

**Corresponding author*: abimran@chem.buet.ac.bd; imransust@yahoo.com

Suvardhan Kanchi, Ayyappa Bathinapatla, Anitha Varghese and Phumlane Selby Mdluli (eds.) Nano-Biosensor Technologies for Diagnosis of Infectious Diseases, (365–400) © 2025 Scrivener Publishing LLC

high performance, sensitivity, and biocompatibility. This chapter discusses nanobiosensors for pandemic management on a global scale, including rapid pathogen detection and identification, POC diagnostics, multiplexed detection, real-time monitoring, integration with digital technologies, environmental monitoring, vaccine and treatment development, surveillance, and outbreak prediction. Detection technologies based on nanobiosensors outperform pandemic instruments compared to conventional diagnostic techniques. This chapter provides a comprehensive understanding of nanobiosensor technology, its applications, and its relevance in detecting infectious diseases such as SARS-CoV-2, COVID-19, influenza, MERS, human immunodeficiency virus (HIV)/acquired immunodeficiency syndrome (AIDS), hepatitis, Ebola, Zika, and dengue. It also discusses the selection and optimization of nanomaterials to create efficient nanobiosensors, identifies current issues and proposes prospective solutions, and concludes with recommendations for researchers, physicians, and policymakers to develop and implement innovative infectious disease diagnostic technology. The primary goal is to facilitate the advancement of nanobiosensor technology research for the benefit of society.

***Keywords*:** COVID-19, SARS-CoV-2, swine flu, Middle East respiratory syndrome, electrochemical biosensors, optical biosensors, plasmonic biosensors, aptameric biosensors

12.1 Introduction

The COVID-19 pandemic has underscored the critical importance of rapid, accurate, and reliable diagnostic tools in detecting and monitoring infectious diseases. Unlike other pandemics in the past, COVID-19 has spread rapidly and affected a large number of people worldwide, making the need for effective diagnostic tools even more pressing. One of the main challenges in controlling the spread of COVID-19 has been the asymptomatic transmission of the virus, which means that infected individuals may not show any symptoms but can still spread the virus to others. This highlights the importance of diagnostic tools in identifying infected individuals and preventing further transmission. Early diagnosis and isolation of infected individuals can help reduce the spread of the virus, thereby reducing the burden on healthcare systems and the overall number of cases. Another challenge in controlling the spread of COVID-19 has been the emergence of new virus variants. These variants may be more transmissible or resistant to current treatments and vaccines, making it even more important to have diagnostic tools that can quickly and accurately identify them. Rapid identification of these variants can help to inform public health interventions and guide the development of new treatments and vaccines.

The COVID-19 pandemic has also highlighted the need for diagnostic tools that are accessible, affordable, and scalable. In many parts of the world, including low- and middle-income countries, there are limited resources for diagnostic testing, which can hinder efforts to control the spread of the virus. Rapid diagnostic tests that can be administered at POC, such as antigen tests, can help to overcome these challenges by providing a quick and affordable diagnosis in settings with limited resources. In addition to diagnostic tools, the COVID-19 pandemic has also highlighted the need for reliable surveillance systems to track the spread of the virus. Surveillance systems that can quickly detect and report new virus cases can help inform public health interventions and prevent further transmission. Wastewater-based epidemiology is an emerging technique used to detect the virus in wastewater, providing an early warning system for outbreaks and enabling targeted interventions. The COVID-19 pandemic has highlighted the critical need for rapid, accurate, and reliable diagnostic tools to detect and monitor infectious diseases. Traditional diagnostic methods, such as PCR and enzyme-linked immunosorbent assays (ELISA), have been widely used but often have limitations regarding their sensitivity, specificity, and cost. As a result, researchers have turned to biosensor technology as a promising solution to these challenges. Biosensors are analytical devices that combine biological recognition elements, such as antibodies or enzymes, with transducer elements, such as optical or electrochemical sensors, to detect and quantify a wide range of analytes. Biosensors have several advantages over traditional diagnostic methods, including their high sensitivity, specificity, and speed, as well as their ease of use and portability.

Nanomaterials are unique compared to bulk materials because of their size and properties at the nanoscale. At the nanoscale, materials exhibit properties significantly different from the same material at the macro scale. For instance, the high surface area to volume ratio of nanomaterials leads to a higher number of surface atoms than bulk materials, making their surface properties more dominant. Additionally, their small size results in quantum confinement effects, which cause electronic, optical, and magnetic properties to be significantly different from bulk materials. Another factor that makes nanomaterials unique is their ability to interact with biological systems in ways that bulk materials cannot. For example, nanoparticles with specific sizes, shapes, and surface chemistries can penetrate cell membranes and interact with proteins and other biomolecules. This unique interaction of nanomaterials with biological systems opens up new opportunities for drug delivery, biosensing, and imaging applications. Nanomaterials can also be engineered with specific properties and structures that allow them to perform novel functions. For example, nanoparticles can be designed to bind to specific molecules, such as cancer cells selectively, and can be used for targeted drug delivery. Similarly, nanocomposites can be engineered with high mechanical strength and thermal stability, making them ideal for applications in fields such as aerospace and electronics. Nanotechnology and nanomaterials have become increasingly important in modern times due to their unique properties and potential applications in various fields, including medicine, electronics, and energy. Nanomaterials are defined as materials with at least one dimension between 1 and 100 nm, while nanotechnology refers to the science and engineering of materials at the nanoscale. The development of engineered nanomaterials, such as nanocomposite gels [1–4], nanocomposite films [5–7], structural colored nanomaterials [8–10], organometallic nanomaterials [11], molecular machines [12], nanomaterial-based biosensors [13], and others have replaced conventional bulk materials in various industrial sectors [14–17] and theoretical investigations [18]. Improvements in nanotechnology, such as the ability to modify and control microscopic shapes and properties, have allowed for developing these designed nanomaterials. Concerns about environmental and health implications, as well as ethical problems, are raised by the production of such materials. Because of their compatibility with nanotechnology, a plethora of engineered nanomaterials with specific characteristics and prospective uses have been created. Medical, technology, and energy sectors all stand to benefit from new materials, but there are legitimate questions about their safety and ethical implications that must be answered. More study is needed to completely comprehend the advantages and disadvantages of engineered nanomaterials, and to guarantee their safe and responsible application.

Nanobiosensors employ nanomaterials, such as nanoparticles or nanotubes, into their design. Nanomaterials are well-suited for biosensors due to their unique physical and chemical properties, such as their large surface area, high reactivity, and ability to enhance signal amplification. They have emerged as a promising analytical instrument for detecting, diagnosing, and monitoring pandemics. They can promptly and affordably detect viral antigens [13, 19–21], antibodies [22–24], and nucleic acids [25–27] with high sensitivity and specificity. In addition, they can be readily incorporated into existing diagnostic platforms, such as lateral flow assays, microarrays, and lab-on-a-chip devices, making them versatile tools for various applications. A nanobiosensor system is a highly effective analytical instrument that integrates nanotechnology and biotechnology to detect and quantify the presence of biological or chemical entities in a sample. A nanomaterial-based sensing platform interacts with the target analyte, and a transducer converts the resulting signal into a measurable output.

Nanobiosensor systems are suitable for a variety of applications, such as medical diagnostics [28–30], environmental monitoring [31–33], and food safety [34–38], due to their high sensitivity [39], specificity [40], and quick response time [41]. In addition, the small size of the nanomaterials used in the sensing platform enables miniaturization and integration with electronic devices, facilitating the development of portable, handheld nanobiosensor systems for POC settings. The capability to engineer nanobiosensor systems with specific properties and structures enables the development of highly selective and versatile sensors that can simultaneously detect multiple analytes. As a result, nanobiosensor systems have the potential to revolutionize analytical chemistry and resolve some of the most pressing issues in healthcare, the environment, and industry.

This chapter presents a thorough review of nanobiosensors for addressing global pandemics. It discusses various strategies, such as rapid pathogen detection and identification, POC diagnostics, multiplexed detection, real-time monitoring, integration with digital technologies, environmental monitoring, vaccine and treatment development, surveillance, and outbreak prediction that can be implemented through the use of nanobiosensors. The significance of nanobiosensors in managing pandemics has been highlighted through a comparative analysis of the benefits and drawbacks of conventional diagnostic methods and nanobiosensor-based detection techniques. The chapter provides a summary of the various sections detailing the utilization of newly developed nanobiosensors for detecting SARS-CoV-2 and diagnosing COVID-19, influenza, MERS, HIV/AIDS, hepatitis, Ebola, Zika, and dengue. The chapter has been successfully finalized by incorporating the process of selecting and optimizing nanomaterials for developing nanobiosensors, along with an analysis of the current challenges and potential solutions. This resource aims to provide valuable insights for academic researchers, medical practitioners, and government officials keen on creating and implementing cutting-edge diagnostic technologies to address the challenges posed by infectious diseases and to enhance the understanding of the principles and applications of nanobiosensor technology.

12.2 Nanobiosensors for Global Pandemics

Nanobiosensors are a significant asset to the management of global pandemics because of their ability to facilitate early detection, diagnosis, monitoring, and containment of pandemics. These sensors make it possible to quickly identify infections, conduct testing at the POC, do multiplexed detection,

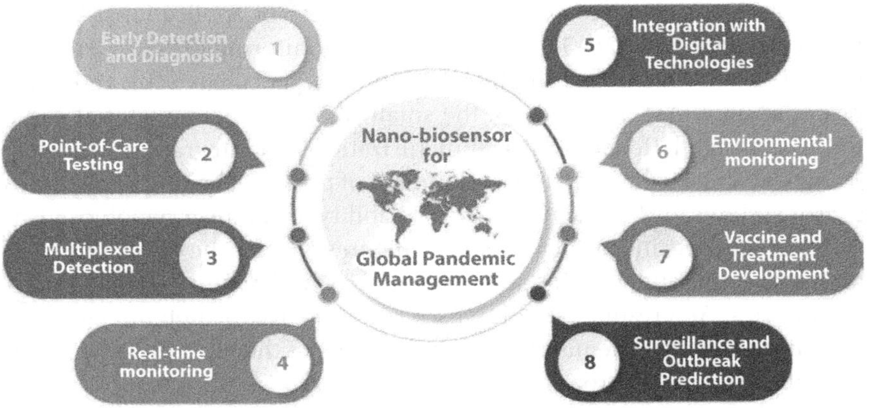

Figure 12.1 Multiple strategies to manage the global pandemics by utilizing nanobiosensors as analytical tools.

carry out real-time monitoring, monitor the environment, and integrate with digital technology. In addition, nanobiosensors contribute to the research and development of treatments and vaccines and the monitoring and forecasting of outbreaks. It is vital to remember that although they play a key part in pandemic response, they are just one component of a holistic plan that also involves effective public health measures, international coordination, and continuous research into prevention and treatment. The multiple strategies involved in managing global pandemics by utilizing nanobiosensors as an analytical tool have been listed in Figure 12.1.

12.2.1 Rapid Pathogen Detection and Identification

Developing nanobiosensors with high sensitivity and specificity for rapidly detecting and identifying pathogens is of utmost importance in effectively managing worldwide pandemics. The timely identification of disease allows for timely intervention, resulting in better patient outcomes and decreased rates of disease transmission. Nanobiosensors can detect pathogens at low concentrations, which presents a potential advantage over conventional techniques that may necessitate higher pathogen loads. To attain optimal sensitivity and specificity, it is common practice for researchers to modify nanobiosensors' surfaces with biomolecules, such as antibodies, aptamers, or molecularly imprinted polymers, which exhibit selective binding to target analytes. Nanomaterials such as gold nanoparticles, carbon nanotubes, and quantum dots can significantly increase sensitivity through signal amplification following analyte binding. Optimizing the transduction mechanism is crucial in achieving rapid pathogen detection as it enables the conversion of the biomolecular recognition event into a measurable signal. Electrochemical, optical, and surface plasmon resonance-based transduction mechanisms provides rapid response times and high signal-to-noise ratios. The expeditious identification methods hold significant importance during pandemics. The promptly identifying infected individuals enables timely quarantine, treatment, and contact tracing, which can effectively curb the spread of the infection. Furthermore, identifying and tracking nascent mutations enables policymakers to make knowledgeable determinations regarding public health interventions, such as limitations on travel and immunization initiatives.

12.2.2 POC Diagnostics

POC testing enables healthcare professionals to swiftly diagnose patients in various settings, including hospitals, clinics, and residences, without requiring specialized laboratory equipment or personnel. POC nanobiosensors can potentially mitigate the strain on healthcare systems and reduce the transmission of infectious diseases by enabling timely diagnosis and treatment. The devices frequently exhibit uncomplicated designs, minimal sample processing requirements, and intuitive interfaces, rendering them easily accessible to a diverse spectrum of users. The latest progressions in microfluidics, lab-on-a-chip, and paper-based devices have facilitated the reduction in size of nano-biosensors, thereby augmenting their mobility and convenience. Integrating sample processing techniques, such as cell lysis or nucleic acid amplification, can optimize the diagnostic workflow and significantly decrease the time required to obtain results. Nanobiosensors with POC have the potential to provide significant value in resource-constrained environments where conventional laboratory infrastructure may not be readily available. The timely and precise diagnoses offered by these tools can aid in treatment decision-making and promote efficient disease management at the community level. Moreover, the extensive implementation of POC diagnostics in pandemics can facilitate the screening of a vast population and provide valuable insights into focused containment strategies.

12.2.3 Multiplexed Detection

Nanobiosensors with multiplexed detection can identify numerous analytes concurrently, providing a robust approach for comprehensive screening in pandemics. It is advantageous in identifying co-infections, monitoring the spread of various diseases or variants, and assessing the effectiveness of vaccines against diverse strains. In order to attain multiplexed detection, it is possible to devise nanobiosensors that incorporate arrays of recognition elements, with each element being tailored to target a specific analyte. The arrays can comprise diverse biomolecules such as antibodies, aptamers, or other relevant biomolecules customized to target specific pathogens. Using nanomaterials or enzymatic reactions as signal amplification strategies can significantly improve the sensitivity and specificity of multiplexed detection. Diverse transduction mechanisms, including electrochemical, optical, and piezoelectric, can transform biomolecular recognition occurrences into quantifiable signals. Through the differentiation of signals attributed to individual analytes, the nanobiosensor can furnish a comprehensive depiction of the infection status. The utilization of multiplexed nanobiosensors can potentially optimize the diagnostic process by reducing the number of required tests, conservation of time, resources, and sample volume. Moreover, the capacity to identify numerous pathogens concurrently amplifies the comprehension of disease dynamics, such as the interplay between co-infections and their influence on patient outcomes. In addition, multiplexed detection techniques can aid in the surveillance of newly arising viral strains and provide valuable insights for creating focused interventions, such as antiviral therapies or vaccines.

12.2.4 Real-Time Monitoring

Monitoring infection status in real-time facilitates healthcare practitioners' customization of treatment approaches and provides policymakers with valuable insights for informed decision-making regarding public health measures. Nanobiosensors have the potential to aid in disease monitoring and treatment evaluation by offering uninterrupted, instantaneous information on the existence, density, or biomarkers of pathogens. Advancements in wireless communication and data processing technologies can further enhance the potential of real-time monitoring. Incorporating nanobiosensors with smartphones, wearable devices, or Internet of Things (IoT) platforms facilitates the retrieval and examination of data from a distance, enabling healthcare providers and authorities to continuously monitor patient populations or geographical regions. The real-time data can also facilitate the execution and assessment of public health interventions, such as quarantines, lockdowns, and resource distribution. Through the surveillance of disease patterns and modes of transmission, policymakers can modify their approaches to mitigate the effects of pandemics on healthcare infrastructure and the economy.

12.2.5 Integration With Digital Technologies

Integrating nanobiosensor data with digital technologies, including IoT platforms, wearable devices, and smartphone apps, facilitates large-scale data collection, analysis, and dissemination. Various healthcare systems can potentially enhance pandemic management, facilitate comprehension of disease transmission patterns, and promote better collaboration among healthcare providers and authorities. Mobile applications can improve data visualization and interpretation, enabling individuals to track their infection status and receive tailored health advice. Nanobiosensors with wearable devices enable the monitoring of vital signs and biomarkers, thereby facilitating the provision of real-time health insights and the potential identification of early signs of infection. IoT platforms enable the aggregation of data from numerous nanobiosensors that are strategically deployed in diverse locations. This approach facilitates the creation of a comprehensive depiction of disease prevalence and transmission. Through the analysis of this data, academic researchers and public health authorities can discern patterns, forecast potential epidemics, and create focused interventions to mitigate the effects of pandemics.

12.2.6 Environmental Monitoring

Nanobiosensors in environmental monitoring and surveillance can yield significant findings regarding the prevalence and dispersion of pathogens in densely populated locations such as public spaces and transportation hubs. The collected data can serve as a valuable tool in directing infection control measures and providing insights for public health interventions. Nanobiosensors intended for environmental monitoring necessitate durability, resilience to environmental factors, and the ability to detect pathogens at low concentrations. Integrating nanobiosensors into air or water monitoring systems can furnish uninterrupted information on the levels of pathogens in the environment. Researchers can execute specific cleaning and disinfection procedures by identifying regions with elevated pathogen levels, modifying ventilation systems, or mandating social distancing protocols. In addition, environmental surveillance has the potential to facilitate the monitoring of infectious disease transmission patterns across various geographic locations, thereby facilitating the implementation of timely alert mechanisms and proactive containment strategies.

12.2.7 Vaccine and Treatment Development

Nanobiosensors have the potential to significantly contribute to the development of vaccines and treatments by detecting and analyzing new pathogens, thus expediting the creation of efficient interventions. Nanobiosensors have the potential to assess vaccine efficacy and safety during clinical trials and post-marketing surveillance by detecting and monitoring biomarkers linked to the immune response. Using nanobiosensors can expedite the process of high-throughput screening for potential drug candidates by providing a swift and precise evaluation of the interaction between target proteins and drug molecules. The data provided can serve as a valuable resource for optimizing drug candidates, potentially resulting in reduced time and cost associated with drug development. Furthermore, utilizing nanobiosensors can facilitate the investigation of drug resistance mechanisms, thereby providing a basis for developing innovative approaches to address resistant strains. Nanobiosensors can offer significant contributions to understanding pathogen biology and host-pathogen interactions, which can be utilized to develop precise and efficient therapeutic interventions.

12.2.8 Surveillance and Outbreak Prediction

Collecting and analyzing data from nanobiosensors deployed in diverse locations allows academic researchers and public health authorities to observe patterns in disease prevalence and transmission. The data gathered can be utilized to forecast possible epidemics, enabling preemptive execution of control strategies and allocation of resources. The integration of data obtained from nanobiosensors with other sources, including electronic health records, travel data, and social media, can provide a holistic understanding of disease dynamics. Sophisticated data analysis methodologies such as advanced data analytics, machine learning, and artificial intelligence can be utilized to detect patterns and associations, facilitating the creation of prognostic models for the propagation of illnesses. Through proactive measures to predict and mitigate outbreaks, policymakers can strategically allocate resources and interventions to mitigate the spread of infectious diseases in high-risk regions. This approach can help minimize pandemics' negative impact on healthcare systems and the economy. In addition, using surveillance data can facilitate the creation of focused public health campaigns and educational programs to promote disease prevention and control.

12.3 Nanobiosensors for COVID-19

COVID-19, caused by the SARS-CoV-2 virus, which falls under the β-coronavirus category, is currently recognized as one of the most contagious and lethal illnesses. Individuals infected with a particular pathogen may exhibit either asymptomatic or symptomatic presentations yet still possess the ability to transmit the pathogen to uninfected individuals. Swift and extensive diagnostic testing are crucial for virus diagnosis and surveillance on a large scale. Healthcare practitioners currently employ various diagnostic tools, such as reverse transcriptase-PCR (RT-PCR), CT imaging, chest X-ray radiography, serological immunoassay, and ELISA, to detect COVID-19. Traditional diagnostic methods necessitate skilled technicians and advanced laboratory infrastructure. The complex laboratory configuration and time lag between sample collection, diagnosis, and the decline in viral load due to the body's natural immune response have motivated researchers and technologists to expedite the development of dependable diagnostic techniques for the large-scale screening of COVID-19. Using nanobiosensors for virus identification and disease confirmation presents several advantages over traditional diagnostic methods, including enhanced sensitivity, faster response times, increased reliability, and improved portability. Nanobiosensors have the potential to detect samples as low as $10^{-18\,M}$, exhibiting high sensitivity. They offer a promising solution to address the limitations of false negatives and positives that are associated with RT-PCR tests. Additionally, these devices are convenient and can be easily utilized in POC centers.

Various categories of biological molecules such as proteins, antibodies, aptamers, enzymes, nucleic acids, and cells are commonly immobilized by means of chemisorption and/or physisorption on the transducer surface. The study quantifies the biochemical interactions between antibody, antigen, and protein, as well as the nucleic acid hybridization that occurs on the transducer's surface. These interactions are measured through various methods, including electrochemical, optical, piezoelectric, magnetic, mechanical, and thermal changes. Nanoparticles is a common strategy in biosensor research to improve biomarker detection and identification sensitivity in various samples. Nanobiosensors have been instrumental in the development of diagnostic and detection platforms due to their exceptional electronic and optical properties, high chemical reactivity, superior stability, increased surface area, effective adsorption, robustness, biocompatibility, and ability to facilitate quick and highly sensitive diagnosis with minimal sample volume (such as patient's nose swab, throat swab, saliva sputum, and blood). These features are particularly crucial for COVID-19 detection. Various nanomaterials, including lanthanide-doped polystyrene nanoparticles, gold-based nanostructures, iron oxide, magnetic particles, carbon black, and graphene nanoparticles, have been employed in creating electrochemical, optical, and magnetic nanobiosensors for COVID-19 prognosis [13]. The advantages and disadvantages of different diagnostic methods have been summarized in Table 12.1. For last several years, there has been tremendous development in the sector of nanobiosensor fabrication for detecting different virus strains. Figure 12.2 illustrates the chronological development of nanobiosensors for detecting and diagnosis of global pandemic caused by different types of viruses.

Nanobiosensors for detecting SARS-CoV-2 (COVID-19) can be classified into various categories based on their working principles and mechanisms. These categories include electrochemical, optical, magnetic, aptameric, field effect transistor (FET), and plasmonic biosensors. Recently, there has been a growing interest in utilizing transistor-based tools as efficient platforms for quickly and economically detecting diverse species. Organic transistors have been identified as having potential applications in detecting COVID-19 due to their ability to conjugate with biological systems. Optical nanobiosensors utilize techniques such as fluorescence, luminescence, surface plasmon, plasmon-enhanced fluorescence, surface-enhancing Raman scattering, colorimetry, reflectance, and absorbance to investigate alterations resulting from the interplay between bioreceptor and analyte. The aforementioned factors indicate the target substance's levels and the nature of the utilized nanoscale materials. One potential benefit of utilizing a magnetic nanobiosensor as opposed to an electrochemical or optical nanobiosensor is the potential for reduced background noise, which may be attributed to the non-magnetic properties of the biological environment. The incorporation of aptamers has the potential to improve both the sensitivity and selectivity of nanobiosensors. An aptamer is a brief oligonucleotide strand of 25–90 bases. The significant role played by the three-dimensional (3D) spatial arrangements of aptamers in detecting viral components is attributed to their high combining ability. The aptamer-conjugated nanobiosensors present significant benefits in the detection of COVID-19. Aptameric nanobiosensors can conduct instantaneous

Table 12.1 The advantages and disadvantages of different diagnostic methods applied in spotting SARS-CoV-2 and COVID-19. The table is reproduced with the permission from Reference [13].

Sl#	Test name	Advantages	Disadvantages
1.	Molecular test: nucleic acid amplification test (NAAT), reverse transcriptase PCR (RT-PCR), loop-mediated isothermal amplification (LAMP)	Gold standard diagnosis test, highly accurate with greater specificity and sensitivity	Require sophisticated sample collection, preparation, and storage, slow (3 h–6 h), complicated, expensive, false negative result.
2.	Antigen test	Fast (15–30 min), inexpensive, simple, offers quick detection and screening.	Potential for false positives, average sensitivity, poor accuracy, qualitative test only
3.	Antibody test: ELISA	Simple, specific, sensitive, efficient, safe, and eco-friendly	Labor-intensive, expensive antibody and culture media required, false negative result, difficulties in transport and storage
4.	Nanobiosensors	High sensitivity and selectivity, fast, suitable for automation, label free detection, better stability and reproducibility, low detection limit, wide range of response limit.	Susceptible to sample matrix, inferior shelf life, need extensive research for commercialization and mass production.

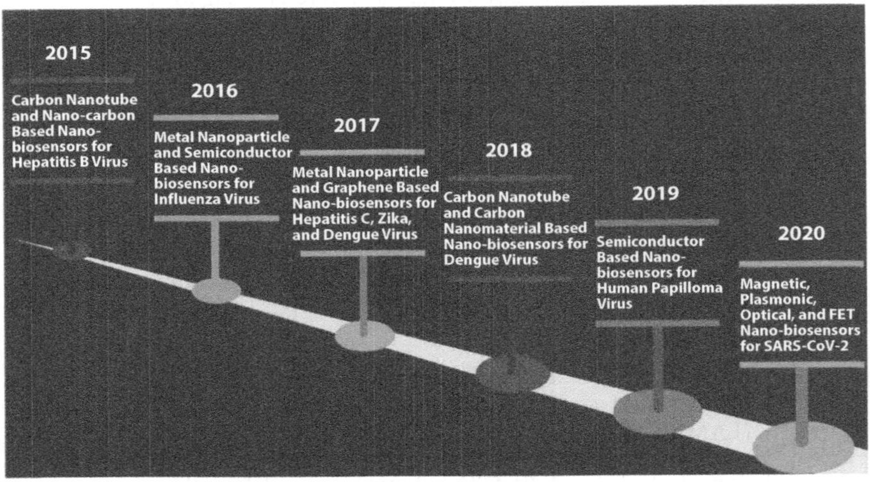

Figure 12.2 Chronological development of nanobiosensors for detecting and diagnosing global pandemic caused by different virus types.

analysis and exhibit multi-target responsiveness. Certain deoxyribonucleic acid (DNA) aptamers, including MSA1 and MSA5, exhibit a notable binding affinity, typically within the few nM range, towards the S1 subunit of the spike protein found in SARS-CoV-2. Plasmons arise from the interaction between electromagnetic radiation and the free electrons on the surface of metal nanoparticles. The plasmonic metallic film comprises a layer of metal nanoparticles measuring 50 nm in diameter, which have been coated onto a prism. The specimen utilized in commercially available plasmonic sensors. The judicious utilization of plasmonic nanobiosensors that are readily available in the market could potentially yield favorable outcomes in terms of instantaneous and unmarked identification, as well as improved consistency and recyclability of sensor chips. Recently developed nanobiosensors, employed for the detection of SARS-CoV-2 and diagnosis of COVID-19, have been summarized in Table 12.2.

12.4 Nanobiosensors for Influenza

The potential of nanobiosensors for detecting and managing influenza H1N1 (swine flu) and H5N1 (avian flu) is significant, given the substantial outbreaks and public health concerns caused by these two strains the influenza virus. Employing nanobiosensors presents many benefits for precise

Table 12.2 The summary of recently developed nanobiosensors, employed for the detection of SARS-CoV-2 and diagnosis of COVID-19.

Sl#	Type	Nanomaterial	Supporting	Target	Detection limit	Ref
1	Electrochemical	AuNPs	WO_3	Virus ribonucleic acid (RNA)	298 fM	[42]
2	FET	Reduced graphene oxide (rGO)	Cu NP, Ag NP	Spike protein		[43]
3	Electrochemical			Spike protein	0.4 BAU/mL	[44]
4	Electrochemical	Carbon black (CB) and AuNPs	Polyethylene terephthalate (PET)	Spike S1 antibody (anti-SP Ab)	101 fg mL^{-1}	[45]
5	Plasmonic	AuNPs	Ta_2C-MXene	N and OFR1a	0.2 pg/mL	[46]
6	Aptameric	AuNPs		Angiotensin-converting enzyme-2 (ACE-2)	----	[47]
7	Optical	AuNPs	Ti_3C_2Tx/Ru (II)	SARS-CoV-2 RdRp gene	7.39 aM	[48]
8	Aptamer-fluorescent	Designer DNA nanostructures (DDNs)	DNA aptamers	Virion particles	3.92x10 Viral-genome-copies/mL	[49]
9	Aptameric	AgNPs	DNA aptamers	Spike protein	100 fg/mL (240 aM)	[50]
10	Optical	SiO_2 NPs	DNA coating	SARS-CoV-2 RNA	10 aM	[51]
11	Nanoplasmonic	Au nanograting		COVID-19-specific receptor binding domain (RBD) protein	0.50 ng/mL to 50 μg/mL	[52]
12	Electrochemical	Antifouling peptides	Peptide nucleic acid (PNA)	viral RNA	0.38 fM	[53]

diagnosis, continuous monitoring, and effective disease control. They can detect distinct viral components or biomarkers linked to H1N1 and H5N1 infections, including viral antigens, nucleic acids, and antibodies. The sensors exhibit a remarkable sensitivity and specificity, facilitating the swift identification of viral strains even when present in low concentrations. This capability enables timely diagnosis and intervention by medical professionals. In addition, nanobiosensors can distinguish between different subtypes of influenza viruses, thereby enabling the implementation of focused therapeutic approaches.

The utilization of nanobiosensors in managing H1N1 and H5N1 infections presents a noteworthy benefit in terms of POC testing. Nanobiosensor devices that are portable and user-friendly have the potential to be utilized in a wide range of settings, including hospitals, clinics, airports, and residential environments. The implementation of rapid, on-site diagnosis can effectively reduce the duration of laboratory testing, thereby accelerating treatment and alleviating the strain on healthcare systems. Furthermore, the utilization of POC testing facilitates the examination of populations with elevated risk factors, such as individuals who travel frequently or those employed in the poultry industry, to promptly detect and isolate infected persons. The detection of multiple analytes or pathogens presents a valuable approach for conducting comprehensive screening of viral strains such as H1N1, H5N1, and others. The utilization of multiplexed detection has the potential to identify co-infections, track the spread of multiple diseases during an outbreak, and monitor emerging variants. The data obtained from this study can provide valuable insights for public health interventions and aid in creating specific vaccines or antiviral therapies.

Utilizing nanobiosensors for real-time monitoring of infection status can prove to be highly beneficial in effectively managing H1N1 and H5N1 outbreaks. The acquisition of continuous and real-time data pertaining to the presence, concentration, or biomarkers of a virus can aid in monitoring the advancement of a disease and evaluating the effectiveness of treatment. Nanobiosensors with wireless communication technologies and data processing platforms facilitate the remote retrieval and analysis of real-time data. This feature enables healthcare providers and authorities to monitor patient populations or geographical regions continuously. Using nanobiosensors for environmental monitoring and surveillance can offer significant contributions in determining the prevalence and dispersion of H1N1 and H5N1 viruses in densely populated locations such as public spaces and transportation hubs. Collecting continuous data on pathogen levels in the environment can provide valuable guidance for infection control efforts.

This data can also inform public health interventions aimed at mitigating disease transmission and enable the implementation of early warning systems for proactive containment measures. Recently developed nanobiosensors, employed for detecting influenza H1N1 (swine flu) and H5N1 (avian flu), have been summarized in Table 12.3.

12.5 Nanobiosensors for MERS

The MERS is a viral respiratory illness first reported in Saudi Arabia in 2012. It is caused by a coronavirus called MERS-CoV. The virus is primarily transmitted from animals to humans, but human-to-human transmission can also occur. Symptoms of MERS include fever, cough, and shortness of breath, and the illness can progress to pneumonia and kidney failure. There is currently no specific treatment or vaccine for MERS, and the mortality rate is estimated to be around 35%. The MERS-CoV, also known as the corona virus, is classified as a highly pathogenic virus. The incidence of severe illnesses and human fatalities in Saudi Arabia and globally is significant. Respiratory infections in humans attributed to corona viruses (CoV) have been documented since the 1960s, with HCoV-229E and HCoV-OC43 being among the identified strains. Subsequently, additional strains of CoV have been identified, including SARS-CoV, HCoV NL63, and HCoV HKU1. The initial documentation of severe respiratory illness in humans caused by MERS-CoV was recorded in June 2012 at a hospital in Jeddah, Saudi Arabia. In a recent development, a new case of acute kidney injury and acute respiratory syndrome has been reported in Qatar within a brief timeframe. In April of 2012, a pair of patients were identified as having contracted MERS-CoV within Jordan. Since the identification of MERS-CoV, it has been reported to have spread to 23 countries. As of April 27, 2017, the World Health Organization (WHO) has reported 1936 cases of MERS-CoV, resulting in 690 fatalities and a case fatality rate of 35.6%. In May and June 2015, there was an outbreak of MERS-CoV in South Korea, which resulted in 180 confirmed cases of infection and 36 fatalities. The incubation period of MERS-CoV exhibits a range of 2–14 d. Individuals in the age groups of children, young adults, and pregnant women are more vulnerable to MERS-CoV. The clustering studies provide compelling evidence for human-to-human transmission of the virus. The transmission of the pathogen in question has the potential to result in various health outcomes ranging from mild illness to severe pneumonia, respiratory syndrome, and multi-organ failure. Additional clinical manifestations

Table 12.3 The summary of recently developed nanobiosensors, employed for detecting influenza H1N1 (swine flu) and H5N1 (avian flu).

Sl#	Type	Nanomaterial	Supporting materials	Target	Diagnosis	Detection limit	Ref
1	Conductive thread-based			HA protein	H1N1		[54]
2	Electrochemical	Multi-walled carbon nanotube (MWCNT)	Au-Pt hybrid nanomaterial	Anti-hemagglutinin (anti-H1)	H1N1	3.54 μg/mL	[55]
3	Electrochemical	AuNPs	Ag/AgCl Carbon Paper (CP) and Pt	RNA of influenza H1N1 virus	H1N1	0.25 pg/mL	[56]
4	Colorimetric	Magnetic beads		RNA of influenza H1N1 virus	H1N1	0.152 pM	[57]
5	Electrochemical	AuNPs	Micro islands	Spike protein	H1N1 and SARS- CoV-2	5.89 pg/mL	[58]
			(NMIs) sensor				
6	Electrochemical	Nickel oxide (NiO)-reduced graphene oxide (rGO)	MXene nanocomposite	H1N1 (QMGFMTSPKHSV) and H5N1 (GHPHYNNPSLQL)	H1N1 and H5N1	3.63 nM of H1N1 and 2.39 nM for H5N1	[59]
7	Optical	Ag nanoclusters (Ag NCs)	Graphene oxide (GO)	H1N1 and H5N1 (tDNA1 and tDNA2)	H1N1 and H5N1	0.25 HAU	[60]
8	Electrochemical	AgNPs		HA protein	H5N1	3.26 nM	[61]
9	Aptameric	Au-IDA microelectrode	H5N1 aptamer	H5N1	H5N1	0.25 HAU	[62]
10	Genosensors	Cysteine modified Au- SPE	ss-DNA H1N1	ss-cDNA H1N1 (HA)	H1N1	0.667 ng/mL	[63]

associated with MERS-CoV infections comprise anorexia, fever, nausea, vomiting, diarrhea, abdominal pain, and disseminated intercellular coagulation. Confirming MERS-CoV cases necessitates the involvement of the WHO and the Ministry of Health. The recommended laboratory tests in Saudi Arabia should be conducted accordingly. According to the WHO, real-time reverse transcription PCR (rRT-PCR) is a recommended method. Subjecting the suspected cases to screening via two quantitative real-time reverse-transcription PCR (qRT-PCR) tests targeting two distinct genome regions is imperative. Furthermore, it is essential to reconfirm the results obtained through the aforementioned tests by conducting one qRT-PCR test targeting the positive genome. Numerous alternative methods for detecting viruses, including immunofluorescence assay, protein microarray assay, reverse transcription loop-mediated isothermal amplification assay (RT-LAMP), viral plaque assay, hemagglutination assay, viral flow cytometry (FCM), and ELISA, are also viable options. Although these methods exhibit high sensitivity, their applicability for large-scale screening of multiple samples is limited due to their high cost and lengthy analysis time. Furthermore, these methodologies require personnel with specialized expertise to execute and are inappropriate for POC diagnostics. Several nanobiosensors have been documented in academic literature to detect viruses through the utilization of diverse transducers, which are considered superior alternatives to conventional assays. Table 12.4 presents a summary of the newly developed nanobiosensors that have been utilized for the detection of MERS-CoV.

12.6 Nanobiosensors for HIV/AIDS

The first report of AIDS was made by the US Center for Diseases Control in 1981. In 1983, the identification of HIV as the causative agent of the disease was established. The prevalence of AIDS, being one of the significant obstacles of the current century, is increasing at a rapid pace. Based on the statistical data gathered, it can be inferred that there existed an estimated 36.9 million individuals who were afflicted with HIV by the end of the year 2017. Furthermore, the worldwide prevalence of HIV among adults was calculated to be 0.8%. The AIDS syndrome has been acknowledged as a significant peril to humanity. Thus, the timely detection of HIV infection holds significant value for researchers globally. There exists a significant need to rapidly, specifically, and precisely detect HIV at present. The ELISA method is commonly utilized for the diagnosis of HIV infection by detecting the presence or absence of HIV antibodies, including IgM/IgG Abs.

Table 12.4 The summary of recently developed nanobiosensors, employed for the detection of MERS-CoV.

Sl#	Type	Nanomaterial	Supporting materials	Target	Diagnosis	Detection limit	Ref
1	Electrochemical	AuNPs	Carbon NP	S protein	MERS-CoV	0.001ng/mL	[64]
					HCoV	0.01ng/mL	
2	Electrochemical	Carbon nanofiber	Cotton fiber	S protein	MERS-CoV	0.07 pg/mL	[65]
3	Aptameric	GO-MoS2	MF DNA aptamer	MERS-NV	MERS- CoV-2	0.4049 pg/mL	[66]
4	Optical	AgNPs	Paper-based colorimetric DNA sensor	PNA probes	MERS-CoV	1.53 nM	[67]
5	Plasmonic	AgNPs	Dye molecules	E-protein and open reading frames (ORF) gene	MERS-CoV	6×10^{11} copies/μL (1 pmol/μL)	[68]

The present study employs a diagnostic approach that involves the examination of blood or saliva samples to detect the presence of virus-specific antibodies. Since the introduction of the HIV antibody test in Canada in 1985, HIV testing has posed a significant policy challenge for health authorities. The test under consideration exhibits high sensitivity; however, it necessitates a specific duration of time, commonly referred to as the "window period," following the contraction of the HIV. The duration of the window period following infection can range from several weeks to several months. It is worth noting that the human immune system requires a certain amount of time to generate antibodies to HIV, and this process does not occur immediately following infection. Furthermore, the ELISA entails multiple incubation and washing procedures, succeeded by spectrophotometric identification utilizing a chromogenic substrate. It is widely acknowledged that the promotion and implementation of HIV testing methods that can detect the virus during the early stages of infection are imperative.

Over three decades since the onset of the HIV/AIDS epidemic, numerous scholars have proposed the utilization of ultrasensitive biosensing techniques to rapidly detect $CD4^+$ cells, HIV genes, and HIV biomarkers. In contrast to immunoassays that require a prolonged period for antibody generation, the identification of HIV-related proteins or genes offers an alternative approach for early detection of the infection. Additionally, biosensing strategies exhibit none of the drawbacks associated with HIV detection methods that rely on antibodies, such as their high cost and complexity. In the past few years, a number of nanobiosensors have been developed with the aim of detecting HIV infection. While biosensors may present themselves as viable substitutes for ELISA kits, a number of them are subject to certain constraints, such as diminished sensitivity. Moreover, the development of multiplexed nanobiosensors for the concurrent detection of diverse analytes has continued to pose a significant obstacle in the bioanalytical domain [69–72]. Table 12.5 presents a summary of the newly developed nanobiosensors that have been utilized for the detection of HIV.

12.7 Nanobiosensors for Other Human Viruses

Nanobiosensors are also effective for the detection of multiple human diseases causing viruses such as hepatitis, Ebola, Zika, and dengue. Table 12.6 presents a summary of the newly developed nanobiosensors that have been utilized for the detection of hepatitis, Ebola, Zika, and dengue.

Table 12.5 The summary of recently developed nanobiosensors, employed for the detection of HIV.

Sl#	Type	Nanomaterial	Supporting materials	Target	Diagnosis	Detection limit	Ref
1	Electrochemical	CVD-grown graphene sheets		HIV-2 Ab	HIV	12 pM	[73]
2	Electrochemical	AuNP	SWCNT	HIV-1 Protease	HIV	0.8 pM	[74]
3	Electrochemical	AuNP	Ferrocene- labeled lipoic acid (Fc-LA)	HIV-1 reverse transcriptase	HIV type I and II	0.7 fM	[75]
4	Electrochemical	MWCNTs	Molecularly imprinted polymers (MIPs)	HIV-p24	HIV	0.083 pg/mL	[76]
5	Electrochemical	AuNP	Electrochemically reduced graphene oxide (ERGO)	HIV-1 gene	HIV	37 aM	[77]
6	Optical	Immunochromatographic strips (ICS)	ICS-based sensor	$CD4^+$ cells	HIV	34 cells/μL	[78]
7	Optical	Graphene oxide (GO)	Peptide molecules	HIV-1 protease	HIV type I and II	1.18 ng/mL	[79]
8	Optical	Silver nanoclusters (AgNCs)	G-rich DNA_1 G-rich DNA_2	Gene	HIV type I and II	3.26 nM	[80]

12.8 Selection and Optimization of Nanomaterials for Nanobiosensors

Previous studies have shown that carbon nanoparticles pose no significant environmental risks when present in lower concentrations. Carbon-based nanomaterials exhibit low toxicity towards microorganisms and relatively low toxicity towards aquatic organisms such as algae, fish, and other arthropods. Carbon nanotubes and graphene are promising carbonaceous materials that are sustainable, cost-effective, and environmentally friendly [92]. These nanoparticles have the potential to be utilized in the design and development of nanobiosensors. Various categories of nanomaterials utilized in the production of nanobiosensors are synthesized through diverse techniques, including top-down and bottom-up methodologies. Carbon nanomaterials such as nanotubes, nanofibers, and graphenes with diameters ranging from 0.4 to 500 nm are synthesized through the chemical vapor deposition (CVD) method. This involves the decomposition of gaseous hydrocarbon over transition metal-catalyst particles [93]. Alternative synthetic methods, such as arc discharge, flame synthesis, and laser ablation, may not be optimal for large-scale production due to their high cost, energy consumption, and lower efficiency. Numerous corporations worldwide are engaged in the production of carbon nanoparticles. Pyrograf Products, Inc. (USA) is a widely recognized supplier. The cost of the material fluctuates depending on its intended use, with prices ranging from $85 per pound for electrical applications to $170 per pound for thermal applications. Several well-known manufacturers of carbon nanomaterials, such as Carbon NT&F 21, Catalyx Nanotech, Inc., Catalytic Materials LLC, and CoMoCat®, offer a range of prices for their products based on factors such as quality and purity. The utilization of particular catalysts, such as metals, has the potential to modify the chemical, physical, optical, electronic, and magnetic characteristics of nanomaterials in a direct manner.

Inorganic nanomaterials, including metal nanoparticles such as gold, silver, platinum, and palladium, as well as metal/carbon nanotube hybrids and metal oxides like TiO_2, ZnO, ZrO_2, CuO, Al_2O_3, Fe_2O_3, and Fe_3O_4, exhibit promise for the development of nanobiosensors. The prevalent method utilized for the production of Au NPs is through the reduction of a metal salt in a solution that comprises a stabilizing agent. The solvated metal atom dispersion technique (SMAD) can be utilized for the large-scale production of Au NPs. Spherical Au NPs ranging from 1 to 6 nm in size were synthesized through Au vaporization in the presence of organic solvents within a vacuum chamber. The production of Ag NPs can be achieved through

Table 12.6 The summary of recently developed nanobiosensors, employed for the detection of hepatitis, Ebola, Zika, and dengue.

Sl#	Type	Nanomaterial	Supporting materials	Target	Diagnosis	Detection limit	Ref
1	Electrochemical	AuNPs	CRISPR/Cas1 2a-based viral DNA	HBV DNA	Hepatitis B (HBV)	0.1pM	[81]
2	Aptameric	Graphene		Hepatitis C virus core protein	Hepatitis C (HCV)	15.6 aM	[82]
3	Optical	MnO_2	MF DNA aptamer	HEV DNA	Hepatitis E (HEV)	3.26 fM	[83]
4	Electrochemical			HAV DNA	Hepatitis A (HAV)	0.65 pM	[84]
5	Optical	GO	Magnetic beads	Ebola virus DNA	Ebola	4.89 pM	[85]
6	Optical	Silicon photonic microring		EBOV and SUDV sGP	Ebola	1.00 ng/mL	[86]
7	Electrochemical	AuNPs	ZV aptamer	ZV DNA	Zika virus (ZV)	38.49 pM	[87]
8	Electrochemical	AgNPs	MXene nanoparticle	ZV DNA	ZV	38.14 pM	[88]
9	Electrochemical	Au thin-film	Ni thin-film	DENV DNA	Dengue virus (DENV)	0.12 pM	[89]
10	Electrochemical	Carbon NM		DENV DNA	DENV	0.11 nM	[90]
11	Multiplex electrochemical	AgNPs		Non- structural protein 1 (NS1)	ZV	0.54 ng/mL	[91]
					DENV	1.17 ng/mL	

the reduction of a silver salt utilizing a reducing agent such as β-d-glucose, sodium borohydride (NaBH4), and colloidal stabilizers including polyvinyl alcohol (PVA), bovine serum albumin (BSA), polyvinyl pyrrolidone (PVP), citrate, starch, and cellulose. Platinum nanoparticles (Pt NPs) can be synthesized by reducing potassium hexachloroplatinate with trisodium citrate and sodium dodecyl sulfate stabilizer while subjecting the mixture to rigorous and uninterrupted stirring. Pt nanoflowers were synthesized using a facile and scalable approach involving ethanol reduction at 85°C in the presence of PVP-10 as a capping agent. The precise control of the structure and order of Pt nanoflowers can be achieved by regulating reaction conditions and parameters. The thermal decomposition of a metal acetate solution in a surfactant at 300°C can produce palladium nanoparticles (Pd NPs) bulk. The attainment of monodispersity and particle size of Pd NPs can be accomplished through the modification of the surfactant mole ratio. Using oleylamine as a surfactant enables the synthesis of Pd NPs with an average size exceeding 10 nm. This method exhibits a comparative advantage over alternative techniques, as it allows facile production of high-quality nanoparticles without necessitating additional selective measures. Metal-carbon nanotube hybrids can be synthesized through various methods, including spontaneous reduction, electrochemical deposition, physical evaporation, and solid state reaction with metal salts at high temperatures. Metal oxide nanoparticles can be synthesized through various methods, including vapor-liquid-solid (VLS) phase growth, solution-liquid-solid (SLS) phase growth, and vapor-solid (VS) phase growth processes. These methods can be employed with or without the assistance of a catalyst. Quantum dots (QDs) are monodispersed colloidal semiconducting nanocrystals with diameters ranging from 2-10 nm. They can be synthesized using various methods such as vapor phase reactions, hydrothermal and solvothermal techniques, ultrasound irradiation, reactions in confined solids, nanolithography, and microwave irradiation method.

In order to enhance the efficacy of nanoparticles, it is necessary to implement pre-treatment or post-treatment procedures. The procedures for modification, which encompass metal-oxide infusion, acid/base activations, and steam/CO_2 activation, are subject to certain limitations, such as their complexity and protracted nature, the need for supporting processes, and elevated operational expenses. Various methods, such as physical, chemical, and biological modifications can be utilized to functionalize nanomaterials. Physical modifications in treatments include steam/gas, microwave, and magnetic force, whereas chemical modifications involve the use of H_2O_2, HCl, H_2SO_4, HNO_3, H_3PO_4, KOH, NaOH, impregnation, and coating. The physical and chemical modification techniques exhibit both benefits and drawbacks. The utilization of steam or gas treatment is a viable method for the large-scale production of engineered nanomaterials that exhibit enhanced porosity, surface area, and surface reactivity, making them commercially valuable. Nevertheless, this procedure poses a challenge in terms of regulation. In contrast, microwave technology is characterized by its ease of control, cost-effectiveness, and enhanced functionality in producing nanomaterials with increased surface area and functional groups. However, it is essential to note that this process is energy-intensive. The magnetic separation method is a viable option for eliminating surface contaminants, although it should be noted that the magnetic attraction may affect the particle's surface area. Using H_2O_2 as a treatment method is both cost-effective and environmentally sustainable. This approach results in the production of enhanced nanoparticles with a surface area that contains oxygen. However, it is essential to note that the yield of this process is relatively low. The utilization of acid and alkali treatments produces particles that exhibit an increased presence of carboxyl functional groups on their surface. However, these methods have been found to negatively impact the environment. The enhancement of surface area, anion–cation exchange capacity, negative surface charge density, and hydrophobicity of nanomaterials can be effectively achieved through biological modifications such as bacterial conversion and anaerobic digestion.

Modernizing conventional waste management strategies is necessary to investigate the field of nano-ecotoxicology. The rapid increase in demand for nanomaterials necessitates large-scale industrial production. From this standpoint, it is imperative to analyze the advantages and disadvantages of laboratory and industrial synthetic processes to determine the most cost-effective and environmentally sustainable technique for the large-scale production of nanomaterials. Implementing environmentally conscious operating conditions, such as utilizing an optimum process temperature, solvent-free or environmentally benign solvents, and auxiliaries, can contribute to the overall sustainability of nanomaterials used to develop nanobiosensors. Additionally, utilizing non-toxic catalysts and safer chemicals, promoting a higher atom economy, utilizing renewable feedstock, and implementing energy-efficient operating conditions can further enhance the sustainability of the manufacturing process. The selection and design of biodegradable raw materials and finished products, as well as real-time analysis for pollution prevention, can also contribute to the overall sustainability of the process. Furthermore, the selection and utilization of inherently safer chemicals can aid in developing a sustainable technique for the selection and optimization of nanomaterials for nanobiosensor development.

12.9 Current Challenges and Prospective Solutions

Similar to other technological fields, nanotechnology encompasses various technical disciplines such as physics, chemistry, biology, material science, electronics, and self-assembly. However, the use of nanomaterials or nanoparticles in the creation and advancement of nanotechnology products is associated with certain limitations. The unique physicochemical properties of nanoparticles give rise to distinct toxicological and environmental hazards that differ from those of their bulk counterparts. Furthermore, the rate of decomposition of nanoparticles is relatively slow, which increases the likelihood of their accumulation at the site of administration. Nanoparticles are subject to various aging mechanisms within both environmental and biological contexts, ultimately resulting in adverse ecological and physiological effects. The challenge may be tackled by utilizing biodegradable and safer nanoparticles. These may include clay, cellulose nanomaterials, halloysite nanotubes, starch nanocrystals, nanocarbon, and carbonaceous nanomaterials.

The limited selectivity and specificity of nanobiosensors pose a significant obstacle in diagnosing viral diseases. This can lead to errors in the detection process of samples, particularly in cases where there are mutations in virus proteins and mRNA. The simultaneous utilization of multiple diagnostic factors, including ORF1ab and RNA-dependent RNA polymerase (RdRp) genes, is a common practice. Based on the available reports, utilizing multiple diagnostic factors, including ORF 1a/b, RdRp genes, and N- and S-proteins, can significantly improve the specificity and selectivity of sensors. The provenance of the samples under analysis can potentially introduce uncertainties in the sensitivity and specificity of the nanobiosensors utilized for disease detection.

The lack of a suitable platform for virus detection poses a significant obstacle in creating and advancing nanobiosensors. The advancements in smart phone technology have the potential to provide a cost-effective means of swiftly detecting viruses. Developing efficient nanobiosensors is of utmost importance in light of the global prevalence of viral diseases. These sensors, similar to blood sugar sensors, must be designed and fabricated with enhanced accuracy and performance. However, the production of nanobiosensors for the detection of viral diseases involves multiplexing and intricate sample preparation procedures. The development and widespread availability of a nanobiosensor with high sensitivity and specificity pose a significant challenge in manufacturing and distribution. A comprehensive comprehension of the molecular mechanism and interactions among the constituents of a nanobiosensor, coupled with a lucid understanding of their molecular structures, would represent a significant advancement in tackling this fundamental obstacle. The potential utilization of machine learning algorithms and artificial intelligence in the future could improve the precision and accuracy of nanobiosensors for detecting viral diseases. Implementing extensive clinical trials and intricate sample processing methodologies is imperative in comprehending sensors' stability and viability for commercial purposes.

12.10 Conclusion

The implementation of nanobiosensors has the potential to significantly transform the diagnostic field through the provision of prompt, precise, and high-sensitivity detection of pathogens, even at low levels. The ability to promptly initiate medical interventions, lower disease transmission rates, and enhance patient outcomes is paramount. The creation of compact and easily operable nanobiosensors enables POC examination in various environments, reducing the strain on medical infrastructures and contributing to the containment of contagious illnesses. Incorporating multiplexed detection in nanobiosensors is a crucial aspect that facilitates extensive screening for diverse pathogens. This feature also allows for detecting co-infections, tracking multiple diseases during a pandemic, and monitoring emerging variants. The utilization of nanobiosensors for real-time monitoring of infection status is a valuable tool for healthcare professionals. This technology can guide treatment strategies and inform public health measures, including quarantines, lockdowns, and resource allocation.

Using nanobiosensors for environmental monitoring can substantially contribute to the prevention of infections by identifying pathogens in densely populated areas such as public spaces and transportation hubs. The data obtained from this study inform public health interventions aimed at reducing disease transmission rates and facilitates the creation of early warning systems for proactive containment measures. The amalgamation of nanobiosensors and digital technologies, including smartphone applications, wearable gadgets, and IoT platforms, can potentially enhance pandemic management efficacy. The implementation of extensive data gathering, examination, and dissemination facilitates enhanced collaboration among healthcare professionals, scholars, and governing bodies, culminating in a more efficient reaction to pandemics. Furthermore, nanobiosensors have the potential to greatly assist in the advancement of vaccines and treatments by expediting the detection and analysis of emerging pathogens. The sensors can also monitor the effectiveness and safety of interventions during clinical trials and post-marketing surveillance.

The utilization of nanobiosensors holds great promise in revolutionizing the management of pandemics. This is due to their ability to offer prompt, precise, and up-to-the-minute information that can aid in the early detection, diagnosis, monitoring, and containment of such outbreaks. Acknowledging that these sensors constitute a solitary constituent of a more extensive strategy for responding to pandemics is imperative. The successful management of pandemics necessitates a comprehensive strategy that integrates technological advancements such as nanobiosensors with strong public health interventions, global cooperation, and continuous investigation into preventing and curing infectious illnesses. The utilization of nanobiosensors in these domains can enhance the durability and agility of the worldwide healthcare system, thereby enabling it to effectively address the difficulties posed by pandemics in the times ahead.

References

1. Harun-Ur-Rashid, M. and Imran, A.B., Superabsorbent Hydrogels from Carboxymethyl Cellulose, in: *Carboxymethyl Cellulose*, M.I.H. Mondal (Ed.), p. 159, Nova Science Publisher, Inc., New York, U.S.A., 2019.
2. Harun-Ur-Rashid, M., Seki, T., Takeoka, Y., Structural colored gels for tunable soft photonic crystals. *Chem. Rec.*, 9, 2, 87–105, 2009.
3. Karim, M.R., Harun-Ur-Rashid, M., Imran, A.B., Effect of sizes of vinyl modified narrow-dispersed silica cross-linker on the mechanical properties of acrylamide based hydrogel. *Sci. Rep.*, 13, 1, 5089, 2023.
4. Rezaul Karim, M., Harun-Ur-Rashid, M., Imran, A.B., Highly Stretchable Hydrogel Using Vinyl Modified Narrow Dispersed Silica Particles as Cross-Linker. *ChemistrySelect*, 5, 34, 10556–10561, 2020.
5. Harun-Ur-Rashid, M., Foyez, T., Imran, A.B., Fabrication of Stretchable Composite Thin Film for Superconductor Applications, in: *Sensors for Stretchable Electronics in Nanotechnology*, K. Pal (Ed.), pp. 63–78, CRC Press, Boca Raton, 2021.
6. Huda, M.N. *et al.*, Characteristics of High-Density Poly (N-isopropylacrylamide)(PNIPA) Brushes on Silicon Surface by Atom Transfer Radical Polymerization. *Trans. Mater. Res. Soc. Jpn.*, 35, 4, 845–848, 2010.
7. Huda, N., Rashid, M., Harun-Ur-Rashid, M., A Review on Stimuli-responsive grafted membranes Based on Facile Synthesis Process and Extensive Applications. *Int. J. Innov. Appl. Stud.*, 8, 3, 1296–1312, 2014.
8. Harun-Ur-Rashid, M. *et al.*, Angle-independent structural color in colloidal amorphous arrays. *ChemPhysChem*, 11, 3, 579–583, 2010.

9. Harun-Ur-Rashid, M. *et al.*, Template synthesis for stimuli-responsive angle independent structural colored smart materials. *Trans. Mater. Res. Soc. Jpn.*, 34, 2, 333–337, 2009.
10. Takeoka, Y. *et al.*, Structurally coloured secondary particles composed of black and white colloidal particles. *Sci. Rep.*, 3, 1, 2371, 2013.
11. Harun-Ur-Rashid, M. and Imran, A.B., Organometallic Nanomaterials Synthesis and Sustainable Green Nanotechnology Applications, in: *Green Nanoarchitectonics*, K. Pal (Ed.), pp. 249–270, Jenny Stanford Publishing, New York, 2022.
12. Imran, A.B., Harun-Ur-Rashid, M., Takeoka, Y., Polyrotaxane Actuators, in: *Soft Actuators: Materials, Modeling, Applications, and Future Perspectives*, K. Asaka and H. Okuzaki (Eds.), pp. 81–147, © Springer Nature: 152 Beach Road, #21-01/04 Gateway East, Singapore 189721, Singapore, 2019.
13. Harun-Ur-Rashid, M. *et al.*, Rapid diagnosis of COVID-19 via nano-biosensor- implemented biomedical utilization: a systematic review. *RSC Adv.*, 12, 15, 9445–9465, 2022.
14. Harun-Ur-Rashid, M. *et al.*, Green Polymer Nanocomposites in Automotive and Packaging Industries. *Current Pharm. Biotechnol.*, 24, 1, 145–163, 2023.
15. Harun-Ur-Rashid, M., Pal, K., Imran, A.B., Hybrid Nanocomposite Fabrication of Nanocatalyst with Enhanced and Stable Photocatalytic Activity. *Top. Catal.*, 67, 17–45, 2024.
16. Harun-Ur-Rashid, M. and Imran, A.B., Nanomaterials in the Automobile Sector, in: *Emerging Applications of Nanomaterials*, N.B. Singh, M.A.B.H. Susan, R.G. Chaudhary (Eds.), pp. 124–150, Materials Research Foundations: 105 Springdale Lane, Millersville, PA 17551, U.S.A., 2023.
17. Harun-Ur-Rashid, M., Imran, A.B., Susan, M., Prospective Nanomaterials for Food Packaging and Safety, in: *Emerging Applications of Nanomaterials, Materials Research Forum LLC*, N.B. Singh, M.A.B.H. Susan, R.G. Chaudhary (Eds.), pp. 327–352, Materials Research Foundations, USA, 2023.
18. Khair, A. *et al.*, Diffusimetry renounces graham's law, achieves diffusive convection, concentration gradient induced diffusion, heat and mass transfer. *Defect Diffus. Forum*, 407, 173–184, 2021.
19. Feng, L. *et al.*, Nano-Biosensors Based on Noble Metal and Semiconductor Materials: Emerging Trends and Future Prospects. *Metals*, 13, 792, 2023, https://doi.org/10.3390/met13040792.
20. Abdelhamied, N. *et al.*, Bacteriophage-based nano-biosensors for the fast impedimetric determination of pathogens in food samples. *Sci. Rep.*, 13, 1, 3498, 2023.
21. Aquino, A. *et al.*, Updating the use of nano-biosensors as promising devices for the diagnosis of coronavirus family members: A systematic review. *J. Pharm. Biomed. Anal.*, 211, 114608, 2022.
22. Ahangari, A., Mahmoodi, P., Mohammadzadeh, A., Advanced nano biosensors for rapid detection of zoonotic bacteria. *Biotechnol. Bioeng.*, 120, 41–56, 2023.

23. Christopher, F.C. *et al.*, Recent advancements in rapid analysis of pesticides using nano biosensors: A present and future perspective. *J. Clean. Prod.*, 269, 122356, 2020.
24. Hassan, R.Y.A., Advances in Electrochemical Nano-Biosensors for Biomedical and Environmental Applications: From Current Work to Future Perspectives. *Sensors*, 22, 19, 7539, 2022, https://doi.org/10.3390/s22197539.
25. Kwakye, S., Goral, V.N., Baeumner, A.J., Electrochemical microfluidic biosensor for nucleic acid detection with integrated minipotentiostat. *Biosens. Bioelectron.*, 21, 12, 2217–2223, 2006.
26. Choi, S. *et al.*, Microfluidic-based biosensors toward point-of-care detection of nucleic acids and proteins. *Microfluid. Nanofluid.*, 10, 2, 231–247, 2011.
27. Huertas, C.S. *et al.*, Advanced Evanescent-Wave Optical Biosensors for the Detection of Nucleic Acids: An Analytic Perspective. *Front. Chem.*, 7, 2019, https://doi.org/10.3389/fchem.2019.00724.
28. Zhang, J. *et al.*, Wearable biosensors for human fatigue diagnosis: A review. *Bioeng. Transl. Med.*, 8, 1, e10318, 2023.
29. T, G.A.D. and Pekyardimci, S., Nanostructured Electrochemical Biosensors for Medical Applications, in: *Emerging Technologies in Biophysical Sciences: A World Scientific Reference*, pp. 157–174, World Scientific, Singapore, 2020.
30. Choi, H.K. and Yoon, J., Nanotechnology-Assisted Biosensors for the Detection of Viral Nucleic Acids: An Overview. *Biosensors*, 13, 2, 208, 2023, doi: DOI: 10.3390/bios13020208.
31. Wanjari, V.P. *et al.*, Laser-induced graphene-based electrochemical biosensors for environmental applications: a perspective. *Environ. Sci. Pollut. Res.*, 30, 15, 42643–42657, 2023.
32. Moorthy, M.S., Nagappan, S., Ha, C.-S., Mesoporous Silica and Organosilica Biosensors for Water Quality and Environmental Monitoring, in: *Surface-Functionalized Ceramics*, pp. 369–406, 2023.
33. Herrera-Domínguez, M. *et al.*, Optical Biosensors and Their Applications for the Detection of Water Pollutants. *Biosensors*, 13, 3, 370, 2023, DOI: 10.3390/bios13030370.
34. Meira, D.I. *et al.*, Recent advances in nanomaterial-based optical biosensors for food safety applications: Ochratoxin-A detection, as case study. *Crit. Rev. Food Sci. Nutr.*, 64, 18, 6318–6360, 2024.
35. Ravindran, N. *et al.*, Recent advances in Surface Plasmon Resonance (SPR) biosensors for food analysis: a review. *Crit. Rev. Food Sci. Nutr.*, 63, 8, 1055–1077, 2023.
36. Mahmood Khan, I. *et al.*, Surface functionalized AuNCs optical biosensor as an emerging food safety indicator: Fundamental mechanism to future prospects. *Coord. Chem. Rev.*, 474, 214842, 2023.
37. Kakkar, S. *et al.*, Progress in Fluorescence Biosensing and Food Safety towards Point-of- Detection (PoD) System. *Biosensors*, 13, 2, 249, 2023, DOI: 10.3390/bios13020249.

38. Wu, S. *et al.*, Chitosan-Based Hydrogels for Bioelectronic Sensing: Recent Advances and Applications in Biomedicine and Food Safety. *Biosensors*, 13, 1, 93, 2023, DOI: 10.3390/bios13010093.
39. Jiao, L. *et al.*, On the Road from Single-Atom Materials to Highly Sensitive Electrochemical Sensing and Biosensing. *Anal. Chem.*, 95, 1, 433–443, 2023.
40. Guo, X. *et al.*, Highly selective fluorescent probe in aqueous solution based on coumarin Schiff base for detecting Cu2+ and specific biosensing of glutathione in mitochondria. *J. Photochem. Photobiol., A*, 435, 114350, 2023.
41. Devraj, L. *et al.*, CHAPTER5 - Nanotechnology-oriented sensors for the quick recognition of foodborne microbes and pathogens, in: *Nanotechnology Applications for Food Safety and Quality Monitoring*, A. Sharma, *et al.* (Eds.), pp. 93–112, Academic Press, Netherlands, 2023.
42. Hussein, H.A. *et al.*, Designing and fabrication of electrochemical nanobiosensor for the fast detection of SARS-CoV-2-RNA. *Sci. Rep.*, 13, 1, 5139, 2023.
43. Alnaji, N., Wasfi, A., Awwad, F., The design of a point of care FET biosensor to detect and screen COVID-19. *Sci. Rep.*, 13, 1, 4485, 2023.
44. Shoute, L.C.T. *et al.*, Label-free impedimetric immunosensor for point-of-care detection of COVID-19 antibodies. *Microsyst. Nanoeng.*, 9, 1, 3, 2023.
45. Ferreira, L.M.C. *et al.*, Using low-cost disposable immunosensor based on flexible PET screen-printed electrode modified with carbon black and gold nanoparticles for sensitive detection of SARS-CoV-2. *Talanta Open*, 7, 100201, 2023.
46. Yang, W. *et al.*, Ultra-sensitive specific detection of nucleic acids in pathogenic infections by Ta2C-MXene sensitization-based ultrafine plasmon spectroscopy combs. *Sens. Actuators, B*, 387, 133785, 2023.
47. Jaisankar, A., Krishnan, S., Ramalingam, M., Design and Development of Nanoscale Aptasensors for Viral Diagnostics. *Lett. Appl. NanoBioScience*, 12, 2, 7, 2022.
48. Wei, J. *et al.*, Entropy-driven assisted T7 RNA polymerase amplification-activated CRISPR/Cas13a activity for SARS-CoV-2 detection in human pharyngeal swabs and environment by an electrochemiluminescence biosensor. *J. Hazard. Mater.*, 452, 131268, 2023.
49. Lee, H. *et al.*, Rapid detection of intact SARS-CoV-2 using designer DNA Nets and a pocket-size smartphone-linked fluorimeter. *Biosens. Bioelectron.*, 229, 115228, 2023.
50. Park, K.S. *et al.*, Ultra-sensitive label-free SERS biosensor with high-throughput screened DNA aptamer for universal detection of SARS-CoV-2 variants from clinical samples. *Biosens. Bioelectron.*, 228, 115202, 2023.
51. Robin, P. *et al.*, A DNA biosensors-based microfluidic platform for attomolar real-time detection of unamplified SARS-CoV-2 virus. *Biosens. Bioelectron.: X*, 13, 100302, 2023.

52. Jiang, W. *et al.*, Label-free integrated microfluidic plasmonic biosensor from vertical- cavity surface-emitting lasers for SARS-CoV-2 receptor binding domain protein detection. *Opt. Express*, 31, 8, 12138–12149, 2023.
53. Li, Y. *et al.*, Peptide nucleic acid and antifouling peptide based biosensor for the non- fouling detection of COVID-19 nucleic acid in saliva. *Biosens. Bioelectron.*, 225, 115101, 2023.
54. Son, S.U. *et al.*, Conductive Thread-Based Immunosensor for Pandemic Influenza A (H1N1) Virus Detection. *ACS Appl. Mater. Interfaces*, 15, 6, 7759–7766, 2023.
55. Büyüksünetçi, Y.T. and Anık, Ü., Electro-Nano Diagnostic Platform Based on Antibody–Antigen Interaction: An Electrochemical Immunosensor for Influenza A Virus Detection. *Biosensors*, 13, 2, 176, 2023, doi: DOI: 10.3390/bios13020176.
56. Bao, Q. *et al.*, Electrochemical biosensor based on antibody-modified Au nanoparticles for rapid and sensitive analysis of influenza A virus. *Ionics*, 29, 5, 2021–2029, 2023.
57. Zhou, H. *et al.*, CRISPR/Cas13a combined with hybridization chain reaction for visual detection of influenza A (H1N1) virus. *Anal. Bioanal. Chem.*, 414, 29, 8437–8445, 2022.
58. Siavash Moakhar, R. *et al.*, A Versatile Biomimic Nanotemplating Fluidic Assay for Multiplex Quantitative Monitoring of Viral Respiratory Infections and Immune Responses in Saliva and Blood. *Adv. Sci.*, 9, 33, 2204246, 2022.
59. Manohara Reddy, Y.V. *et al.*, Fine-tuning of MXene-nickel oxide-reduced graphene oxide nanocomposite bioelectrode: Sensor for the detection of influenza virus and viral protein. *Biosens. Bioelectron.*, 214, 114511, 2022.
60. Feng, D.-Q. and Liu, G., Target-Activating and Toehold Displacement Ag NCs/GO Biosensor-Mediating Signal Shift and Enhancement for Simultaneous Multiple Detection. *Anal. Chem.*, 93, 48, 16025–16034, 2021.
61. Kim, J.H. *et al.*, Biomimetic isolation of affinity peptides for electrochemical detection of influenza virus antigen. *Sens. Actuators, B*, 343, 130161, 2021.
62. Karash, S. *et al.*, Rapid detection of avian influenza virus H5N1 in chicken tracheal samples using an impedance aptasensor with gold nanoparticles for signal amplification. *J. Virol. Methods*, 236, 147–156, 2016.
63. Ravina, *et al.*, Hemagglutinin gene based biosensor for early detection of swine flu (H1N1) infection in human. *Int. J. Biol. Macromol.*, 130, 720–726, 2019.
64. Layqah, L.A. and Eissa, S., An electrochemical immunosensor for the corona virus associated with the Middle East respiratory syndrome using an array of gold nanoparticle- modified carbon electrodes. *Microchim. Acta*, 186, 4, 224, 2019.
65. Eissa, S., Al-Kattan, K., Zourob, M., Combination of Carbon Nanofiber-Based Electrochemical Biosensor and Cotton Fiber: A Device for the Detection of the Middle- East Respiratory Syndrome Coronavirus. *ACS Omega*, 6, 47, 32072–32080, 2021.

66. Kim, G. *et al.*, Fabrication of MERS-nanovesicle biosensor composed of multi-functional DNA aptamer/graphene-MoS2 nanocomposite based on electrochemical and surface- enhanced Raman spectroscopy. *Sens. Actuators, B*, 352, 131060, 2022.
67. Teengam, P. *et al.*, Multiplex Paper-Based Colorimetric DNA Sensor Using Pyrrolidinyl Peptide Nucleic Acid-Induced AgNPs Aggregation for Detecting MERS-CoV, MTB, and HPV Oligonucleotides. *Anal. Chem.*, 89, 10, 5428–5435, 2017.
68. Kim, H. *et al.*, Development of Label-Free Colorimetric Assay for MERS-CoV Using Gold Nanoparticles. *ACS Sens.*, 4, 5, 1306–1312, 2019.
69. Nandi, S. *et al.*, Chapter One - Biosensor platforms for rapid HIV detection, in: *Advances in Clinical Chemistry*, G.S. Makowski (Ed.), pp. 1–34, Elsevier, Netherlands, 2020.
70. Zhang, J. *et al.*, 3-D DNA walker-mediated ultrasensitive dual-signal biosensor for HIV DNA detection. *Sens. Actuators, B*, 385, 133637, 2023.
71. Li, Z. *et al.*, Bioinspired CRISPR-Mediated Cascade Reaction Biosensor for Molecular Detection of HIV Using a Glucose Meter. *ACS Nano*, 17, 4, 3966–3975, 2023.
72. Farzin, L. *et al.*, HIV biosensors for early diagnosis of infection: The intertwine of nanotechnology with sensing strategies. *Talanta*, 206, 120201, 2020.
73. Kwon, O.S. *et al.*, Large-Scale Graphene Micropattern Nano-biohybrids: High- Performance Transducers for FET-Type Flexible Fluidic HIV Immunoassays. *Adv. Mater.*, 25, 30, 4177–4185, 2013.
74. Mahmoud, K.A. and Luong, J.H.T., Impedance Method for Detecting HIV-1 Protease and Screening For Its Inhibitors Using Ferrocene–Peptide Conjugate/Au Nanoparticle/Single- Walled Carbon Nanotube Modified Electrode. *Anal. Chem.*, 80, 18, 7056–7062, 2008.
75. Labib, M. *et al.*, Electrochemical analysis of HIV-1 reverse transcriptase serum level: Exploiting protein binding to a functionalized nanostructured surface. *Talanta*, 85, 1, 770–778, 2011.
76. Ma, Y. *et al.*, A multi-walled carbon nanotubes based molecularly imprinted polymers electrochemical sensor for the sensitive determination of HIV-p24. *Talanta*, 164, 121–127, 2017.
77. Shamsipur, M. *et al.*, Sub-femtomolar detection of HIV-1 gene using DNA immobilized on composite platform reinforced by a conductive polymer sandwiched between two nanostructured layers: A solid signal-amplification strategy. *Anal. Chim. Acta*, 1055, 7–16, 2019.
78. Xiao, W. and Rapid, A. *et al.*, Simple, and Low-Cost CD4 Cell Count Sensor Based on Blocking Immunochromatographic Strip System. *ACS Sens.*, 4, 6, 1508–1514, 2019.
79. Zhang, Y. *et al.*, Graphene oxide-based biosensing platform for rapid and sensitive detection of HIV-1 protease. *Anal. Bioanal. Chem.*, 410, 24, 6177–6185, 2018.

80. Zou, R. *et al.*, DNA-programming multicolor silver nanoclusters for sensitively simultaneous detection of two HIV DNAs. *Sens. Actuators, B*, 296, 126608, 2019.
81. Du, Y. *et al.*, Amplification-free detection of HBV DNA mediated by CRISPR-Cas12a using surface-enhanced Raman spectroscopy. *Anal. Chim. Acta*, 1245, 340864, 2023.
82. Palacio, I. *et al.*, Attomolar detection of hepatitis C virus core protein powered by molecular antenna-like effect in a graphene field-effect aptasensor. *Biosens. Bioelectron.*, 222, 115006, 2023.
83. Alam, N. *et al.*, Label-free ultra-sensitive colorimetric detection of hepatitis E virus based on oxidase-like activity of MnO_2 nanosheets. *Anal. Bioanal. Chem.*, 415, 4, 703–713, 2023.
84. Manzano, M. *et al.*, Rapid and label-free electrochemical DNA biosensor for detecting hepatitis A virus. *Biosens. Bioelectron.*, 100, 89–95, 2018.
85. Gao, Y. *et al.*, One-step self-assembly of multilayer graphene oxide via streamlined click reactions for sensitive colorimetric assays. *Anal. Chim. Acta*, 1241, 340806, 2023.
86. Qavi, A.J. *et al.*, Rapid detection of an Ebola biomarker with optical microring resonators. *Cell Rep. Methods*, 2, 6, 100234, 2022.
87. Jang, M. *et al.*, Fabrication of Rapid Electrical Pulse-Based Biosensor Consisting of Truncated DNA Aptamer for Zika Virus Envelope Protein Detection in Clinical Samples. *Materials*, 16, 6, 2355, 2023, DOI: 10.3390/ma16062355.
88. Park, G. *et al.*, Selection of DNA aptamer and its application as an electrical biosensor for Zika virus detection in human serum. *Nano Convergence*, 9, 1, 41, 2022.
89. Bangruwa, N., Bhartiya, P.K., Mishra, D., A novel spin-based label-free electrochemical dna hybridization biosensor and its applications for dengue virus detection. *Sens. Actuators, B*, 382, 133447, 2023.
90. Alhazmi, H.A. *et al.*, Development of screen- printed carbon electrode-based immunosensors for the electrochemical detection of dengue virus antigen. *J. King Saud Univ. Sci.*, 35, 3, 102568, 2023.
91. Sampaio, I. *et al.*, Electrochemical detection of Zika and Dengue infections using a single chip. *Biosens. Bioelectron.*, 216, 114630, 2022.
92. Kumar, M. *et al.*, Ball milling as a mechanochemical technology for fabrication of novel biochar nanomaterials. *Bioresour. Technol.*, 312, 123613, 2020.
93. Charitidis, C.A. *et al.*, Manufacturing nanomaterials: from research to industry. *Manuf. Rev.*, 1, 11, 2014, https://doi.org/10.1051/mfreview/2014009.

13

Biosensing Technologies to Improve Neurological Disease Management

Poojith Nuthalapati[1,2*], Arjun Singh[2,3], Brinda Niravkumar Desai[4], Preeti Reddy Yendapalli[5], Reethika Gongireddy[6], Karan Singh[2], Bhaswanth Bollu[7] and Dheeraj K. Pinninty[8]

[1]*Department of Neurology, Yale University School of Medicine, New Haven, Connecticut, USA*
[2]*Department of Neurology, Massachusetts General Hospital, Harvard Medical School, Boston, MA, USA*
[3]*Department of Neurology, University of Kansas Medical Center, Kansas City, KS, USA*
[4]*Department of General Medicine, G.M.E.R.S. Medical College, Valsad, India*
[5]*Department of Internal Medicine, Saint Clare's Health, Denville, NJ, USA*
[6]*Department of General Medicine, Bhaskar Medical Collage, NTR University of Health Sciences, India*
[7]*Department of Internal Medicine, Texas Tech University Health Sciences Centre, El Paso, TX, USA*
[8]*Department of Internal Medicine, RWJ Barbnabas Health, Jersey City, NJ, USA*

Abstract
Neurological diseases present a global health challenge, demanding innovative methods to diagnose and treat efficiently. Biosensors, because of their capacity to deliver precise, and sensitive biomarker detection, are asserted as influential assets within the sphere of neurological diseases. This field has shown great promise, owing to its nanoscience and technology. Various advances in nanoscience can create sensors that can detect chronic diseases in the initial stages. Biosensors are a promising frontier for the monitoring of neurological diseases. This chapter discusses the various applications of biosensors in the context of neurological diseases. It covers areas identifying biomarkers of neurodegenerative diseases, detecting neurotransmitters, and identifying genetic mutations linked to hereditary

**Corresponding author*: poojith.nuthalapati@yale.edu

Suvardhan Kanchi, Ayyappa Bathinapatla, Anitha Varghese and Phumlane Selby Mdluli (eds.) Nano-Biosensor Technologies for Diagnosis of Infectious Diseases, (401–422) © 2025 Scrivener Publishing LLC

neurological conditions. Additionally, it addresses how biosensors can monitor disease progression, assess drug effectiveness, and aid in developing patient-specific treatment strategies.

***Keywords*:** Nanobiosensors, neurologogical disorders, CNS disorders, Alzheimer's, Parkinson's disease, multiple sclerosis

13.1 Introduction

Biosensors are devices that help to study the interactions between probes and targets on their surfaces. These interactions can be converted into the measured signals. These signals can be in different forms, such as light, bending of a small beam, or electricity [1, 2]. Nanobiosensors have proven beneficial in medical diagnostics and have also shown their role in the management of diverse neurological disorders. Recently, substantial attempts to identify biomarkers linked to central nervous system (CNS) disorders with clinically significant results have been ongoing [3]. Several genomic, proteomic, and metabolic indicators of neurological disorders can be identified quickly, sensitively, and accurately using these biomarkers (Figure 13.1). A definitive diagnosis of most neurological disorders often requires multiple approaches. An In depth medical history & neurological examination, and often neuroimaging techniques such as magnetic resonance imaging (MRI), electromyogram (EMG), and electroencephalogram (EEG) to come to a diagnosis [4]. However, the drawbacks of the traditional methods include limited accuracy, time consumption, and challenges in automating these techniques (Figure 13.1). Immunosensing, microarray use, and amplification performed in a non-enzymatic manner result in significant alleviation of these drawbacks [4]. While functioning as diagnostics, biosensors also facilitate the identification of disease signatures by enabling the discovery of biomarkers from tissues and genomes. Hence, biosensors serve dual purposes in the field of neurological disorders. They function as tools for identifying biomarkers and as diagnostic aids.

Numerous neurological studies have traditionally used cerebrospinal fluid (CSF) for biochemical analysis. Although CSF is the gold standard physiological sample used for the diagnosis of neurological disorders, similar biomarkers can also be found in readily accessible bodily fluids. The availability of biomarkers in easily obtainable biological fluids offers the advantage of a less invasive sample acquisition process [5]. Modern disease biosensors, across different platforms for neurological disorder analysis,

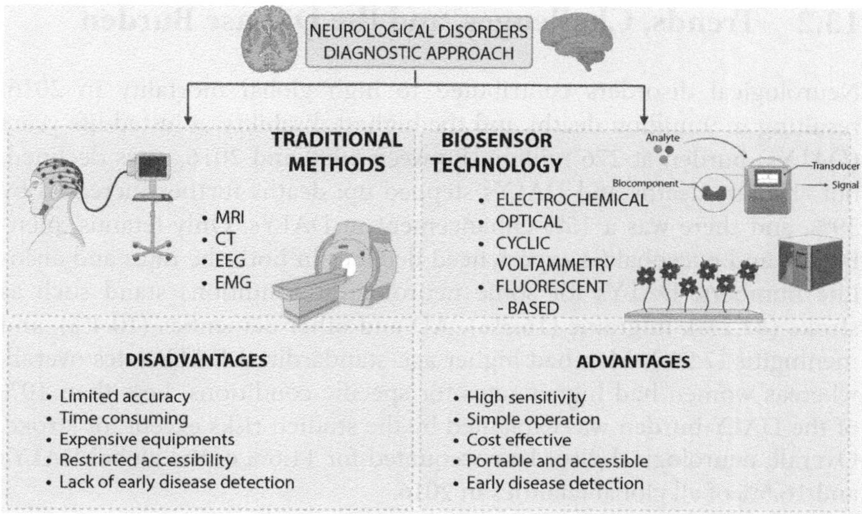

Figure 13.1 Traditional vs. biosensor-based approaches for diagnosing neurological conditions. **Legend:** A comparison of traditional diagnostic methods with biosensor technology for neurological disorders highlights several differences. Traditional technologies involving MRI, computed tomography (CT), EEG, and EMG are often time-consuming, costly, and, sometimes, lack the necessary sensitivity and specificity for early detection and ongoing monitoring of neurological conditions. On the other hand, biosensors present a minimally invasive or non-invasive alternative by identifying specific biomarkers related to neurological diseases. Utilizing nanomaterials, biomolecules, and electrical or optical transducers, biosensors facilitate the rapid, sensitive, and continuous monitoring of biomarkers in body fluids such as blood, tears, and saliva. With their extreme sensitivity, specificity, affordability, and user-friendliness, biosensors exhibit considerable promise for prompt diagnosis, treatment selection, and disease tracking in various neurological disorders.

bind genetic targets or biochemical substances in body fluids and tissues [3]. Many studies have underscored the capacity to identify biomarkers associated with neurological disorders, such as Alzheimer's disease (AD), including amyloid beta and tau proteins. Additionally, biosensors have facilitated the continuous monitoring of neurotransmitter concentrations in diseases such as Parkinson's disease, offering invaluable information on disease advancement and the need for medication modifications [3]. Although the potential of biosensors for advancing diagnostics is clear, obstacles remain. Issues such as maintaining the stability of biomarkers and navigating the transition from laboratory experimentation to practical clinical deployment remain as an ongoing challenge.

13.2 Trends, Challenges, and the Disease Burden

Neurological disorders contributed to high global mortality in 2016, resulting in 9 million deaths and the highest disability-adjusted life years (DALYs) burden at 276 million. Between 1990 and 2016, rates declined, but absolute deaths and DALYs stepped up: deaths further increased by 39%, and there was a 15% enhancement in DALYs. Only tetanus, meningitis, and encephalitis experienced declines in both the rates and absolute numbers. DALYs for some neurological conditions stand such as stroke (42.2%), migraine (16.3%), AD and other dementias (10.4%), and meningitis (7.9%). Men had higher age-standardized DALY rates overall, whereas women had higher rates for specific conditions. Less than 10% of the DALY burden was explained by the studied risks except for stroke. Overall, neurological disorders accounted for 11.6% of the global DALYs and 16.5% of all global fatalities in 2016.

13.3 CNS Diseases

Neurological diseases contribute significantly to global illness and death primarily because of the absence of rapid and sensitive diagnostic tools. These include neurodegenerative, neuroinflammatory, and neuroinfectious diseases. In the US, an estimated 6.2 million individuals could be affected by Alzheimer's as per a 2022 report by the Alzheimer's Disease Association [6]. Additionally, the Parkinson's Foundation states that almost one million Americans currently contend with Parkinson's disease [7]. However, the extent of undiagnosed CNS infection has been underestimated. Even in the era following advancements in polymerase chain reaction, approximately 30% of individuals suspected of having a CNS infection do not receive a definitive diagnosis. This lack of diagnostic accuracy contributes to the death of nearly one-third of patients. CNS infections have several origins, symptoms, and prognosis. They can cause considerable morbidity and mortality and manifest suddenly. An inherent clinical hurdle in dealing with CNS infections arises from the relatively isolated environment of the CNS and its protective mechanisms. The primary safeguard of the CNS is the barrier between the blood and the brain, which does not allow pathogens and sizable molecules to enter the CSF from the blood. One challenge that emerges owing to these protective mechanisms is the task of pinpointing the precise neurovascular compartment housing the infection [8]. The scope of the neurovascular space affected by infections frequently spans from the protective meningeal layers to the innermost regions of the brain, creating a spectrum of involvement [9]. As CNS infections have a non-specific clinical presentation, clinicians frequently order and concurrently evaluate many diagnostic tests. Given the abundance of possible pathogens and variety of available testing techniques, this can be a difficult task. Early and accurate diagnosis is essential for effective treatment and management of these diseases. Additionally, individuals may display symptoms, despite having a very small number of organisms. The most effective approaches for diagnostic testing vary depending on the specific pathogens involved, which results from the diverse ways in which these pathogens cause infections and differences in how immune systems respond. Several elements influence the effectiveness of tests, such as disease prevalence and intricacies linked to the specific testing platform.

13.4 Utility of Neurobiosensors

Neurobiosensors encompass a diverse array of technologies ranging from commercially available kits to experimental prototypes. Table 13.1 shows the list of some commercially available biosensors platforms. They demonstrate detection capabilities equivalent to and can even surpass traditional assays [4]. Current diagnostic procedures for neurological disorders in clinical laboratories often require multistep immunoassays. Microarrays are being actively employed for biomarker discovery. Immunosensors function by detecting the interactions between antibodies and antigens. These interactions can be identified using immobilized or immunogenic peptides. Many immunosensors use surface plasmon resonance (SPR) to detect and quantify neurological biomarkers. Nanoscale optical biosensors based on SPR have been utilized for the detection of AD. Instances of using Biacore™ SPR systems include their application in characterizing regular and elongated polyglutamine sequences. This study was conducted to ascertain the pathological threshold of Huntington's disease [10]. SPR has also been used for the detection of complex targets. This involves the incorporation of selective peptides into the SPR system, which can bind to cells, bacteria, and viruses associated with neurological disorders. Alternatively, aptamers can substitute for peptides for greater selectivity and cost-effectiveness. Similarly, chips that utilize proteins or peptides are instrumental in biomarker discovery. Their use has helped identify AD determination [11, 12] biomarker panels. Protein arrays from Ciphergen have proposed early stage AD biomarkers obtained from CSF samples. For this purpose, they employed surface-enhanced laser desorption/ionization time-of-flight mass spectrometry.

Table 13.1 Some commercially available biosensor platforms for the detection of neurological diseases.

Biosensor platform	**Company**
Microarray technology (DNA-based)	Cepheid
DNA-based microarray technology	Applied Biosystems
xMAP technology	Luminex
Protein and Peptide microarray technology	Ciphergen
Microarray technology (DNA-based)	Affymetrix

13.5 The Technology Behind Biosensor Development

13.5.1 Electrochemical Biosensors

Neurobiosensors can be of different types and are based on the principles of biosensing, selective detection of specific neural biomolecules, or changes in neural activity (Figure 13.2). Electrochemical neurobiosensors

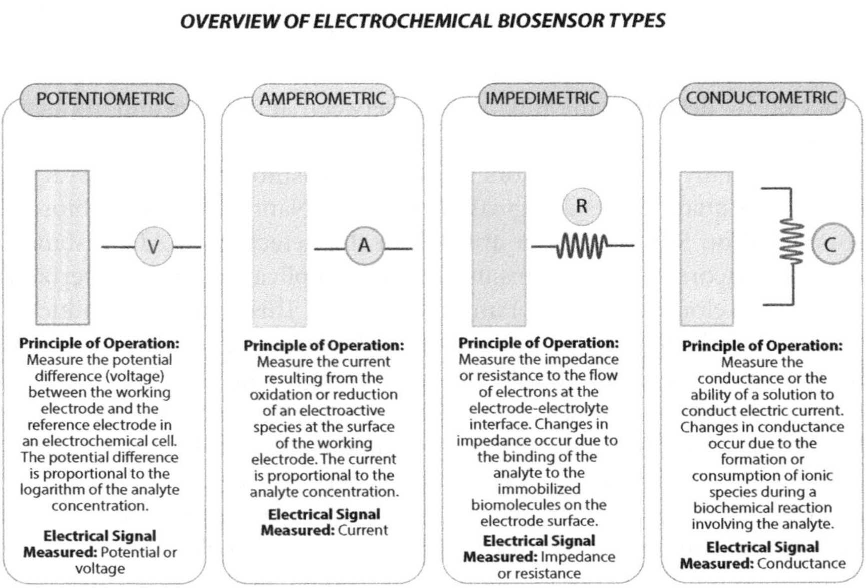

Figure 13.2 Legend: The figure depicts the four primary types of electrochemical biosensors: potentiometric, amperometric, impedimetric, and conductometric, along with their respective principles of operation and the types of electrical signals they measure.

are based on the core concept of electrochemistry (processes involving electron transfer). The outcome is a measurable electrical signal that allows the quantification of target analytes [13, 14]. They are categorized into four primary groups, each corresponding to a specific electrochemical technique employed to quantify the electrical signal generated by a biochemical process (Figure 13.2). These categories include potentiometric, amperometric, impedimetric, and conductometric biosensors [15] (Figure 13.2). Amperometric biosensors are designed to indirectly assess analytes through the inhibition of specific enzymes [16]. These inhibition-based biosensors can achieve remarkable sensitivity, particularly when the target enzyme is inhibited even at minute concentrations of its inhibitor [17] (Figure 13.2). Dysregulation of acetylcholine (Ach) in the brain leads to the development of Parkinson's disease and AD, progressive dementia, schizophrenia, and Myasthenia Gravis [18]. However, existing techniques for assessing ACh levels in bodily fluids are cumbersome, costly, time-consuming, and unsuitable for on-site monitoring. Consequently, substantial research efforts have given rise to biosensors for detecting markers associated with neurological disease pathways that involve the inhibition of neural esterases, such as acetylcholinesterase (AChE) [19, 20]. Although AChE is predominantly located in cholinergic neurons, it is worth noting that significant AChE activity, responsible for the degradation of circulating ACh, is also present in the blood cells. Blood-based biomarkers offer clear advantages, owing to their easy and painless collection, high efficiency, and cost-effectiveness. They enable convenient and repeatable measurements, particularly for frail elderly individuals. However, a significant challenge lies in the indirect connection between the peripheral blood and CNS diseases.

Primary electrochemical techniques employed to detect biomarkers related to Alzheimer's and Parkinson's disease include cyclic voltammetry (CV), differential pulse voltammetry, and electrochemical impedance spectroscopy (EIS). Numerous biosensor devices operate by forming a recognition complex between the biosensing material and analyte, typically arranged in a single-layer or thin-film format. With the biosensor material gradually developing on conductor surfaces, the properties of capacitance and resistance change at the interface of the surface and electrolyte. These changes were accurately represented by EIS. Consequently, EIS is a highly favored technique for discerning surface-level changes and obtaining detailed insights into the activities of biosensing elements at interface [21, 22].

Improving the diagnostic precision in AD can involve identifying reduced levels of amyloid beta and an elevation in total tau protein, which can be phosphorylated within the CSF. Traditionally, the pathology of

AD is associated with the presence of amyloid plaques and the buildup of hyperphosphorylated tau aggregates. Enzyme-linked immunosorbent assay is the predominant technique used to measure T-tau and P-tau levels. It can identify all isoforms of tau regardless of their degree of phosphorylation [23, 24]. In 2008, Kerman *et al.* introduced a highly sensitive method for detecting tau protein by employing a nanoparticle chip plated with gold, using localized SPR (LSPR).

The method showed exceptional sensitivity, detecting tau at concentrations as low as 10 pg/mL in CSF [25].

13.5.2 Optical Biosensors

Optical biosensors constitute the frequently described category of biosensors. Detecting them involves utilization of the optical field and biorecognition element [26] interactions (Figure 13.3). Different biological components, such as antibodies, enzymes, antigens, nucleic acids, receptors, entire cells, and tissues, are utilized as biorecognition components (Figure 13.3). Aptamers have been extensively employed in optical biosensors for biomarker analyses. They are favored because of their cost-effectiveness compared to antibodies, synthetic production without ethical concerns, and ability to achieve a low detection limit for biomarkers. Ziu *et al.* [27] highlighted tau protein's importance and developed an optical biosensor using an aptamer to detect the Tau-441 protein. By combining it with a streptavidin surface, they achieved a low detection threshold of 6.7

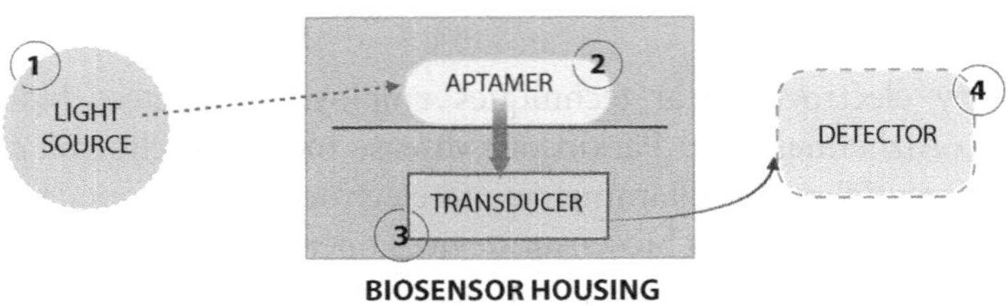

Figure 13.3 Schematic of Optical Biosensors. **Legend:** The optical biosensor consists of a light source (1) that provides an incident light beam directed toward the biorecognition element (2), which is typically an immobilized biomolecule like an antibody or aptamer that specifically binds to the target analyte. Upon binding, an optical interaction occurs, causing a change in the optical properties that is detected by the transducer (3). The transducer transforms this optical signal into a quantifiable output, which is subsequently examined by the part of the detector (4). The flow of the signal represents the transmission of the generated signal from the transducer to the detector for further analysis and quantification of the target analyte. The entire setup is housed within a biosensor housing that encloses and integrates all the necessary components.

nanomolar, which was achieved through the application of biolayer interferometry. Amyloid-β and α-synuclein are two neurodegenerative proteins included in the selectivity studies that they conducted. The high selectivity of the biosensor developed specifically for Tau-441 is also emphasized. Additionally, they stressed the benefits of optical biosensors, like their simplicity, speed, and exceptional selectivity, proposing their application in biological fluids. The most common optical biosensing method is surface plasmon resonance analysis (SPRA). A metallic surface or materials that conduct is employed at the junction linking two media: glass and liquid. Polarized light is utilized for illumination at a particular angle, causing surface plasmons to generate. This is accompanied by a decrease in the light intensity at the resonance angle, further being proportional to the surface mass [26].

A study by Lisi *et al.* [28]. showed the clinical importance of tau and the need for its determination at picomolar levels, although the SPR method can measure it in nanomoles. Thus, the detection level of Tau was further reduced to 125 pM by the development of an amplification strategy by this research group. This approach involved enhancing multi-walled carbon nanotubes by attaching a secondary antibody specific to the protein (tau). This enhanced sandwich assay significantly amplified the SPR signal, exceeding the signals of the conventional method by a factor of hundred.

Although target biomarkers can be detected in real time and directly from various body fluids using optical biosensors, the ability of these biosensors can be impaired if there is any change in the sensor surface. Hence, biomarkers for neurodegeneration are preferred to be detected using electrochemical biosensors because they have high sensitivity, can resist alterations in the matrix, and can be utilized in different types of biological fluids [26]. Several electrochemical biosensors have been developed for these purposes. One such method, developed by Wang *et al.*, relies on establishing a robust antibody-antigen complex capable of detecting tau-441 with a detection limit of 0.03 pM [29].

13.5.3 Cyclic Voltammetry (CV)

By repeatedly oscillating the potential, CV produces cyclic current feedback, revealing the redox properties of neurotransmitters. The reaction encompasses both oxidation and reduction reactions, allowing for specific quantitative evaluation and detection of neurotransmitters. When the neurotransmitters undergo oxidation at the electrode surface, an oxidation peak is generated in the voltammogram. The neurotransmitter concentration is closely linked to the size of this oxidation peak, allowing for direct

Table 13.2 Variety of biomarkers and biosensors devised for neurological disorder diagnosis. AD, Alzheimer's disease; MS, multiple sclerosis; PD, Parkinson's disease; SPR, surface plasmon resonance; LSPR, localized surface plasmon resonance; DPV, differential pulse voltammetry; CV, cyclic voltammetry; CSF, cerebrospinal fluid; (R)-Sal, 1(R)-methyl-6,7-dihydroxy-1,2,3,4-tetrahydroisoquinoline; (R)-NMSal-1, (R),2(N)-dimethyl-6,7-dihydroxy-1,2,3,4-tetra-hydroisoquinoline.

Disease	Biomarker	Platform/biosensor type	Application	Source of sample	Limit of detection (LOD)	References
AD	Tau protein and its isoforms	Electrochemical, SPR, and EIS	Diagnosis	Human serum, plasma, and CSF	10 Fm to 125 pM	[29, 35–37]
	Amyloid beta isoforms	CV and DPV	Diagnosis	Human serum, artificial CSF, and plasma	0.2 pg/mL to 100 pg/mL	[38–40]
	Alpha-synuclein	Optical, EIS, CV, DPV, and SPR	Diagnosis	Human serum and artificial CSF	0.13 pg/mL to 3.5×10^{-5} ng/mL	[41, 42]
	Amyloid beta isoforms	EIS, CV, DPV, and SPR	Diagnosis	Human serum, artificial CSF, and plasma	0.2 pg/mL to 100 pg/mL	[38, 40, 43]
	Acetylcholine	Electrochemical	Diagnosis	Spiked serum	4 nM to 10 μM	[44, 45]
	Amyloid beta peptides and Tau proteins	Optical	Diagnosis	Artificial plasma	34.9 fM (Aβ), 26 fM (Aβ), and 23.6 fM (Tau protein)	[46]
	Apolipoprotein E	Electrochemical	Diagnosis	Spiked in buffer	286 nM	[48]

(Continued)

Table 13.2 Variety of biomarkers and biosensors devised for neurological disorder diagnosis. AD, Alzheimer's disease; MS, multiple sclerosis; PD, Parkinson's disease; SPR, surface plasmon resonance; LSPR, localized surface plasmon resonance; DPV, differential pulse voltammetry; CV, cyclic voltammetry; CSF, cerebrospinal fluid; (R)-Sal, 1(R)-methyl-6,7-dihydroxy-1,2,3,4-tetrahydroisoquinoline; (R)-NMSal-1, (R),2(N)-dimethyl-6,7-dihydroxy-1,2,3,4-tetra-hydroisoquinoline. (*Continued*)

Disease	Biomarker	Platform/biosensor type	Application	Source of sample	Limit of detection (LOD)	References
MS	Myelin-based protein	Electrochemical	Diagnosis and patient monitoring	CSF and spiked serum	0.30 nM	[47]
	Tau protein	Electrochemical	Diagnosis	CSF and spiked serum	0.15nM	[47]
	Autoantibodies	Optical and electrochemical	Diagnosis and patient monitoring	Serum from patient and CSF	0.149 ng/mL	[49, 50]
PD	Alpha-synuclein	LSPR, LC optical EIS, and DPV	Research	Human serum and artificial CSF	70 nM to 1.13 ng/mL	[51, 52, 57, 58]
	Dopamine	Optical and electrical	Diagnosis	Spiked in buffer and CSF samples	40 nM (buffer), 1 nM (serum), and 0.830 nM (CSF)	[53–55]
	Acetylcholinesterase	Photoelectrochemical	Screening of drug	(R)-Sal (R)-NMSal	-	[56]

measurement of peak current values to determine the neurotransmitter concentration. This technique has proven to be highly valuable in neuroscience, providing crucial insights into the basic brain processes related to addiction and various neurological disorders. Table 13.2 shows the list of variety of biomarkers and biosensors devised for neurological disorder diagnosis.

13.5.4 Biosensors Based on Fluorescence

Fluorescence-based technology is extremely responsive to the detecting of several biomolecules [30, 31]. Table 13.2 Variety of biomarkers and biosensors devised for neurological disorder diagnosis. Biomarker detection is also based on the use of several fluorescent modalities that are economical, compact, and portable [31]. In the case of neurological diseases such as AD, fluorescent biosensors utilizing several biomarkers such as neurotoxin, Tau, dopamine, and Apolipoprotein E4 (ApoE4) DNA have been used [32]. AβO is the main target analyte detected by these biosensors as it is associated with the initiation and progression of AD. Hence, it is important to quantify this biomarker in plasma and CSF to determine the severity [33, 34]. A biosensor utilizing shape-coding, incorporating gold nanoparticles with unique shape-dependent properties within distinct LSPR, was created for detecting multiple biomarkers associated with AD in a simulated blood sample [46]. The Q-body technique is an innovative approach for the detection of diverse arrays of biomolecules. It utilizes the fluorescence-reducing effect of dye(s) associated with antibody fragment [59]. Although fluorescence techniques are effective for detecting AD biomarkers, they have certain drawbacks. Challenges include being time-consuming, costly, and leading to blockage of active binding sites. Moreover, it may also affect the bioreceptor-biomarker interaction [60]. Nevertheless, it has been widely used in clinical settings, specifically to determine the initial phase of Aβ nucleation, owing to its strong staining capabilities and covinience [61, 62].

13.5.5 Microelectrode Arrays

These are advanced tools used in drug screening, in addition to studying neural activity at cellular and network levels. These specialized devices have multiple microscale electrodes that can record electrical activity from multiple points within the neural tissue or cultured neurons. Microelectrode arrays (MEAs) can have numerous electrodes, ranging from a few to thousands of electrodes, depending on their design and purpose. The electrodes were arranged in a grid-like pattern to record the electrical activity from multiple points simultaneously. When neurons are cultured on an array surface, connections and communication are established through electrical impulses. The electrodes on the arrays record these electrical signals, which represent the neural activity. The key requirements for tools measuring neurobiological events include their small size and rapid responsiveness, thus making microelectrodes highly suitable for such purposes, given that their dimensions typically fall within the micron scale [63]. The use of microelectrodes for electrochemical detection requires the detection of substances with electroactive properties. Numerous neurochemical transmitters, including dopamine and serotonin, fall under this category and can be electrochemically identified. The true advantage of microelectrodes in combination with fast-scan cyclic voltammetry (FSCV) is the direct detection of electroactive neurotransmitters directly [64]. Coupling of the FSCV with microelectrodes enables real-time monitoring of rapid alterations in neurotransmitter levels, occurring within sub-second intervals during pharmacological interventions or behavioral interventions. MEA serves two primary types of measurements: extracellular recordings, which capture neural activity outside cells, and intracellular recordings, which focus on activity within neural cells. Extracellular measurements are less invasive and can be recorded for a long time; however, their quality is poor compared to that of intracellular measurements. On the other hand, intracellular recordings are comparatively more invasive and involve cell damage, show higher sensitivity, and utilize synaptic potential [65, 66]. This technique has also been used to study the effects of drugs on neural activity. Exposure of neurons that have been cultured to various drugs and measurement of changes in electrical patterns can aid in drug development and toxicity screening. Additionally, it has been employed to study numerous neurological diseases. By utilizing neurons derived from patients or animal models, disease conditions can be mimicked, and the aberrant neural activity associated with these disorders can be studied [67].

13.6 Clinical Applications

Diagnosing temporal lobe epilepsy clinically primarily relies on EEG or video-EEG recordings combined with the history of seizures in patients. The main reason for this is the lack of proven molecular biomarkers that are detectable in blood. However, microRNAs (miRs) have potential as biomarkers as they exhibit specific expression patterns in various cell types, are associated with brain excitability, and are present in CSF and sweat in addition to blood [68]. Aidan *et al.* [69] and Raoof *et al.* [70] observed that miR-27a-3p, miR-328-3p, and miR-654-3p are promising biomarkers in patients with psychogenic non-epileptic seizures. Using electrocatalytic nanoparticles as a label, a point-of-care device, TORNADO, was established by Aidan *et al.* [69] for the quick identification of these microRNAs in fluids like plasma. The signal-to-noise ratio was optimized by selecting an electrode with a low electrocatalytic activity for the substrate, such as hydrogen peroxide. Thus, even one particle with increased catalytic activity linked to a particular target molecule-binding reaction potentially yields a detectable current, thereby amplifying the resulting signal. Here, high diagnostic accuracy was obtained by quantifying miRNAs within the exosome-enriched fraction. Nevertheless, numerous miRNAs tend to bind with proteins, and, specifically, miR-328-3p bound to argonaute showed a significant selective increase in patients following seizures. Thus, point-of-care devices can be highly effective in diagnosing and understanding the underlying pathomechanisms and prognoses of several neurological diseases.

Furthermore, early diagnosis of AD remains a difficult challenge, primarily because of the absence of reliable biomarkers. However, blood-based microRNAs, such as Let-7b and miR-206, have been detected to an increased extent in patients with AD and moderate cognitive damage. Rusling *et al.* [68] validated the miR-206 biomarker and sample-to-answer disk by quantitative polymerase chain reaction (qPCR). Elevated miR-206 levels are strongly linked to cognitive decline and memory deficits that precede the onset of dementia. Microfluidic devices such as this could efficiently stratify patients with slight cognitive loss due to Alzheimer's risk, aiding early intervention.

Multiple sclerosis (MS) stands as the primary cause of non-traumatic disability in young adults, marked by the demyelination of neurons. Assessments for patients with MS usually occur during neurological appointments at 3- to 12-month intervals, in addition to emergency visits for relapse. However, these appointments may not effectively capture the symptom fluctuations, relapse onset, or disability progression among visits [71, 72]. Wearable biosensors are gaining attention as tools to measure disability in neurological conditions, such as MS. They offer the potential to provide more measurable, empirical, and important evaluations of neurological functionality, not only during clinic visits, but also in the intervals between them [73–75]. Studies on people living with MS have mostly used commercial wearable biosensors that focus on steps and energy. A recent study compared these and found that they were good for overall energy but less accurate for specific activities, especially resistance exercises [76]. Chitnis *et al.* [72] evaluated the viability, adherence, and association of metrics obtained from biosensor utilization with conventional clinical disability assessments conducted by neurologists. Additionally, the study group applied algorithms to extract various motion and dexterity features and correlated them with conventional neurological disability assessments. The study discovered that advanced wearable

Table 13.3 Details of clinical trials for various biosensors for neurological diseases.

Reference	Disease	Biosensor
NCT02177877	Epilepsy	Wristwatch Accelerometer
NCT03369171	Multiple sclerosis	MYO Armband (electroMYOgraphy sensor)
NCT05588115	Traumatic brain Injury	Glial fibrillary acidic protein–based biosensor
NCT03103919	Parkinson's disease	Kinesia-ONE
NCT05370105	Stroke	SPRi biosensor
NCT04201834	Huntington's disease	BioStamp nPoint device
NCT03882671	Epilepsy	EpiTel Epilog, Empatica E4, and GENEActiv
NCT02308605	Stroke	SMARTCap
NCT05627661	Autoimmune epilepsy	-
NCT02545166	Stroke	SMARTChip
NCT02358252	Stroke	Intelligent biosensor system
NCT04731311	Alzheimer's disease	BrightGo

biosensors, which record accelerometry and gyroscopic movements, yielded metrics closely aligned with complicated MS scales typically evaluated by neurologists. Biosensor metrics that correlated well with conventional neurological measurements were primarily observed during inactive walking, stance of mobility, velocity of turn angles, and postural shift. Table 13.3 details of clinical trials for various biosensors for neurological diseases. These metrics can be consistently obtained both within clinical settings and during everyday activities, offering the potential to evaluate and track neurological impairment between appointments at the clinic.

13.7 Conclusion

Biosensor systems enable the definition of important genomic and proteomic signatures and facilitate a better understanding of neurological diseases. Numerous advanced biosensor technologies for detecting and quantifying diverse biomarkers are rapidly advancing, such that future devices deliver significant advantages in diagnostics and patient monitoring. These advancements will further reduce the requirements for extensive biological sample processing. Integrating biosensing technologies into portable microfluidic systems enables point-of-care utilization for real-time and earlier neurological disease diagnosis.

References

1. Li, J. and Wu, N. (Eds.), Biosensors Based on Nanomaterials and Nanodevices, CRC Press, Boca Raton, FL, 2014.
2. Banica, F.G., Chemical Sensors and Biosensors, Wiley, New York, 2012.
3. Abreu, C.M., Soares-dos-Reis, R., Melo, P.N. *et al.*, Emerging Biosensing Technologies for Neuroinflammatory and Neurodegenerative Disease Diagnostics. *Front. Mol. Neurosci.*, 11, 1–13, 2018.
4. Bell, K.M. and Kornguth, S.E., Biosensors for Neurological Disease, in: Handbook of Biosensors and Biochips, John Wiley & Sons, Ltd, Chichester, England, 2007.
5. Çomoglu, S.S., Güven, H., Acar, M., Öztürk, G. *et al.*, Tear levels of tumor necrosis factor-alpha in patients with Parkinson's disease. *Neurosci. Lett.*, 553, 63–67, 2013.
6. https://www.alz.org/media/documents/alzheimers-facts-and-figures.pdf, Accessed 29th September 2023.
7. https://www.parkinson.org/understanding-parkinsons/statistics, Accessed 29th September 2023.
8. Tan, K., Patel, S., Gandhi, N., Chow, F. *et al.*, Burden of neuroinfectious diseases on the neurology service in a tertiary care center. *Neurology*, 71, 15, 1160–1166, 2008.
9. Bowers, K.M. and Mudrakola, V.V., Neuroinfections: presentation, diagnosis, and treatment of meningitis and encephalitis. *EMJ Neurol.*, 8, 1, 93–102, 2020.
10. Bennett, M.J., Huey-Tubman, K.E., Herr, A.B. *et al.*, Inaugural article: a linear lattice model for polyglutamine in CAG expansion diseases. *Proc. Natl. Acad. Sci. U.S.A.*, 99, 18, 11634–11639, 2002.
11. Choe, L.H., Dutt, M.J., Relkin, N. *et al.*, Studies of potential cerebrospinal fluid molecular markers for Alzheimer's disease. *Electrophoresis*, 23, 14, 2247–2251, 2002.
12. Carrette, O., Demalte, I., Scherl, A. *et al.*, A panel of cerebrospinal fluid potential biomarkers for the diagnosis of Alzheimer's disease. *Proteomics*, 3, 8, 1486–1494, 2003.

13. Zhang, S., Wright, G., Yang, Y., Materials and techniques for electrochemical biosensor design and construction. *Biosens. Bioelectron.*, 15, 273–282, 2000.
14. Uniyal, S. and Sharma, R.K., Technological advancement in electrochemical biosensor based detection of Organophosphate pesticide chlorpyrifos in the environment: A review of status and prospects. *Biosens. Bioelectron.*, 116, 37–50, 2018.
15. Grieshaber, D., MacKenzie, R., Vörös, J., Reimhult, E., Electrochemical biosensors—Sensor principles and architectures. *Sensors*, 8, 1400–1458, 2008.
16. Honeychurch, K.C. and Piano, M., Electrochemical (bio) sensors for environmental and food analyses. *Biosensors*, 8, 57, 2018.
17. Upadhyay, L.S.B. and Verma, N., Enzyme inhibition based biosensors: A review. *Anal. Lett.*, 46, 225–241, 2013.
18. Tandon, R., Cholinergic aspects of schizophrenia. *Br. J. Psychiatry*, 174, 7–11, 1999.
19. Kumaran, S. and Morita, M., Application of a cholinesterase biosensor to screen for organophosphorus pesticides extracted from soil. *Talanta*, 42, 649–655, 1995.
20. Dhull, V., Gahlaut, A., Dilbaghi, N., Hooda, V., Acetylcholinesterase biosensors for electrochemical detection of organophosphorus compounds: a review. *Biochem. Res. Int.*, 2013, 731501, 2013, doi: 10.1155/2013/731501. Epub 2013 Dec 9. PMID: 24383001; PMCID: PMC3872028.
21. Bertok, T., Lorencova, L., Chocholova, E. et al., Electrochemical Impedance Spectroscopy Based Biosensors: Mechanistic Principles, Analytical Examples and Challenges towards Commercialization for Assays of Protein Cancer Biomarkers. *ChemElectroChem*, 6, 989–1003, 2019.
22. Zamfir, L.-G., Puiu, M., Bala, C., Advances in electrochemical impedance spectroscopy detection of endocrine disruptors. *Sensors*, 20, 6443, 2020.
23. Hampel, H., Blennow, K., Shaw, L.M. *et al.*, Total and phosphorylated tau protein as biological markers of Alzheimer's disease. *Exp. Gerontol.*, 45, 30–40, 2010.
24. Lu, H., Zhu, X.C., Jiang, T., Yu, J.T., Tan, L., Body fluid biomarkers in Alzheimer's disease. *Ann. Transl. Med.*, 3, 5, 70, 2015. https://doi.org/10.3978/j.issn.2305-5839.2015.02.13
25. Kerman, K., Kim, D.-K., Hiep, H.M., Tamiya, E., Detection of Alzheimer's tau protein using localised surface plasmon resonance-based immunochip. *Talanta*, 74, 1038–1042, 2008.
26. Karaboğa M.N.S. and Sezgintürk M.K., Biosensor approaches on the diagnosis of neurodegenerative diseases: Sensing the past to the future. *J. Pharm. Biomed. Anal.*, 209, 114479, 2022 Feb 5. doi: 10.1016/j.jpba.2021.114479. Epub 2021 Nov 24. PMID: 34861607.
27. Ziu, I., Laryea, E.T., Alashkar, F., Wu, C.G., Martic, S., A dip-and-read optical aptasensor for detection of tau protein. *Anal. Bioanal. Chem.*, 412, 1193–1201, 2020.

28. Lisi, S., Scarano, S., Fedeli, S., Pascale, E., Cicchi, S., Ravelet, C., Peyrin, E., Minunni, M., Toward sensitive immuno-based detection of tau protein by surface plasmon resonance coupled to carbon nanostructures as signal amplifiers. *Biosens. Bioelectron.*, 93, 289–292, 2017.
29. Wang, S.X., Acha, D., Shah, A.J., Hills, F., Roitt, I., Demosthenous, A., Bayford, R.H., Detection of the tau protein in human serum by a sensitive four-electrode electrochemical biosensor. *Biosens. Bioelectron.*, 92, 482–488, 2017.
30. Lichtman, J.W. and Conchello, J.A., Fluorescence microscopy. *Nat. Methods.*, 2, 12, 910–9, 2005 Dec, doi: 10.1038/nmeth817. PMID: 16299476.
31. Zhu, H., Isikman, S.O., Mudanyali, O, Greenbaum, A., Ozcan, A., Optical imaging techniques for point-of-care diagnostics. *Lab. Chip.*, 13, 1, 51–67, 2013 Jan 7. doi: 10.1039/c2lc40864c. Epub 2012 Oct 9. Erratum in: Lab Chip. 2013 Dec 21;13(24):4890. PMID: 23044793; PMCID: PMC3510351.
32. Murti, B.T., Putri, A.D., Huang, Y.J., Wei, S.M., Peng, C.W., Yang P.K., Clinically oriented Alzheimer's biosensors: expanding the horizons towards point-of-care diagnostics and beyond. *RSC Adv.*, 11, 33, 20403–20422, 2021 Jun 8. doi: 10.1039/d1ra01553b. PMID: 35479927; PMCID: PMC9033966.
33. Wang, M.J., Yi, S., Han, J.-Y., Park, S.Y., Jang, J.-W., Chun, I.K. *et al.*, Oligomeric forms of amyloid-β protein in plasma as a potential blood-based biomarker for Alzheimer's disease. *Alzheimer's Res. Ther.*, 9, 98, 2017.
34. Wang-Dietrich, L., Funke, S.A., Kühbach, K., Wang, K., Besmehn, A. *et al.*, The amyloid-β oligomer count in cerebrospinal fluid is a biomarker for Alzheimer's disease. *J. Alzheimers Dis.*, 34, 985–94, 2013.
35. Dai, Y., Molazemhosseini, A., Liu, C., *In vitro* quantified determination of β-amyloid 42 peptides, a biomarker of neuro-degenerative disorders, in PBS and human serum using a simple, cost-effective thin gold film biosensor. *Biosensors*, 17, 29, 2017.
36. Li, X., Jiang, M., Cheng, J., Ye, M., Zhang, W. *et al.*, Signal multi-amplified electrochemical biosensor for voltammetric determination of tau-441 protein in biological samples using carbon nanomaterials and gold nanoparticles to hint dementia. *Microchim. Acta*, 187, 1–8, 2020.
37. Tao, D., Shui, B., Gu, Y., Cheng, J. et al., Development of a label-free electrochemical aptasensor for the detection of Tau381 and its preliminary application in AD and non-AD patients' sera. *Biosensors*, 9, 84, 2019.
38. Le, H.T.N., Park, J., Chinnadayyala, S.R., Cho, S., Sensitive electrochemical detection of amyloid beta peptide in human serum using an interdigitated chain-shaped electrode. Biosens. Bioelectron., 114, 111694, 2019.
39. Sun, L., Zhong, Y., Gui, J., Wang, X., Zhuang, X., Weng, J., A hydrogel biosensor for high selective and sensitive detection of amyloid-beta oligomers. *Int. J. Nanomed.*, 13, 843, 2018.
40. Negahdary, M. and Heli, H., An electrochemical peptide-based biosensor for the Alzheimer biomarker amyloid-β (1–42) using a microporous gold nanostructure. *Microchim. Acta*, 186, 1–8, 2019.

41. Ma, Y., Hu, Q., Liu, C., Wang, L., A nanospherical conjugated microporous polymer- graphene nanosheets modified molecularly imprinted electrochemical sensor for high sensitivity detection of α-Synuclein. *J. Electroanal. Chem.*, 862, 113994, 2020.
42. Karaboğa, M.N.S. and Sezgintürk, M.K., Cerebrospinal fluid levels of alpha-synuclein measured using a poly-glutamic acid-modified gold nanoparticle-doped disposable neuro-biosensor system. *Analyst*, 144, 611–621, 2019.
43. You, M., Yang, S., An, Y., Zhang, F., He, P., A novel electrochemical biosensor with molecularly imprinted polymers and aptamer-based sandwich assay for determining amyloid-β oligomer. *J. Electroanal. Chem.*, 862, 114017, 2020.
44. Chauhan, N., Chawla, S., Pundir, C.S., Jain, U., An electrochemical sensor for detection of neurotransmitter-acetylcholine using metal nanoparticles, 2D material and conducting polymer modified electrode. *Biosens. Bioelectron.*, 89, 377–383, 2017.
45. Moreira, F.T.C., Sale, M.G.F., Di Lorenzo, M., Towards timely Alzheimer diagnosis: a self-powered amperometric biosensor for the neurotransmitter acetylcholine. *Biosens. Bioelectron.*, 87, 607–614, 2017.
46. Kim, H., Lee, J.U., Song, S., Kim, S., Sim, S.J. et al., A shape-code nanoplasmonic biosensor for multiplex detection of Alzheimer's disease biomarkers. *Biosens. Bioelectron.*, 101, 96–102, 2018.
47. Derkus, B., Acar Bozkurt, P., Tulu, M., Emregul, K.C., Yucesan, C., Emregul, E., Simultaneous quantification of myelin basic protein and tau proteins in cerebrospinal fluid and serum of multiple Sclerosis patients using nanoimmunosensor. *Biosens. Bioelectron.*, 89, 781–788, 2017.
48. Cheng, X.R., Hau, B.Y., Endo, T., Kerman, K. *et al.*, Au nanoparticle-modified DNA sensor based on simultaneous electrochemical impedance spectroscopy and localized surface plasmon resonance. *Biosens. Bioelectron.*, 53, 513–518, 2014.
49. Real-Fernández, F., Passalacqua, I., Peroni, E., Chelli, M. *et al.*, Glycopeptide-based antibody detection in multiple sclerosis by surface plasmon resonance. *Sensors*, 12, 5596–5607, 2012.
50. Derkus, B., Emregul, E., Yucesan, C., Cebesoy Emregul, K., Myelin basic protein immunosensor for multiple sclerosis detection based upon label-free electrochemical impedance spectroscopy. *Biosens. Bioelectron.*, 46, 53–60, 2013.
51. Khatri, A., Punjabi, N., Ghosh, D., Maji, S.K., Mukherji. Detection, S., and differentiation of α-Synuclein monomer and fibril by chitosan film coated nanogold array on optical sensor platform. *Sens. Actuators B Chem.*, 255, 692–700, 2018.
52. Yang, X., Zhao, X., Liu, F., Li, H., Zhang, C.X., Yang, Z., Simple, rapid and sensitive detection of Parkinson's disease related alpha-synuclein using a DNA aptamer assisted liquid crystal biosensor. *Soft. Matter.*, 17, 18, 4842–4847, 2021 May 12, doi: 10.1039/d1sm00298h. PMID: 33889925.

53. Yildirim, A. and Bayindir, M., Turn-on fluorescent dopamine sensing based on *in situ* formation of visible light emitting polydopamine nanoparticles. *Anal. Chem.*, 86, 5508–5512, 2014.
54. Park, S.J., Song, H.S., Kwon, O.S., Chung, J.H., Lee, S.H., An, J.H. *et al.*, Human dopamine receptor nanovesicles for gate-potential modulators in high-performance field-effect transistor biosensors. *Sci. Rep.*, 4, 4342, 2014.
55. Govindaraju, S., Ankireddy, S.R., Viswanath, B., Kim, J., Yun, K., Thirumalraj, B. et al., Fluorescent gold nanoclusters for selective detection of dopamine in cerebrospinal fluid. *Sci. Rep.*, 7, 1–12, 2017.
56. Huang, Q., Chen, H., Xu, L., Lu, D., Tang, L., Jin, L. *et al.*, Visible-light-activated photoelectrochemical biosensor for the study of acetylcholinesterase inhibition induced by endogenous neurotoxins. *Biosens. Bioelectron.*, 45, 292–299, 2013.
57. Ge, C.-Y., Rahman, M., Zhang, W., Lopa, N.S., Jin, L., Yoon, S., Jang, H. *et al.*, An electrochemical immunosensor based on a self-assembled monolayer modified electrode for label-free detection of α-synuclein. Sensors, 20, 617, 2020.
58. Yin, Z., Cheng, X., Wang, G., Chen, J., Jin, Y., Q. *et al.*, SPR immunosensor combined with Ti 4+@ TiP nanoparticles for the evaluation of phosphorylated alpha-synuclein level. *Microchim. Acta*, 187, 1–9, 2020.
59. Abe, R., Ohashi, H., Iijima, I., Ihara, M., Takagi, H. *et al.*, "Quenchbodies": quench-based antibody probes that show antigen-dependent fluorescence. *J. Am. Chem. Soc.*, 133, 17386–17394, 2011.
60. Schöning, M.J. and Poghossian, A., Label-free biosensing: advanced materials, devices and applications, Springer, Cham, Switzerland, 2018.
61. Biancalana, M. and Koide, S., Molecular mechanism of Thioflavin-T binding to amyloid fibrils. *Biochim. Biophys. Acta*, 1804, 1405–1412, 2010.
62. Xue, C., Lin, T.Y., Chang, D., Guo, Z., Thioflavin T as an amyloid dye: fibril quantification, optimal concentration and effect on aggregation. *R. Soc. Open Sci.*, 4, 160696, 2017.
63. Wightman, R.M., Detection technologies. Probing cellular chemistry in biological systems with microelectrodes. *Science*, 311, 1570–1574, 2006.
64. Kita, J.M. and Wightman, R.M., Microelectrodes for studying neurobiology. *Curr. Opin. Chem. Biol.*, 12, 491–6, 2008.
65. Spira, M.E. and Hai, A., Multi-electrode array technologies for neuroscience and cardiology. *Nat. Nanotechnol.*, 8, 83–94, 2013.
66. Robinson, J.T., Jorgolli, M., Park, H., Nanowire electrodes for high-density stimulation and measurement of neural circuits. *Front. Neural. Circuits*, 7, 38, 2013 Mar 12, doi: 10.3389/fncir.2013.00038. PMID: 23486552; PMCID: PMC3594763.
67. Görtz, P., Siebler, M., Ihl, R., Hennin, U., Multielectrode array analysis of cerebrospinal fluid in Alzheimer's disease versus mild cognitive impairment: A potential diagnostic and treatment biomarker. *Biochem. Biophys. Res. Commun.*, 434, 293–297, 2013.

68. Rusling, J.F. and Forster, R.J., Biosensors Designed for Clinical Applications. *Biomedicines*, 9, 702, 2021.
69. Kenny, A., McArdle, H., Calero, M., Rabano, A., Madden, S.F., Adamson, K., Forster, R.J., Spaine, E., Prehn, J.H.M., Henshall, D.C. *et al.*, Elevated Plasma microRNA-206 Levels Predict Cognitive Decline and Progression to Dementia from Mild Cognitive Impairment. *Biomolecules*, 9, 734, 2019.
70. Raoof, R., Bauer, S., El Naggar, H., Connolly, N.M.C., Brennan, G.P., Brindley, E., Hill, T., McArdle, H., Spain, E., Forster, R.J. *et al.*, Dual-Center, Dual-Platform MicroRNA Profiling Identifies Potential Plasma Biomarkers of Adult Temporal Lobe Epilepsy. *EBioMedicine*, 38, 127–141, 2018.
71. Reich, D.S., Lucchinetti, C.F., Calabresi, P.A., Multiple Sclerosis. *N. Engl. J. Med.*, 378, 169–180, 2018.
72. Chitnis, T., Glanz, B.I., Gonzalez, C., Healy, B.C., Saraceno, T.J., Sattarnezhad, N. *et al.*, Quantifying neurologic disease using biosensor measurements in-clinic and in free-living settings in multiple sclerosis. *NPJ Digit. Med.*, 11, 123, 2019.
73. Block, V.A. *et al.*, Remote physical activity monitoring in neurological disease: a systematic review. *PLoS One*, 11, e0154335, 2016.
74. Bradshaw, M.J., Farrow, S., Motl, R.W., Chitnis, T., Wearable biosensors to monitor disability in multiple sclerosis. *Neurol. Clin. Pract.*, 7, 354–362, 2017.
75. Kos, D., Nagels, G., D'Hooghe, M.B., Duquet, W., Ilsbroukx, S., Delbeke, S., Kerckhofs, E., Measuring activity patterns using actigraphy in multiple sclerosis. *Chronobiol. Int.*, 24, 2, 345–56, 2007, doi: 10.1080/07420520701282364. PMID: 17453852.
76. Sandroff, B.M. and Motl, R.W., Comparison of ActiGraph activity monitors in persons with multiple sclerosis and controls. *Disabil. Rehabil.*, 35, 725–731, 2013.

14

Nanotechnology-Based Strategies for Improvement of Disease Diagnostic Systems for Future Outbreaks

Busiswa Dyan[1*], Tintswalo N. Mgwenya[1], Kamogelo S. Setlolamathe[1], Phumlane S. Mdluli[1,2] and Nicole R.S. Sibuyi[1,3†]

[1]*Diagnostic Unit, Health Platform, Advanced Materials Division, Mintek, Randburg, South Africa*
[2]*Faculty of Applied Sciences, Durban University of Technology, Durban, South Africa*
[3]*Department of Science and Innovation/Mintek Nanotechnology Innovation Centre, Biolabels Research Node, Department of Biotechnology, University of the Western Cape, Bellville, South Africa*

Abstract
In the last decade, the world has faced major outbreaks due to viral infections such as severe acute respiratory syndrome coronavirus 2 (SARS-Cov-2) and Ebola virus. The severe health impact that these viruses caused led to pandemics and instability in the world. The recent Coronavirus disease 2019 (COVID-19) pandemic caused by SARS-Cov-2 infections, demonstrated that there is an urgent need to improve diagnosis of infectious and chronic diseases; including tuberculosis and cancer. The lesson learned during the pandemic is that simple, rapid, and affordable point-of-care (PoC) diagnostic tools can prevent transmissions and manage disease outbreaks. Nanotechnology have shown a huge potential in developing rapid and cost-effective diagnostic tools that can improve diagnosis of diseases at PoC and low resource areas, thus controlling and preventing the outbreaks. Nano-based sensors are highly favored because of the unique properties of nanomaterials used in these systems such as small size, surface adaptability and multifunctionality. These unique properties resulted in the development of novel and nano-smart devices

Corresponding author: BusiswaD@mintek.co.za
†*Corresponding author*: NicoleS@mintek.co.za

Suvardhan Kanchi, Ayyappa Bathinapatla, Anitha Varghese and Phumlane Selby Mdluli (eds.) Nano-Biosensor Technologies for Diagnosis of Infectious Diseases, (423–448) © 2025 Scrivener Publishing LLC

with enhanced activity, high sensitivity and specificity which can improve disease diagnosis and intervention. This chapter will discuss the use and role of various nanomaterials such as gold nanoparticles (AuNPs) in development of nano-smart diagnostic tools in relation to the conventional systems.

***Keywords*:** Infectious diseases, nanotechnology, point-of-care, diagnostics, gold nanoparticles, paper-based immunoassays

14.1 Introduction

Infectious diseases can cause widespread repercussions and even death, especially in resource-limited areas. This is mostly escalated by late diagnostics, followed by lack of effective treatments or vaccines. In the midst of outbreaks, the successful control and prevention of disease transmission rely on the rapid disease diagnosis [1]. Current diagnostics methods for infectious disease include molecular methods and immunoassays.

In order to reduce the spread of the pathogens during a global pandemic, it becomes crucial to look for quick diagnostic techniques. Numerous companies use reverse-transcriptase polymerase chain reaction (RT-PCR) as a gold standard method for diagnosis of infectious diseases. Despite being an accurate method, RT-PCR has a number of drawbacks that make it challenging to control the spread of infectious disease during an outbreak. In an effort to expedite diagnosis, serological tests based on the detection of antibodies and antigens, such as the lateral flow immunochromatographic assay (LFA), are introduced [2]. Serology tests are known to be less time consuming, have a low risk linked to contamination of samples, and requires less complicated equipment. In addition to serological testing, there is still an opportunity to create sensitive, rapid, and precise diagnostic methods. The use of biosensors in disease diagnosis is also promising, for example gold nanoparticles (AuNPs)-based electrochemical biosensor for SARS-CoV-2 spike protein [3], LFA to detect SARS-CoV-2 nucleocapsid protein [4], and SARS-CoV-2 Induced reactive oxygen species electrochemical sensor [5] have been developed. Biosensors have many benefits, including high sensitivity and specificity, low analysis costs, rapid execution time, a low limit of detection, and the ability to build small platforms that can be used at the PoC [6].

14.2 Infectious Disease Outbreaks

Pathogens can be transmitted from infected animals or animal products into the human population in a number of different ways. Some examples of how a disease can spread from animals to humans include a hunter touching diseased animal blood, a farmer breathing in air exhaled by sick animals, and contaminated meat being served at an open-air market. Influenza, Ebola, Zika, and the human immunodeficiency virus (HIV) SARS-CoV-2 are just a few of the animal viruses that infect humans [7]. These viruses cause illnesses that are easily disseminated among animals reared in overcrowded conditions. viruses are able to multiply only inside living animals, and they can quickly adjust to a new host [8]. A single infection can quickly become a widespread pandemic. Connectivity between countries and regions has increased dramatically, making it easy for infections to spread rapidly across continents, as witnessed with COVID-19. Many epidemics, such as influenza, are expected each year. Although influenza epidemics are uncommon, they do occur frequently. Several pandemics are more severe than others. A single case of an infectious disease may sometimes be considered an outbreak. This may be the case if the disease is uncommon, such as foodborne botulism, or has significant public health repercussions, such as anthrax [9, 10].

The phrases “endemic,” “outbreak,” “epidemic,” and “pandemic” as described in Figure 14.1, all refer to the presence and spread of a health problem in relation to their expected rates and geographical locations [11]. An outbreak is a sudden rise in the incidence of a disease when the number of cases exceeds what would be expected for the place or season [12]. The incidence rate shifts depending on the pathogen responsible for the illness, as well as the duration and nature of the patient’s exposure to the

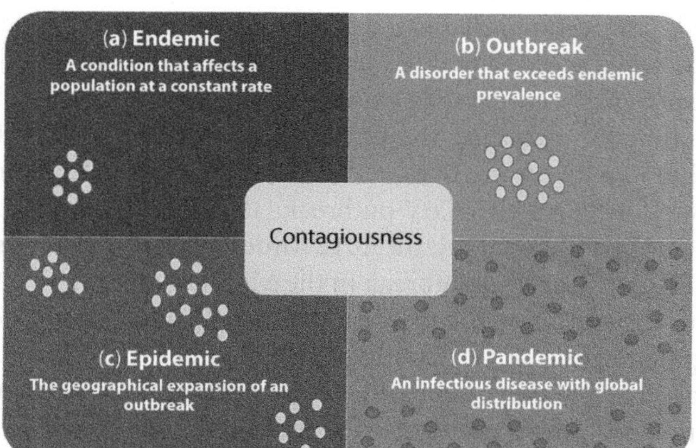

Figure 14.1 Interchangeable terms used to define a contagious disease in relation to a geographical region: endemic (a), outbreak (b), epidemic (c), and pandemic (d).

pathogen. An outbreak might affect hundreds of people in one town or thousands across a whole continent [12]. Diseases with an environmental source, such as those that spread through contaminated water or food, are also included in the category of outbreaks, which is more commonly used to refer to infectious disease epidemics [12, 13]. In a single population, an epidemic is the rapid transmission of a disease to a large number of hosts in a short period of time [14]. An example of an epidemic is the sudden, severe outbreak of a disease, or the occurrence of an abnormally high number of instances of a disease in a community or region. A pandemic, on the other hand, is an outbreak of an infectious illness that has spread across a vast region, such as multiple continents or the entire world, and affected a significant number of people [14]. In a pandemic, a wide range of countries get infected rapidly by the same illness [13]. The recent SARS-CoV-2 outbreak, which claimed millions of lives worldwide, is an example of a pandemic. The pandemic was tracked to have started in Huanan South China seafood market in Wuhan, China. The SARS-CoV-2 initially spread in China and quickly spread to other countries causing a pandemic. A historical perspective on pandemics that occurred in the 20th century sheds light on the "mother of pandemics," the H1N1 influenza pandemic that hit the world from 1918 to 1919. The root cause of H1N1 was a highly infectious strain of the influenza-A virus [15]. The H1N1 pandemic consisted of two waves of infection; the first wave, which occurred in early 1918, caused moderate sickness. A second wave occurred later in the year and claimed between 20 and 50 million lives [16]. Five hundred million people were estimated to have been infected by the 1918 influenza pandemic [16]. Due to the high death rates caused by the H1N1 pandemic, the global life expectancy decreased by a number of years. The death rate is believed to have exceeded that of the entire First World War [17].

The twenty-first century has experienced a surge of severe infectious disease outbreaks, including the COVID-19 pandemic, which has had a catastrophic impact on the lives and livelihoods of people all around the world [18]. The 2003 SARS-CoV outbreak [19], the 2009 influenza A virus subtype H1N1 pandemic [20], the 2012 Middle East respiratory syndrome coronavirus (MERS-CoV) outbreak in the Middle East [21], the 2013-2016 Ebola virus disease epidemic in West Africa [22], and the 2015 Zika virus disease epidemic [23] are all examples of epidemics that have occurred in recent years. Each of these pandemics caused widespread death and illness as they spread or were transmitted from country to country. Neglected tropical diseases, HIV infection, tuberculosis, and malaria continue to be major causes of death and disability in nations with poor and lower-middle incomes [24].

Infections like seasonal influenza and the new COVID-19 are examples of infectious conditions with which the global community has learnt to coexist. According to the World Health Organization (WHO), between 3 and 5 million individuals have severe flu each year, with up to 650,000 succumbing to its effects [25, 26]. As of the 21st of February 2023, the WHO had received reports of 757,264,511 confirmed cases of COVID-19 worldwide, with 6,850,594 fatalities [27]. As of the 25th of February 2023, a total of 13,224,955,795 doses of vaccine will have been distributed [27]. During times of peak infections, healthcare systems were put under extreme strain, and those working in the medical field had a higher chance of contracting the illness, and there was a ripple effect on the economy as schools and businesses came to a total shutdown.

14.2.1 Significant Epidemics and Pandemics Across the Globe

A number of diseases that have rapidly spread and resulted in epidemics throughout human history could have been averted if individuals had access to vital information on how these diseases propagate in addition to improved availability of health care resources [14]. There were several other pandemics that occurred in the 20th century, which claimed thousands of lives, and had far-reaching consequences for human society, agriculture, and the economy. The influenza pandemic had the highest recurrence rate and persisted into the twenty-first century. Since 1918, there have been other pandemics of influenza, including those in 1957 and 1968, as well as H1N1 in 2009 and other influenza-related pandemics such as the Russian flu pandemic of 1989 and the recent COVID-19 pandemic, all of which were caused by influenza viruses that were not H1N1. Table 14.1 provides a brief overview of recent and historical pandemics.

There are a number of important infectious diseases with a high prevalence across the globe, but they are not yet categorized as epidemics or pandemics due to a lack of definitive data, including duration and death toll. Tuberculosis (TB), even though curable and preventable, has been reported by the WHO as the leading infectious cause of death worldwide, with an estimated 10 million new cases and 1.5 million deaths annually (before the COVID-19 pandemic) with infections caused by a single infectious agent (*Mycobacterium*) [24]. Unfortunately, there was a dearth of resources that could- provide dates and death tolls during significant TB pandemics. The global pandemic of TB that began in Europe in the 18th and 19th centuries and followed a seasonal pattern persists today [41].

Table 14.1 Summary of pandemics that took place worldwide.

Outbreak	Cause	Duration	Statistics	Sign and symptoms	Preventative measures	Refs
COVID-19	SARS-Cov-2	2019–Ongoing	> 500 million infections and > 6.5 million deaths	Pneumonia, respiratory illnesses, high fever, dry cough, headache, sore throat, runny nose, chills, fatigue, loss of appetite	Vaccination, avoiding public spitting, protective clothing, isolation, hygiene, global shutdown	[28]
SARS	SARS-Cov	2002–2004	8098 infected and 774 deaths	High fever, cough, shortness of breath headaches, feeling of discomfort, body pains	Hygiene, cover mouth when sneezing or coughing, isolation	[19]
H1N1 influenza	(H1N1) pdm09 virus	2009	60.8 million cases and 151,700–575,400 deaths	Fever, cough, headache, muscle and joint pain, severe malaise, sore throat, runny nose	Quarantine, hand hygiene, isolation, personal protective equipment, school closures, social distancing, workplace closures, and vaccines	[29]

(Continued)

Table 14.1 Summary of pandemics that took place worldwide. (*Continued*)

Outbreak	Cause	Duration	Statistics	Sign and symptoms	Preventative measures	Refs
HIV/AIDS	HIV	1981–Ongoing	84.2 million infections and 40.1 million deaths	Flu-like symptoms, swollen glands, skin rashes, non-healing lips sore, thrush, TB, bad cough and pneumonia, pains in hands and feet, non-stop diarrhea, weakness, tiredness, skin tumors, weight loss, headaches, seizure, difficulty in concentrating and in swallowing	Antiretroviral therapy, use medicines to prevent HIV (PrEP, PEP), choosing less risky sexual behaviors, getting tested for HIV, testing and treatment of sexually transmitted diseases	
Asian flu	Influenza A (H2N2) virus	1957–1958	1.1 million deaths	Cough and mild fever, vomiting, pneumonia, otitis	Antibiotics and intramuscular injection against virus A2	[30]

(*Continued*)

Table 14.1 Summary of pandemics that took place worldwide. (*Continued*)

Outbreak	Cause	Duration	Statistics	Sign and symptoms	Preventative measures	Refs
H3N2/Fujian flu	Influenza A(H3N2) virus	1968	1 million deaths	Malaise, fever, myalgia, cough, headache, coryza, and sore throat	Antiviral medications, influenza vaccines	[31]
Spanish influenza	H1N1 virus	1918– 1919	500 million infected and 20–50 million deaths	Pneumonia, respiratory illnesses, high fever, dry cough, headache, sore throat, runny nose, chills, fatigue, loss of appetite	Avoiding public spitting, Using protective gear gloves & masks, Isolation, washing hands thoroughly and frequently, closing schools and gyms	[32]
Russian flu	*Myxovirus influenzae*	1989–1990	1 million deaths	Multivisceral and neurological involvement, skin rash, early iterative relapses	Closing of schools and factories, hand hygiene, isolation	[33, 34]

(*Continued*)

Table 14.1 Summary of pandemics that took place worldwide. *(Continued)*

Outbreak	Cause	Duration	Statistics	Sign and symptoms	Preventative measures	Refs
Cholera	*Vibrio cholerae* bacteria	1817–1824	Total deaths remain unknown	Acute watery diarrhea with severe dehydration	Surveillance, water, sanitation and hygiene, social mobilization, treatment, and oral cholera vaccines	[35]
Small pox	Variola virus	1870–1875	500,000 deaths	High fever, vomiting, mouth sores, fluid-filled lesions	Vaccines	[36, 37]
Black death	*Yersinia pestis*	1347–1352	75–200 million deaths	Fever, chills, swellings leaking blood and pus	Mouths, noses with scented handkerchiefs, and isolation	[38, 39]
Antonine plague	Small pox, measles	165AD–180 AD	2000 deaths/day	Abdominal pain, diarrhea, vomiting, bleeding from the mouth, nose, rectum, or skin	Never treated successfully	[40]

TB/HIV syndetic refers to the interconnectedness of TB and HIV/Acquired immunodeficiency syndrome (AIDS) epidemics [42].

Hepatitis, like TB, has a high frequency but is not yet recognized as an epidemic or pandemic. As of 2019, the WHO estimated that 296 million people worldwide would have chronic hepatitis B, with an additional 1.5 million new cases emerging each year [43]. Almost 820,000 fatalities are attributed to hepatitis B each year (primary liver cancer), according to the WHO. Hepatitis B is widespread in parts of Asia and Africa [43]. In addition, nearly 1% of all hepatitis B virus (HBV) infections are found in roughly 2.7 million people who have been infected with both HBV and HIV [44]. About 1.5 million new cases of chronic hepatitis C are reported each year, making the total number of people living with the virus to be around 58 million [45]. Cirrhosis and hepatocellular carcinoma accounted for around 80% of 2019 deaths from the condition [46]. Over the years, hepatitis C virus (HCV) has caused a number of pandemics. Other diseases, such as HIV, also had their origins in West Africa before becoming pandemic by the end of the 20th century. Although HIV was not considered an endemic disease before the turn of the century, it is reasonable to argue that it is currently in some regions of the world after two decades in the 21st century.

14.3 Pandemic-Potential Priority Diseases for Future Outbreaks

Viruses having pandemic potential arise on a regular basis, although not all of them trigger a pandemic. The WHO monitors viruses such as influenza viruses having pandemic potential and prepares for future pandemics. The WHO has recognized diseases that represent the greatest threat to public health due to their potential for epidemic transmission and/or the availability of weak or non-existent solutions. COVID-19, Crimean-Congo hemorrhagic fever, Ebola virus disease and Marburg virus disease, Lassa fever, MERS-CoV and SARS, Nipah and henipaviral diseases, Rift Valley fever, and Zika are among the identified diseases [47, 48].

There are minimal resources available for disease research and development, despite the fact that there are several diseases in the world that can cause pandemics. For a rapid and effective response to an epidemic or pandemic, research and the development of countermeasures are necessary.

During the 20th century, the 1918 influenza epidemic was most likely to strike the young and healthy, swiftly killing those aged 15 to 45 years of age.

This was owing to their robust immune systems, which unleashed a 'cytokine storm' of virus-fighting chemicals in an immunological response. These chemicals attached themselves to lung tissue and caused fatal damage [15–17]. Targeting cytokines during the management of disease in patients could have improved survival rates and reduced mortality. Nevertheless, due to a lack of suitable technologies, medicines, and resources to enable the research and development of proper diagnosis and preventative measures, there was no strategy designed to avoid the global spread of infectious illnesses.

Improvements in technology and increased investment in research and development in the 21st century have led to greater success in halting the spread of infections and establishing effective strategies for managing and treating existing outbreaks. Many people are still getting infected and dying from outbreaks, for example there are many people who died from COVID-19, and most of these patients had hyper-inflammation with features of cytokine storm syndrome and associated acute respiratory distress syndrome [28]. Only anti-inflammatory treatments have improved survival in these patients [28]. Young individuals, in general, have robust immune systems; therefore, they have felt safe from COVID-19. Still, there are many aspects of the virus and its effects on humans that remain unknown. Cytokine storms, which pose a significant health risk, are being countered by newly available medications. It is clear that more work needs to be done to identify new ways to combat the infections. Opposed to the old ways of combating the infections, the discovery of vaccines provide protection against many diseases. But to some extent, for example, COVID-19, despite evidence that vaccinated people can be re-infected by variants like Omicron and Delta [49], the vaccine is still quite successful in protecting people from severe cases of COVID-19 [50].

Diagnostics, vaccinations, and antivirals have become increasingly successful in preventing and responding to infectious disease threats as a result of scientific and technological advancements. It is crucial to note that progress has been achieved in combating these diseases and the new SARS-CoV-2 prevention efforts, even if HIV/AIDS, TB, and malaria remain major killers and disablers in low- and middle-income nations. The ability of modern science to rapidly respond to risks from new illnesses was exemplified by the rapid development of a vaccine against SARS-CoV-2.

The symptoms of the recurring seasonal flu infections are similar, and their treatment is also related. If researchers can establish a connection between viruses like the Russian flu and COVID-19, they may be able to develop effective counter measures against future pandemics. Scientists can now identify the type of virus that infected the lungs of Russian flu patients who had their tissue preserved. Effective preventive measures must be developed to avoid the global spread of infectious diseases. One possible way to get insight and contain infections is through early diagnostics.

14.4 Combating Infectious Diseases Through Diagnostics

Despite progress in medical science, which include the development of vaccines and therapeutic interventions; infectious diseases continue to affect millions of people around the world [51]. Most infectious diseases show similar symptoms such as dyspnea, headache, loss of appetite, loss of taste and smell, panting, sore throat, vomiting, diarrhea, rhinorrhea, and abdominal pain [52]. Fever (85.6%), cough (68.7%), and fatigue (39.4%) are among the most reported symptoms. This makes it difficult to properly diagnose diseases based on symptoms. Proper diagnosis of diseases is necessary to prescribe appropriate and suitable interventions and therapy.

Developing countries have many initiatives to reduce the burden of diseases such as vaccination, improved water supply, screening-based mass treatment, use of vector control and intermittent preventive treatment [53–55].

14.4.1 Clinical Methods for Diagnosis of Infectious Diseases

Microbiological (microscopy or culture), immuno- (ELISA) and molecular (nucleic acid amplification) assays are the most employed tests during outbreaks of infectious diseases, as shown in Figure 14.2. All these methods are performed in laboratories under strict regulations to avoid disease transmissions [56]. In clinical laboratories the culture method is still the core method to characterize and identify unknown pathogens. It offers important information about the viability of the pathogen, isolates can be fingerprinted (by phenotyping or genotypic methods) and susceptibility to antibiotics can be screened at the same time [57].

Even though the culture method has stood the test of time, it has its disadvantages such as long turnaround time (takes 3 to 5 d) for microbial growth and identification [58]. These limitations delays diagnosis and treatment of patients with suspected infectious disease. Disease diagnosis at early stages is very crucial as wrong prescriptions can lead to antibiotic resistance [59]. Drug resistance threatens the treatment and control of outbreaks and increases the health care costs making difficult to contain

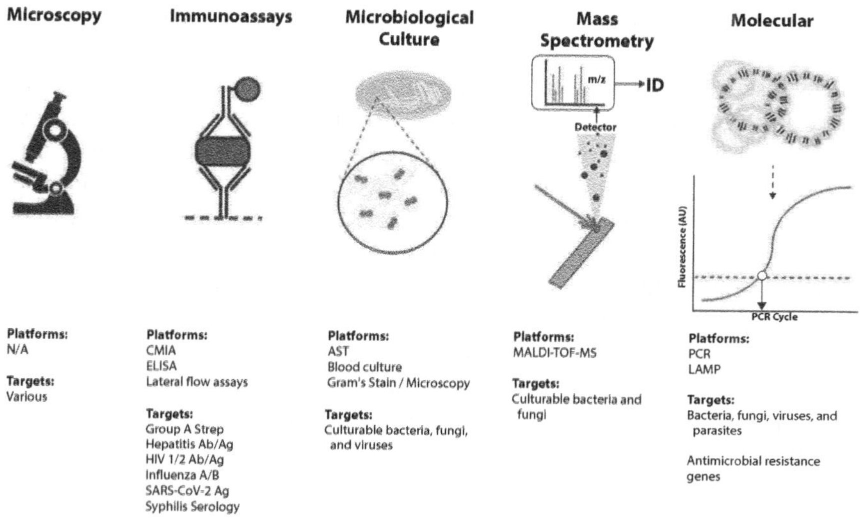

Figure 14.2 Conventional laboratories methods used for diagnosis of infectious diseases. Infections from unknown source starts with microbial culture for fingerprinting, immunoassays and molecular assays are used as confirmatory tests. Reprinted with permission from Elsevier [56].

it [60]. Over the last decade, molecular methods have been introduced to compensate for the shortcomings of microbiology tests [61]. The introduction of nucleic acid-based assays such as PCR, ligase chain reaction (LCR), nucleic acid sequence-based amplification (NASBA), and strand displacement amplification (SDA) as routine tests for diagnostic offered sensitivity and specificity [62, 63]. It is important to mention the initial step of PCR requires extraction of the deoxyribonucleic acid and ribonucleic (DNA/RNA) from desired pathogen [64]. In 2019, a novel coronavirus strain posed a health threat globally and still an ongoing pandemic [65]. Although the science world had advanced with technologies and have guidelines on outbreaks, there was still a struggle to contain the Coronavirus.

It became urgent to accurately diagnose the virus as it was spreading at high rates all over the world threatening people's lives. Nucleic acid amplification test (specifically real-time PCR, RT-PCR) of nasopharyngeal or oropharyngeal swabs were used as preferred method for diagnosis of the virus [66]. RT-PCR is a known standard method that is still being used to test for the virus. As accurate and sensitive as RT-PCR is, it has limitations such as the use of highly specialized equipment, trained personnel and a long turnaround time (4–8 h) to run a sample [67]. Because COVID-19 is a contagious disease and spread at an alarming rate, the PCR method was not a suitable method as results took longer. Due to the global pandemic it become important to search for other quick diagnostic methods to minimize disease transmission. Serological tests based on antigen-antibody detection [2] are used to determine exposure history or immune status of patients. Immunoglobin M (IgM) appears in the patient's blood after 3 to 8 d of infection, and immunoglobin G (IgG) after 8 d [2]. In order to speed up diagnosis instead of antibody-based tests, the antigen tests can be used to detect the presence of the pathogens within blood [68], serum, and plasma samples [69]. The antigen-based test detects protein fragments of the pathogens. The serological tests are able to give results within 15 min compared to RT-PCR which takes hours [70]. However, serological assays also need to be performed by experts and in a controlled environment. To minimize the impact of outbreaks and prevent transmissions, diagnostic at a point of need is desirable.

14.4.2 LFAs for PoC Diagnostics (PoCD)

Emerging viral outbreaks are a perpetual problem, and each one requires a unique rapid response for managing the epidemic transmission. The demand for rapid tests is growing, which encourages the creation of novel methods for clinical diagnosis [71]. In comparison to traditional analytical techniques, LFAs have a number of benefits outlined in Table 14.2, including affordability, ease of use, on-site response, speed, and results that can be seen with the naked eye. The lengthy turnaround times of the diagnostic techniques currently in use make it challenging to deliver treatment on schedule. It also takes competent individuals to conduct the test and

Table 14.2 Advantages and disadvantages of LFAs. Adapted from reference [71].

Advantages	Disadvantages
• Less interference • User friendly • Rapid • Portable • Good reproducibility • Long shelf life • Stable • Inexpensive • Do not require an expert • Adequate specificity	• Reproducibility varies a lot • Low signal intensity • Poor quantitative discrimination • Pretreatment of sample maybe required • Mostly qualitative or semi quantitative

analyze the data at centralized laboratories [72]. Paper-based immunoassays offer a great technique that will achieve the requirements needed for PoCD [73]. The availability of a simple, accurate, inexpensive and ideally PoCD is very important to control the infections at its early stages as disease can have a fatal result if not treated [74]. LFAs allow for rapid PoCD of diseases by detecting associated antigens, proteins, antibodies and small molecules [75]. LFAs can be used to examine a wide range of biological samples including urine [76], saliva [77], plasma [78], whole blood [79], and other fluids.

The paper-based PoC LFAs are made up of three components, *i.e.*, paper as the substrate, biomolecules (antibodies or antigens) as the detection element, and reporter molecules as the signal transforming element [80]. The detection element (*e.g.*, antibodies) are attached to colored or fluorescent labels [81], the end-user will interpret the visual effect that these labels provide as test results. Colored lines in both control and test lines represent a positive test, as shown in Figure 14.3. Nowadays, colloidal gold is the most widely used label in commercial LFA. Due to their strong affinity for biomolecules easily modifiable surface and outstanding durability in optimized experiments, AuNPs are widely used. They can also be readily conjugated with biomolecules such as antibodies to achieve a desired biosensor design [82]. Many other labeling substances, such as carbon, selenium, fluorescent labels, up-converting phosphor, magnetic nanoparticles, and enzymes have been used in LFIAs. They however have their limitation. For instance, carbon nanotubes have been shown to exhibit a limit of detection that is 10-fold lower than that of gold [83]. Other limitations include lower sensitivity, weaker signals, delicate to pH and higher costs [84].

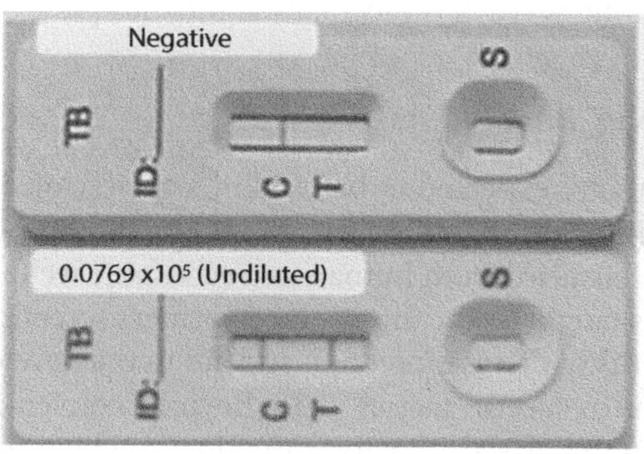

Figure 14.3 Detection of TB antigens using LFAs. One line at the control (C) line represents a negative test, two lines at both the C and test (T) lines indicates a positive test. Reprinted with permission from MDPI [85].

14.5 Nanotechnology in Diagnostics

Nanotechnology is the branch of science and engineering that design and produce structures with dimensions in the nanoscale (1–100 nm) range [86]. Nanomaterials have remarkable physical, chemical, electrical, optical, or biological properties compared to their bulk or macro-scale materials [87]. Because of their small size and large surface area, they exhibit enhanced electrical conductivity, surface reactivity, and magnetic characteristics, making them suitable tools for development of novel diagnostic systems [88]. These properties could advance the POCD, enhance the performance of the existing technologies such as microfluidic devices, in vitro assay, or lab-on-a-chip devices [89]. Nanotechnology can contribute vastly in the development of cost-effective, sensitive and rapid diagnostic systems. A number of nanotechnology-based diagnostic systems that are able to detect disease-specific biomarkers, pathogens and toxins have been explored. This is especially relevant during outbreaks, where instant results are required for better clinical decision making and interventions [90].

14.5.1 Rapid AuNPs-Based Diagnostics

AuNPs, among the various nanomaterials (organic and inorganic NPs) used in theranostics, can serve in multiple sensors due to their distinct properties such as optical, fluorescent, and conductivity [91]. As a result, they have been widely used in LFAs and in-solution in colorimetric assays as detection probes. These type of assays offer a broad and distinctive set of features that may be used to enhance POC tests. In reality, POC tests utilizing nanotechnology, in particular AuNPs, have altered how we think about the upcoming wave of molecular diagnostics [92].

14.5.1.1 AuNPs-Based LFAs

Nano-based tools are expected to be used in a variety of situations around the world, ranging from handling present (COVID-19, Ebola) and future health threats to ensuring both human and economic welfare. The COVID-19 pandemic has taught us the importance for mass screening in an attempt to obtain the status of disease infection and aid in the development of effective and targeted response tactics [93]. Beyond conventional diagnostic methods, the AuNPs-based LFA for COVID-19, which was developed, is currently the best PoC testing approach. It has the remarkable advantages of being quick, affordable, simple to use, portable, and fast acting, which can help to contain the virus's spread throughout the world [94]. Many other

AuNP-based LFA are currently being developed to help detect infectious diseases rapidly and approved by the WHO. These include LFA for detection of HIV [95], TB [85], Ebola [1], and malaria [96], to name a few.

The lateral-flow devices can be enhanced by incorporating nanopaper and nano-channel based sensors that enable the users to observe the results with smartphones or the human eye [97]. Smartphones are now becoming a platform for imaging and bioanalysis. Many of the smartphones can now be supported to behave like a traditional benchtop spectroscopic and imaging devices thanks to their built-in light sources, batteries, cameras, network connectivity, and processing power [98]. Smartphones are connected to the sensor system, which then conducts analysis while using NPs as an add-on. The smartphone will then interpret the findings [99]. This battery operated and smartphone camera-based amplifications with inorganic nanomaterials such as AuNPs and quantum dots are coming to be the next generations for detection of infectious diseases [99].

14.5.1.2 AuNPs-Based In-Solution Assays

The inability of the global health system to quickly and accurately diagnose diseases is a significant issue. It is challenging to control the pandemic in terms of which persons to quarantine and the capability to track the infection across communities due to the lack of quick diagnostic tools, especially in remote locations [100]. The ability to conduct widespread testing has already demonstrated benefits to nations like South Korea and Singapore, providing precise information about strict contact tracing and mandatory quarantine for virus carriers, which results in greater control and slows the spread of disease [101].

As it was previously mentioned above, nanotechnology-based colorimetric bioassays are appealing and attractive in biosensor design for their simplicity, visual output, and no need for complex instruments [102]. Nano-enabled biosensors with highly selective optical detection techniques have been widely developed for the quick detection of a variety of targets [103]. Recently, a group developed a colorimetric assay based on AuNPs, capped with suitably designed thiol-modified antisense oligonucleotides (ASOs) for SARS-CoV-2 N-gene (nucleocapsid phosphoprotein), as shown in Figure 14.4. The assay utilizes the intrinsic optical properties of plasmonic AuNPs and targeting ability of the ASOs to develop a selective, and "naked-eye" detection platform for the COVID-19 causative virus, SARSCoV-2, without the requirement of any sophisticated instrumental techniques. Results are obtained within 10 min and assessed through the change in color of a solution [100].

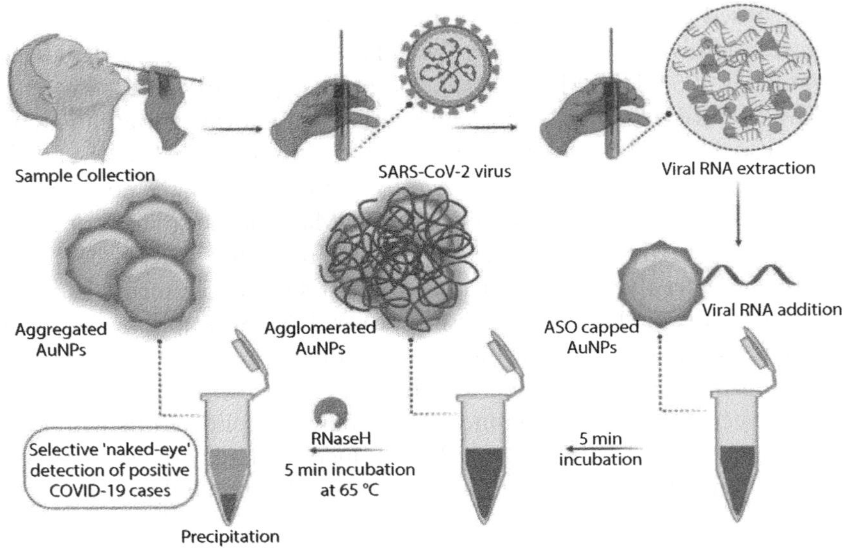

Figure 14.4 Detection of SARS-CoV-2 using gold AuNPs-based in a solution. Reprinted with permission from ACS [100].

The rapid response of such test make it possible for mass screening of infections during future outbreak, as seen with COVID-19 pandemic. The ability to quickly diagnose a disease will be a key part in handling future outbreak thus "flattening the curve" and reducing the rate of new infections.

14.6 Conclusion

Because outbreaks have an impact on nations' economies, they pose a serious threat to civilization. This is due to the fact that countries may implement measures like travel restrictions, school closings, therapy, and prophylaxis that may allow us to limit disease epidemics. Outbreaks can be controlled by developing vaccinations and having a reliable diagnosis in place. How important it is to have quick diagnostic technologies available to contain the outbreak has been made clear by the COVID-19 pandemic. PCR-based techniques were used as diagnostic methods due to the dependability of the technology. Although PCR is a reliable method, there are a few disadvantages, including the requirement for highly skilled people and the amount of time it takes to get results. Nanotechnology has emerged as useful tool that can be used in diagnosis of disease analytes. The application and development of biosensors using nanotechnology offers massive advantages in the diagnosis of viral diseases. Because it offers low cost, rapid diagnostic, user friendly methods that can help contain and manage the epidemics.

References

1. Phan, J.C. *et al.*, Lateral Flow Immunoassays for Ebola Virus Disease Detection in Liberia. *J. Infect. Dis.*, **214**, S222–S228, 2016.
2. Li, Z. *et al.*, Development and clinical application of a rapid IgM-IgG combined antibody test for SARS-CoV-2 infection diagnosis. *J. Med. Virol.*, **92**, 1518–1524, 2020.
3. Karakuş, E., Erdemir, E., Demirbilek, N., Liv, L., Colorimetric and electrochemical detection of SARS-CoV-2 spike antigen with a gold nanoparticle-based biosensor. *Anal. Chim. Acta*, **1182**, 2021, 10.1016/j.aca.2021.338939.
4. Kim, H.Y. *et al.*, Development of a SARS-CoV-2-specific biosensor for antigen detection using scFv-Fc fusion proteins. *Biosens. Bioelectron.*, **175**, 112868, 2021.
5. Miripour, Z.S. *et al.*, Real-time diagnosis of reactive oxygen species (ROS) in fresh sputum by electrochemical tracing; correlation between COVID-19 and viral-induced ROS in lung/respiratory epithelium during this pandemic. *Biosens. Bioelectron.*, **165**, 112435, 2020.
6. Yasri, S. and Wiwanitkit, V., Sustainable materials and COVID-19 detection biosensor: A brief review. *Sens. Int.*, **3**, 100171, 2022.
7. Priyadarsini, S.L., Suresh, M., Huisingh, D., What can we learn from previous pandemics to reduce the frequency of emerging infectious diseases like COVID-19? *Global Transitions*, **2**, 202–220, 2020.
8. Brier, J. and Jayanti, L.D., Introduction to Animal Viruses. *Viruses*, **21**, 1–9, 2020.
9. Rasetti-Escargueil, C., Lemichez, E., Popoff, M.R., Public health risk associated with botulism as foodborne zoonoses. *Toxins (Basel)*, **12**, 2019, doi: 10.3390/toxins12010017.
10. Alebie, A., Misgie, F., Atnaf, A., Surafel, K., A Review on Anthrax and its Public Health and Economic Importance. *Acad. J. Anim. Dis.*, **4**, 196–204, 2015.
11. Singer, B.J., Thompson, R.N., Bonsall, M.B., The effect of the definition of 'pandemic' on quantitative assessments of infectious disease outbreak risk. *Sci. Rep.*, **11**, 1–13, 2021.
12. Maciel, E.L., Oliveira, W.K., Siqueira, P.C., Croda, J., Are we near the end of the pandemic? *Rev. Soc. Bras. Med. Trop.*, **55**, 1–4, 2022.

13. O'Quin, J., Outbreak Management, in: *Infect. Dis. Manag. Anim. Shelter*, Second Ed., pp. 113–142, 2021, doi:10.1002/9781119294382.ch6.
14. Jamison, D.T., *Disease Control Priorities: improving health and reducing poverty*, 3rd edition, vol. 391, The Lancet, Washington (DC), 2018.
15. Taubenberger, J.K., The origin and virulence of the 1918 "Spanish" influenza virus. *Proc. Am. Philos. Soc.*, **150**, 86–112, 2006.
16. Agrawal, A. *et al.*, A Comparative Analysis of the Spanish Flu 1918 and COVID-19 Pandemics. *Open Public Health J.*, **14**, 128–134, 2021.
17. Noymer, A. and Garenne, M., The 1918 Influenza Epidemic's Effects on Sex Differentials in Mortality in the United States. *Bone*, **23**, 1–7, 2008.
18. Fadaka, A.O. *et al.*, Understanding the epidemiology, pathophysiology, diagnosis and management of SARS-CoV-2. *J. Int. Med. Res.*, **48**, 030006052094907, 2020.
19. Schneider, E., Severe Acute Respiratory Syndrome (SARS), in: *Netter's Infect. Dis.*, pp. 537–543, 2012, doi:10.1016/B978-1-4377-0126-5.00089-6.
20. World Health Organization, Evolution of a pandemic, Health (San Francisco), 2010.
21. Al Hajjar, S., Memish, Z.A., McIntosh, K., Middle east respiratory syndrome coronavirus (MERS-CoV): A perpetual challenge. *Ann. Saudi Med.*, **33**, 427–436, 2013.
22. Ohimain, E.I., and Silas-Olu, D., The 2013–2016 Ebola virus disease outbreak in West Africa. *Curr. Opin. Pharmacol.*, **60**, 360–365, 2021.
23. Lowe, R. *et al.*, The zika virus epidemic in brazil: From discovery to future implications. *Int. J. Environ. Res. Public Health*, **15**, 2018, DOI 10.3390/ijerph15010096.
24. Moshinsky, M., The ImpactofInfectious Diseases on the Development of Africa. *Nucl. Phys.*, **13**, 104–116, 2010.
25. Nair, H. *et al.*, Global burden of respiratory infections due to seasonal influenza in young children: A systematic review and meta-analysis. *Lancet*, **378**, 1917–1930, 2011.
26. Paget, J. *et al.*, Global mortality associated with seasonal influenza epidemics: New burden estimates and predictors from the GLaMOR Project. *J. Glob. Health*, **9**, 1–12, 2019.
27. Emro, W. H. O., Who emro | covid-19, pp. 6–8, 2023.
28. Güner, R., Hasanoğlu, I., Aktaş, F., Covid-19: Prevention and control measures in community. *Turk. J. Med. Sci.*, **50**, 571–577, 2020.
29. Baldo, V. *et al.*, The new pandemic influenza A/(H1N1)PDM09 virus: Is it really 'new'? *J. Prev. Med. Hyg.*, **57**, E19–E22, 2016.
30. Zhdanov, V., The USSR. 1957 Influ. Pandemic USSR, pp. 489–494, 1957, doi: 10.4324/9781315882574.
31. Jester, B.J., Uyeki, T.M., Jernigan, D.B., Fifty years of influenza A(H3N2) following the pandemic of 1968. *Am. J. Public Health*, **110**, 669–676, 2020.

32. Martini, M., Gazzaniga, V., Bragazzi, N.L., Barberis, I., The Spanish Influenza Pandemic: A lesson from history 100 years after 1918. *J. Prev. Med. Hyg.*, **60**, E64–E67, 2019.
33. Yoneyama, T. and Krishnamoorthy, M.S., Influence of the cold war upon influenza pandemic of 1957-1958. *Proc. - 2010 6th IEEE Int. Conf. e-Science, eScience 2010*, pp. 9–16, 2010, doi:10.1109/eScience.2010.25.
34. Berche, P., Since January 2020 Elsevier has created a COVID-19 resource centre with free information in English and Mandarin on the novel coronavirus COVID-19 . The COVID-19 resource centre is hosted on Elsevier Connect , the company ' s public news and information, 2020.
35. Steere-Williams, J., Cholera Pandemics, 1816-1861, in: *Encycl. Pestilence, Pandemics, Plagues*, pp. 95–105, 2008.
36. Sköld, P., The history of smallpox and its prevention in Sweden. *Asclepio*, **54**, 71–91, 2002.
37. Barquet, N. and Domingo, P., Smallpox: The triumph over the most terrible of the ministers of death. *Ann. Intern. Med.*, **127**, 635–642, 1997.
38. Frith, J., The history of plague - Part 1. The three great pandemics. *J. Mil. Vet. Health*, **20**, 11–16, 2012.
39. Green, M., *Taking 'Pandemic' Seriously: Making the Black Death GlobalThe Medieval Globe*, The Medieval globe, Arizona State University, vol. 1, 2015.
40. Littman, R.J. and Littman, M.L., Galen and the Antonine Plague. *Am. J. Philol.*, **94**, 3 (Autumn , 1973), 243–255, 2016, Published by: The Johns Hopkins University Press Stable URL : http://www.jstor.
41. Barberis, I., Bragazzi, N.L., Galluzzo, L., Martini, M., The history of tuberculosis: From the first historical records to the isolation of Koch's bacillus. *J. Prev. Med. Hyg.*, **58**, E9–E12, 2017.
42. Ojo, T. *et al.*, HIV, Tuberculosis, and Food Insecurity in Africa—A Syndemics-Based Scoping Review. *Int. J. Environ. Res. Public Health*, **19**, 1–31, 2022.
43. Sonderup, M.W. and Spearman, C.W., Global Disparities in Hepatitis B Elimination—A Focus on Africa. *Viruses*, **14**, 2022, DOI: 10.3390/v14010082.
44. Platt, L. *et al.*, Prevalence and burden of HBV co-infection among people living with HIV: A global systematic review and meta-analysis. *J. Viral Hepat.*, **27**, 294–315, 2020.
45. Jefferies, M., Rauff, B., Rashid, H., Lam, T., Rafiq, S., Update on global epidemiology of viral hepatitis and preventive strategies. *World J. Clin. Cases*, **6**, 589–599, 2018.
46. Chidambaranathan-Reghupaty, S., Fisher, P.B., Sarkar, D., Hepatocellular carcinoma (HCC): Epidemiology, etiology and molecular classification. *Adv. Cancer Res.*, **149**, 1–61, 2021.
47. Neumann, G. and Kawaoka, Y., Which Virus Will Cause the Next Pandemic? *Viruses*, **15**, 199, 2023.
48. Jonkmans, N., D'Acremont, V., Flahault, A., Scoping future outbreaks: A scoping review on the outbreak prediction of the WHO Blueprint list of priority diseases. *BMJ Glob. Health*, **6**, 1–13, 2021.

49. Buchan, J., Catton, H., Schaffer, F.A., Sustain and Retain in 2022 and Beyond: The Global Nursing Workforce and the Covid-19 pandemic, Int. Counc. Nurses, pp. 1–71, 2022.
50. Zheng, C. *et al.*, Real-world effectiveness of COVID-19 vaccines: a literature review and meta-analysis. *Int. J. Infect. Dis.*, **114**, 252–260, 2022.
51. Mori, Y. and Notomi, T., Loop-mediated isothermal amplification (LAMP): A rapid, accurate, and cost-effective diagnostic method for infectious diseases. *J. Infect. Chemother.*, **15**, 62–69, 2009.
52. Martinez, A.W., Phillips, S.T., Whitesides, G.M., Carrilho, E., Diagnostics for the developing world: Microfluidic paper-based analytical devices. *Anal. Chem.*, **82**, 3–10, 2010.
53. Baird, J.K., Malaria control by commodities without practical malariology. *BMC Public Health*, 17, 2017, doi: 10.1186/s12889-017-4454-x.
54. Huttly, S.R.A., Morris, S.S., Pisani, V., Reviews/Analyses Prevention of diarrhoea in young children in developing countries. *Bull. World Health Organ.*, **75,** 163–174, 1997.
55. Turner, H.C., Ottesen, E.A., Bradley, M.H., A refined and updated health impact assessment of the Global Programme to Eliminate Lymphatic Filariasis (2000–2020). *Parasites Vectors*, **15**, 2022, doi.org/10.1186/s130071-002-05268-w.
56. Tran, N.K., Albahra, S., Rashidi, H., May, L., Innovations in infectious disease testing: Leveraging COVID-19 pandemic technologies for the future. *Clin. Biochem.*, **117,** 10–15, 2023 Jul, doi: 10.1016/j.clinbiochem.2021.12.011.
57. Poxton, I.R., Molecular Techniques in the Diagnosis and Management of Infectious Diseases: Do They Have a Role in Bacteriology?, *Med. Princ. Pract.*, **14,** Suppl 1, 20–6, 2005, doi: 10.1159/000086181.
58. Medicine, G. G.-T. @ in M., Molecular diagnostics in infectious diseases and public health microbiology: cottage industry to postgenomics. *Trends Mol. Med.*, 8, 280–7, Elsevier, 2002.
59. Becker, K., Prinzipien und stellenwert von nukleinsaure-nachweistechniken (NNT) in der modernen mikrobiologischen diagnostik. *Anasthesiol. Intensivmed. Notfallmed. Schmerzther.*, **35**, 35–40, 2000.
60. Mabey, D., Peeling, R.W., Ustianowski, A., Perkins, M.D., The identification of asymptomatic or poorly symptomatic cases of a disease or infection. DIAGNOSTICS FOR THE DEVELOPING WORLD. *Nat. Rev. Microbiol.*, **2**, 3–4, 2004.
61. van Belkum, A., Molecular diagnostics in medical microbiology: Yesterday, today and tomorrow. *Curr. Opin. Pharmacol.*, **3**, 497–501, 2003.
62. Farkas, K., Hassard, F., McDonald, J.E., Malham, S.K., Jones, D.L., Evaluation of molecular methods for the detection and quantification of pathogen-derived nucleic acids in sediment. *Front. Microbiol.*, **8**, 2017, doi: 10.3389/fmicb.2017.00053.
63. Versalovic, J. and Lupski, J.R., Molecular detection and genotyping of pathogens: More accurate and rapid answers. *Trends Microbiol.*, **10**, 15–21, 2002.

64. Saiki, R.K., Enzymatic amplification of beta-globin genomic sequences and restriction site analysis for diagnosis of sickle cell anemia. *Science (80)*, **230**, 1350–1355, 1985.
65. Ramakrishnan, S.G. *et al.*, Nanotechnology based solutions to combat zoonotic viruses with special attention to SARS, MERS, and COVID 19: Detection, protection and medication. *Microb. Pathogen.*, **159**, 2021, doi: 10.016/j.micpath.2021.105133.
66. Cheng, M.P. *et al.*, Diagnostic testing for severe acute respiratory syndrome-related coronavirus 2: A narrative review. *Ann. Intern. Med.*, **172**, 726–734, 2020.
67. Kashir, J. and Yaqinuddin, A., Loop mediated isothermal amplification (LAMP) assays as a rapid diagnostic for COVID-19. *Med. Hypotheses*, **141**, 2020, doi: 10.1016/j.mehy.2020.109786.
68. Lambert-Niclot, S. *et al.*, Evaluation of a rapid diagnostic assay for detection of sars-cov-2 antigen in nasopharyngeal swabs. *J. Clin. Microbiol.*, **58**, 2020, doi: 10.1128/JCM.00977-20.
69. Vengesai, A. *et al.*, A systematic and meta-analysis review on the diagnostic accuracy of antibodies in the serological diagnosis of COVID-19. *Syst. Rev.*, **10**, 1–23, 2021.
70. Smithgall, M.C., Dowlatshahi, M., Spitalnik, S.L., Hod, E.A., Rai, A.J., Types of Assays for SARS-CoV-2 Testing: A Review. *Lab. Med.*, **51**, E59–E65, 2020.
71. Mao, X., Wang, W., Du, T.E., Rapid quantitative immunochromatographic strip for multiple proteins test. *Sens. Actuators, B Chem.*, **186**, 315–320, 2013.
72. García-Basteiro, A.L. *et al.*, Point of care diagnostics for tuberculosis. *Rev. Port. Pneumol. Engl. Ed.*, **24**, 73–85, 2018.
73. Eddabra, R. and Ait Benhassou, H., Rapid molecular assays for detection of tuberculosis. *Pneumonia*, **10**, 1–12, 2018.
74. Heidt, B. *et al.*, Point of care diagnostics in resource-limited settings: A review of the present and future of PoC in its most needed environment. *Biosensors*, **10**, 2020, doi: 10.3390/bios10100133.
75. Koczula, K.M. and Gallotta, A., Lateral flow assays. *Essays Biochem.*, **60**, 111–120, 2016.
76. Moreno, M.D.L. *et al.*, Detection of gluten immunogenic peptides in the urine of patients with coeliac disease reveals transgressions in the gluten-free diet and incomplete mucosal healing. *Gut*, **66**, 250–257, 2017.
77. Torjesen, I., Covid-19: How the UK is using lateral flow tests in the pandemic. *BMJ*, **372**, 1–3, 2021.
78. Petrou, L. *et al.*, Lateral Flow Test (LFT) Detects Cell-Free MicroRNAs Predictive of Preterm Birth Directly from Human Plasma. *Adv. NanoBiomed Res.*, **2**, 2200026, 2022.
79. Li, H., Han, D., Hegener, M.A., Pauletti, G.M., Steckl, A.J., Flow reproducibility of whole blood and other bodily fluids in simplified no reaction lateral flow assay devices. *Biomicrofluidics*, **11**, 1–15, 2017.

80. Bahadır, E.B. and Sezgintürk, M.K., Lateral flow assays: Principles, designs and labels. *TrAC, Trends Anal. Chem.*, **82**, 286–306, 2016.
81. Chiao, D.J., Shyu, R.H., Hu, C.S., Chiang, H.Y., Tang, S.S., Colloidal gold-based immunochromatographic assay for detection of botulinum neurotoxin type B. *J. Chromatogr. B Anal. Technol. Biomed. Life Sci.*, **809**, 37–41, 2004.
82. Ardekani, L.S. and Thulstrup, P.W., Gold Nanoparticle-Mediated Lateral Flow Assays for Detection of Host Antibodies and COVID-19 Proteins. *Nanomaterials*, **12**, 1–17, 2022.
83. Qiu, W. *et al.*, Carbon nanotube-based lateral flow biosensor for sensitive and rapid detection of DNA sequence. *Biosens. Bioelectron.*, **64**, 367–372, 2015.
84. Parolo, C. *et al.*, Tutorial: design and fabrication of nanoparticle-based lateral-flow immunoassays. *Nat. Protoc.*, **15**, 3788–3816, 2020.
85. Seele, P.P., Dyan, B., Skepu, A., Maserumule, C., Sibuyi, N.R.S., Development of Gold-Nanoparticle-Based Lateral Flow Immunoassays for Rapid Detection of TB ESAT-6 and CFP-10. *Biosensors*, **13**, 1–15, 2023.
86. Bardhan, N., Nanomaterials in diagnostics, imaging and delivery: Applications from COVID-19 to cancer. *MRS Commun.*, **12**, 1119–1139, 2022.
87. Nasrollahzadeh, M., Sajadi, S.M., Sajjadi, M., Issaabadi, Z., *An Introduction to Nanotechnology. Interface Science and Technology*, vol. 28, Elsevier Ltd., Iran, 2019.
88. Thwala, L.N., Ndlovu, S.C., Mpofu, K.T., Lugongolo, M.Y., Mthunzi-Kufa, P., Nanotechnology-Based Diagnostics for Diseases Prevalent in Developing Countries: Current Advances in Point-of-Care Tests. *Nanomaterials*, **13**, 1–19, 2023.
89. Tyagi, P.K., Tyagi, S., Kumar, A., Ahuja, A., Gola, D., Contribution of nanotechnology in the fight against covid-19. *Biointerface Res. Appl. Chem.*, **11**, 8233–8241, 2021.
90. Gupta, R. *et al.*, Nanotechnology-Based Approaches for the Detection of SARS-CoV-2. *Front. Nanotechnol.*, **2**, 6, 2020.
91. Sibuyi, N.R.S. *et al.*, Multifunctional Gold Nanoparticles for Improved Diagnostic and Therapeutic Applications: A Review. *Nanoscale Res. Lett.*, **16**, 174, 2021.
92. Cordeiro, M., Carlos, F.F., Pedrosa, P., Lopez, A., Baptista, P.V., Gold nanoparticles for diagnostics: Advances towards points of care. *Diagnostics*, **6**, 2016, doi: 10.3990/diagnostic 6040043.
93. Mistry, D.A., Wang, J.Y., Moeser, M.E., Starkey, T., Lee, L.Y.W., A systematic review of the sensitivity and specificity of lateral flow devices in the detection of SARS-CoV-2. *BMC Infect. Dis.*, **21**, 2021, doi: 10.1186/s12879-021-06528-3.
94. Yakoubi, A. and Dhafer, C.E.B., Advanced Plasmonic Nanoparticle-Based Techniques for the Prevention, Detection, and Treatment

of Current COVID-19. *Plasmonics*, 18, Springer US, 2023, doi.org/10.1007//511468-022-01754-0.

95. Martiskainen, I. *et al.*, Double-antigen lateral flow immunoassay for the detection of anti-hiv-1 and-2 antibodies using upconverting nanoparticle reporters. *Sensors (Switzerland)*, **21**, 1–17, 2021.
96. Guirgis, B.S.S. *et al.*, Gold nanoparticle-based fluorescence immunoassay for malaria antigen detection. *Anal. Bioanal. Chem.*, **402**, 1019–1027, 2012.
97. Yayehrad, A.T. *et al.*, Could Nanotechnology Help to End the Fight Against COVID-19? Review of Current Findings, Challenges and Future Perspectives. *Int. J. Nanomed.*, **16**, 5713, 2021.
98. Zhang, M., Cui, X., Li, N., Smartphone-based mobile biosensors for the point-of-care testing of human metabolites. *Mater. Today Bio*, **14**, 100254, 2022.
99. Pang, R., Zhu, Q., Wei, J., Meng, X., Wang, Z., Enhancement of the Detection Performance of Paper-Based Analytical Devices by Nanomaterials. *Molecules*, **27**, 1–31, 2022.
100. Moitra, P. *et al.*, Selective Naked-Eye Detection of SARS-CoV-2 Mediated by N Gene Targeted Antisense Oligonucleotide Capped Plasmonic Nanoparticles. *ACS Nano*, **14**, 7617–7627, 2020.
101. COVID-19: extending or relaxing distancing control measures. www.thelancet.co./public-health, 5, 2020, doi: 10.1016/S2468-2667(20)30072-4.
102. Pan, D. *et al.*, A facile synthesis of novel self-assembled gold nanorods designed for near-infrared imaging. *J. Nanosci. Nanotechnol.*, **10**, 8118–8123, 2010.
103. Miyagawa, A., Okada, Y., Okada, T., Aptamer-Based Sensing of Small Organic Molecules by Measuring Levitation Coordinate of Single Microsphere in Combined Acoustic-Gravitational Field. *ACS Omega*, **5**, 3542–3549, 2020.

15

Biocompatibility and Toxicity of Nanomaterials in the Designing of Tools for the Diagnosis of Infectious Diseases

Manju Manuel

Department of Chemistry, CHRIST (Deemed to be University), Bengaluru, India

Abstract

Nanomaterials have revolutionized the landscape of infectious disease diagnostics by offering unparalleled advantages in terms of sensitivity, specificity, and rapidity. However, their integration into diagnostic tools necessitates a profound understanding of their biocompatibility and toxicity profiles to ensure both diagnostic accuracy and patient life. The successful translation of these nanomaterials into practical diagnostic tools hinges on a thorough understanding of their biocompatibility and toxicity. Biocompatibility, a fundamental characteristic of nanomaterials refers to their ability to coexist with biological systems without causing harm or triggering immune responses. This chapter unfolds the pivotal role of biocompatibility assessment, examining the compatibility of nanomaterials with biological systems. *In vitro* and *in vivo* evaluation methods, and adherence to regulatory standards, are emphasized as essential components of biocompatibility analysis.

Simultaneously, the exploration of nanotoxicity and its hazardous effect highlights the significance of establishing safe exposure limits. Toxicity is a pressing concern when dealing with nanomaterials. The chapter explores the factors that contribute to nanomaterials toxicity, including size-dependent effects, surface modifications, and the route of exposure. It also delves into the mechanisms by which nanomaterials can exert toxicity, such as reactive oxygen species, reaction with surface expanded group, and penetration into the cell. To mitigate the potential risk associated with nanomaterials, the chapter discusses strategies for improving biocompatibility. Finally, it gives a glance into the various tools prepared highlighting the successful integration of biocompatible nanotoxicity of nanomaterials

Email: manju.manuel@res.christuniversity.in

Suvardhan Kanchi, Ayyappa Bathinapatla, Anitha Varghese and Phumlane Selby Mdluli (eds.) Nano-Biosensor Technologies for Diagnosis of Infectious Diseases, (449–468) © 2025 Scrivener Publishing LLC

into infectious disease diagnostics. Ultimately, this chapter emphasizes the need for comprehensive biocompatibility and toxicity evaluation as integral components of designing effective and safe diagnostic tools thereby contributing to the advancement of healthcare through cutting-edge nanotechnology.

Keywords: Biocompatibility, nanomaterials, infectious diseases, toxicity

15.1 Introduction

Nanomaterials have gained significant importance in infectious diseases due to their unique properties and versatile applications. They can be engineered to carry and deliver drugs more effectively to the site of infection. Their small size allows for targeted delivery, reducing the dose required and minimizing side effects [1]. Some nanomaterials, such as silver nanoparticles (NPs), have inherent antimicrobial properties which can be used to coat medical devices to prevent the spread of infectious agents [2]. The biosensors with high sensitivity and specificity incorporating the nanomaterials play a crucial role in early detection and treatment. In vaccine development, nanomaterials are used as carriers for antigens and enhance the immune response. Most of the nanomaterials are designed to exhibit highly specific interactions with biomolecules, such as viruses, bacteria, or specific biomarkers associated with infectious diseases. The faster results of the nanomaterial-based diagnostic tests are especially crucial in case of rapidly spreading infectious diseases [3]. Moreover, nanomaterials can be used to enhance various imaging techniques such as magnetic resonance imaging (MRI) or fluorescence imaging. This can provide detailed insights into the localization and progress of infectious diseases [4].

Evaluating the biocompatibility and toxicity of nanomaterials when using them for diagnosing infectious diseases is essential. This is critical in diagnostic methods that involve direct contact with patients such as *in vitro* diagnostics, medical imaging, or nanoscale drug delivery [5]. Evaluating biocompatibility helps to identify adverse effects and allows for the selection of nanomaterials that are less likely to provoke side effects. In some cases, nanomaterials may accumulate in the body over time. Chronic exposure to toxic nanomaterials can lead to long-term health problems. Toxic nanomaterials may produce false positive or negative results, which can seriously affect infectious disease diagnosis [6]. Nanomaterials that are toxic to the biological system may cause risk to the environment when they are disposed of improperly. The safety of nanomaterial-based diagnostics is vital for maintaining public trust in the healthcare system.

This chapter aims to explore the critical intersection of biocompatibility and toxicity as they pertain to the development of nanomaterial-based tools for the diagnosis of infectious diseases. As the field of nanomedicine continues to advance, integrating nanomaterials into diagnostic tools offers unprecedented opportunities for early disease detection, monitoring, and treatment. However, the biocompatibility of these nanomaterials is of paramount importance, as it directly influences their safe and effective use in biological systems. Furthermore, understanding the potential toxicity of these materials is essential for mitigating any adverse effects on both the diagnostic tools themselves and the patients they are designed to aid. To address these crucial concerns, this chapter explores the objectives and structure of this chapter, starting with an overview of nanomaterial-based diagnostic tools, followed by a comprehensive examination of biocompatibility and toxicity assessments. Subsequently, delves into case studies and emerging technologies, demonstrating how these considerations are integrated into the design and implementation of tools for infectious disease diagnosis. This chapter hopes to provide a comprehensive understanding of the challenges and opportunities in this rapidly evolving field and foster a deeper appreciation of the intricate balance between nanomaterial functionality, biocompatibility, and safety.

15.2 An Overview of Nanomaterials in Infectious Disease Diagnosis

Nanomaterials have revolutionized the field of infectious disease diagnosis by offering unique properties and capabilities that enhance the sensitivity, specificity, and versatility of diagnostic tools. Several nanomaterials are commonly used in this context, each with distinct advantages. Some key aspects of nanomaterials in infectious diseases include:

15.2.1 Bacteria-Targeting Nanomaterials

These nanomaterials have been widely used to treat bacterial, fungal, viral, and parasitic infectious diseases in recent years. They show great potential as antimicrobial agents due to their broad-spectrum activity [7].

15.2.2 Nanotechnology in Clinical Infectious Diseases

Nanotechnology has enhanced our ability to diagnose, treat, and prevent infectious diseases [8]. It has significantly benefited healthcare in resource-rich as well as resource-poor parts of the world.

15.2.3 Nanomaterial Vaccines

Nanomaterials are uniquely positioned to address challenges in developing prophylactic strategies for infectious diseases such as acquired immuno deficiency syndrome (AIDS), malaria, and tuberculosis [9]. They can produce immune responses that are distinct from those of other vaccine-preventable diseases, which is crucial for these diseases.

Nanotechnology has significantly benefited healthcare in resource-rich and resource-poor parts of the world, and it continues to influence changes in the practice of medicine. Figure 15.1 depicts an overview of some of the most frequently employed nanomaterials for infectious disease diagnosis.

Nanomaterials have an exceptionally high surface-to-volume ratio, which allows for a greater interaction with target molecules. This property is advantageous for immobilizing recognition molecules (*e.g.*, antibodies or aptamers) and increasing the chances of capturing low-concentration biomarkers [4]. Nanomaterials can be precisely engineered to have specific sizes and shapes, which can be tailored for optimal interactions with target biomolecules [10]. This tunability is crucial for designing sensors with enhanced selectivity and sensitivity. Quantum-sized nanomaterials,

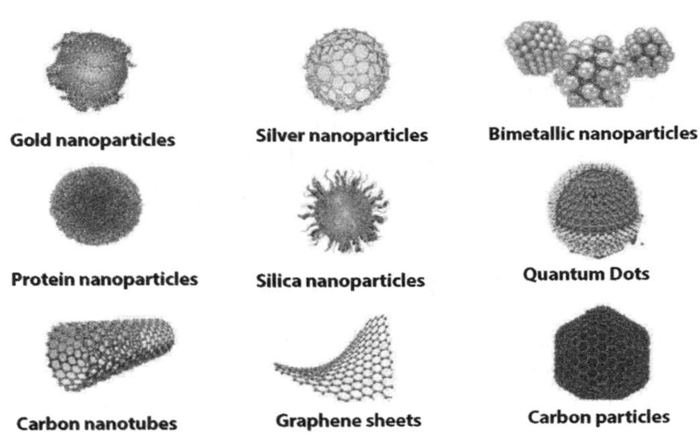

Figure 15.1 Most frequently used nanomaterials in the biomedical level with unique structure and synthesis method.

such as quantum dots, exhibit size-dependent optical and electronic properties. This property enables the creation of highly sensitive fluorescent or luminescent labels for biomolecular recognition and detection [11]. Nanomaterials can amplify detection signals through various mechanisms, including surface-enhanced Raman scattering (SERS) for enhanced vibrational spectroscopy, plasmon resonance for colorimetric detection, and electrochemical amplification for sensitive electrochemical assays [12]. Many nanomaterial-based assays require small sample volumes, making them suitable for point-of-care diagnostics and reducing the invasiveness of sample collection. Nanomaterials can provide real-time monitoring of dynamic biological processes, such as the release of biomarkers, changes in pH, or alterations in ion concentrations, enabling continuous disease assessment. Overall, the unique properties of nanomaterials have led to significant advancements in diagnostic technologies, enabling the development of more sensitive, accurate, and efficient tools for the early detection and management of various diseases.

15.3 Biocompatibility Assessment

Biocompatibility refers to the ability of a material to coexist harmoniously with living organisms without causing harm, adverse reactions, or immunological responses. In the context of medical and biological applications, it specifically relates to the compatibility of biomaterials, medical devices, or substances with the tissues, cells, and biological systems they come into contact with. Biocompatible materials are designed to interact with the human body or other living systems in a way that does not induce toxic or injurious responses and allows for their intended function.

Assessing biocompatibility in nanomaterials is a pivotal undertaking in advancing the diagnosis of infectious diseases. This critical evaluation ensures that nanomaterials seamlessly integrate with biological systems, minimizing potential risks and optimizing efficacy. Signifying a delicate balance between effectiveness and potential harm, this assessment guarantees that selected nanomaterials enhance diagnostic accuracy without compromising the biological environment. Tailoring nanomaterials for optimal compatibility becomes paramount, allowing for seamless integration into diagnostic tools. Ultimately, the goal is to advance diagnostic precision, providing clinicians and researchers with a powerful tool to combat infectious diseases effectively. This meticulous scrutiny marks a significant leap forward in elevating diagnostic capabilities and shaping the future of

Table 15.1 Nanomaterials which are successfully incorporated with the biological system for diagnosing various diseases.

Case study	Nanomaterials	Applications/outcome	References
Gold nanoparticles in the cancer study	Gold nanoparticles	Enhanced imaging contrast for precise cancer cell targeting without harm.	[13]
Liposomal nanocarriers for drug delivery.	Liposomal nanocarriers.	Targeted drug delivery, minimizing side effects in infectious disease treatment.	[14]
Quantum dots for infectious disease imaging.	Quantum dots	Early detection and monitoring of infectious diseases with minimal tissue impact.	[15]
Polymeric nanomaterials for gene therapy	Polymeric nanomaterials	Biocompatible carriers for precise delivery of genetic materials in gene therapy.	[16]
Magnetic nanoparticles in hyperthermia treatment	Magnetic nanoparticles	Controlled hyperthermia for targeted destruction of pathogens while sparing healthy tissues.	[17]
Carbon nanotubes for antibacterial coatings	Carbon nanotubes	Creation of antibacterial coatings on medical devices, preventing infections.	[18]
Silica nanoparticles in antiviral therapies	Silica nanoparticles	Creation of antibacterial coatings on medical devices, preventing infections.	[19]
Graphene oxide for immunomodulation	Graphene oxide	Potential immunomodulation applications in infectious diseases.	[20]
Chitosan nanoparticles in oral vaccines	Chitosan nanoparticles	Oral vaccine delivery, protecting components for effective mucosal delivery.	[20]
Silver nanoclusters for wound healing	Silver nanoclusters	Antimicrobial properties for sterile wound environments, accelerating healing.	[21]

healthcare. Table 15.1 contains the successful incorporation of nanomaterials at the medical level.

15.4 Mechanism of Nanoparticles in the Infectious Disease Diagnosis

The reactive mechanism of the nanomaterials not yet be fully known but believed to have the following mechanisms which ultimately suppress/irradicate the of infectious diseases.

15.4.1 Reactive Oxygen Species (ROS) Formation

Nanoparticles (NPs) possess the capability to generate reactive oxygen species (ROS), which can induce damage to the cell membrane and deoxyribo nucleic acid (DNA) of pathogenic cells. The bactericidal or bacteriostatic properties of metal NPs are contingent upon factors such as their particle size, capping method, and concentration. Extensive analysis of available evidence reveals that NPs exhibit antibacterial activity against both gram-positive and gram-negative strains. In the context of gram-negative strains, the cell wall composition features a thin layer of peptidoglycan polymer, measuring approximately 7–8 nm. This unique structural attribute is directly implicated in the observed antibacterial activity of NPs. The thin cell wall facilitates enhanced penetration of NPs into bacterial cells, while the negatively charged surface of the gram-negative strains establishes a robust electrostatic interaction with NPs [22]. This interaction, in turn, promotes the generation of ROS and induces oxidative stress within the bacterial cells as shown in Figure 15.2.

15.4.2 Interaction With Surface-Exposed Groups

Nanoparticles can interact with various surface-exposed groups on the cell membrane, leading to microbial destruction and inactivation as represented in Figure 15.3. These interactions often involve specific binding to receptors or functional groups present on the cell surface, enabling NPs to effectively target and engage with microbial entities. The unique physicochemical properties of NPs, such as size, shape, and surface charge, play a crucial role in determining the extent and specificity of their interaction with the cell membrane [24]. By exploiting these properties, NPs can disrupt key cellular processes and compromise the integrity of the microbial

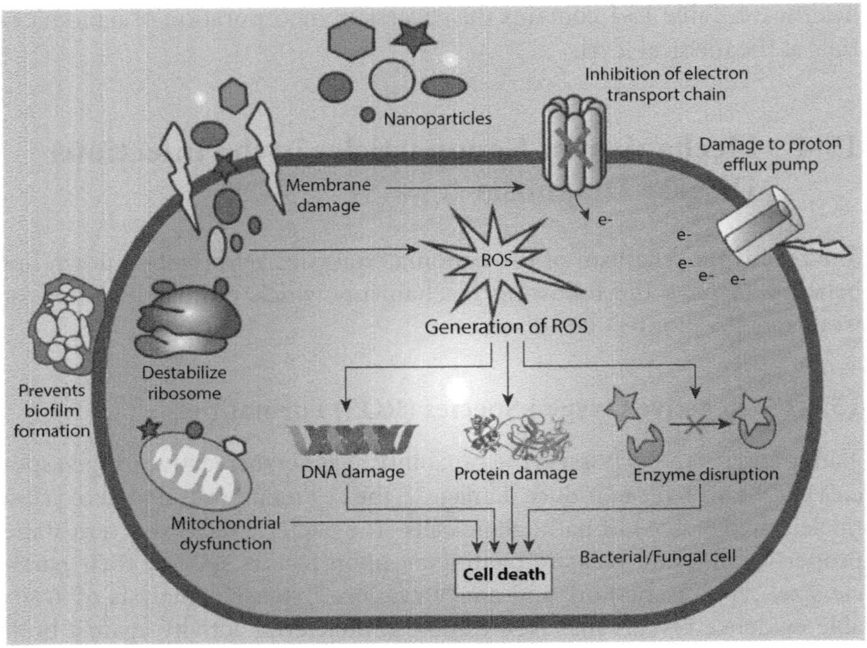

Figure 15.2 Schematic illustration of antibacterial mechanism of NPs and the production toxic reaction species (ROS). Adapted from reference [23], published by PMC PubMed Central.

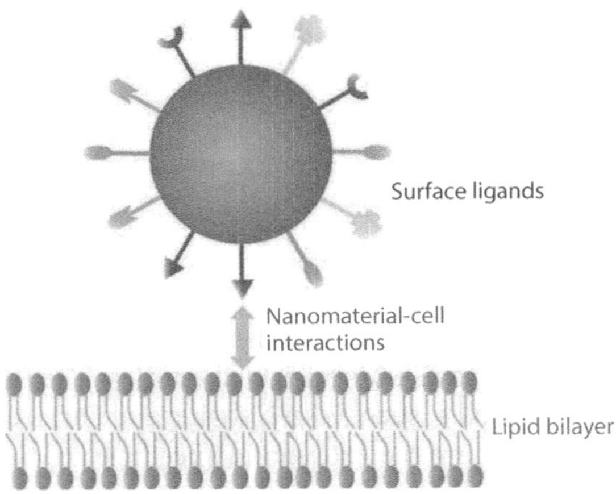

Figure 15.3 Schematic representation of the interaction of the nanomaterials with the surface molecules [25].

cell envelope. This targeted approach holds promise for the development of advanced antimicrobial strategies with potential applications in medicine and environmental protection.

15.4.3 Penetration Into the Cell

The small particle size of NPs allows them to penetrate into the bacteria, disrupting their internal processes. In the realm of combating infectious diseases stemming from anticipated and unforeseen viruses and bacteria, nanotechnology plays a pivotal role. Antimicrobial agents, crucial for treatment, encompass chemical compounds comprising bioactive polymers or synthesized polymers, either in isolation or coupled with NPs (NPs) [26]. Nanoparticles, characterized by their elevated surface-to-volume ratio and well-established antibacterial properties, exhibit significant efficacy even at low concentrations [27]. Their principal function involves serving as antimicrobial agents, thwarting and eradicating pathogenic viruses, fungi, and bacteria. Recently, a spectrum of organic substances, encompassing polymers and biopolymers, has displayed promising attributes as antiviral and antibacterial agents for treating diseases induced by hazardous bacteria and viruses [28].

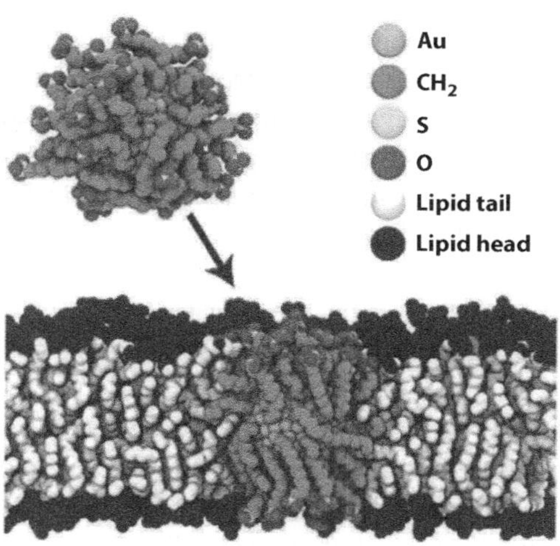

Figure 15.4 Illustration shows the entry of a gold nanoparticle (in orange) wrapped with a monolayer of hydrophobic/hydrophilic material (shown in blue-green, yellow and red), moving through a cell membrane consist of lipids (white and blue) [29].

15.5 *In Vitro* and *In Vivo* Evaluation Methods of Biocompatibility Analysis

Biocompatibility analysis, a critical aspect of evaluating the safety and efficacy of medical devices and materials, places significant emphasis on employing rigorous *in vitro* and *in vivo* evaluation methods while adhering to stringent regulatory standards. *In vitro* assessments involve testing within controlled laboratory environments using biological systems such as cell cultures. *In vivo* evaluations, on the other hand, involve testing within living organisms, allowing for a more comprehensive understanding of the potential biological interactions and responses to the medical device or material. These studies enable researchers to gauge the compatibility of materials at a cellular level, assessing factors like cytotoxicity and inflammatory responses. Table 15.2 underlines the NPs which are extensively used in the *in vitro* and *in vivo* experiments to evaluate its biocompatibility.

Adherence to regulatory standards is paramount to ensuring the reliability and relevance of biocompatibility assessments. Regulatory bodies, such as the U.S. Food and Drug Administration (FDA) and the European Medicines Agency (EMA), establish guidelines and standards that must be followed to guarantee the safety and efficacy of medical products [37]. Compliance with these standards is crucial for obtaining regulatory approval and bringing medical devices to market. Biocompatibility analysis is particularly crucial in the development of implantable medical devices, prosthetics, and other healthcare products. Ensuring that these devices are biocompatible helps mitigate the risk of adverse reactions, such as inflammation or rejection, when introduced into the human body. By integrating *in vitro* and *in vivo* evaluation methods and adhering to rigorous regulatory standards, researchers and manufacturers it is possible to enhance the overall safety and performance of medical devices, fostering confidence in their application within clinical settings.

15.6 Toxicity of Nanomaterials

Nanotoxicity refers to the adverse effects caused by nanomaterials on biological systems. These effects arise due to the unique properties of nanomaterials, such as their small size and high surface area, which can lead to interactions with biological structures. Types of nanotoxicity include cytotoxicity, where nanomaterials harm cells, and genotoxicity, affecting genetic material. Additionally, immunotoxicity involves interference with

Table 15.2 The *in vivo* and *in vitro* assessment of the nanomaterials in the biomedical diagnosis.

Nanoparticle	**Composition**	**Biocompatibility features**	**Applications**	**References**
Liposomes	Phospholipid bilayers	Mimics cell membranes, widely used in drug delivery.	*In vivo* drug delivery, *in vitro* cell culture studies.	[30]
Polymeric nanoparticles	Biodegradable polymers	Tailored for controlled release, reduced toxicity.	*In vivo* drug delivery, *in vitro* controlled release studies.	[31]
Gold nanoparticles	Gold	Inert, stable, low toxicity.	*In vivo* imaging, photothermal therapy; *in vitro* imaging and diagnostics.	[32]
Silver nanoparticles	Silver	Broad-spectrum antimicrobial properties.	*In vivo* wound dressings, antimicrobial coatings; *in vitro* antimicrobial studies	[33]
Iron oxide nanoparticles	Iron oxide	Biodegradable, magnetic properties.	*In vivo* hyperthermia for localized infections; *in vitro* magnetic resonance imaging (MRI) studies	[34]
Quantum dots	Semiconductor nanocrystals	Size-tunable, versatile imaging capabilities.	*In vivo* tracking and imaging of infectious agents; *in vitro* cellular imaging studies	[35]
Dendrimers	Repeated branching structures	Precise control over size and surface properties.	*In vivo* antiviral drug delivery, gene therapy; *in vitro* targeted drug delivery studies.	[36]

the immune system, while oxidative stress results from an imbalance in cellular antioxidants. Nanotoxicity, the potential adverse effects of NPs on human health and the environment, is influenced by various factors. Research has shown that the following factors can affect the toxicity of NPs.

15.6.1 Particle Size and Surface Area

Smaller NPs tend to have greater acute toxicities in animal models. The ratio of surface area to volume of NPs increases exponentially with decreasing size, leading to an increase in their biological and chemical reactivities. For example, the number of expressed surface molecules increased from 5% to 60% when the NP's size shrank from 40 to 3 nm. Therefore, depending on surface area and particle size, NPs with the same chemical composition can have radically varying levels of cytotoxicity [38].

15.6.2 Chemical Composition

The biological effects of nanomaterials are influenced by their chemical composition. The chemical composition of nanomaterials can significantly influence their toxicity. Different types of NPs, such as metal, metal oxide, and carbon-based NPs, have been studied for their potential toxic effects. The chemical composition affects nanotoxicity through various mechanisms, including the release of toxic ions, surface reactivity, and interactions with biological systems [39].

15.6.3 Shape and Crystallinity

The shape or crystallinity of an NP can influence its toxicity. For example, needle-shaped NPs may exhibit increased toxicity due to their improved multiple endocytic mechanisms, while "sharp" shapes are often associated with higher toxicity and induction of cell death [40]. Additionally, the crystallinity of NPs can affect their nanotoxicological aspects, with different crystalline structures influencing their toxicity. Understanding these relationships is important for assessing the potential risks associated with the use of NPs [41].

15.6.4 Dose

Nanomaterials are known to have dose-dependent toxic effects. Nanotoxicity often follows a dose-response relationship, meaning that as the dose of NPs increases, the biological effects become more pronounced.

A study was done using silver nanoparticle (AgNP) uses for medicinal purpose and was performed to explore the effect of various doses of AgNPs in rats. Four different doses of AgNP (4, 10, 20, and 40 mg/kg) were injected intravenously and found that <10 mg/kg is safe for the biomedical applications [42].

15.6.5 Exposure Routes

The route of exposure to NPs, such as inhalation or dermal contact, can affect their toxic effects. Research conducted on rats has demonstrated that NPs have a greater ability to cause lung toxicity than microparticles [43]. Through the food chain, they can enter the human body from the environment. Furthermore, after being inhaled, some of the NPs may be ejected from the lung's mucociliary system and swallowed into the gastrointestinal tract.

15.7 The Environmental and Health Hazards Caused by Nanoparticles

The utilization of NPs in the diagnosis of infectious diseases has undoubtedly brought about innovative advancements in medical technology. However, it is imperative to acknowledge and carefully address the potential environmental and health hazards associated with their use. Figure 15.5 showcases various concerns regarding NPs in the context of infectious disease diagnosis [44].

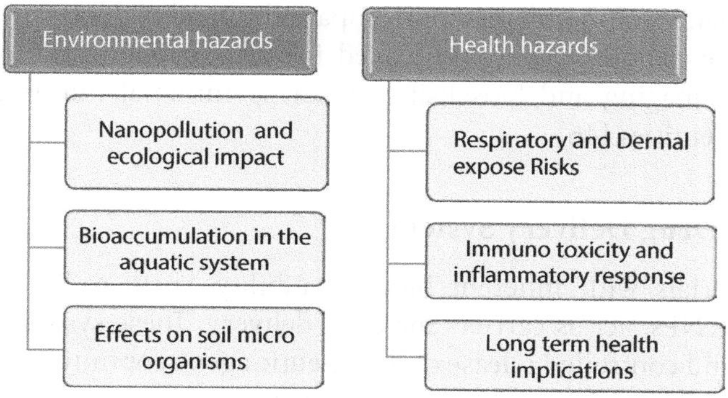

Figure 15.5 The environmental and health hazards caused by the nanomaterials used in the diagnosis of infectious diseases.

Engineering NPs with appropriate encapsulation and surface modifications can enhance biocompatibility and reduce potential health risks. Proper ventilation, personal protective equipment, and adherence to established safety protocols can minimize the risk of exposure to healthcare workers and researchers [45]. Developing effective waste management protocols for the disposal of NP-containing diagnostic tools is crucial. This includes ensuring proper containment and disposal methods to prevent environmental contamination. Regulatory bodies should continuously evaluate and update safety standards, taking into account advancements in nanotechnology and potential environmental and health risks.

15.8 The Tools Developed for the Diagnosis of Infectious Diseases

The advancement of nanotechnology has led to the development of various tools, taking into careful consideration the biocompatibility and toxicity of nanomaterials. These tools play a crucial role in diverse fields, including medicine, diagnostics, and environmental monitoring. Here, we explore some notable tools designed with a focus on biocompatibility and toxicity considerations.

15.8.1 Nanosensors

Nanoparticles, engineered for biocompatibility, serve as effective sensing elements in nanosensors. These sensors can detect specific biomarkers associated with diseases, offering rapid and precise diagnostic capabilities while minimizing adverse effects on biological systems. Surface modifications of nanoprobes enhance biocompatibility and reduce potential toxicity. These nanoprobes are employed in various diagnostic techniques, including imaging and biosensing, ensuring minimal interference with biological entities [46].

15.8.2 Drug Delivery System

Nanomaterials with inherent biocompatibility, such as liposomes and polymeric NPs, act as carriers for drug delivery. These systems allow for targeted and controlled release of therapeutic agents, optimizing treatment efficacy while minimizing systemic toxicity. The development of smart NPs incorporates controlled release mechanisms. These systems respond to specific stimuli, ensuring precise drug delivery to the target site while mitigating potential toxic effects on healthy tissues [47].

15.8.3 Biosensors

Biosensors utilize biocompatible nanomaterials to detect pollutants and contaminants in the environment. These sensors, designed with a focus on toxicity reduction, offer real-time monitoring capabilities without adversely affecting the ecosystems being studied. Nanoprobes equipped with toxicity-responsive elements enable the detection of environmental toxins [48]. These sensors provide valuable insights into the ecological impact of pollutants, contributing to effective environmental management.

15.8.4 Biomedical Imaging Tools

Biomedical imaging tools, such as magnetic resonance imaging (MRI) and computed tomography (CT), employ contrast agents with enhanced biocompatibility. These agents, often based on nanomaterials, improve imaging quality while minimizing potential harm to patients. Fluorescence imaging utilizes biocompatible nanomaterials for enhanced contrast and improved imaging resolution [48]. These materials enable non-invasive visualization of biological structures, offering valuable insights in medical diagnostics.

15.8.5 *In Vitro* Diagnostic Platform

In vitro diagnostic platforms leverage biocompatible nanomaterials for assay development. These platforms, including immunoassays and nucleic acid-based tests, ensure accurate and sensitive detection of diseases with minimal impact on biological samples [49]. Lab-on-a-chip systems incorporate cellular-compatible nanomaterials, allowing for the miniaturization of diagnostic processes. These systems offer portability and efficiency, making them valuable tools for point-of-care diagnostics [50].

The tools designed with a keen focus on the biocompatibility and toxicity of nanomaterials represent a significant stride in harnessing the potential of nanotechnology for various applications. From disease diagnosis to drug delivery and environmental monitoring, these tools exemplify the responsible integration of nanomaterials, ensuring their positive impact on science and healthcare while minimizing potential risks to biological systems.

15.9 Conclusion

This book chapter serves as a comprehensive guide to the intricate interplay between biocompatibility and toxicity of nanomaterials in the design of tools for diagnosing infectious diseases. It begins with a thought-provoking introduction, navigating through the pivotal role of NPs in revolutionizing infectious disease diagnosis. The overview of NPs establishes their significance in transforming diagnostic tools, setting the stage for a deeper understanding. The inclusion of a case study table enriches the discussion, offering practical insights into biocompatibility assessment and real-world applications. Delving into the mechanisms of NP involvement in infectious disease diagnosis, the chapter clarifies how reactive oxygen species, surface interactions, and penetration contribute to tool efficacy. This mechanistic understanding is crucial for optimizing design and mitigating potential toxic effects. Exploration of *in vitro* and *in vivo* evaluation methods forms a robust foundation for researchers, providing valuable insights into designing safer nanomaterials for diagnostics. Addressing environmental and health hazards associated with nanomaterials emphasizes responsible development and deployment, crucial for ensuring the sustainability of nanotechnology applications in medicine. Concluding with an overview of diverse tools developed for infectious disease diagnosis, the chapter highlights remarkable progress in nanotechnology. From biosensors to imaging tools, these innovations exemplify the nuanced understanding of biocompatibility and toxicity. In essence, this chapter is a forward-looking guide, weaving theoretical insights and practical applications. By addressing each facet with depth and clarity, it contributes significantly to the evolving landscape of nanotechnology in infectious disease diagnosis, paving the way for future advancements that are safer, more effective, and ethically responsible.

References

1. Fries, C.N., Curvino, E.J., Chen, J.-L., Permar, S.R., Fouda, G.G., Collier, J.H., Advances in nanomaterial vaccine strategies to address infectious diseases impacting global health. *Nat. Nanotechnol.*, 16, 4, 1–14, 2021, doi: 10.1038/s41565-020-0739-9.
2. Xie, M. *et al.*, Antibacterial Nanomaterials: Mechanisms, Impacts on Antimicrobial Resistance and Design Principles. *Angew. Chem. Int. Ed.*, 62, 17, e202217345, Apr. 2023, doi: https://doi.org/10.1002/anie.202217345.

3. Sharma, A. *et al.*, Optical Biosensors for Diagnostics of Infectious Viral Disease: A Recent Update. *Diagnostics*, 11, 11, 2021, doi: 10.3390/diagnostics11112083.
4. Smith, B.R. and Gambhir, S.S., Nanomaterials for *In Vivo* Imaging. *Chem. Rev.*, 117, 3, 901–986, Feb. 2017, doi: 10.1021/acs.chemrev.6b00073.
5. Huang, R. *et al.*, Advances of functional nanomaterials for magnetic resonance imaging and biomedical engineering applications. *WIREs Nanomed. Nanobiotechnol.*, 14, 4, e1800, Jul. 2022, doi: https://doi.org/10.1002/wnan.1800.
6. Staroń, A., Długosz, O., Pulit-Prociak, J., Banach, M., Analysis of the Exposure of Organisms to the Action of Nanomaterials. *Materials*, 13, 2, 2020, doi: 10.3390/ma13020349.
7. Hung, Y.-P., Chen, Y.-F., Tsai, P.-J., Huang, I.-H., Ko, W.-C., Jan, J.-S., Advances in the Application of Nanomaterials as Treatments for Bacterial Infectious Diseases. *Pharmaceutics*, 13, 11, 1913, 2021.
8. Tobin, E. and Brenner, S., Nanotechnology fundamentals applied to clinical infectious diseases and public health. *Open Forum Infect. Dis.*, 8, 12, ofab583, 2021.
9. Kirtane, A.R., Verma, M., Karandikar, P., Furin, J., Langer, R., Traverso, G., Nanotechnology approaches for global infectious diseases. *Nat. Nanotechnol.*, 16, 4, 369–384, 2021, doi: 10.1038/s41565-021-00866-8.
10. Yadav, S.K., Yadav, R.D., Tabassum, H., Arya, M., Recent Developments in Nanotechnology-Based Biosensors for the Diagnosis of Coronavirus. *Plasmonics*, 18, 3, 955–969, 2023.
11. Castro, R.C., Ribeiro, D.S.M., Santos, J.L.M., Visual detection using quantum dots sensing platforms. *Coord. Chem. Rev.*, 429, 213637, 2021, doi: https://doi.org/10.1016/j.ccr.2020.213637.
12. Pilot, R., Signorini, R., Durante, C., Orian, L., Bhamidipati, M., Fabris, L., A Review on Surface-Enhanced Raman Scattering. *Biosensors*, 9, 2, 2019, doi: 10.3390/bios9020057.
13. Sztandera, K., Gorzkiewicz, M., Klajnert-Maculewicz, B., Gold Nanoparticles in Cancer Treatment. *Mol. Pharm.*, 16, 1, 1–23, Jan. 2019, doi: 10.1021/acs.molpharmaceut.8b00810.
14. Tenchov, R., Bird, R., Curtze, A.E., Zhou, Q., Lipid Nanoparticles—From Liposomes to mRNA Vaccine Delivery, a Landscape of Research Diversity and Advancement. *ACS Nano*, 15, 11, 16982–17015, Nov. 2021, doi: 10.1021/acsnano.1c04996.
15. Mousavi, S.M. *et al.*, The Pivotal Role of Quantum Dots-Based Biomarkers Integrated with Ultra-Sensitive Probes for Multiplex Detection of Human Viral Infections. *Pharmaceuticals*, 15, 7, 880, 2022.
16. Mangraviti, A. *et al.*, Polymeric Nanoparticles for Nonviral Gene Therapy Extend Brain Tumor Survival *in Vivo*. *ACS Nano*, 9, 2, 1236–1249, Feb. 2015, doi: 10.1021/nn504905q.

17. Vassallo, M. *et al.*, Improvement of Hyperthermia Properties of Iron Oxide Nanoparticles by Surface Coating. *ACS Omega*, 8, 2, 2143–2154, Jan. 2023, doi: 10.1021/acsomega.2c06244.
18. Hadidi, N. and Mohebbi, M., Anti-Infective and Toxicity Properties of Carbon Based Materials: Graphene and Functionalized Carbon Nanotubes. *Microorganisms*, 10, 12, 2022, doi: 10.3390/microorganisms10122439.
19. LaBauve, A.E. *et al.*, Lipid-coated mesoporous silica nanoparticles for anti-viral applications via delivery of CRISPR-Cas9 ribonucleoproteins. *Sci. Rep.*, 13, 1, 6873, 2023, doi: 10.1038/s41598-023-33092-4.
20. Peng, G. *et al.*, Graphene oxide elicits microbiome-dependent type 2 immune responses via the aryl hydrocarbon receptor. *Nat. Nanotechnol.*, 18, 1, 42–48, 2023, doi: 10.1038/s41565-022-01260-8.
21. Singh, M. *et al.*, Silver Nanoparticles and Its Mechanistic Insight for Chronic Wound Healing: Review on Recent Progress. *Molecules*, 27, 17, 2022, doi: 10.3390/molecules27175587.
22. Yang, B., Chen, Y., Shi, J., Reactive Oxygen Species (ROS)-Based Nanomedicine. *Chem. Rev.*, 119, 8, 4881–4985, Apr. 2019, doi: 10.1021/acs.chemrev.8b00626.
23. Sharmin, S., Rahaman, M.M., Sarkar, C., Atolani, O., Islam, M.T., Adeyemi, O.S., Nanoparticles as antimicrobial and antiviral agents: A literature-based perspective study. *Heliyon*, 7, 3, 2021.
24. Baer, D.R. *et al.*, Characterization challenges for nanomaterials. *Surf. Interface Anal.*, 40, 3–4, 529–537, Mar. 2008, doi: https://doi.org/10.1002/sia.2726.
25. Verma, A. and Stellacci, F., Effect of Surface Properties on Nanoparticle–Cell Interactions. *Small*, 6, 1, 12–21, Jan. 2010, doi: https://doi.org/10.1002/smll.200901158.
26. Collignon, P.C. *et al.*, World Health Organization ranking of antimicrobials according to their importance in human medicine: a critical step for developing risk management strategies to control antimicrobial resistance from food animal production. *Clin. Infect. Dis.*, 63, 8, 1087–1093, 2016.
27. Mitchell, M.J., Billingsley, M.M., Haley, R.M., Wechsler, M.E., Peppas, N.A., Langer, R., Engineering precision nanoparticles for drug delivery. *Nat. Rev. Drug Discov.*, 20, 2, 101–124, 2021, doi: 10.1038/s41573-020-0090-8.
28. Balasubramaniam, B. *et al.*, Antibacterial and Antiviral Functional Materials: Chemistry and Biological Activity toward Tackling COVID-19-like Pandemics. *ACS Pharmacol. Transl. Sci.*, 4, 1, 8–54, Feb. 2021, doi: 10.1021/acsptsci.0c00174.
29. Penetration of nanomaterials into the cell, [Online]. Available: https://news.mit.edu/2013/the-gold-standard-for-cell-penetration-0823.
30. Torchilin, V.P., Recent advances with liposomes as pharmaceutical carriers. *Nat. Rev. Drug Discov.*, 4, 2, 145–160, 2005.
31. Danhier, F., Ansorena, E., Silva, J.M., Coco, R., Le Breton, A., Préat, V., PLGA-based nanoparticles: an overview of biomedical applications. *J. Control. Release*, 161, 2, 505–522, 2012.

32. Dreaden, E.C., Alkilany, A.M., Huang, X., Murphy, C.J., El-Sayed, M.A., The golden age: gold nanoparticles for biomedicine. *Chem. Soc. Rev.*, 41, 7, 2740–2779, 2012.
33. Durán, N. and Seabra, A.B., Metallic oxide nanoparticles: state of the art in biogenic syntheses and their mechanisms. *Appl. Microbiol. Biotechnol.*, 95, 275–288, 2012.
34. Laurent, S. *et al.*, Magnetic iron oxide nanoparticles: synthesis, stabilization, vectorization, physicochemical characterizations, and biological applications. *Chem. Rev.*, 108, 6, 2064–2110, 2008.
35. Michalet, X. *et al.*, Quantum dots for live cells, *in vivo* imaging, and diagnostics. *Science (80-.)*, 307, 5709, 538–544, 2005.
36. Svenson, S. and Tomalia, D.A., Dendrimers in biomedical applications—reflections on the field. *Adv. Drug Deliv. Rev.*, 64, 102–115, 2012.
37. Halamoda-Kenzaoui, B., Holzwarth, U., Roebben, G., Bogni, A., Bremer-Hoffmann, S., Mapping of the available standards against the regulatory needs for nanomedicines. *WIREs Nanomed. Nanobiotechnol.*, 11, 1, e1531, Jan. 2019, doi: https://doi.org/10.1002/wnan.1531.
38. Egbuna, C. *et al.*, Toxicity of Nanoparticles in Biomedical Application: Nanotoxicology. *J. Toxicol.*, 2021, 9954443, 2021, doi: 10.1155/2021/9954443.
39. Akçan, R., Aydogan, H.C., Yildirim, M.Ş., Taştekin, B., Sağlam, N., Nanotoxicity: A challenge for future medicine. *Turk. J. Med. Sci.*, 50, 4, 1180–1196, 2020.
40. Kladko, D.V., Falchevskaya, A.S., Serov, N.S., Prilepskii, A.Y., Nanomaterial Shape Influence on Cell Behavior. *Int. J. Mol. Sci.*, 22, 10, 2021, doi: 10.3390/ijms22105266.
41. Fu, P.P., Xia, Q., Hwang, H.-M., Ray, P.C., Yu, H., Mechanisms of nanotoxicity: generation of reactive oxygen species. *J. Food Drug Anal.*, 22, 1, 64–75, 2014.
42. Tiwari, D.K., Jin, T., Behari, J., Dose-dependent *in-vivo* toxicity assessment of silver nanoparticle in Wistar rats. *Toxicol. Mech. Methods*, 21, 1, 13–24, Jan. 2011, doi: 10.3109/15376516.2010.529184.
43. Oberdörster, G., Oberdörster, E., Oberdörster, J., Nanotoxicology: an emerging discipline evolving from studies of ultrafine particles. *Environ. Health Perspect.*, 113, 7, 823–839, 2005.
44. Som, C., Wick, P., Krug, H., Nowack, B., Environmental and health effects of nanomaterials in nanotextiles and façade coatings. *Environ. Int.*, 37, 6, 1131–1142, 2011, doi: https://doi.org/10.1016/j.envint.2011.02.013.
45. Saleh, T.A., Trends in the sample preparation and analysis of nanomaterials as environmental contaminants. *Trends Environ. Anal. Chem.*, 28, e00101, 2020, doi: https://doi.org/10.1016/j.teac.2020.e00101.
46. Deng, J., Zhao, S., Liu, Y., Liu, C., Sun, J., Nanosensors for diagnosis of infectious diseases. *ACS Appl. Bio Mater.*, 4, 5, 3863–3879, 2020.
47. Esim, O., Kurbanoglu, S., Savaser, A., Ozkan, S. A., Ozkan, Y., Chapter 9 - Nanomaterials for Drug Delivery Systems, in: *New Development in*

Nanosensors for Pharmaceutical Analysis, S. A. Ozkan and A. B. T.-N. D. in N. for P. A. Shah (Eds.), pp. 273–301, Academic Press, Cambridge, MA, USA, 2019, doi: https://doi.org/10.1016/B978-0-12-816144-9.00009-2.

48. Rodrigues Ribeiro Teles, F.S., Pires de Távora Tavira, L.A., Pina da Fonseca, L.J., Biosensors as rapid diagnostic tests for tropical diseases. *Crit. Rev. Clin. Lab. Sci.*, 47, 3, 139–169, Jun. 2010, doi: 10.3109/10408363.2010.518405.
49. Markwalter, C.F., Kantor, A.G., Moore, C.P., Richardson, K.A., Wright, D.W., Inorganic Complexes and Metal-Based Nanomaterials for Infectious Disease Diagnostics. *Chem. Rev.*, 119, 2, 1456–1518, Jan. 2019, doi: 10.1021/acs.chemrev.8b00136.
50. Cruz, C., Miranda, A., Santos, T., 17 - Aptamer-based microfluidics for circulating tumor cells, in: *Woodhead Publishing Series in Biomaterials*, P. B. T.-A. E. N. and C.T. Kesharwani (Eds.), pp. 403–432, Woodhead Publishing, Cambridge, United Kingdom, 2023, doi: https://doi.org/10.1016/B978-0-323-85881-6.00004-X.

16

Strengthening the Health System of the Communities in the Battle Against Infectious Diseases

Dinoy Mathew[1]*, Anu P. Mathew[2], Bobby Simon[3] and Ancy Joseph[4]

[1]*Deva Matha College, Kuravilangad, Kottayam (Dt), Kerala, India*
[2]*Department of Commerce, Deva Matha College Kuravilangad, Kerala, India*
[3]*Department of Commerce, St. Thomas College Palai, Kerala, India*
[4]*Govt. T.D. Medical College, Alappuzha, Kerala, India*

Abstract

This chapter explores the imperative role of reinforcing community health systems in combating infectious diseases. The global prevalence of infectious diseases underscores the necessity for robust health systems to effectively address these threats. The chapter highlights the significance of a comprehensive approach to health security, emphasizing its role as a foundation for disease prevention, early detection, and treatment. Health information system has an indispensable role in promoting efficient data management and decision-making. Education and awareness emerge as essential tools for empowering communities with knowledge to mitigate disease risks. The critical importance of vaccination in preventing the spread of infectious diseases is also underscored, alongside the need for accessible and equitable vaccination programs. Preventive mechanisms are explored as a proactive strategy to minimize disease transmission, followed by a discussion on surveillance for timely identification and response. The chapter emphasizes the significance of networking and collaboration among diverse stakeholders for collective and effective disease control efforts. The profound impact of the coronavirus disease 2019 (COVID-19) pandemic on infectious diseases and health systems is also touched upon, proposing strategies for strengthening community health systems to mitigate these effects and ensure resilient and effective healthcare delivery. The interconnectedness of the components discussed pinpoints the necessity of a

**Corresponding author*: frdinoy.kavalammackal@devamatha.ac.in

Suvardhan Kanchi, Ayyappa Bathinapatla, Anitha Varghese and Phumlane Selby Mdluli (eds.) Nano-Biosensor Technologies for Diagnosis of Infectious Diseases, (469–484) © 2025 Scrivener Publishing LLC

holistic approach to strengthen community health systems in the face of infectious diseases. In fostering a resilient healthcare framework, this approach can contribute significantly to global health security and mitigate the impact of infectious diseases.

Keywords: Health information system, vaccination, preventive mechanism, surveillance, networking and collaboration, COVID-19, community health systems

16.1 Introduction

The health system of communities plays a crucial role in preventing, managing, and controlling infectious diseases. Infectious diseases are illnesses caused by pathogenic microorganisms, such as bacteria, viruses, parasites, or fungi [1]. These diseases can spread rapidly within communities and have far-reaching consequences for individuals and society as a whole [2]. Infectious diseases pose significant threats to public health worldwide, with outbreaks often resulting in severe illness, mortality, and economic burden [3]. Effective healthcare systems are essential to mitigate the impact of infectious diseases on communities [4]. By examining various aspects such as access to healthcare services, disease surveillance mechanisms, preventive measures, and response strategies, we can gain valuable insights into how communities can better prepare for and respond to infectious disease outbreaks. Understanding the strengths and weaknesses of community health systems is crucial for policymakers, healthcare providers, researchers, and other stakeholders involved in disease prevention and control efforts [5]. Infectious diseases exert a profound influence on the health systems of communities, posing significant challenges to both individuals and healthcare providers [6]. These diseases not only cause immense suffering and loss of life but also strain the resources and infrastructure of healthcare systems. The burden imposed by infectious diseases can overwhelm the capacity of community health systems to respond effectively [7]. One crucial impact is the increased demand for healthcare services, including diagnostic tests, treatments, and hospitalizations [8]. This surge in demand often leads to overcrowded hospitals, limited access to care, and stretched healthcare personnel [9]. Additionally, infectious diseases can disrupt routine healthcare services as resources are redirected towards outbreak control measures. Furthermore, infectious diseases have far-reaching consequences beyond physical health. They can trigger social unrest, economic instability, and psychological distress within communities [10]. The fear associated with outbreaks often leads to stigmatization and discrimination against affected individuals or groups.

Infectious diseases arise from various pathogenic microorganisms, including viruses, bacteria, protozoa, fungi, and other parasites. These ailments can spread through animals, humans, insects, and other vectors. Infectious diseases continue to be a leading global cause of death. The study on the threat of infectious diseases to health workers suggested an interdisciplinary approach for comprehensive and innovative strategies to address infectious disease outbreaks [3]. Another study elaborates on reducing the risk of infectious disease transmission to healthcare workers by adopting robotics, wearables and other intelligent systems [4]. Infection prevention control (IPC) initiatives play a decisive role in averting infections that can be prevented, ultimately saving lives. Successful IPC efforts result in a reduction of over 30% in rates of healthcare-associated infections (HAIs), which in turn prevents sickness and death. These programs also deter the emergence and spread of antimicrobial resistance (AMR), all the while notably decreasing healthcare expenses [11].

16.2 Primary Healthcare

Since the inception of human civilization, infectious diseases have posed a continuous threat to the security of public health. A strong healthcare system is essential for the early detection and control of infectious diseases. The significance of primary healthcare as the first point of contact for people with health concerns has been already established [5, 12]. It provides a range of services, including preventive care, chronic disease management, and treatment for common illnesses. This makes it well-positioned to identify and control infectious diseases early on before they can spread. In healthcare facilities, implementing measures and practices for infection prevention and control weakens the chances for drug-resistant pathogens to propagate. One way that primary healthcare can do this is through early detection and surveillance. This involves regular check-ups and screening to identify individuals with early symptoms of infectious diseases. By detecting diseases early, healthcare providers can take steps to prevent them from spreading, such as providing treatment or isolating patients. Furthermore, these systems facilitate coordination among various stakeholders such as local governments, healthcare providers, non-profit organizations, and community members [13]. This collaborative approach ensures a comprehensive response to infectious diseases through effective communication channels.

However, these systems often face numerous challenges that hinder their effectiveness in addressing such public health threats. One key challenge is the limited funding allocated to community health infrastructure, resulting in inadequate resources and personnel to adequately respond to outbreaks [14]. Additionally, there is a lack of coordination and collaboration between different stakeholders within the community health system, including healthcare providers, local authorities, and community members [15]. This leads to fragmented efforts and an ineffective response to infectious diseases. Furthermore, the accessibility and affordability of healthcare services remain major obstacles for vulnerable populations within communities, limiting their ability to seek timely care or preventive measures [16]. Lastly, the ongoing issue of vaccine hesitancy poses a significant challenge in achieving community-wide immunization coverage against infectious diseases [17].

The various strategies for strengthening community health systems in the context of infectious diseases can be outlined in Figure 16.1.

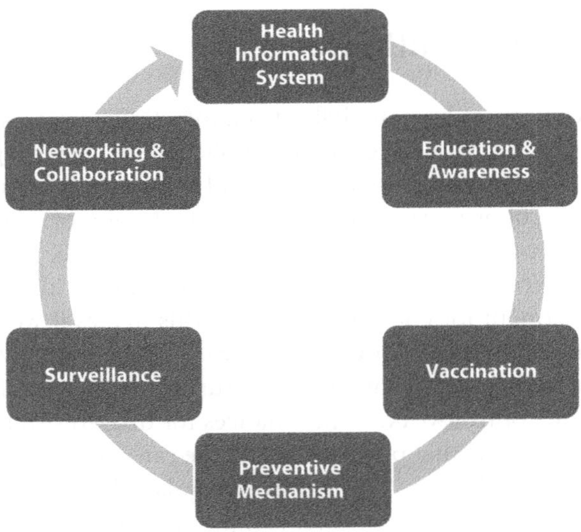

Figure 16.1 Strategies for strengthening public health systems.

16.2.1 Health Information System

The current approach of the health systems revolves around swiftly identifying these occurrences and promptly taking action in response. A technology-based health information system can help to improve the competency of the healthcare system [6]. This type of system can collect and store data on diseases, which can be used to track the incidence and spread of diseases. This information can then identify areas where preventive measures are needed. The context of developing a health information system in developing country settings is challenging [18]. Timely and accurate diagnosis of infectious diseases is also essential for controlling them. This is because it allows healthcare providers to initiate treatment promptly, which can help to prevent the spread of the disease. Timely collection and analysis of epidemiological data can aid in the early understanding of disease patterns. This information can be used to develop targeted interventions to prevent the spread of diseases. Rapid diagnostic tests can be used to quickly identify specific pathogens, which can help to guide treatment.

16.2.2 Education and Awareness

Social awareness and education are essential for the success of any health system. Emphasizing health education within communities is vital for preventing infectious diseases. Implementing targeted campaigns on hygiene practices, vaccination awareness, and disease prevention measures can empower individuals to protect themselves and their communities [19].

The social repercussions and stigma linked to infectious disease pandemics reveal striking similarities across different health conditions, healthcare systems, and cultural contexts. The analysis of ignorance and stigma associated with COVID-19 suggested awareness campaigns [7]. When people are aware of the risks of disease and how to prevent them, they are more likely to take steps to protect themselves and their communities. This can be done through education and awareness campaigns that teach people about the mode of transmission of diseases, preventive measures, and the importance of seeking early treatment.

Empowered individuals with knowledge of the mode of transmission and preventive measures can lead health promotion campaigns to raise awareness about appropriate precautionary measures. These individuals can be doctors, nurses, community health workers, or even ordinary citizens who are passionate about public health. They can use their knowledge and skills to educate others about the importance of health and how to stay healthy. Continuous training makes the system alert to handle various scenarios. Healthcare workers need to be constantly trained on new diseases and emerging threats. This will help them to stay up-to-date on the latest information and be prepared to respond to any health crisis.

16.2.3 Vaccination

Vaccination is one of the most effective ways to prevent the spread of infectious diseases. Vaccines work by exposing the body to a weakened or inactive form of a virus or bacteria, which helps the body to develop immunity to the disease. When enough people in a community are vaccinated, it creates herd immunity, which protects even those who are not vaccinated [8]. Herd immunity works because the virus or bacteria has a harder time spreading when fewer people are susceptible to it. This is because vaccinated people are less likely to get sick and spread the disease to others.

Routine vaccination of children and adults is essential for creating herd immunity. By getting vaccinated, we can protect ourselves, our families, and our communities from infectious diseases. Increasing the availability of vaccination is essential for realizing the objectives of the Sustainable Development Goals (SDGs). Illnesses like polio and smallpox, which used to claim the lives of numerous children, have been eradicated, and similar progress is being made with other diseases, largely due to the success of reliable and efficient vaccines. It is of utmost importance to raise consciousness regarding the significance of immunization and to guarantee that parents and caretakers are well-informed about the vaccination process. They should be educated about the appropriate timing and location for vaccinating their children, the intervals between doses, and the critical nature of not missing any scheduled appointments. Enhanced endeavors to convey the advantages and potential drawbacks of vaccines, along with tackling concerns through evidence-backed data, will contribute to enhancing and upholding global confidence in both vaccines and healthcare systems. UNICEF has a key role in materializing this goal by actively engaging in collaborations with various entities, particularly radio stations, faith-based groups, community leaders, and influential figures. The aim is to disseminate accurate information regarding the merits of immunization and foster partnerships that aid in spreading reliable knowledge.

16.2.4 Preventive Mechanism

A strong preventive mechanism is essential for controlling the spread of infectious diseases. This includes ensuring access to safe drinking water, adequate sanitation facilities, and proper waste management. These measures help to reduce the number of germs in the environment, which can help to prevent the spread of diseases. Investing in healthcare infrastructure at the community level enables prompt diagnosis, treatment, and containment of infectious diseases [20]. This includes ensuring adequate healthcare facilities, laboratory capacity, medical supplies, and trained personnel are available in remote or marginalized areas [2].

Establishing skillful teams of infection control personnel will not only aid in the enhanced and lasting execution of interventions for infection prevention and control, along with HAI surveillance, but it will also enhance the capacity to identify and manage potential outbreaks of such infections. These skilled experts will additionally establish connections between initiatives addressing HAIs and infection prevention/control, and more comprehensive programs aimed at overseeing antimicrobial use and monitoring AMR [9]. There is an urgent need for tracking infectious diseases and an efficient surveillance and response system for controlling outbreaks [21]. This system should be able to track the incidence and spread of diseases and respond quickly to outbreaks. It should also be able to communicate effectively with the public and with other agencies.

16.2.5 Surveillance

The management of infectious diseases can be regarded as a global public good, and public health surveillance serves as a means to attain this goal [22]. The landscape of public health surveillance is intricate, involving various actors from both governmental and non-governmental sectors who play roles in contributing, receiving, and facilitating the exchange of data [10]. Developing robust disease surveillance mechanisms at the community level is crucial for early detection and response to infectious diseases. This involves training community health workers to identify and report suspected cases promptly, establishing reliable communication channels, and integrating local knowledge into surveillance activities [23].

Monitoring healthcare-related infections and IPC initiatives focused on preventing and controlling these infections serve a dual purpose: addressing AMR and bolstering overall patient safety [24]. Evidence from numerous nations demonstrates that when governments and healthcare leadership make HAI surveillance and prevention a priority, substantial transformation can occur [22]. The timely exchange of public health surveillance data is crucial for enhancing coordinated and efficient risk management in public health reactions [25]. Collaborative sharing of such data across nations enhances the ability to detect and respond to diseases [14] and aids in pinpointing outbreak sources when national data falls short [26]. This practice also diminishes the potential or real consequences of a global health emergency. By systematically implementing surveillance for HAIs, engaging in infection prevention and control initiatives, and simultaneously striving to compile and communicate precise AMR data while refining antimicrobial stewardship practices, nations can enhance their ability to detect and prevent AMR effectively.

16.2.6 Networking and Collaboration

In the realm of community health systems, addressing infectious diseases requires a comprehensive approach that emphasizes collaboration and partnerships [27]. The universal healthcare community must holistically tackle upcoming challenges to health security by delving deeper into the prevention and resolution of human security issues. There remains a chance to react with greater efficiency and thoroughness in response to the signals that demand our attention [8]. The resilience of the community can be enhanced by promoting social cohesion, social networking, and trust. This can help to build a sense of community and encourage people to work together to prevent the spread of diseases. Collaboration enables different stakeholders, such as healthcare providers, public health agencies, and community organizations, to pool their resources and expertise [28]. Various networks and collaborations exist for disease surveillance, such as the World Health Organization (WHO) Global Outbreak Alert and Response Network (GOARN) and the European Centre for Disease Prevention and Control (ECDC) network. Collaboration among all healthcare providers and other agencies is important [29] for timely reporting of notifiable diseases and participation in outbreak investigations. This will help to ensure that diseases are detected and controlled quickly. Effective disease prevention often requires collaboration at the grassroots level, involving local healthcare providers, community leaders, and public health authorities [30]. Additionally, collaborations between the public and private sectors can enhance resources and expertise for disease prevention efforts. Collaborative efforts can yield beneficial and inventive outcomes in pursuit of specific public health objectives. These partnerships can establish potent frameworks for tackling challenging issues, harnessing diverse partners' ideas, resources, and skills [31].

16.3 Impact of COVID-19 on Infectious Diseases and Health Systems

The emergence of the novel coronavirus severe acute respiratory syndrome coronavirus 2 (SARS-CoV-2) and the subsequent COVID-19 pandemic has brought into sharp focus the critical role of community health systems in mitigating the impact of public health emergencies globally [32]. Beyond its direct impact on morbidity and mortality, COVID-19 has catalyzed significant shifts in the management of infectious diseases and the functioning of health systems [33]. Community health systems serve as the frontline defense against infectious diseases, providing essential healthcare services, promoting preventive measures, and facilitating community engagement. Community health systems have faced challenges in terms of capacity, resources, and workforce management [34]. The pandemic has highlighted the need for innovative solutions such as telehealth services and remote monitoring to ensure continuity of care. Additionally, it has exposed existing disparities in healthcare access and reinforced the importance of community-based healthcare models [35].

The COVID-19 pandemic has influenced the transmission patterns, prevalence, and management of various infectious diseases. Studies have reported both positive and negative impacts on infectious disease dynamics. For instance, the implementation of non-pharmaceutical interventions such as mask-wearing and social distancing has led to a decrease in the transmission of respiratory viruses like influenza [36]. Conversely, disruptions in routine healthcare services, including vaccination campaigns and disease surveillance, have contributed to the resurgence of vaccine-preventable diseases and other infections [37].

The strain exerted by COVID-19 on healthcare systems has highlighted vulnerabilities and gaps in healthcare delivery. Overwhelmed hospital capacities, shortages of medical supplies, and workforce exhaustion have impeded the provision of essential healthcare services, including those for infectious diseases [38]. Health systems have adapted by implementing telemedicine, triage systems, and makeshift facilities to manage the influx of COVID-19 patients while maintaining continuity of care for other illnesses. However, the long-term sustainability and resilience of these adaptations remain uncertain.

The COVID-19 pandemic has yielded valuable lessons for strengthening community health systems. Community health systems encompass a network of primary healthcare providers, community health workers, and local health facilities that deliver essential health services to populations at the grassroots level. These systems play a pivotal role in disease surveillance, outbreak response, and health promotion, particularly in resource-constrained settings. The COVID-19 pandemic has underscored the importance of community-based approaches in reaching marginalized populations, addressing health disparities, and building trust within communities. Key insights include the need for decentralized healthcare delivery models, community empowerment, and multisectoral collaboration.

Effective communication strategies, community engagement initiatives, and investment in digital health technologies have emerged as essential components of pandemic response efforts. Furthermore, the pandemic has highlighted the importance of prioritizing the well-being of frontline healthcare workers and integrating mental health services into community-based care.

Building on lessons learned from COVID-19, several strategies can be employed to strengthen community health systems, as evidenced in Figure 16.2.

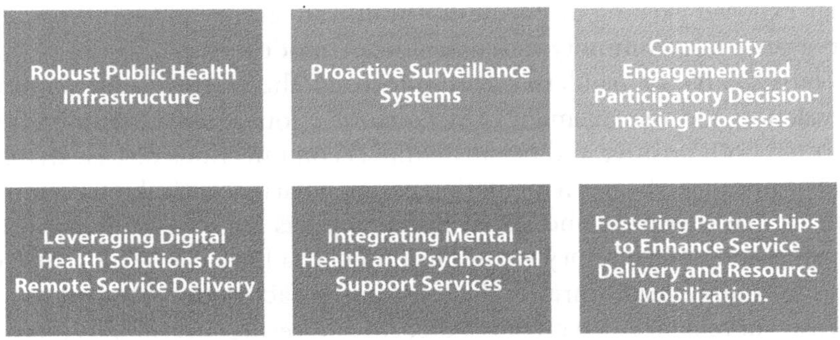

Figure 16.2 Strategies for strengthening community health systems.

Investments in vaccine development, healthcare workforce training, and pandemic preparedness are imperative to mitigate the impact of future pandemics on infectious diseases and health systems. Furthermore, international collaboration and data sharing are crucial for effective pandemic response and containment efforts.

Table 16.1 illustrates the key changes and developments in community health systems before and after the COVID-19 pandemic.

Table 16.1 Community health systems before and after COVID-19 pandemic.

Aspects	Before COVID-19	After COVID-19
Healthcare access	▪ Moderate access to healthcare ▪ Readily available routine check-ups and non-urgent appointments	▪ Limited access to healthcare ▪ Strained healthcare facilities ▪ Increased demand for critical care

(*Continued*)

Table 16.1 Community health systems before and after COVID-19 pandemic. (*Continued*)

Aspects	Before COVID-19	After COVID-19
Healthcare infrastructure facilities	▪ Adequate infrastructure and resources in place ▪ Hospitals operating at normal capacity	▪ Overwhelmed healthcare facilities ▪ Shortages of beds, equipment, and medical supplies ▪ Increased reliance on telemedicine
Public health education	▪ General awareness about common illnesses and preventive measures ▪ Moderate adherence to hygiene and vaccination protocols	▪ Heightened awareness about infectious diseases and hygiene ▪ Emphasis on vaccination campaigns ▪ Increased compliance with preventive measures
Mental health support	▪ Mental health services available but underutilized ▪ Stigma around mental health issues persists	▪ Increased demand for mental health support services ▪ Heightened awareness about mental health importance ▪ Expansion of mental health programs
Economic impact	▪ Stable employment and economic growth ▪ Limited impact on the healthcare sector	▪ Economic downturn and job loss ▪ Financial strain on individuals and healthcare systems ▪ Redistribution of healthcare funds to pandemic response

16.4 Conclusion

By prioritizing the development of resilient community health systems worldwide, we can enhance our collective ability to combat infectious diseases effectively [39]. Thorough investing in community health systems with strong government and professional leadership is the need of the hour to combat infectious diseases meticulously [40]. This can enhance the capacity to swiftly identify and manage infectious diseases, thereby fortifying the healthcare infrastructure. To achieve these goals, it is important to utilize varied strategies for implementing IPC programs effectively, operating at the levels of nations, regions, and healthcare facilities. Moreover, it is prudent to pair these strategies with financial support, like a stimulus package, to enhance the adoption of efficient infection control measures. This comprehensive approach will aid in addressing present and future challenges posed by infectious diseases, eventually leading to the establishment of sustainable healthcare facilities.

References

1. Edubirdie, Impact of Infectious Diseases on Society: Opinion Essay, 2022, December 27, https://edubirdie.com/examples/impact-of-infectious-diseases-on-society-opinion-essay/.
2. GGI Insights, Global Health Security: Safeguarding Humanity's Wellbeing, 2024, January 31, https://www.graygroupintl.com/blog/global-health-security.
3. Musau, J., Baumann, A., Kolotylo, C., O'Shea, T., Bialachowski, A., Infectious disease outbreaks and increased complexity of care. *Int. Nurs. Rev.*, 62, 3, 404–11, December 2015, doi: 10.1111/inr.12188.
4. Tavakoli, M., Robotics, Smart Wearable Technologies, and Autonomous Intelligent Systems for Healthcare During the COVID-19 Pandemic: An Analysis of the State of the Art and Future Vision. *Adv. Intell. Syst.*, 2, 1–7, 2020.
5. Pandve, H.T. and Pandve, T.K., Primary healthcare system in India: Evolution and challenges. *Int. J. Health Syst. Disaster Manage.*, *1*, 3, 125–128, 2013.
6. Li, X., Lu, J., Hu, S., Cheng, K., Maeseneer, J.D., Meng, Q., Mossialos, E., The primary health-care system in China. *Lancet*, *390*, 10112, 2584–2595, 2017.
7. Saeed, F., Mihan, R., Mousavi, S.Z., Reniers, R.L., Bateni, F.S., Alikhani, R., Mousavi, S.B., A Narrative Review of Stigma Related to Infectious Disease Outbreaks: What Can Be Learned in the Face of the Covid-19 Pandemic? *Front. Psychiatry*, 11, 1–8, 2020, https://doi.org/10.3389/fpsyt.2020.565919.

8. Heymann, D.L., Chen, L., Takemi, K., Fidler, D.P., Tappero, J.W., Thomas, M.J., Rannan-Eliya, R.P., Global health security: the wider lessons from the west African Ebola virus disease epidemic. *Lancet, 385*, 9980, 1884–1901, 2015, doi: 10.1016/s0140-6736(15)60858-3.
9. Walia, K., Ohri, V.C., Mathai, D., Antimicrobial Stewardship Programme of ICMR. Antimicrobial stewardship programme (AMSP) practices in India. *Indian J. Med. Res.*, 358, 130–8, 2015, 2014. doi:10.4103/0971-5916.164228 pmid:26354210.
10. Edelstein, M., Lee, L.M., Herten-Crabb, A., Heymann, D.L., Harper, D.R., Strengthening global public health surveillance through data sharing. *Emerg. Infect. Dis.*, 24, 7, 1324–1330, 2018, https://doi.org/10.3201/eid2407.151830.
11. World Health Organization, Core components of infection prevention and control programmes: assessment tools for IPC programmes, 2011.
12. Khoo, H.S., Lim, Y.W., Vrijhoef, H.J., Primary healthcare system and practice characteristics in Singapore. *Asia Pac. Fam. Med.*, 13, 1, 8, 2014, https://doi.org/10.1186/s12930-014-0008-x.
13. Roy, B., Stanojevich, J., Stange, P., Jiwani, N., King, R., Koo, D., Development of the Community Health Improvement Navigator Database of Interventions. *MMWR Suppl.*, 65, 1–9, 2016, place_Holder_For_Early_Release:http://dx.doi.org/10.15585/mmwr.su6502a1external icon.
14. Phommasack, B., Jiraphongsa, C., Ko Oo, M., Bond, K.C., Phaholyothin, N., Suphanchaimat, R., Ungchusak, K., Macfarlane, S.B., Mekong Basin Disease Surveillance (MBDS): a trust-based network. *Emerg. Health Threats J.*, 6, 1–9, 2013, https://doi.org/10.3402/ehtj.v6i0.19944.
15. USAID, Strengthening Community Health Systems, Global Health. U.S. Agency for International Development, 2023, https://www.usaid.gov/global-health/health-systems-innovation/health-systems/strengthening-community-health-systems.
16. Tulane University, Community Health Promotion in Rural Areas, 2023, https://publichealth.tulane.edu/blog/community-health-promotion-rural-areas/.
17. Ellwanger, J.H., Veiga, A.B.G., Kaminski, V.L., Valverde-Villegas, J.M., Freitas, A.W.Q., Chies, J.A.B., Control and prevention of infectious diseases from a One Health perspective. *Genet. Mol. Biol., 44*, 1 Suppl 1, e20200256, 2021, https://doi.org/10.1590/1678-4685-GMB-2020-0256.
18. Nguyen, T.N., Developing health information systems in developing countries: Lessons learnt from a longitudinal action research study in Vietnam. *Electron. J. Inf. Syst. Dev. Countries*, 89, 5, 1–27, 2023, https://doi.org/10.1002/isd2.12268.
19. USA Health, Infectious Disease, https://www.usahealthsystem.com/specialties/infectious-disease.
20. Warren, E., Preventing, Containing, and Treating Infectious Disease Outbreaks at Home and Abroad, https://elizabethwarren.com/plans/combating-infectious-disease-outbreaks.

21. M'ikanatha, N.M., Lynfield, R., Julian, K.G., Beneden, C.A., de Valk, H., *Infectious disease surveillance*, Wiley-Blackwell, John Wiley & Sons, Ltd., UK, 2013, DOI:10.1002/9781118543504.
22. Pittet, D., Allegranzi, B., Storr, J. *et al.*, Infection control as a major World Health Organization priority for developing countries. *J. Hosp. Infect.*, 358, 285–92, 2008, doi: 10.1016/j.jhin.2007.12.013 pmid:18329137.
23. Quinn, S.C. and Kumar, S., Health inequalities and infectious disease epidemics: a challenge for global health security. *Biosecur. Bioterror.: biodefense strategy, practice, and science, 12*, 5, 263–273, 2014, https://doi.org/10.1089/bsp.2014.0032.
24. O'Neill, J., Infection prevention, control and surveillance: limiting the development and spread of drug resistance, The Review on Antimicrobial Resistance, 2016, https://amr-review.org/sites/default/files/Health%20 infrastructure%20and%20 surveillance%20final%20version_LR_NO%20 CROPS.pdf.
25. International Organization of Public Health Institutes, Public health surveillance: a call to share data, 2016 May, http://www.ianphi.org/news/2016/datasharing1.html.
26. Nordic Outbreak Investigation Team C, Joint analysis by the Nordic countries of a hepatitis A outbreak, October 2012 to June 2013: frozen strawberries suspected. *Euro Surveill.*, 18, 20520, 2013. 10.2807/1560-7917.ES2013.18.27.20520.
27. CDC, Our Partners, NCEZID, https://www.cdc.gov/ncezid/what-we-do/our-partners.html.
28. Danbury, CT, Community Health, https://www.danbury-ct.gov/202/Community-Health.
29. Davies, G.P., Williams, A.M., Larsen, K., Perkins, D., Roland, M., Harris, M.F., Coordinating primary health care: an analysis of the outcomes of a systematic review. *Med. J. Aust.*, *188*, S8, 565–568, 2008.
30. Salam, R.A., Maredia, H., Das, J.K. *et al.*, Community-based interventions for the prevention and control of helminthic neglected tropical diseases. *Infect. Dis. Poverty*, 3, 23, 2014, https://doi.org/10.1186/2049-9957-3-23.
31. Reich, M., Public–private partnerships for public health. *Nat. Med.*, 6, 617–620, 2000, https://doi.org/10.1038/76176.
32. World Health Organization, Coronavirus disease (COVID-19) pandemic, 2020, https://www.who.int/emergencies/diseases/novel-coronavirus-2019/advice-for-public?adgroupsurvey={adgroupsurvey}&gad_source=1&gclid=CjwKCAiA_OetBhAtEiwAPTeQZ3clmxPPEuJZWg-FMMB1MlyePNbRW TOanFro74wnLIr2 KohuN XlGnRoC33UQAvD_BwE.
33. McKinsey & Company, COVID-19's potential impact on healthcare, 2020, https://www.mckinsey.com/industries/healthcare/our-insights/understanding-the-hidden-costs-of-covid-19s-potential-impact-on-us-healthcare.
34. Haileamlak, A., The impact of COVID-19 on health and health systems. *Ethiop. J. Health Sci.*, *31*, 6, 1073–1074, 2021, https://doi.org/10.4314/ejhs.v31i6.1.

35. Healthy People 2030, Access to Health Services, https://health.gov/healthypeople/priority-areas/social-determinants-health/literature-summaries/access-health-services.
36. Jüni, P., Rothenbühler, M., Bobos, P., Thorpe, K.E., da Costa, B.R., Fisman, D.N., Egger, M., Impact of climate and public health interventions on the COVID-19 pandemic: a prospective cohort study. *Can. Med. Assoc. J.*, 192, 21, E566–E573, 2020.
37. Reintjes, R., Lessons in contact tracing from Germany. *BMJ*, 369, m2522, 2020, doi:10.1136/bmj.m2522.
38. Wang, C.J., Ng, C.Y., Brook, R.H., Response to COVID-19 in Taiwan: Big Data Analytics, New Technology, and Proactive Testing. *JAMA*, 323, 14, 1341–1342, 2020, https://doi.org/10.1001/jama.2020.3151.
39. Mpofu, E., *Sustainable Community Health Systems and Practices in Diverse Settings*, Palgrave Macmillan, Camden, London, 2020, ISBN 978-3-030-59686-6 ISBN 978-3-030-59687-3 (eBook) https://doi.org/10.1007/978-3-030-59687-3.
40. Hutchison, B., Levesque, J.-F., Erin Strumpf, N.C., Primary Health Care in Canada: Systems in Motion. *Milbank Q.*, *89*, 2, 256–288, 2011.

Index